Methods in Cell Biology

Cytokinesis

Volume 137

Series Editors

Leslie Wilson
Department of Molecular, Cellular and Developmental Biology
University of California
Santa Barbara, California

Phong Tran
University of Pennsylvania
Philadelphia, USA &
Institut Curie, Paris, France

Methods in Cell Biology

Cytokinesis

Volume 137

Edited by

Arnaud Echard
Institut Pasteur, Paris, France

ELSEVIER

ACADEMIC PRESS
An imprint of Elsevier
elsevier.com

Academic Press is an imprint of Elsevier
50 Hampshire Street, 5th Floor, Cambridge, MA 02139, USA
525 B Street, Suite 1800, San Diego, CA 92101-4495, USA
125 London Wall, London EC2Y 5AS, UK
The Boulevard, Langford Lane, Kidlington, Oxford OX5 1GB, UK

First edition 2017

ISBN: 978-0-12-809673-4
ISSN: 0091-679X

For information on all Academic Press publications visit our website at https://www.elsevier.com

Publisher: Zoe Kruze
Acquisition Editor: Zoe Kruze
Editorial Project Manager: Colford Hannah
Production Project Manager: Surya Narayanan Jayachandran
Senior Designer: Matthew Limbert

Typeset by TNQ Books and Journals

Contents

Contributors

M. Abe
RIKEN, Wako, Saitama, Japan

S. Adar
Ben Gurion University of the Negev, Beer Sheva, Israel

N. Baranova
Institute for Science and Technology Austria (IST Austria), Klosterneuburg, Austria

Y. Bellaïche
Institut Curie, PSL Research University, Paris, France; Sorbonne Universités, Paris, France

G. Belloni
Istituto di Biologia e Patologia Molecolari del CNR, Università di Roma Sapienza, Roma, Italy

M. Bennett
Boston College, Chestnut Hill, MA, United States

P. Bhatia
University of Lausanne, Lausanne, Switzerland

E. Bi
University of Pennsylvania, Philadelphia, PA, United States

J. Borrego-Pinto
European Molecular Biology Laboratory (EMBL), Heidelberg, Germany

K.G. Bourdages
Université de Montréal, Montréal, QC, Canada

D. Burgess
Boston College, Chestnut Hill, MA, United States

J.C. Canman
Columbia University, New York, NY, United States

S. Celton-Morizur
Inserm, U1016, Paris, France; CNRS, UMR 8104, Paris, France; Paris Descartes University, Paris, France

Y. Chai
Tsinghua University, Beijing, China

A. Chaigne
MRC Laboratory for Molecular Cell Biology, University College London, London, United Kingdom

T. Davies
Columbia University, New York, NY, United States

C. Desdouets
Inserm, U1016, Paris, France; CNRS, UMR 8104, Paris, France; Paris Descartes University, Paris, France

J. Dumont
Institut Jacques Monod, Paris, France

A. Echard
Institut Pasteur, Paris, France; Centre National de la Recherche Scientifique, Paris, France

N. Elia
Ben Gurion University of the Negev, Beer Sheva, Israel

A. Ettinger
Institute of Epigenetics and Stem Cells, Munich, Germany

C.M. Field
Harvard Medical School, Boston, MA, United States; Marine Biological Laboratory, Woods Hole, MA, United States

M. Fortier
Inserm, U1016, Paris, France; CNRS, UMR 8104, Paris, France; Paris Descartes University, Paris, France

A. Frappaolo
Istituto di Biologia e Patologia Molecolari del CNR, Università di Roma Sapienza, Roma, Italy

S. Frémont
Institut Pasteur, Paris, France; Centre National de la Recherche Scientifique, Paris, France

O. Gershony
Ben Gurion University of the Negev, Beer Sheva, Israel

M.G. Giansanti
Istituto di Biologia e Patologia Molecolari del CNR, Università di Roma Sapienza, Roma, Italy

I. Goliand
Ben Gurion University of the Negev, Beer Sheva, Israel

R.A. Green
University of California, San Diego, CA, United States

G.R.X. Hickson
Sainte-Justine Hospital Research Center, Montreal, QC, Canada; Université de Montréal, Montréal, QC, Canada

W.B. Huttner
Max Planck Institute of Molecular Cell Biology and Genetics, Dresden, Germany

J.-R. Huynh
Institut Curie, Paris, France; CNRS UMR3215, Inserm U934, Paris, France

S.N. Jordan
Columbia University, New York, NY, United States

J. Kashiwazaki
Gakushuin University, Tokyo, Japan

A. Kechad
Sainte-Justine Hospital Research Center, Montreal, QC, Canada; Université de Montréal, Montréal, QC, Canada

T. Kobayashi
RIKEN, Wako, Saitama, Japan; CNRS, Illkirch, France

J. König
Technische Universität Dresden, Dresden, Germany

Y. Kosodo
Korea Brain Research Institute, Daegu, Korea

P. Kothari
Johns Hopkins University, Baltimore, MD, United States

J. Lafaurie-Janvore
Institut Curie, Paris, France

C. Lafaurie
Versailles, France

P. Lenart
European Molecular Biology Laboratory (EMBL), Heidelberg, Germany

W. Li
Tsinghua University, Beijing, China

M. Loose
Institute for Science and Technology Austria (IST Austria), Klosterneuburg, Austria

I. Mabuchi
Gakushuin University, Tokyo, Japan

A.S. Maddox
University of North Carolina at Chapel Hill, Chapel Hill, NC, United States

S.G. Martin
University of Lausanne, Lausanne, Switzerland

J. Mathieu
Institut Curie, Paris, France; CNRS UMR3215, Inserm U934, Paris, France

J. Meaders
Boston College, Chestnut Hill, MA, United States

M. Mishima
University of Warwick, Coventry, United Kingdom

M. Mishra
Tata Institute of Fundamental Research, Mumbai, India

T.J. Mitchison
Harvard Medical School, Boston, MA, United States; Marine Biological Laboratory, Woods Hole, MA, United States

T. Müller-Reichert
Technische Universität Dresden, Dresden, Germany

M. Munzig
Technische Universität Dresden, Dresden, Germany

D. Nachmias
Ben Gurion University of the Negev, Beer Sheva, Israel

S. Okada
University of Pennsylvania, Philadelphia, PA, United States; Kyushu University, Fukuoka, Japan

G. Ou
Tsinghua University, Beijing, China

E.K. Paluch
MRC Laboratory for Molecular Cell Biology, University College London, London, United Kingdom

J.F. Pelletier
Harvard Medical School, Boston, MA, United States; Marine Biological Laboratory, Woods Hole, MA, United States

E. Peterman
University of Colorado Anschutz Medical Campus, Aurora, CO, United States

M. Piel
Institut Curie, Paris, France

R. Piergentili
Istituto di Biologia e Patologia Molecolari del CNR, Università di Roma Sapienza, Roma, Italy

D. Pinheiro
Institut Curie, PSL Research University, Paris, France; Sorbonne Universités, Paris, France; Abel Salazar Biomedical Sciences Institute, University of Porto, Porto, Portugal

R. Prekeris
University of Colorado Anschutz Medical Campus, Aurora, CO, United States

K. Rehain
University of North Carolina at Chapel Hill, Chapel Hill, NC, United States

D.N. Robinson
Johns Hopkins University, Baltimore, MD, United States

S. Ruvolo
Boston College, Chestnut Hill, MA, United States

E.S. Schiffhauer
Johns Hopkins University, Baltimore, MD, United States

S. Sechi
Istituto di Biologia e Patologia Molecolari del CNR, Università di Roma Sapienza, Roma, Italy

I. Segal
Ben Gurion University of the Negev, Beer Sheva, Israel

S. Sherman
Ben Gurion University of the Negev, Beer Sheva, Israel

M. Shirasu-Hiza
Columbia University, New York, NY, United States

M.B. Smith
MRC Laboratory for Molecular Cell Biology, University College London, London, United Kingdom

D. Streichert
Technische Universität Dresden, Dresden, Germany

S. Sundaramoorthy
Columbia University, New York, NY, United States

D. Tian
Tsinghua University, Beijing, China

P. Ullal
University of Lausanne, Lausanne, Switzerland

C. Wloka
University of Pennsylvania, Philadelphia, PA, United States; University of Groningen, Groningen, The Netherlands

Preface

Cytokinesis is one of the most fundamental processes in cell biology and has fascinated biologists since the invention of microscopes. Generations of zoologists and cell biologists have been working on how cells physically divide into independent daughter cells in a variety of organisms, from echinoderms to salamanders. Years of experimental embryology and cell micromanipulation intimately linked to the names of Walther Flemming and Ray Rappaport revealed how the cytokinetic furrow is positioned and demonstrated the crucial role of microtubules in activating contraction in animal cells. Cells in culture as well as cells in tissues adopt stereotyped cell shape changes culminating with cleavage furrow ingression driven by actin and myosin II cytoskeletons. Bridge stability then depends on the septin cytoskeleton. The final cut or abscission is ultimately driven by the constriction of ESCRT-III helices on the side of the Flemming body. Failure at any of these steps produces binucleated, unstable tetraploid cells that can turn into a starting point for tumorigenesis. The fact that almost 40% of tumors arise from tetraploid cells indicates that failure in cytokinesis is a fundamental cause of tumor initiation.

After 35 years of seminal discoveries, Ray Rappaport confessed: "When I began working on cytokinesis, I thought I was tinkering with a beautifully made Swiss watch, but what I was really working on was an old Maine fishing boat engine: overbuilt, inefficient, never-failed and repaired by simple measures." The aim of this volume of Methods in Cell Biology is to provide state-of-the-art protocols and a variety of modern approaches to study cytokinesis. The molecular mechanisms of cytokinesis are addressed in multiple organisms (echinoderms, flies, worms, yeasts, mice), in individual cultured cells and using in vitro assays. The methods presented in this volume will cover all these experimental models, using genetics, cell biology, microscopy, biochemistry, physics, modeling, and reconstitution assays.

The first two chapters deal with the role of lipids in echinoderm embryos (Chapter 1) and in mammalian cells (Chapter 2), describing the latest advances using probes to follow lipid dynamics during cytokinesis. Protein dynamics throughout cell division in budding yeast will be detailed in Chapter 3. Chapters 4 and 5 will cover the strength of RNAi and genetics in *Drosophila* to describe cytokinesis at the cellular level and at the tissue level, highlighting the importance of neighboring cells during cytokinesis. *Drosophila* is also a remarkable model for studying the evolutionarily conserved phenomenon of incomplete cytokinesis in male and female germ lines (Chapters 6 and 7). Chapters 8 and 9 are dedicated to incomplete cytokinesis and postcytokinetic events including midbody release in mice. The role of the midbody as a platform for abscission and polarity establishment will be addressed in mammalian cells cultured in 3D (Chapter 10) and in 2D human cells using micropatterning techniques (Chapter 11) or superresolution microscopy (Chapter 12). The importance of novel approaches in EM tomography and correlative light-scanning EM for understanding cytokinetic abscission and postabscission events will be highlighted in Chapters 13 and 14. Postabscission events will then be described in vivo in *Caenorhabditis elegans* using high-resolution live cell imaging (Chapter 15). Furrow ingression and stable rings, as well as the use of thermosensitive mutants, to dissect the temporal requirement of particular proteins in *C. elegans* will also be described in Chapters 16 and 17. The next four chapters will cover the latest methodologies used to probe cytokinesis at a physical levels, including physical measurements and modelization in *Dictyostelium* (Chapter 18), active 3D contour for monitoring cytokinesis in mammalian cells (Chapter 19), FRAP measurements in fission yeast (Chapter 20), and single molecule measurements to study polymerization of FtsZ-FtsA copolymers in bacteria (Chapter 21). Finally, the last section is dedicated to reconstitution assays, including the preparation of the centralspindlin complex from human cells and its use in functional assays (Chapter 22), in vitro reactivation of the cytokinetic contractile ring in fission yeast (Chapter 23), and *Xenopus* extract approaches for the study of microtubule organization and signaling in cytokinesis (Chapter 24).

I would like to sincerely thank all my colleagues for their enthusiasm in contributing to this volume and for sharing their experimental secrets, which will be of great help for the community. I would also like to thank Dr. Phong Tran who contacted me for editing this volume as well as the editorial team for their great efficacy. I hope you will enjoy reading this 137th volume of Methods in Cell Biology and that the presented protocols will inspire new discoveries.

Arnaud Echard
Institut Pasteur, Paris

CHAPTER

Labeling membrane domains in dividing echinoderm eggs

1

M. Bennett, J. Meaders, S. Ruvolo, D. Burgess[1]

Boston College, Chestnut Hill, MA, United States

[1]*Corresponding author: E-mail: david.burgess@bc.edu*

CHAPTER OUTLINE

Methods in Cell Biology, Volume 137, ISSN 0091-679X, http://dx.doi.org/10.1016/bs.mcb.2016.04.004

Abstract

Sea urchins have long been used as a model organism to address questions in cell biology and development. Work utilizing this model has elucidated processes of fertilization, the description of centrosomes, the separation of chromosomes during mitosis, and numerous other landmark discoveries. They have been a favorite among researchers for a variety of reasons: Eggs are large and malleable, making them relatively easy to work with; the embryos have synchronized, rapid divisions; they are easy to microinject and manipulate; and their genome sequence is available. In this Chapter, we focus on methods used to study the membrane dynamics during cytokinesis using sea urchin embryos. The focus of this chapter is to provide methods for live-cell imaging of membrane domains in sea urchin embryos.

1. GENERAL CARE AND USE OF SEA URCHINS

Beginning in the 1800's, the first experiments on cell division utilized sea urchins (Wilson, 1928). Since then, they have played an enduring and significant role in pushing the science of cell division and developmental biology forward. For the study of cytokinesis, early research on echinoderm eggs still frames the discussion of models and mechanisms of cytokinesis (Rappaport, 1996).

1.1 GAMETE COLLECTION AND FERTILIZATION PROTOCOL

Detailed methods of care and use of sea urchin eggs for study in cell and developmental biology are found in a methods book (Ettensohn, Wessel, & Wray, 2004). Eggs and sperm can both be retrieved by injecting the urchin with potassium chloride into the gonad wall, causing contraction of the gonads, and release of the gametes. Potassium chloride (0.5 M) can be injected into the urchin body cavity by turning the urchin over. The needle should be tilted to be at an angle away from the mouth during insertion and injection. The KCl serves to induce the gonad to contract and release either sperm or egg. The eggs and sperm released from the gonopores at the top of the urchin, with sperm being white and eggs being a yellowish color.

Sperm should be collected "dry" by micropipetting the shed white sperm from the gonopore directly into an Eppendorf tube. Females can be inverted on a small beaker full of seawater so that the top of the urchin has contact with the water. From there, the female will release eggs for ~10 min into the beaker below.

Fertilization is straightforward but a few steps must be taken. First the egg jelly coat must be swelled to facilitate fertilization by pipetting up and down, then letting the eggs settle. This step should be repeated 2–3 times. This hydrates the jelly coat of the eggs. Once the eggs have settled, an aliquot can be added to a 15 mL conical tube of *para*-aminobenzoate (PABA) water (described later). A small amount ($\sim$8 μL) of collected sperm can be added to a separate 15 mL conical tube of seawater to form a dilution. Next, a few drops of this sperm dilution should be added to the tube of eggs in PABA water, and the combination of eggs and sperm inverted multiple times to ensure adequate mixing. The mixture of sperm and egg should be left for 1 min before using a hand centrifuge to spin the embryos down. The seawater should then be aspirated off, and the embryos should be drawn into a pipet before putting the fertilized eggs though an appropriate seize Nytex mesh into a small beaker of seawater (or modified seawater) to mechanically strip the fertilization envelope off the eggs.

1.2 PREVENTING DI-TYROSINE CROSSLINKING OF THE FERTILIZATION ENVELOPE USING *PARA*-AMINOBENZOATE

Di-tyrosine bonds that form to harden the fertilization envelope can be an impediment to working with clear, naked eggs, which is required for most membrane labeling protocols. The presence of the fertilization envelope blocks access by most of the probes used to study the membrane compartments of the egg. Therefore, PABA can be a powerful tool in preventing the crosslinking. Consisting of a benzene ring substituted with an amino group and a carboxyl group, PABA is an effective nontoxic inhibitor of ovoperoxidase, which is an enzyme that forms these crosslinking bonds. Without the crosslinking of bonds, PABA in combination with Nytex can strip the egg of the fertilization envelope. PABA performs a sort of chemical stripped while Nytex performs a mechanical stripped by choosing a Nytex with a diameter slightly smaller than that of the egg.

1.2.1 para*-Aminobenzoate seawater*

0.062 g/100 mL seawater
Once combined, PABA dilution stays for approximately one week.

1.2.2 Artificial seawaters

Urchins must be fertilized in seawater, but can be raised in modified seawaters such as the calcium-free seawater (CaFSW) listed below. Seawaters should be prepared at room temperature, then chilled or warmed to the temperature that it will be used before bringing water to the correct pH with NaOH. In addition, deionized water should be used. Components can be added to a water volume that is less than the final desired volume, as volume should be adjusted last.

1.2.3 Calcium-free seawater

26.5 g NaCl
0.7 g KCl
5.81 g $MgSO_4$ or 11.9 g $MgSO_4$ $7H_2O$
0.5 g NaHCO3
pH to $\sim$8.2
Salinity to 32–35 ppt or specific gravity to $\sim$1.025

1.2.4 Artificial seawater

28.3 g NaCl
0.77 g KCl
5.41 g $MgCl_2$ $6H_2O$
3.42 g $MgSO_4$ or 7.13 g $MgSO_4$ $7H_2O$
0.2 g $NaHCO_3$
1.56 g $CaCl_2$ $2H_2O$
pH to ~8.2
Salinity to 32–35 ppt or specific gravity to ~1.025

2. PLASMA MEMBRANE LABELS

Labeling the plasma membrane (PM) can be an extremely powerful tool for imaging cytokinesis. Membrane labeling is necessary for visualizing membrane dynamics, tracking cell movement, and identification of unique cells. There exist numerous dyes with differing strengths and weaknesses that also label different components of the PM.

2.1 CELLMASK

CellMask (Life Technologies) is an amphipathic molecule with a negatively charged hydrophilic dye, which aids in anchoring the CellMask probe into the PM. CellMask can be used for both live-cell imaging and in fixed cells for immunofluorescence, and comes in different fluorophores such as orange, deep red, and green. The best procedure for utilizing CellMask involves incubating fertilized demembranated embryos and imaging at a desired developmental stage. It provides a fairly even membrane staining; however, it can be endocytosed relatively quickly so the best images are captured within an hour and a half of addition of the dye.

2.1.1 Incubation procedure for CellMask

1. Fertilize embryos, strip fertilization membrane, and raise embryos according to procedures listed previously.
2. Transfer 1 mL of concentrated embryos to an Eppendorf tube.
3. Add 2 μL of CellMask (Life Technologies) to embryos and incubate shaking **in the dark** for 15 min.
4. Wash embryos 3 times for 1 min with seawater.
5. Prepare a coverslip or chamber slide.
6. Pellet embryos and then pipette up and down to mildly disperse embryos in solution.
7. Transfer embryos in solution to a coverslip for fluorescence imaging.

2.2 FM 4-64

FM 4-64 is a lipophilic styryl dye that inserts into the outer leaflet of the PM (ThermoFisher). Useful for live-cell imaging, FM 4-64 is virtually nontoxic to cells and has been widely used to visualize the PM of sea urchin embryos. However, FM 4-64, like CellMask can begin to be endocytosed relatively quickly. Therefore the best practice is to minimize time between incubation of the sample in the dye and imaging. Incubation of embryos close to the desired developmental stage is preferred. Overall, FM 4-64 is endocytosed more quickly than CellMask by sea urchin eggs. FM 4-64 excited at 515 nm and emits at 640 nm.

2.2.1 Incubation procedure for FM 4-64

1. Fertilize embryos, strip fertilization membrane, and raise embryos according to procedures listed previously.
2. Transfer 1 mL of concentrated embryos to an Eppendorf tube.
3. Add 2 μL of FM 4-64 (ThermoFisher) to embryos and incubate shaking **in the dark** for 15 min.
4. Wash embryos 3 times by gentle hand centrifugation, aspirating the supernatant, and resuspending the embryos in fresh seawater.
5. Prepare a coverslip or chamber slide.
6. Pellet embryos and then pipette up and down to mildly disperse embryos in solution.
7. Transfer embryos in solution to a coverslip for fluorescence imaging.

2.3 CHOLERA TOXIN B AND C_{12} FLUORESCEIN

Cholera toxin B (CTB) and C_{12} fluorescein (5-dodecanoylaminofluorescein) are nontoxic fluorescence probes for investigating the PM. C_{12} fluorescein uniformly labels the PM, while CTB specifically binds to the ganglioside GM1, a marker for lipid rafts. These rafts are a dynamic microdomain within the PM that serve various cellular functions from intracellular signaling to membrane remodeling (Gudejko, Alford, & Burgess, 2012). By labeling cells with CTB and C_{12} fluorescein, further insight into membrane architecture during cytokinesis can be understood.

There are several methods for experimentally investigating lipid raft dynamics using CTB and C_{12} fluorescein labeling. A simple experiment involves incubating fertilized embryos with CTB and C_{12} fluorescein followed by imaging at a desired developmental stage. The position of CTB-tagged lipid rafts can then be determined by comparing CTB fluorescence against the uniform labeling of the PM by C_{12} fluorescein. Our lab has used this type of experiment to examine lipid raft dynamics in the cleavage furrow during cytokinesis (Fig. 1) (Gudejko et al., 2012). Furthermore, dual-labeling with two CTB-conjugated fluorophores (eg, CTB-488 and CTB-647) at distinct time points can allow research into distinct membrane events through observations of colocalization.

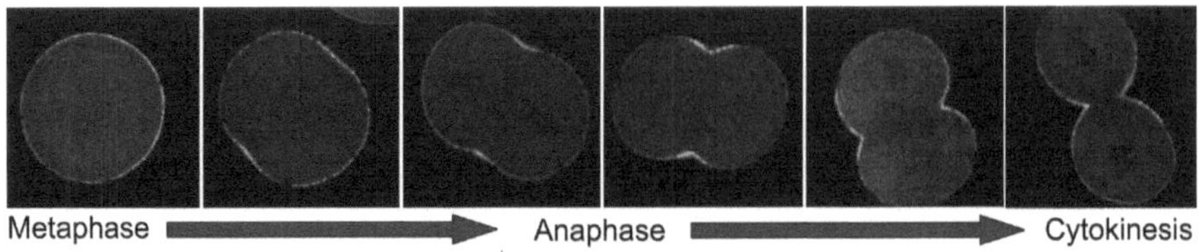

FIGURE 1 Cholera toxin B (CTB)-488 labels GM1 rich microdomains.

Fertilized sea urchin embryos were labeled with CTB and imaged throughout mitosis confocal microscopy. CTB binds to ganglioside GM1, a component of lipid raft microdomains in fertilized sea urchin embryos. During cytokinesis, CTB labeling is observed to migrate to the cleavage furrow.

Live-cell imaging with CTB can be obtained by simply incubating the fluorophore with sea urchin embryos for a desired period of time. Appropriate incubation times can range from 15 to 60 min. CTB actively undergoes endocytosis and recycling when added to embryos, therefore managing incubation periods according to your experiments will be important to monitor.

2.3.1 Incubation procedure for cholera toxin B and/or C_{12} fluorescein

1. Fertilize embryos, strip fertilization membrane, and raise embryos according to procedures provided in the general care section.
2. Transfer 1 mL of concentrated embryos to an Eppendorf tube.
3. Add 1 μL of CTB (5 mg/mL diluted in artificial seawater, Molecular Probes) and/or 1 μL of C_{12} fluorescein (0.5 μg/mL diluted in artificial seawater, Life Technologies) to embryos and incubate shaking in the dark for 15−60 min.
4. Wash embryos 3 times for 1 min with CaFSW.
5. Prepare a coverslip or chamber slide for imaging (Wessel et al., 2004).
6. Pellet embryos and then pipette up and down to mildly disperse embryos in solution.
7. Transfer embryos in solution to a coverslip for fluorescence imaging.

3. INTRACELLULAR LABELING

Intracellular labeling techniques can provide helpful information regarding localization, potential interactions, as well as trafficking patterns during the cell cycle. To visualize these events, many different intracellular probes with specific binding sites can be utilized. For instance, there is the option of labeling various organelles and compartments such as the endoplasmic reticulum, mitochondria, and nucleus. Alternatively, incubating cells with endocytic/exocytic markers can be utilized to analyze trafficking patterns between these various systems, as well as between endosome and lysosome compartments. Thus, a variety of techniques are available to ensure that the intracellular environment is well researched.

3.1 ENDOCYTOSED FLUORESCENT DEXTRAN

Dextran is a low molecular weight polysaccharide bead that can be used for indirect intracellular labeling. It is actively taken up by sea urchin embryos to visualize trafficking events within the cell (Alford, Ng, & Burgess, 2009). Variability between dextran-conjugated fluorophore parameters provides researchers with a wide array of products that can be used in accordance to experimental setups.

The primary benefit of dextran labeling is for visualizing intracellular dynamics (Fig. 2). Cytokinesis is a highly dynamic process requiring the trafficking of proteins to and from the ingressing cleavage furrow, thus making dextran a valuable tool for investigating trafficking. Dextran that is incubated with cells will be taken up through endocytosis and trafficked to various compartments and organelles depending on developmental stage. Furthermore, microinjection of dextran incorporates the label directly into the cytoplasm and can serve as a fluorescent marker for successful injections. These two methods for introducing dextran-conjugated fluorophores only add to the numerous parameters that can be adjusted when commercially ordering it.

Size exclusion properties of dextran molecules are exceptional for experiments designed to examine membrane diffusion and size-specific transport. Various dextran molecules of molecular weights ranging from 10,000 to 500,000 Da can be purchased for the purpose of size-dependent transport research. Furthermore, while the ease of live-cell imaging is an obvious advantage, fixable dextran can be purchased to image fixed cells. Similar to other conjugated probes, dextran can be ordered with a variety of different fluorophores allowing for colabeling. Experiments with multiple intracellular fluorophores can be used to identify whether endocytosis and trafficking events direct cargo to the same or unique compartments (as identified through colocalization). Thus, the incredible variability between dextran-conjugated fluorophores means that dextran can be purchased according to your experimental setup. Interestingly, small latex beads can also be used for biochemical purposes, as they can easily be isolated due to their buoyancy.

As mentioned before, dextran can be introduced to sea urchin embryos through incubation or microinjection. Common incubation periods range from 15 to 60 min depending on the experiment. Regardless of the method, dextran will be actively

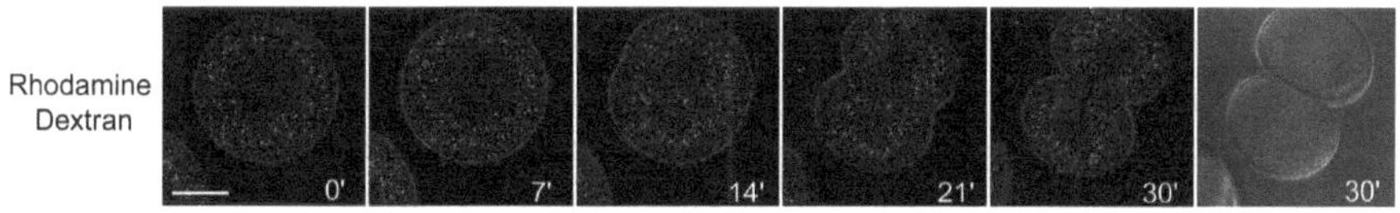

FIGURE 2 Rhodamine dextran undergoes fluid phase endocytosis.

Fertilized sea urchin embryos were labeled with Rhodamine dextran and imaged throughout mitosis using time-lapse fluorescence microscopy. Dextran is actively internalized during mitosis and undergoes intracellular trafficking and endocytosis to different compartments and organelles. After sufficient incubation with embryos, Rhodamine dextran localization can be observed uniformly throughout the cytoplasm within different compartments.

transported throughout the cell as determined by the developmental stage. A longer incubation period will cause this molecule to become internalized and compartmentalized, and thus incubation should be planned according to any imaging constraints.

3.1.1 Incubation procedure for dextran–Texas red

1. Fertilize embryos, strip fertilization membrane, and raise embryos according to procedures provided in the general care section.
2. Transfer 1 mL of concentrated embryos to an Eppendorf tube.
3. Add 5 μL of dextran-Texas red (10 mg/mL diluted in artificial seawater, ThermoFisher) to embryos and incubate shaking in the dark for 15–60 min.
4. Wash embryos 3 times for 1 min with CaFSW.
5. Prepare a coverslip or chamber slide for imaging (Wessel, Voronina, & Brooks, 2004).
6. Pellet embryos and then pipette up and down to mildly disperse embryos in solution.
7. Transfer embryos in solution to a coverslip for fluorescence imaging.

3.1.2 Live-cell microinjection procedure for dextran-Texas red

1. Fertilize embryos, strip fertilization membrane, and raise embryos according to procedures provided in the general care section.
2. Prepare micromanipulator and injection chamber for microinjection (Böttger, Walker, & Unuma, 2004).
3. With fertilized eggs prepared for injection, calibrate the micromanipulator to eject the correct volume of dextran-Texas red (10 mg/mL diluted in artificial seawater) for your sample.
4. Carefully inject several embryos within the field of view, moving between embryos without contacting the coverslip.
5. Incubate cells shaking in the dark for 15–60 min.
6. Pipette off excess seawater from the imaging chamber and carefully replace with fresh CaFSW.
7. Transfer imaging chamber to a fluorescence microscope for imaging.

3.1.3 Fixed cell microinjection procedure for dextran-Texas red (fixable)

1. Fertilize embryos, strip fertilization membrane, and raise embryos according to procedures provided in the general care section.
2. Prepare micromanipulator and coated injection chamber for microinjection (Böttger, Walker, & Unuma, 2004).
3. With fertilized eggs prepared for injection, calibrate the micromanipulator to eject the correct volume of dextran-Texas red (10 mg/mL diluted in artificial seawater) for your sample.
4. Carefully inject several embryos within the field of view, moving between embryos without contacting the coverslip.
5. Incubate cells shaking in the dark for 15–60 min.

6. Pipette off excess seawater from the imaging chamber and carefully replace with fresh CaFSW.
7. Prepare cells for fixation protocol (Strickland, von Dassow, Ellenberg, Foe, Lenart, & Burgess, 2004).
8. Transfer imaging chamber to a fluorescence microscope for imaging.

3.2 RhoA Activity Monitoring With eGFP-rGBD

GTPases are a family of proteins that are activated through GTP binding and are important for regulating different steps of the cell cycle. They are maintained in their active GTP-bound state or inactive GDP-bound state through guanine nucleotide-exchange factors and GTPase activating proteins, respectively. RhoA is a GTPase that localizes to the equatorial cortex of sea urchins prior to cytokinesis and directly controls PM ingression at the division plane. It is an essential component for ingression to occur, and recent studies investigating its dynamics have utilized a reporter system to track RhoA localization in live cells.

Observing the dynamics of RhoA localization during cytokinesis is achievable using a previously reported fluorescent RhoA system designed for *Xenopus*, but applicable to echinoderms (Benink & Bement, 2005). The system utilizes microinjection of mRNA encoding for eGFP bound to the Rho GTPase-binding domain of rhotekin, which binds to active RhoA-GTP (eGFP-rGBD). After microinjection of this reporter construct, tracking RhoA localization throughout mitosis can be achieved with fluorescence microscopy and time-lapse imaging. After injection of approximately 3% of cell volume with eGFP-rGBD, Bement, Benink, and Von Dassow (2005) noticed low levels of RhoA expression prior to anaphase with expression peaking during cytokinesis.

Potential experiments include manipulating upstream RhoA factors and observing changes to the division plane and specifically RhoA confinement. Furthermore, tracking changes to the division plane become simpler, as the recruitment of RhoA to the furrow can predict ingression patterns. Thus, studying the cytoskeletal factors that contribute to RhoA confinement at the cleavage furrow are now achievable and easily obtainable through mRNA microinjection of eGFP-rGBD. Supplemental methods and information regarding the eGFP-rGBD constructs can be found in Benink and Bement (2005), or Bement et al. (2005).

3.3 ENDOPLASMIC RETICULUM LABELING

The endoplasmic reticulum (ER) is a widely distributed membrane compartment that is continuous with the nuclear envelope. It is involved in protein synthesis, lipid formation, and is a primary regulator of intracellular calcium. Visualization of the ER in echinoderm eggs has been achieved using dicarbocyanine dyes and expression of a GFP fusion protein. These experiments have allowed characterization of ER morphology throughout cytokinesis and the rest of the cell cycle in sea urchin embryos (Terasaki, 2000).

3.3.1 DiI method

The first and most commonly used method for labeling the ER is referred to as the DiI method. DiIC$_{18}$(3) and DiIC$_{16}$(3) (D382, D-384; Molecular Probes) are long-chain dicarbocyanine dyes, which become irreversibly incorporated into membrane bilayers through their alkyl chains allowing subsequent lateral diffusion (Axelrod, 1979). DiIC$_{18}$(3) and DiIC$_{16}$(3) are typically dissolved in different types of oil. When small oil droplets are applied to the cell surface using a micromanipulator, the dyes can diffuse into the PM (Fig. 3). DiI$_{18}$(3) in oil has been used to visualize the cleavage furrow in sea urchin embryos by applying a small droplet to the cell surface during metaphase (Ng, Chang, & Burgess, 2005). Within 8–12 min, DiIC$_{18}$(3) is seen enriched in the cleavage furrow.

For ER labeling, DiI dissolved in oil is microinjected into sea urchin eggs (Terasaki & Jaffe, 1991). The oil droplet is immediately seen inside eggs, covering ~5–10% of the internal space (Fig. 4). The dye will diffuse from the droplet to lipid bilayers upon contact, and after ~30 min all of the membranes continuous with and derived from the ER will be labeled. After ER and ER-derived membranes are stained, the oil droplet can be removed using suction from a micropipette if needed.

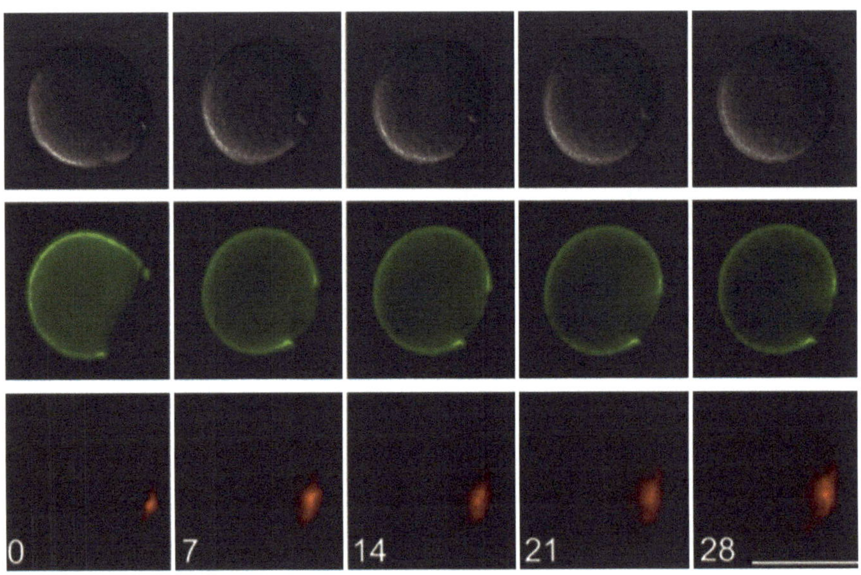

FIGURE 3 Membrane domains label differentially.

Four cell stage sea urchin embryos were dissociated in calcium-free seawater into individual cells. Cholera toxin B (CTB) staining reveals that only the surface containing microvilli is labeled whereas the surface where the blastomeres were in contact are lacking in lipid rafts. Cells were then labeled by application of a droplet of DiIC$_{18}$ to the raft free membrane domain and imaged over time. The DiIC$_{18}$ labeling is excluded from the membrane domain containing lipid rafts.

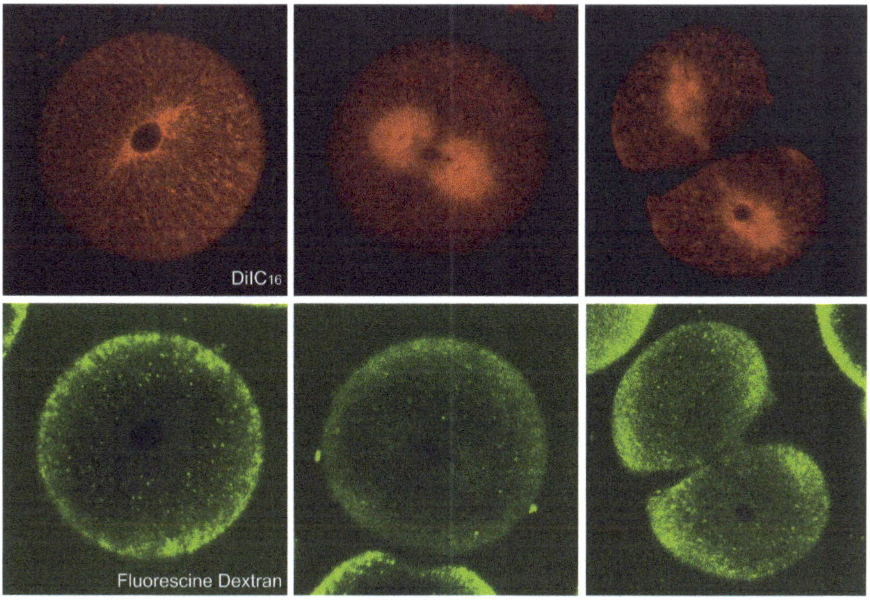

FIGURE 4 DiI injection labels endoplasmic reticulum (ER) and ER-derived lipid bilayers.

$DiIC_{16}$-saturated Wesson oil (red) was microinjected into fertilized sea urchin eggs, which were imaged at prophase, metaphase, and cytokinesis using confocal microscopy. Fluorescein dextran (green) also labels the PM and endocytic vesicles. Overlay of the two channels shows that $DiIC_{16}$ diffuses into ER and ER-derived lipid bilayers but not endocytic lipid bilayers.

To prepare an injection solution, we dissolve several crystals (or several milligrams) in 100–300 μL of Wesson soybean oil using a 1.5 mL microfuge tube. However, olive, canola, and walnut oil can also be used (Jaffe & Terasaki, 2004). DiI dissolves slowly in most oils, so it can either be left for a few hours or overnight, followed by removal of the remaining crystals with centrifugation. Alternatively, we have had success grinding the crystals directly into the oil using a pestle compatible with 1.5 mL microfuge tubes, allowing full dissolution of DiI. The dissolved dye can then be left at room temperature protected from light for at least several months.

Following injection, cells can be cultured until they reach a desired developmental stage and mounted on slides for visualization. $DiI_{16}(3)$ and $DiI_{18}(3)$ both contain a fluorophore with an emission spectra of 570–590 nm. Therefore, a microscope equipped with the appropriate excitation source will allow for detailed imaging of ER dynamics.

3.3.2 GFP-KDEL

The second, more specific method to visualize the ER in echinoderm eggs is with the use of a GFP-KDEL construct (Terasaki, Jaffe, Hunnicutt, & Hammer, 1996). GFP is normally a cytosolic soluble protein, so two key elements were considered when designing GFP-KDEL to ensure it labels the ER. First, proteins being translated in the cytosol can only be brought to the ER when a signal sequence is present. This signal sequence is recognized by the signal recognition particle (SRP), which stops translation until the translation machinery binds to a receptor on the ER. Upon binding, translation resumes and directs newly made protein into the ER. Therefore, the N-terminal portion of GFP-KDEL was engineered with a signal sequence from ECast/PDI, a protein localized to the ER lumen in sea urchins (Lucero, Lebeche, & Kaminer, 1994). The second consideration when engineering GFP-KDEL is that any proteins without ER retention sequences are transported out of the lumen. KDEL is a C-terminal retention sequence that traps proteins in the ER (Munro & Pelham, 1987), and as the name suggests, it keeps GFP-KDEL from exiting the lumenal space.

Because GFP-KDEL must be translated and retained in the ER, microinjection of purified recombinant protein will not work. Therefore, microinjection of purified mRNA encoding the protein is used (Terasaki et al., 1996). To obtain mRNA, GFP-KDEL is cloned into the vector pSP64-RI (Tang, Freeman, O'Rielly, Neel, & Sokol, 1995; Terasaki et al., 1996). pSP64-RI contains a promoter for RNA polymerase SP6, which is useful for making mRNA in vitro. The vector can be replicated in bacteria and later purified for making mRNA using standard techniques. Finally, mRNA can be made from mRNA synthesis kits (Ambion) and stored at −70°C until needed for injection.

Once mRNA is successfully prepared, ~10−20 μg/mL can be injected into sea urchin oocytes or fertilized eggs (Terasaki et al., 1996). In oocytes, GFP signal begins to appear in the ER between 1 and 2 h postinjection and plateaus at around 24 h. In fertilized eggs GFP fluorescence is seen in the ER at around the fourth to fifth cleavage. Furthermore, GFP-KDEL has no apparent transfer of signal to membranes outside the ER, providing researchers with more specific labeling than the DiI method (Terasaki et al., 1996).

ACKNOWLEDGMENTS

This work was supported in part by NSF Grant 1244425 to DB.

REFERENCES

Alford, L. M., Ng, M. M., & Burgess, D. R. (2009). Cell polarity emerges at first cleavage in sea urchin embryos. *Developmental Biology, 330*(1), 12−20.

Axelrod, D. (1979). Carbocyanine dye orientation in red cell membrane studied by microscopic fluorescence polarization. *Biophysical Journal, 26*(3), 557−573.

Bement, W. M., Benink, H. A., & Von Dassow, G. (2005). A microtubule-dependent zone of active RhoA during cleavage plane specification. *Journal of Cell Biology, 170*(1), 91−101.

Benink, H. A., & Bement, W. M. (2005). Concentric zones of active RhoA and Cdc42 around single cell wounds. *Journal of Cell Biology, 168*(3), 429–439.

Böttger, S. A., Walker, C. W., & Unuma, T. (2004). Care and maintenance of adult echinoderms. In C. A. Ettensohn, G. M. Wessel, & G. A. Wray (Eds.), *Methods in cell biology: 74. Development of sea urchins, ascidians, and other invertebrate deuterostomes: experimental approaches* (pp. 17–38).

Ettensohn, C. A., Wessel, G. M., & Wray, G. A. (Eds.). (2004), *Methods cell biology: Vol. 74. Development of sea urchins, ascidians, and other invertebrate deuterostomes: Experimental approaches*. San Diego, CA: Elsevier Academic Press.

Gudejko, H. F. M., Alford, L. M., & Burgess, D. R. (2012). Polar expansion during cytokinesis. *Cytoskeleton, 69*(11), 1000–1009.

Jaffe, L. A., & Terasaki, M. (2004). Quantitative microinjection of oocytes, eggs, and embryos. In C. A. Ettensohn, G. M. Wessel, & G. A. Wray (Eds.), *Methods in cell biology: 74. Development of sea urchins, ascidians, and other invertebrate deuterostomes: experimental approaches* (pp. 219–242).

Lucero, H. A., Lebeche, D., & Kaminer, B. (1994). ER calcistorin/protein difulfide isomerase (PDI). Sequence determination and expression of a cDNA clone encoding a calcium storage protein with PDI activity from endoplasmic reticulum of the sea urchin egg. *Journal of Biological Chemistry, 269*, 23112–23119.

Munro, S., & Pelham, H. R. B. (1987). A C-terminal signal prevents secretion of luminal ER proteins. *Cell, 48*(5), 899–907.

Ng, M. M., Chang, F., & Burgess, D. R. (2005). Movement of membrane domains and requirement of membrane signaling molecules for cytokinesis. *Developmental Cell, 9*(6), 781–790.

Rappaport, R. (1996). *Cytokinesis in animal cells*. Cambridge University Press.

Strickland, L., von Dassow, G., Ellenberg, J., Foe, V., Lenart, P., & Burgess, D. (2004). Light microscopy of echinoderm embryos. In C. A. Ettensohn, G. M. Wessel, & G. A. Wray (Eds.), *Methods in cell biology: 74. Development of sea urchins, ascidians, and other invertebrate deuterostomes: experimental approaches* (pp. 371–409).

Tang, T. L., Freeman, R. M., Oreilly, A. M., Neel, B. G., & Sokol, S. Y. (1995). The Sh2-containing protein-tyrosine-phosphatase Sh-Ptp2 is required upstream of map kinase for early *Xenopus* development. *Cell, 80*(3), 473–483.

Terasaki, M. (2000). Dynamics of the endoplasmic reticulum and golgi apparatus during early sea urchin development. *Molecular Biology of the Cell, 11*(3), 897–914.

Terasaki, M., & Jaffe, L. A. (1991). Organization of the sea urchin egg endoplasmic reticulum and its reorganization at fertilization. *Journal of Cell Biology, 114*(5), 929–940.

Terasaki, M., Jaffe, L. A., Hunnicutt, G. R., & Hammer, J. A. (1996). Structural change of the endoplasmic reticulum during fertilization: evidence for loss of membrane continuity using the green fluorescent protein. *Developmental Biology, 179*(2), 320–328.

Wessel, G. M., Voronina, E., & Brooks, J. M. (2004). Obtaining and handling echinoderm oocytes. In C. A. Ettensohn, G. M. Wessel, & G. A. Wray (Eds.), *Methods in cell biology: 74. Development of sea urchins, ascidians, and other invertebrate deuterostomes: experimental approaches* (pp. 87–114).

Wilson, E. B. (1928). *The cell in development and heredity* (3rd ed.). New York: The Macmillan Company.

CHAPTER

Dynamics of sphingomyelin- and cholesterol-enriched lipid domains during cytokinesis

2

M. Abe*, T. Kobayashi*,§,1

RIKEN, Wako, Saitama, Japan

§CNRS, Illkirch, France

[1]Corresponding author: E-mail: kobayasi@riken.jp

CHAPTER OUTLINE

Abstract

Sphingomyelin (SM) and cholesterol (Chol) are the major lipids in the mammalian cells, which are mainly localized to the plasma membrane. Multiple lines of evidence suggest that these lipids form local lipid domains in the plasma membrane, playing functional roles in the cell. Several observations have suggested that these lipid domains are required for cytokinesis. In this chapter, we show the methods for visualizing SM-rich and/or Chol-rich membrane domains at cytokinesis by using specific lipid-binding proteins. Lysenin, equinatoxin II, perfringolysin O, and pleurotolysin A2 bind specifically to clustered SM-rich domain, dispersed SM-rich domain, Chol-rich domain, and SM/Chol mixtures, respectively. Nontoxic forms of these lipid-binding proteins fused to fluorescent proteins are used for imaging lipid domains in biological membranes at cytokinesis. The image analysis reveals the structures and functions of SM-rich and/or Chol-rich domains at the time of cytokinesis.

Methods in Cell Biology, Volume 137, ISSN 0091-679X, http://dx.doi.org/10.1016/bs.mcb.2016.03.030

INTRODUCTION

Sphingomyelin (SM) is one of the major lipids in the mammalian plasma membrane, accounting for 2–15% of the total phospholipids (Koval & Pagano, 1991). SM has been shown to form local clusters under particular conditions in model membranes (Pandit, Jakobsson, & Scott, 2004; Pandit et al., 2004). One attractive model of the presence of SM-rich domains in biological membranes is the "lipid raft" concept (Simons & Ikonen, 1997). Lipid rafts have been suggested to contain certain characteristic lipids, including SM, glycosphingolipids, and cholesterol (Chol), playing important roles in signal transduction, membrane trafficking, cytoskeleton, and pathogen entry (Pike, 2006). Since most of the evidence for lipid raft relied on indirect methods such as detergent extraction, structure, function, and even existence of raft have been under debate. Imaging the lipid components of the lipid raft may provide conclusive evidence showing the structures and functions.

Several observations have suggested that lipid rafts are involved in cytokinesis (Atilla-Gokcumen et al., 2014; Gudejko, Alford, & Burgess, 2012; Ng, Chang, & Burgess, 2005). Depletion of cholesterol or inhibition of its synthesis impairs cytokinesis (Feng, Schwarz, & Jesuthasan, 2002; Fernandez, Lobo Md Mdel, Gomez-Coronado, & Lasuncion, 2004). Depletion of SM disperses PIP_2 and inhibits the recruitment of the small GTPase RhoA to the cleavage furrow, leading to abnormal cytokinesis (Abe et al., 2012). Together with these observations, imaging the lipid components of lipid raft supports the idea that lipid raft is required for cytokinesis in mammalian cells (Abe et al., 2012; Makino et al., 2015; Ng et al., 2005). In this chapter, we focus on SM-rich and/or Chol-rich binding proteins to observe these lipid domains in biological membrane at cytokinesis. As SM-rich and/or Chol-rich binding proteins, we show here lysenin (Lys), equinatoxin II (EqtII), perfringolysin O (PFO), and pleurotolysin A2 (PlyA2).

Lys is a 297 amino acid protein, which was originally isolated from the coelomic fluid of the earthworm *Eisenia foetida* (Sekizawa, Kubo, Kobayashi, Nakajima, & Natori, 1997). Lys specifically binds to SM (Bruhn, Winkelmann, Andersen, Andra, & Leippe, 2006; Yamaji et al., 1998), especially clustered SM composed of fewer than 10 SM molecules (Makino et al., 2015). Lys has a cytolytic activity to vertebrate spermatozoa, amphibian larvae, and cultured mammalian cells (Hanada et al., 1998; Kobayashi, Sekizawa, Aizu, & Umeda, 2000; Kobayashi, Suzuki, & Ohta, 2006; Yamaji et al., 1998). Due to the cytolytic activity, full-length Lys is not suitable for use in living cells as an SM-specific probe. Study with several truncated Lys indicates that the fragment 161–297 amino acid (NT-Lys) has no pore-forming activity and retains its binding specificity to SM (Kiyokawa et al., 2005).

The EqtII is a member of the actinoporins and was isolated from the sea anemone, *Actinia equina*. The EqtII belongs to a family of pore-forming toxins showing hemolytic activity to mammalian cells. Biochemical assays showed that EqtII specifically binds SM (Bakrac et al., 2008), especially dispersed SM rather than clustered SM (Makino et al., 2015). Pore formation by EqtII is multiple processes, which include binding to SM, intramolecular conformational change, oligomerization on the membrane, and formation of the pore. Inhibition of intramolecular conformational change by disulfide bridge formation in the mutant EqtII(8−69) prevents the hemolytic activity without perturbing SM-binding activity (Hong et al., 2002; Kristan et al., 2004; Rojko et al., 2013).

PFO, also referred as θ toxin, secreted from *Clostridium perfringens* binds to Chol-rich domains both in model and biological membranes (Verherstraeten et al., 2015). PFO exhibits a pore-forming activity in a Chol-dependent manner and causes cytolytic activity to mammalian membranes (Heuck, Savva, Holzenburg, & Johnson, 2007; Hotze & Tweten, 2012). Study with several truncated fragments of PFO indicates C-terminal domain (D4) of PFO is the smallest functional unit that retains the Chol-binding ability without cytolytic activity (Shimada, Maruya, Iwashita, & Ohno-Iwashita, 2002).

Recently, we identified a novel protein, which binds to SM and Chol mixtures (Bhat et al., 2013). We screened for proteins binding specifically to SM/Chol liposomes from edible mushroom *Pleurotus eryngii*. Biochemical analyses indicate that PlyA2, a pleurotolysin ortholog from *P. eryngii*, binds to SM/Chol mixtures but not to phosphatidylcholine/Chol mixtures in model membranes. PlyA2 also binds ceramide phosphoethanolamine, a quite minor sphingolipid in mammalian cells (Bhat et al., 2015). PlyA2 does not have cytolytic activity to mammalian cells.

Using these lipid-binding proteins, SM-rich and/or Chol-rich domains in the outer leaflet of the plasma membrane have been visualized under a confocal microscope. First, the lipid-binding proteins fused to fluorescent proteins are expressed in and purified from *Escherichia coli*. Subsequently, lipid domains in the outer leaflet of the plasma membrane in living mammalian cells are stained by exogenously added nontoxic recombinant proteins. Here we show the methods for visualizing SM-enriched and/or Chol-enriched domains in cell membrane.

1. METHODS

1.1 PREPARATION OF LIPID PROBES

To prepare the lipid probes, the lipid-binding proteins fused to fluorescent proteins are produced in *E. coli*. Since all fusion proteins are tagged with poly histidine, the recombinant proteins can be purified with Ni^{2+} or Co^{2+} columns. In the case of EqtII(8−69)-EGFP, the recombinant protein is oxidized after the purification to prevent its cytolytic activity.

1. Transform the plasmids expressing lipid probes into *E. coli* strain BL-21 (DE3).
2. Grow the bacterial culture in 250 mL lysogeny broth (LB) medium with 50 μg/mL kanamycin until an optical density (600 nm) reaches 0.2 at 30°C.
3. Add 250 mL of LB containing 2M sorbitol, 50 μg/mL kanamycin, 5 mM betaine hydrochloride, and 1 mM isopropyl-D-thiogalactopyranoside, and further incubate overnight at 30°C. Osmotic stress in the presence of sorbitol

and betaine has been shown to increase the yield of recombinant proteins (Blackwell & Horgan, 1991).

4. Harvest the bacterial cells by centrifugation at 5000 × g for 10 min at 4°C. The pellet can be stored at −80°C.
5. Suspend the pellet in binding buffer consisting of 20 mM sodium phosphate buffer (pH 7.4) containing 500 mM NaCl, 20 mM imidazole, and protease inhibitors.
6. Disrupt the cells by sonication on ice. We use a Misonix XL2020 sonicator with a 1/8″ probe (30 × 5-s pulses, a 10-s incubation on ice in between, 40% full power).
7. Remove insoluble protein and cell debris by centrifugation at 9000 × g at 4°C for 20 min. Transfer the supernatant carefully, to avoid disturbing the pellet, to a fresh tube.
8. For purification of recombinant proteins, use Ni^{2+} (GE Healthcare, UK, No. 29048631) or Co^{2+} columns (GE Healthcare, UK, No. 29048565). Elute recombinant proteins from columns by elution buffer consisting of 20 mM sodium phosphate buffer (pH 7.4) containing 500 mM NaCl and 500 mM imidazole.
9. Suspend the obtained fractions in 15 mL of PBS and concentrate them by using an Amicon Ultra-15 Ultracel-30K centrifugal unit.
10. In the case of EqtII(8−69), the fusion protein should be oxidized to form a disulfide bridge between Cys8 and Cys69, preventing the formation of pores (Hong et al., 2002; Kristan et al., 2004). Incubate the EqtII(8−69) fusion protein in 0.5 mM 1,10-phenantroline and 0.1 mM $CuSO_4$ for 1 h at 30°C. Suspend the resulting fractions in 15 mL of PBS and concentrate them by using an Amicon Ultra-15 Ultracel-30K centrifugal unit.
11. Add 30% of glycerol, and store at −20°C.

Notes

- For step 9, stepwise gradient elution can highly purify the recombinant proteins with low yield. However, we give priority to yield over purity since extracts from *E. coli* hardly influence the binding ability of the lipid-binding proteins.
- Here we describe lipid-binding proteins fused to fluorescent proteins. Lipid-binding proteins can also be labeled with a fluorescent dye such as Alexa Fluor 647 using labeling kit (Thermo Fisher Scientific, MA, No. A20173) (Abe et al., 2012).

Materials

Plasmids for expressing lipid-binding proteins fused to EGFP (pET28/His-EGFP-NT-Lys, pET28/EqtII(8−69)-EGFP-His, pET28/His-EGFP-D4 and pET28/PlyA2-EGFP-His) can be obtained from RIKEN BRC (http://dna.brc.riken.jp/).

Ni^{2+} or Co^{2+} columns. We use HisTrap FF crude (GE Healthcare, UK, No. 29048631), HiTrap TALON crude (GE Healthcare, UK, No. 29048565), or Ni-NTA superflow columns (Qiagen, Germany, No. 30622).

Amicon Ultra-15 Ultracel-30K centrifugal unit (Merck Millipore, Germany, No. UFC903008).
Sonicator. Any sonicators for disrupting cells can be used. We use a Misonix XL2020 sonicator with a 1/8″ probe (Misonix, NY).

1.2 SYNCHRONIZING HeLa CELLS

To enrich HeLa cells at the time of cytokinesis, mitotic cells are collected and released from G2/M arrest.

1. Grow HeLa cells in a T-25 flask at 37°C in Dulbecco's modified Eagle's medium (DMEM) supplemented with 10% fetal bovine serum (FBS).
2. For synchronizing HeLa cells, incubate HeLa cells in the presence of 40 ng/mL nocodazole for 3 h, and harvest mitotic cells by mechanical shake-off.
3. Plate the harvested cells in a poly-D-lysine-coated dish, and incubate in the presence of 40 ng/mL nocodazole for additional 30 min.
4. Wash nocodazole out, and the cells are incubated in DMEM supplemented with 10% FBS for 1 h to release the cells from the mitotic arrest.
   Notes
   - During the mechanical shake-off, the efficiency should be monitored under a light microscope.
   - Longer incubation in the presence of nocodazole may take longer to release the cells from mitosis in the fresh medium.

   Materials
   T-25 flask (Thermo Fisher Scientific, MA or Sigma-Aldrich, MO).
   Poly-D-lysine-coated dish (Corning, NY, No. 354077).

1.3 STAINING CELLS WITH LIPID PROBES

1. Replace the medium to DMEM supplemented with 10% lipoprotein deficient serum (LPDS).
2. Stain the cells with 10 μg/mL of the recombinant proteins.
3. Incubate the cells at 37°C for 10 min.
4. Observe the living cells under a confocal microscope (typical images are shown in Fig. 1) without washing.
   Notes
   - Since lipid-binding proteins bind lipoproteins, it is better to use LPDS instead of FBS. Due to this, when FBS is used, a high concentration of lipid-binding protein is needed. Cells do not divide in the absence of serum. Either LPDS or FBS is required for progression of cytokinesis.
   - If you need to fix cells, cells are incubated with 4% paraformaldehyde and 0.2% glutaraldehyde for 30 min after step 3, because chemical fixation does not immobilize lipid molecules (Tanaka et al., 2010). Sometimes fixation before labeling gives different signals compared with living cell staining.

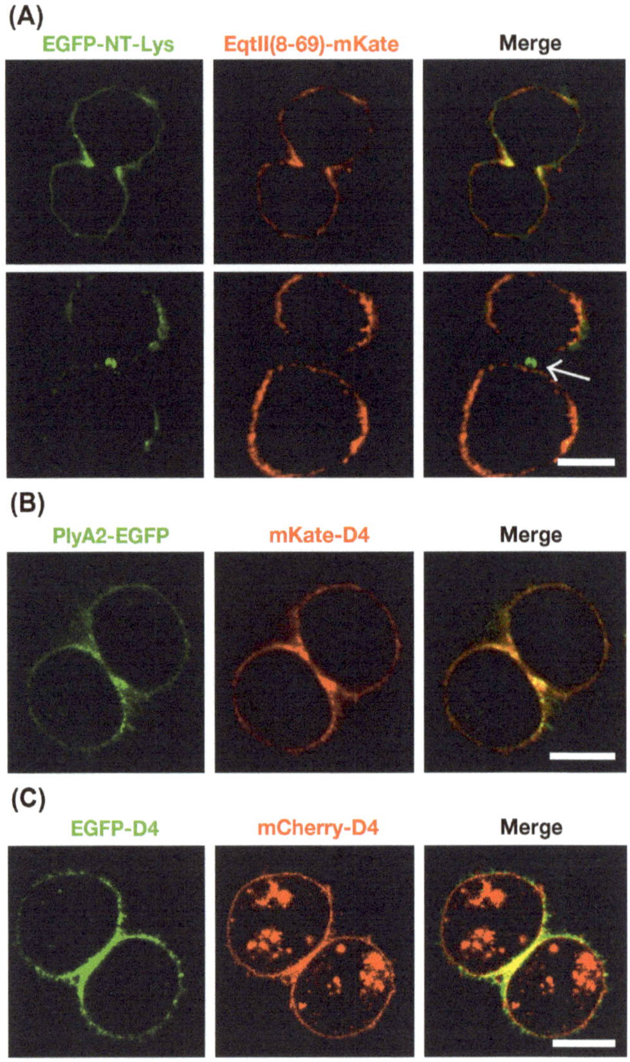

FIGURE 1 Typical images of sphingomyelin (SM)-rich and/or cholesterol (Chol)-rich domains in the cleavage furrow.

(A) Selective accumulation of SM clusters to the midbody. HeLa cells were synchronized, and cells were doubly labeled with EGFP-NT-Lys (green) and EqtII(8–69)-mKate (red). The *arrow* indicates the midbody. (B) Chol-rich domains and SM/Chol domains in the outer leaflet of the plasma membrane at the time of cytokinesis. Synchronized HeLa cells were doubly labeled with PlyA2-EGFP (green) and mKate-D4 (red). (C) Chol-rich domains in the outer but not the inner leaflet of the plasma membrane accumulated in the cleavage furrow. HeLa cells transiently expressing mCherry-D4 (red) were stained with exogenous recombinant EGFP-D4 (green). Bars, 10 μm.

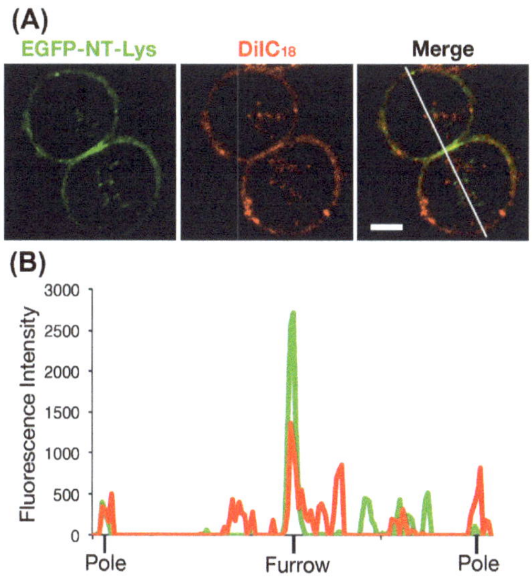

FIGURE 2 Fluorescence intensity of EGFP-NT-Lys and a nonspecific membrane probe.

(A) Clustered sphingomyelin labeled with NT-Lys. HeLa cells were stained with EGFP-NT-Lys (green) and DiIC$_{18}$ (red). Bar, 5 μm. (B) Quantitative analysis of fluorescence intensity. Fluorescence intensity of EGFP-NT-Lys (green) and DiIC$_{18}$ (red) was measured along the white line shown in (A).

- For staining of the Chol-rich domains at the inner leaflet of the plasma membrane, the labeled D4 is transiently expressed in the cells after plasmid transfection (Fig. 1C). It should be noted that artificial aggregates of the labeled D4 are seen in the cells when the labeled D4 is expressed.
- To exclude the possibility that the increased fluorescence intensity is due to the closely apposed membranes in the furrow region, it is important to compare the lipid-binding protein staining with that of a nonspecific membrane probe, such as 1,1′-dioctadecyl-3,3,3′,3′-tetramethylindocarbocyanine perchlorate (DiIC$_{18}$) as shown in Fig. 2.

Materials

LPDS (Sigma-Aldrich, MO, No. S5394).

A confocal microscope equipped with an environmental chamber maintained with humidity, 37°C, and 5% CO_2. We use an FV 1000 confocal microscope with a 60× 1.1 NA PlanApo objective lens (Olympus, Japan).

CONCLUSION

Here we presented methods for visualizing endogenous SM-rich and/or Chol-rich membrane domains at cytokinesis by using specific lipid-binding proteins. Imaging

SM-rich and/or Chol-rich membrane domains has provided evidence showing their structures and functions at the time of cytokinesis. The combined use of lipid-binding proteins and several fluorescent proteins and/or small molecules coupled with advanced microscopy will ultimately reveal accurate dynamics and organization of SM-rich and/or Chol-rich domains in cell membranes.

ACKNOWLEDGMENT

The work of authors' laboratory was supported by the Integrated Lipidology Program of RIKEN, and the Grant-in-Aid for Scientific Research from the Ministry of Education, Culture, Sports, Science and Technology of Japan (25293015 and 15K07957) and Mizutani Foundation.

REFERENCES

Abe, M., Makino, A., Hullin-Matsuda, F., Kamijo, K., Ohno-Iwashita, Y., Hanada, K., ... Kobayashi, T. (2012). A role for sphingomyelin-rich lipid domains in the accumulation of phosphatidylinositol-4,5-bisphosphate to the cleavage furrow during cytokinesis. *Molecular and Cellular Biology, 32*, 1396–1407.

Atilla-Gokcumen, G. E., Muro, E., Relat-Goberna, J., Sasse, S., Bedigian, A., Coughlin, M. L., ... Eggert, U. S. (2014). Dividing cells regulate their lipid composition and localization. *Cell, 156*, 428–439.

Bakrac, B., Gutierrez-Aguirre, I., Podlesek, Z., Sonnen, A. F., Gilbert, R. J., Macek, P., ... Anderluh, G. (2008). Molecular determinants of sphingomyelin specificity of a eukaryotic pore-forming toxin. *The Journal of Biological Chemistry, 283*, 18665–18677.

Bhat, H. B., Ishitsuka, R., Inaba, T., Murate, M., Abe, M., Makino, A., ... Kobayashi, T. (2015). Evaluation of aegerolysins as novel tools to detect and visualize ceramide phosphoethanolamine, a major sphingolipid in invertebrates. *FASEB Journal: Official Publication of the Federation of American Societies for Experimental Biology, 29*, 3920–3934.

Bhat, H. B., Kishimoto, T., Abe, M., Makino, A., Inaba, T., Murate, M., ... Kobayashi, T. (2013). Binding of a pleurotolysin ortholog from *Pleurotus eryngii* to sphingomyelin and cholesterol-rich membrane domains. *Journal of Lipid Research, 54*, 2933–2943.

Blackwell, J. R., & Horgan, R. (1991). A novel strategy for production of a highly expressed recombinant protein in an active form. *FEBS Letters, 295*, 10–12.

Bruhn, H., Winkelmann, J., Andersen, C., Andra, J., & Leippe, M. (2006). Dissection of the mechanisms of cytolytic and antibacterial activity of lysenin, a defence protein of the annelid *Eisenia fetida*. *Developmental and Comparative Immunology, 30*, 597–606.

Feng, B., Schwarz, H., & Jesuthasan, S. (2002). Furrow-specific endocytosis during cytokinesis of zebrafish blastomeres. *Experimental Cell Research, 279*, 14–20.

Fernandez, C., Lobo Md Mdel, V., Gomez-Coronado, D., & Lasuncion, M. A. (2004). Cholesterol is essential for mitosis progression and its deficiency induces polyploid cell formation. *Experimental Cell Research, 300*, 109–120.

Gudejko, H. F., Alford, L. M., & Burgess, D. R. (2012). Polar expansion during cytokinesis. *Cytoskeleton (Hoboken, NJ), 69*, 1000–1009.

Hanada, K., Hara, T., Fukasawa, M., Yamaji, A., Umeda, M., & Nishijima, M. (1998). Mammalian cell mutants resistant to a sphingomyelin-directed cytolysin. Genetic and biochemical evidence for complex formation of the LCB1 protein with the LCB2 protein for serine palmitoyltransferase. *The Journal of Biological Chemistry, 273*, 33787–33794.

Heuck, A. P., Savva, C. G., Holzenburg, A., & Johnson, A. E. (2007). Conformational changes that effect oligomerization and initiate pore formation are triggered throughout perfringolysin O upon binding to cholesterol. *The Journal of Biological Chemistry, 282*, 22629–22637.

Hong, Q., Gutierrez-Aguirre, I., Barlic, A., Malovrh, P., Kristan, K., Podlesek, Z., … Anderluh, G. (2002). Two-step membrane binding by Equinatoxin II, a pore-forming toxin from the sea anemone, involves an exposed aromatic cluster and a flexible helix. *The Journal of Biological Chemistry, 277*, 41916–41924.

Hotze, E. M., & Tweten, R. K. (2012). Membrane assembly of the cholesterol-dependent cytolysin pore complex. *Biochimica et Biophysica Acta, 1818*, 1028–1038.

Kiyokawa, E., Baba, T., Otsuka, N., Makino, A., Ohno, S., & Kobayashi, T. (2005). Spatial and functional heterogeneity of sphingolipid-rich membrane domains. *The Journal of Biological Chemistry, 280*, 24072–24084.

Kobayashi, H., Sekizawa, Y., Aizu, M., & Umeda, M. (2000). Lethal and non-lethal responses of spermatozoa from a wide variety of vertebrates and invertebrates to lysenin, a protein from the coelomic fluid of the earthworm *Eisenia foetida*. *Journal of Experimental Zoology, 286*, 538–549.

Kobayashi, H., Suzuki, H., & Ohta, N. (2006). Exfoliation of the epidermal cells and defecation by amphibian larvae in response to coelomic fluid and lysenin from the earthworm *Eisenia foetida*. *Biomedical Research, 27*, 169–181.

Koval, M., & Pagano, R. E. (1991). Intracellular transport and metabolism of sphingomyelin. *Biochimica et Biophysica Acta, 1082*, 113–125.

Kristan, K., Podlesek, Z., Hojnik, V., Gutierrez-Aguirre, I., Guncar, G., Turk, D., … Anderluh, G. (2004). Pore formation by equinatoxin, a eukaryotic pore-forming toxin, requires a flexible N-terminal region and a stable beta-sandwich. *The Journal of Biological Chemistry, 279*, 46509–46517.

Makino, A., Abe, M., Murate, M., Inaba, T., Yilmaz, N., Hullin-Matsuda, F., … Kobayashi, T. (2015). Visualization of the heterogeneous membrane distribution of sphingomyelin associated with cytokinesis, cell polarity, and sphingolipidosis. *FASEB Journal: Official Publication of the Federation of American Societies for Experimental Biology, 29*, 477–493.

Ng, M. M., Chang, F., & Burgess, D. R. (2005). Movement of membrane domains and requirement of membrane signaling molecules for cytokinesis. *Developmental Cell, 9*, 781–790.

Pandit, S. A., Jakobsson, E., & Scott, H. L. (2004). Simulation of the early stages of nano-domain formation in mixed bilayers of sphingomyelin, cholesterol, and dioleoylphosphatidylcholine. *Biophysical Journal, 87*, 3312–3322.

Pandit, S. A., Vasudevan, S., Chiu, S. W., Mashl, R. J., Jakobsson, E., & Scott, H. L. (2004). Sphingomyelin-cholesterol domains in phospholipid membranes: atomistic simulation. *Biophysical Journal, 87*, 1092–1100.

Pike, L. J. (2006). Rafts defined: a report on the keystone symposium on lipid rafts and cell function. *Journal of Lipid Research, 47*, 1597–1598.

Rojko, N., Kristan, K. C., Viero, G., Zerovnik, E., Macek, P., Dalla Serra, M., & Anderluh, G. (2013). Membrane damage by an alpha-helical pore forming protein equinatoxin II proceeds through succesion of ordered steps. *The Journal of Biological Chemistry, 288*(33).

Sekizawa, Y., Kubo, T., Kobayashi, H., Nakajima, T., & Natori, S. (1997). Molecular cloning of cDNA for lysenin, a novel protein in the earthworm *Eisenia foetida* that causes contraction of rat vascular smooth muscle. *Gene, 191*, 97–102.

Shimada, Y., Maruya, M., Iwashita, S., & Ohno-Iwashita, Y. (2002). The C-terminal domain of perfringolysin O is an essential cholesterol-binding unit targeting to cholesterol-rich microdomains. *European Journal of Biochemistry, 269*, 6195–6203.

Simons, K., & Ikonen, E. (1997). Functional rafts in cell membranes. *Nature, 387*, 569–572.

Tanaka, K. A., Suzuki, K. G., Shirai, Y. M., Shibutani, S. T., Miyahara, M. S., Tsuboi, H., … Kusumi, A. (2010). Membrane molecules mobile even after chemical fixation. *Nature Methods, 7*, 865–866.

Verherstraeten, S., Goossens, E., Valgaeren, B., Pardon, B., Timbermont, L., Haesebrouck, F., … Van Immerseel, F. (2015). Perfringolysin O: the underrated *Clostridium perfringens* toxin? *Toxins (Basel), 7*, 1702–1721.

Yamaji, A., Sekizawa, Y., Emoto, K., Sakuraba, H., Inoue, K., Kobayashi, H., & Umeda, M. (1998). Lysenin, a novel sphingomyelin-specific binding protein. *The Journal of Biological Chemistry, 273*, 5300–5306.

CHAPTER

Analysis of protein dynamics during cytokinesis in budding yeast

3

S. Okada[*,§], **C. Wloka**[*,¶], **E. Bi**[*,1]

*University of Pennsylvania, Philadelphia, PA, United States

[§]Kyushu University, Fukuoka, Japan

[¶]University of Groningen, Groningen, The Netherlands

[1]Corresponding author: E-mail: ebi@mail.med.upenn.edu

CHAPTER OUTLINE

Methods in Cell Biology, Volume 137, ISSN 0091-679X, http://dx.doi.org/10.1016/bs.mcb.2016.04.002

Abstract

Cytokinesis is essential for development and survival of all organisms by increasing cell number and diversity. It is a highly regulated process that requires spatiotemporal coordination of hundreds of proteins functioning in the assembly, constriction, and disassembly of a contractile actomyosin ring, targeted vesicle fusion, and localized extracellular matrix remodeling. Cytokinesis has been studied in multiple systems with a wide range of technologies to learn the common principles. In this chapter, we describe the analysis of protein dynamics during cytokinesis in the budding yeast *Saccharomyces cerevisiae* by several live-cell imaging methods. This, in combination with the power of yeast genetics, has yielded novel insights into the mechanism of cytokinesis. Similar approaches are increasingly used to study this fundamental process in more complex systems.

INTRODUCTION

Cytokinesis is a fundamental process essential for development, tissue homeostasis, and survival of all cellular life. Despite dramatic differences in cell size and geometry, cytokinesis in all animal and fungal cells requires spatiotemporal coordination of a contractile actomyosin ring (AMR), targeted vesicle fusion, and localized extracellular matrix (ECM) remodeling (Balasubramanian, Bi, & Glotzer, 2004; Pollard, 2010; Strickland & Burgess, 2004; Wloka & Bi, 2012). Constriction of the AMR drives cleavage-furrow ingression as well as guides vesicle fusion at the division site (Strickland & Burgess, 2004; Wloka & Bi, 2012). Targeted vesicle fusion not only increases the surface area but also delivers synthetic and hydrolytic enzymatic cargoes for localized ECM remodeling at the division site (Strickland & Burgess, 2004; Wloka & Bi, 2012). Since the early 1990s, cytokinesis research has concentrated on identifying genes, defining their functions, and establishing their interactions in different model systems (Bi et al., 1998; Chang, Woollard, & Nurse, 1996; Echard, Hickson, Foley, & O'Farrell, 2004; Giansanti et al., 1998; Gould & Simanis, 1997; Lippincott & Li, 1998; Maddox, Lewellyn, Desai, & Oegema, 2007; Skop, Liu, Yates, Meyer, & Heald, 2004). It is now clear that cytokinesis is a dynamic and robust process, with its core components conserved between yeast and human (Balasubramanian et al., 2004; Pollard, 2010; Wloka & Bi, 2012). Recent technological developments, especially in the areas of live-cell imaging and gene editing, have reignited tremendous interests in cytokinesis. As a result, this classic problem in cell biology is being studied in multiple systems with a variety of experimental approaches, including superresolution microscopy (Beach et al., 2014; Holden et al., 2014; Laplante, Huang, Bewersdorf, & Pollard, 2016; Ong, Wloka, Okada, Svitkina, & Bi, 2014; Ries, Kaplan, Platonova, Eghlidi, & Ewers, 2012; Strauss et al., 2012), electron microscopy (Guizetti et al., 2011; Maupin & Pollard, 1986; Ong et al., 2014; Sanger & Sanger, 1980; Schroeder, 1972), in vitro reconstitution (Huang, Mishra, Palani, Chew, & Balasubramanian, 2016; Mishra et al., 2013; Nguyen et al., 2014; Young, Buser, & Drubin, 2010), and computational modeling (Bidone, Tang, & Vavylonis, 2014; Vavylonis, Wu, Hao, O'Shaughnessy, & Pollard, 2008; Zhao & Wang, 2016). In this chapter, we focus on various methods for analyzing protein dynamics during cytokinesis in the budding yeast *Saccharomyces cerevisiae*.

Saccharomyces cerevisiae undergoes stereotyped division, has low gene redundancy (eg, one nonmuscle myosin-II instead of two or three in fission yeast and mammalian cells, respectively), and allows highly efficient gene editing via homologous recombination. For these reasons, the budding yeast has become an effective model for elucidating the general principles of cytokinesis. In *S. cerevisiae*, the site of division, which is the site of budding, is specified in late G1 by Cdc42-controlled cell polarization (Bi & Park, 2012). Cdc42 is a small GTPase that plays a central role in polarity establishment from yeast to humans (Etienne-Manneville, 2004; Park & Bi, 2007). Cdc42 controls polarized assembly of cytoskeletal structures, including the formation of a septin ring at the presumptive bud site and targeting actin cables toward that site. The septin ring marks the division site, stays at the bud neck after bud emergence, and serves as the foundation to build a division machine in an ordered manner during the cell cycle (Wloka & Bi, 2012; Wloka et al., 2013). The actin cables, nucleated by the formins Bni1 and Bnr1 at the bud tip and bud neck, respectively, serve as the tracks for myosin-V (Myo2 in budding yeast)–mediated exocytosis to control the size and shape of the bud (Bretscher, 2003). Bnr1 disappears from the bud neck before cytokinesis, while Bni1 localizes to the bud neck during cytokinesis (Buttery, Yoshida, & Pellman, 2007; Pruyne, Gao, Bi, & Bretscher, 2004). Once the septin ring is formed, it recruits Myo1 (the sole myosin-II heavy chain in budding yeast) to form a ringlike structure. Upon bud emergence, the septin ring is expanded into an hourglass structure. In anaphase, Iqg1 (the sole IQGAP in budding yeast) localizes to the bud neck and acts together with Myo1 and the formins to promote actin ring formation (Bi et al., 1998; Epp & Chant, 1997; Lippincott & Li, 1998; Tolliday, VerPlank, & Li, 2002; Vallen, Caviston, & Bi, 2000). At this point, a clearly detectable AMR is assembled at the bud neck, similar to mammalian cells in which an AMR is assembled at the midcell in anaphase. Thus, despite the earlier arrival of myosin-II at the division site in budding yeast, the timing of AMR assembly is the same as that in mammalian cells.

At the onset of cytokinesis, the septin hourglass is triggered by the mitotic exit network to remodel into a double ring that sandwiches the AMR (Lippincott, Shannon, Shou, Deshaies, & Li, 2001). The dramatic morphological change and the characteristic drop in fluorescence intensity of GFP-labeled septins at the bud neck (approximately 25–30%) during this structural reorganization can be used as "a clock" to time molecular events of cytokinesis in individual cells (Dobbelaere, Gentry, Hallberg, & Barral, 2003; Feng, Okada, Cai, Zhou, & Bi, 2015; Wloka et al., 2011, 2013). Once the septin double ring is formed, the AMR starts constricting (Fang et al., 2010; Wloka et al., 2013). Meanwhile, exocytosis is targeted to the bud neck (Finger, Hughes, & Novick, 1998; Wloka et al., 2013). Vesicle fusion at the division site results in localized expansion of the surface area, and also release of the secretory cargo Chs2, a chitin synthase that is essential for primary septum (PS) formation (Oh et al., 2012; Sburlati & Cabib, 1986). Thus, the AMR is spatially coupled to exocytosis. It also guides the centripetal growth of the PS (Fang et al., 2010; Vallen et al., 2000; Wloka et al., 2013). Collectively, the motor activity of Myo1 and PS formation drive cleavage-furrow ingression (Fang et al., 2010; Lord, Laves, & Pollard, 2005). Near the end of PS formation, secondary septa (SS), which are chemically identical to the lateral cell wall, are synthesized at both sides of the PS (Onishi, Ko, Nishihama, & Pringle, 2013). Subsequently, a daughter cell–specific transcription program is activated, leading to the expression and delivery of hydrolytic enzymes to the division site to degrade the PS and part of the SS, which results in cell separation (Weiss, 2012; Yeong, 2005). Here, we describe several methods for analyzing the dynamics of specific proteins and cellular structures involved in cytokinesis in budding yeast, including dual-color live-cell imaging, photoactivation, photoconversion, and fluorescence recovery after photobleaching (FRAP).

1. FIVE-DIMENSIONAL LIVE-CELL IMAGING

Five-dimensional (5D) (xyz + time + color channel) live-cell imaging is essential for directly comparing and quantifying the localization dynamics of protein pairs at the division site during cytokinesis. This is generally done by imaging an individual cytokinetic protein against a common cell cycle marker (eg, cell cycle regulators, mitotic spindle, spindle pole body components, septins, etc.) or pairs of cytokinesis proteins tagged with different fluorophores (usually with GFP and RFP). When performed on specific yeast mutants, 5D imaging can reveal the dependency relationships between different cytokinetic genes. The spatiotemporal maps and the dependency relationships generated by this approach are enormously useful for formulating mechanistic hypotheses to be tested. In the following sections, we describe the devices and methods for this type of imaging.

1.1 DEVICES

1.1.1 Microscope

To perform time-lapse analysis, especially for a long duration, it is better to use an inverted microscope than an upright one, as the former can easily accommodate the use of a dish sample and microfluidics-based perfusion chamber system. These systems, as opposed to the sealed slide sample, provide an abundant supply of nutrients and oxygen, allowing yeast cells to stay healthier for generations during imaging. Oxygen is essential not only for cell proliferation, but also for certain proteins such as GFP to form the fluorophore. To acquire z-stacks of images for 3D reconstruction, the microscope needs a motorized z-axis controller (eg, piezo z stage). For long-term live imaging, it is highly recommended that the microscope is equipped with an automatic focus correction system (eg, Carl Zeiss Definite Focus, Leica Adaptive Focus Control, Nikon Perfect Focus System, and Olympus Zero Drift).

1.1.2 Environmental chamber

Time-dependent parameters of cytokinesis, such as the rate of AMR constriction, are affected by the temperature and other environmental factors such as growth media. Thus, it is highly recommended that the microscope is equipped with a temperature controller and an environmental chamber to ensure stable growth conditions during live-cell imaging.

1.1.3 Objective lens

Specification of an objective lens needs to be carefully considered for each imaging application. For fluorescent microscopy, theoretical resolution δ is determined by the emission wavelength λ and the numerical aperture (NA) of the objective lens: $\delta = 0.61\ \lambda/\mathrm{NA}$. Thus, when a higher spatial resolution is required, an objective lens with a higher NA must be used. The transparency of the objective lens is also an important factor to consider, especially when a limited number of fluorescent molecules are monitored. Thus, it is recommended that the microscope is equipped with an objective lens with high transparency for the wavelength of the fluorescent molecules used in the experiment.

1.1.4 Light source

For quantitative time-lapse analysis, the most important property for a light source is illumination stability over time. Conventional mercury arc lamp or metal halide lamp has been widely used as the excitation light sources, but they have the drawback of electric arc—based intensity fluctuation and gradual decrease in illumination intensity during long-term use (months to years). In contrast, solid-state light sources such as light-emitting diodes (LED) do not have this problem. In addition, the emission of light from the LED and other solid-state light sources is controlled by electric current, which means that switching on/off the excitation light occurs fast, and a physical shutter is no longer needed.

1.1.5 Filter sets

When multichannel imaging is performed, it is important to choose appropriate filter sets. An excessively wide band-pass filter results in poor image quality due to bleed-through between channels. An excessively narrow band-pass filter limits the number of photons to pass through and thus causes difficulties in detecting weak fluorescence signals. It is possible to simulate combinations of filter sets and fluorescent proteins/dyes using online tools provided by optical filter manufacturers such as Semrock SearchLight (http://searchlight.semrock.com) and Omega Optical Curvomatic (http://www.omegafilters.com/curvomatic/).

1.1.6 Camera

Camera devices should be carefully selected based on imaging types and sample properties.

1. Conventional charge-coupled device (CCD) cameras

Conventional CCD cameras are commonly used as area sensors to detect fluorescent signals in the field of live-cell microscopy. Conventional CCD cameras have ~70% quantum efficiency in the range of visible light wavelength. If the samples have relatively high fluorescence intensity, a conventional CCD camera can be used. Compared to Electron Multiplying Charge-Coupled Device (EMCCD) cameras (see the following), conventional CCD cameras have smaller size and higher number of detector cells, and therefore higher resolution.

2. Electron multiplying charge-coupled device (EMCCD) cameras
The main feature of EMCCD cameras is their very high sensitivity. The quantum efficiency of EMCCD cameras can be higher than 90%. If a small number of fluorescent molecules are monitored by highly frequent image acquisitions, an EMCCD camera should be the first choice.
3. Scientific complementary metal-oxide-semiconductor (sCMOS) cameras
Recently, the sensitivity of sCMOS cameras is getting closer to that of EMCCD cameras. The latest models of sCMOS cameras have >80% quantum efficiency. When dealing with samples that have higher background noise (eg, live cells in a culture), the apparent signal/noise (S/N) ratio on sCMOS cameras can surpass that of EMCCD cameras. Comparing to EMCCD cameras, sCMOS cameras have smaller detector cells (typically 6.5 μm) and a larger area of detector array (typically 2048 × 2048), which means that it is possible to take images with higher spatial resolution and covering a larger number of cells than those on EMCCD cameras.

1.1.7 Signal/noise ratio

Biological samples used for time-lapse analysis have a certain thickness, and it is impossible to be free of stray light coming from outside the focal plane. To eliminate the effect of stray light for achieving a higher S/N ratio, confocal and deconvolution systems can be used.

1.1.8 Devices used for the imaging examples shown in this study

- Microscope: IX71 inverted microscope (Olympus)
- Environmental chamber: DH-35 (Warner Instruments)
- Objective lens: 100×/1.4NA Plan S-Apo oil immersion lens (Olympus)
- Light source: Diode lasers (488 and 561 nm)
- Camera: EMCCD camera ImagEM, C9100-13 (Hamamatsu Photonics)
- Confocal system: Spin-disk confocal head CSU10 (Yokogawa)

1.2 SAMPLE PREPARATION IN A GLASS-BOTTOM DISH FOR LONG-TERM LIVE IMAGING

Fig. 1 shows a schematic of the steps involved in sample preparation for live-cell imaging. This glass-bottom "dish" method has the following advantages over the

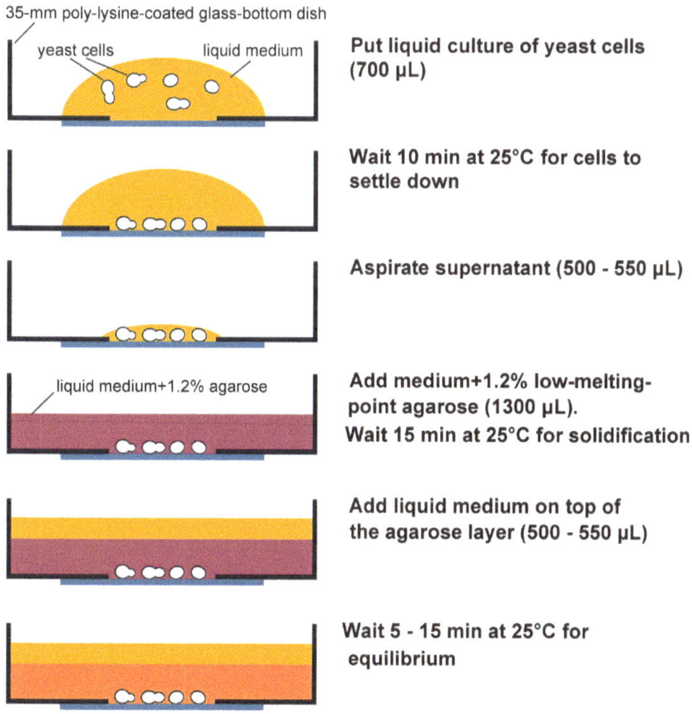

FIGURE 1

Schematic of sample preparation in a poly-lysine–coated glass-bottom dish for long-term live imaging. See text for details.

traditional "slide" method (Fig. 4), which sandwiches yeast cells between a cover glass and a growth medium–containing agarose pad mounted on a slide:

1. With the slide method, the supply of nutrients for cell growth and of oxygen for fluorophore formation could be a problem, especially for long-term live imaging experiments. In our experience, it is difficult to keep cells growing for longer than 5 h when they are sealed between an agarose pad and a cover glass. In contrast, the dish method described below allows time-lapse image acquisition of live cells for more than 8 h without apparent decrease in growth rate. Thus, individual cells can be followed for at least two to three cell division cycles in a single filming session, which would allow visualization of cell lineage–specific phenotypes.
2. With the slide method, the cover glass is sealed along its edges, which prevents the addition of drugs to cells during live imaging. In contrast, the dish method allows this operation, which enables observation of cellular behaviors before and after the drug treatment in a single filming session.

1.2.1 Materials

- To reduce the background fluorescence noise due to fluorescent component contained in the medium, yeast nitrogen base without amino acids and without riboflavin (Formedium Ltd., Cat. # CYN4501) should be used
- MatTek Poly-D-lysine–coated 35-mm dish (Cat. # P35GC-1.5-14.C)
- Lonza SeaPlaque agarose (Cat. # 50101)

1.2.2 Methods

1. Prepare exponentially growing yeast cells. Unbuffered synthetic complete (SC) medium (Guthrie & Fink, 1991) is preferred as the pH of the medium can affect the solidity of agarose. It is better to avoid using buffered medium (eg, YM-1) (Lillie & Pringle, 1980).
2. Apply 700 μL of yeast liquid culture on top of a poly-lysine–coated glass-bottom dish of 35-mm in diameter.
3. Wait for ~10 min for the yeast cells to settle down on the glass surface at 25°C.
4. Gently aspirate 500–550 μL of excessive supernatant using a P1000 micropipette.
5. Add to the dish with 1300 μL of growth medium containing 1.2% low-melting-point agarose that has been melted and kept in a liquid state at 37°C. Gently mix the yeast culture and the agarose-containing medium by a P1000 micropipette. Three strokes of slow pipetting are sufficient for mixing. Wait for ~15 min for the agarose to solidify at 25°C.
   Notes:
   - It is possible to use ordinary agarose (higher melting point), but in that case it is more difficult to handle the melted agarose-containing medium by a micropipette.
   - It is important to avoid making bubbles on the surface of the agarose-containing layer.
   - After solidification of the agarose-containing layer, the layer is stable at higher temperature. For example, the solidified sample can be placed in an environmental chamber set at 37°C for imaging temperature-sensitive yeast cells.
6. Apply 500–550 μL of liquid culture medium on top of the solidified agarose layer. This prevents the sample from drying up, especially for long-term live imaging.
7. Wait for 5–15 min to allow the agarose layer and the liquid medium reach equilibrium.

1.3 IMAGING METHODS

Before image acquisition, the following parameters must be optimized to obtain the best results in image quality and cell healthiness:

1. *Temperature*. Temperature can affect imaging outcome at least in two ways. First, most of the cellular processes are accelerated based on thermodynamic

properties of protein activities. For example, the rate of AMR constriction is higher at 37°C than at 25°C. Second, temperature can also affect maturation efficiency of fluorescent proteins. Most fluorescent proteins derived from the jellyfish *Aequorea victoria* (GFP, CFP, YFP, etc.) mature less efficiently at high temperatures. One way to mitigate this problem is to use the fluorescent proteins from coral species, which form the fluorophore more efficiently at high temperatures. This is especially important when temperature-sensitive mutants are imaged at 37°C.

2. *Duration of filming*. The duration of imaging (seconds, minutes, or hours) depends on what cellular behaviors are being monitored. Generally, shorter duration is used to acquire the average value of single events, while longer duration is used to collect the data on relationships between temporally distinct events (eg, between cytokinesis and the next bud emergence or between consecutive cytokinesis events).
3. *Interval of image acquisition*. The interval between image acquisitions, ie, temporal resolution, is determined by the dynamics of cellular and molecular behaviors being monitored. For a slow process, a longer interval is used (usually in minutes). For proteins with rapid turnover, a shorter interval is required (usually in seconds). However, this interval must be longer by the mechanical response time of the imaging system, which is the sum of many time parameters including those for camera exposure, movement of z-drive, and open/close of the shutter.
4. *Z-stack settings*. To follow a subcellular structure over time, a number of z-sections (=z-stack) for each time point must be acquired for three-dimensional (3D) reconstruction. The thickness of each z-section and the total number of z-sections are mainly determined by the size, shape, and fluorescence intensity of the imaged structure. The total number of image acquisition is also tightly linked to camera exposure setting (see the next section). In general, for 3D reconstruction at a higher resolution, a larger number of thinner z-sections are required.
5. *Exposure settings*. When cells are exposed to excessive light, the light energy causes damage to many cellular components including DNA. This could lead to the activation of cell cycle checkpoints, and ultimately cell death. To avoid phototoxicity, the amount of cell exposure to excitation light should be kept low during the entire imaging period. For each time-lapse analysis, a balanced point of illumination intensity, exposure time, and frequency of image acquisition should be experimentally determined.

As an example of 5D live-cell imaging, we monitored the localization of septins (Cdc3-GFP) and Cdc42 activity (indicated by the biosensor Gic2PBD-RFP) during the cell cycle (Fig. 2) (Okada et al., 2013). Clearly, the septins at the bud neck undergo cell cycle–triggered changes in structural organization. At the onset of cytokinesis, the septin hourglass is instructed to reorganize into a double ring (Fig. 2) (Lippincott et al., 2001). Consequently, the old septin ring at the mother

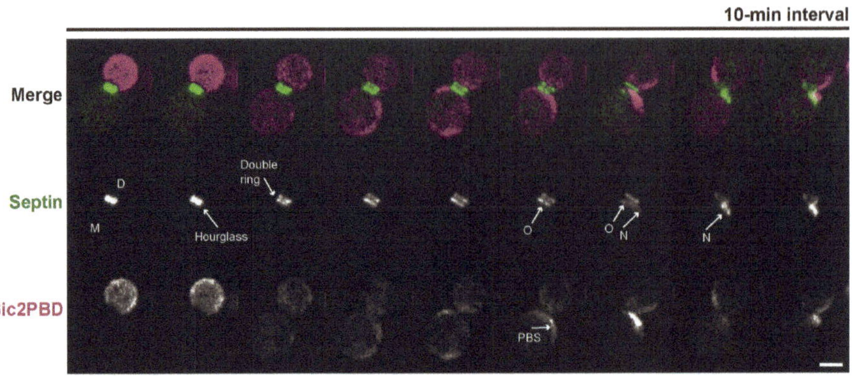

FIGURE 2

Time-lapse analyses of septin and Cdc42-GTP during cytokinesis. Yeast cells harboring Cdc3-GFP (septin) and Gic2PBD-RFP (Cdc42-GTP) were filmed with the dish method. *M*, mother cell; *D*, daughter cell; *O*, old septin ring; *N*, new septin ring; and *PBS*, presumptive bud site. The image sequences were constructed by average projection of 11 z-sections acquired in 0.6-μm increments for each time point. Scale bar = 2 μm.

side disassembles, while a new septin ring forms at an adjacent site (Fig. 2). It is also abundantly clear that Cdc42 is inactivated at the bud cortex prior to cytokinesis and remains inactive during cytokinesis, and becomes activated at the presumptive bud site after cytokinesis (Fig. 2). These dynamic behaviors of septins and Cdc42-GTP during the cell cycle can be quantified as we have done in the past (Okada et al., 2013). Taken together, the 5D live-cell imaging coupled with quantitative analysis can provide significant insights into molecular functions in fundamental processes such as cytokinesis and cell polarization.

2. PHOTOACTIVATION AND PHOTOCONVERSION

Photoactivation and photoconversion are increasingly used for probing protein dynamics in diverse cell types. In the case of photoactivation, target molecules are tagged with a photoactivatable (PA) fluorescent protein. The "dark" molecules within a region of interest (ROI) are activated by a laser to produce fluorescence. This approach allows the tracing of a specific population of molecules in time and space within the cell. In the case of photoconversion, target molecules are tagged with a photoconvertible (PC) fluorescent protein. The molecules that are fluorescent in one channel (eg, the green channel) within a ROI will be converted by a laser to molecules that are now fluorescent in a different channel (eg, the red channel). This approach allows visualization of two population behaviors of the same protein within the same cellular structure or compartment.

The devices for photoactivation and photoconversion experiments are basically the same as those required for 5D live-cell imaging and FRAP experiments. In addition to fluorescence microscope, a system for photostimulation [an ablation laser with an appropriate wavelength for the PA or PC protein or a digital micromirror device (DMD) module] is required.

When photoactivation or photoconversion experiments are performed, several parameters need to be optimized, especially the intensity of photostimulation laser, the number of laser pulses, and the area size for photostimulation. To minimize the phototoxic effect for cells, the combination of a larger number of laser pulses with lower power in a smaller ROI will work better than that of a smaller number of laser pulses with higher power in a larger ROI.

Fig. 3A and B shows an example of 5D live imaging of yeast cells carrying a septin (Cdc3) tagged with a PA-GFP (Patterson & Lippincott-Schwartz, 2002). Fig. 3C and D shows an example of 5D live imaging of yeast cells carrying the same septin tagged with a PC (green to red) fluorescent protein, mEos3.2 (Ong et al., 2014; Zhang et al., 2012). In both cases, a portion of the septin hourglass was photoactivated or photoconverted, and the targeted molecules were followed over time. The results show that septin molecules in the hourglass are largely immobile before the onset of cytokinesis (as indicated by the nuclear position in Fig. 3A and/or by the septin hourglass-to-double ring conversion in Fig. 3A–D). Septin molecules become more dynamic during and after cytokinesis. More specifically, photoactivated or photoconverted septin molecules in the hourglass disappear in situ (without diffusing across the bud neck), and some of these molecules are relocalized to the double ring (Fig. 3B and D). These data suggest that the septin hourglass-to-double ring conversion involves filament disassembly in the hourglass and reassembly at the ends of the hourglass, a clear hypothesis that can be tested at the mechanistic level.

3. FLUORESCENCE RECOVERY AFTER PHOTOBLEACHING

FRAP has been widely used for analyzing protein dynamics involved in diverse cellular functions. In this case, fluorescence of the target molecules in a ROI is bleached by a laser; the behavior of the remaining fluorescent molecules outside the ROI is monitored with respect to the bleached molecules within the ROI. Thus, this approach allows the determination of protein turnover kinetics within the ROI.

Although the dish method can be used for FRAP analysis, the "slide" method (Fig. 4) is preferred for the following reasons: (1) slide assembly is quick and simple, (2) cells are less likely to de-attach during live imaging, and (3) it is compatible with the shorter imaging duration that is often used in FRAP experiments.

Below, we first describe the protocol for sample slide assembly (Fig. 4) and then show examples for FRAP analysis of protein dynamics during cytokinesis in budding yeast.

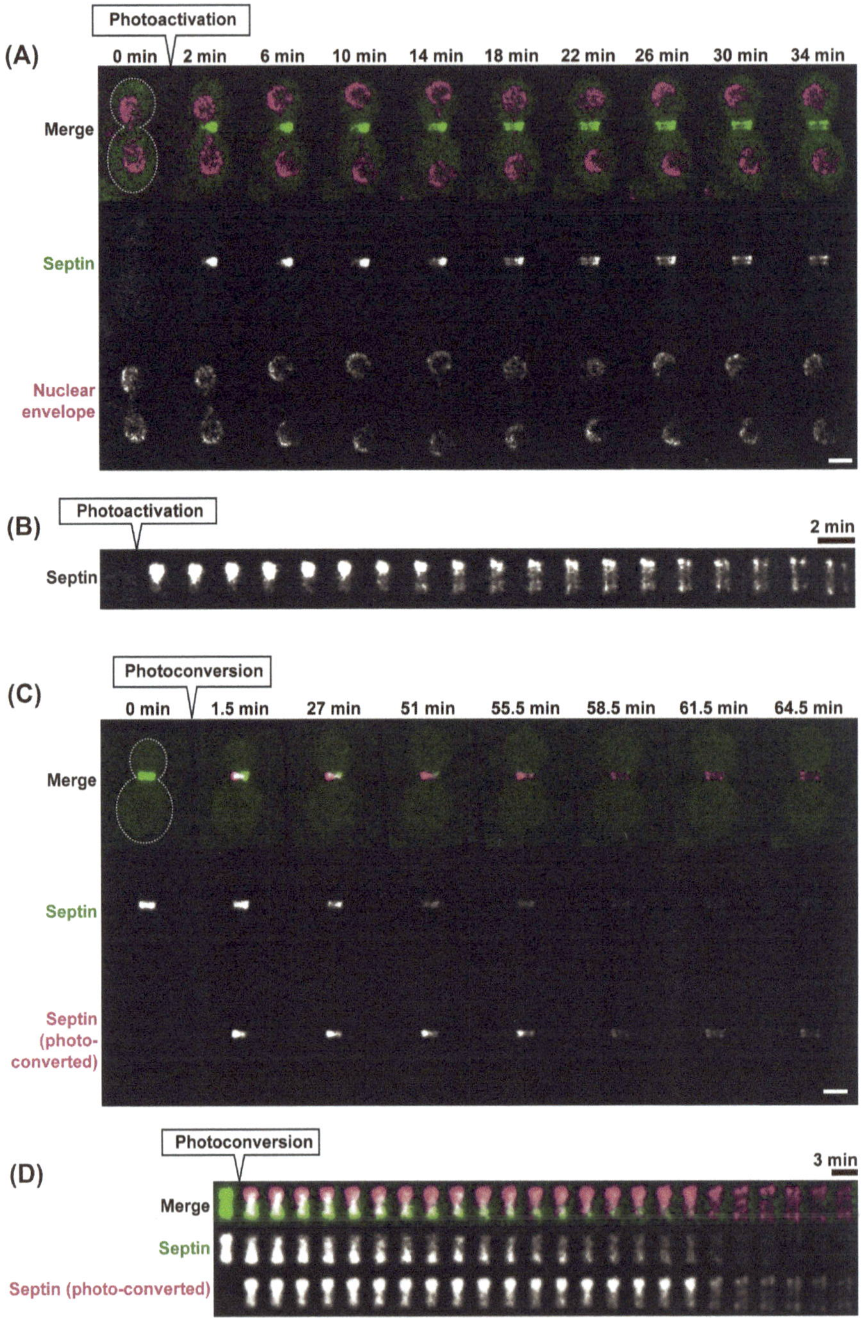

(A)
Photoactivation
0 min
2 min
6 min
10 min
14 min
18 min
22 min
26 min
30 min
34 min
Merge
Septin
Nuclear envelope
(B)
Photoactivation
2 min
Septin
(C)
Photoconversion
0 min
1.5 min
27 min
51 min
55.5 min
58.5 min
61.5 min
64.5 min
Merge
Septin
Septin (photo-converted)
(D)
Photoconversion
3 min
Merge
Septin
Septin (photo-converted)

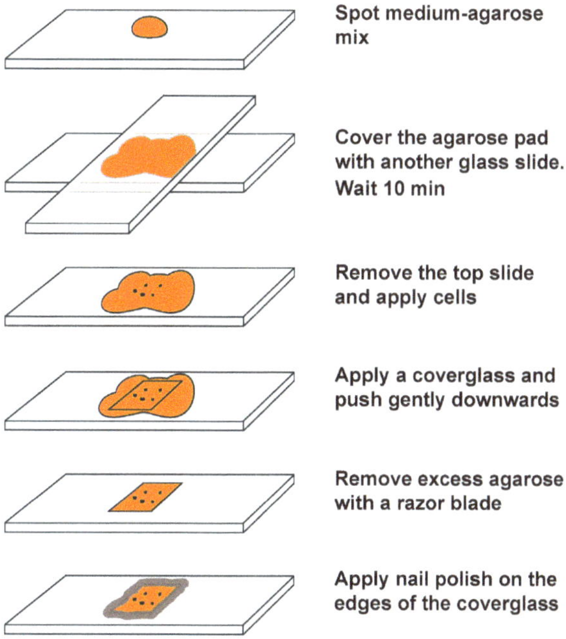

FIGURE 4

Schematic of sample preparation on a glass slide for fluorescence recovery after photobleaching (FRAP) analysis. See text for details.

FIGURE 3

Photoactivation and photoconversion studies of the septin hourglass-to-double ring transition. (A) The budding yeast cells harboring Cdc3-PA-GFP (septin, photoactivatable) and Nup57-mCherry (nuclear envelope) were filmed with the dish method. Part of the septin hourglass was photoactivated between the first and the second time points. (B) Image sequence of the bud-neck region of the cell shown in panel A. (C) The budding yeast cells harboring Cdc3-ymEos3.2 (septin, photoconvertible from green to red) were filmed with the dish method. Part of the septin hourglass was photoconverted between the first and the second time points. (D) Image sequence of the bud-neck region of the cell shown in panel C. Green channel shows the fluorescence intensity of the unconverted Cdc3-ymEos3.2. Red channel (pseudocolored in Magenta) shows the fluorescence intensity of the photoconverted Cdc3-ymEos3.2. The image sequences were constructed by average projection of 11 z-sections acquired in 0.6-μm increments. Scale bar = 2 μm.

3.1 SAMPLE PREPARATION ON A GLASS SLIDE FOR FRAP ANALYSIS

1. Heat an Eppendorf tube containing SC or rich media with 2% low-melting agarose in a boiling water bath for about 5 min.
2. Spot 40 μL of the medium–agarose mix onto a clean glass slide (Corning, 75 × 25 mm). Beware of spill-outs from the tube.
3. Put another glass slide, within 5–10 s, on top of the medium–agarose mix to create a flat pad.
4. Allow cooling and solidification for ~10 min.
5. Remove the top glass slide and spot ~3 μL of a cell culture onto the agarose pad.

   Notes:
   - The media used for the growth of the cell culture should be the same as in the medium–agarose mix.
   - The OD_{600} of the liquid culture should be kept between 0.2 and 0.8 at all times. If too few cells are present, centrifugation for 30 s at ~5000 g and subsequent resuspension in a smaller volume can increase cell number.
6. Put a cover glass (Corning, 10 × 10 mm) on top of the cells and gently push downward with a pipette tip.
7. Cut away the overhanging agarose patch with a clean razor blade and seal the cover glass with nail polish. The glass slide can be used immediately for imaging.

3.2 METHODS

1. Mount the glass slide on a microscope using a 100× objective with oil immersion.
2. Bring a yeast cell in focus and position it in the center of the imaging field. Minimal divergence is only achieved in the center with most FRAP systems.

   Note:
   - Calibration is dependent on the system used and should be done at least weekly.
3. Acquire an image of the cell in time-lapse mode and pause after acquisition of the first time point. Stacks of 12 × 0.4 μm z-sections are routinely used for live imaging and FRAP analysis of budding yeast cells in our laboratory. Exposure times and gain depend on the proteins studied. Generally speaking, a 125-ms exposure time provides a good starting point. Shorter exposure is used for longer imaging since less general photobleaching and phototoxicity would occur.
4. Zoom into a ROI so individual pixels become clearly distinguishable and add pixel-sized dots on the cellular structure to be bleached. Not necessarily every pixel of the cellular structure has to be covered, as bleaching will also occur in the surrounding area (see the next note). The method of bleaching might vary from system to system. Using the MicroPoint (Photonic Instruments) system controlled by Metamorph (Molecular Devices), pixel-sized areas can be selected for bleaching.

Notes:

- The actual bleach is likely to spread beyond the area selected and depends on the intensity used in the system. Thus, a clear idea where the bleach occurred will be only obtained after acquisition of the next image.
- Usually, a quite low intensity of ~10% laser power together with the attenuator slide set on a high attenuation is sufficient for bleaching proteins in budding yeast.
- The laser power is somewhat arbitrary and slightly changing over time as the laser loses power and has to be determined empirically for each protein.
- Too much laser power can be toxic to the cell. If the cell continues to grow, ideally completing a whole new cell cycle, which takes about 2–3 h on the agarose pad, one can reasonably assume that cellular damage was acceptable.

5. Immediately after the bleaching (audible with the MicroPoint system), restart the image-acquisition process.

   General notes:

   Appropriate controls and investigations into the resource(s) for the fluorescence recovery should be performed. For example, the recovery could originate from the cytosol or a cellular structure. The protein could flux within the structure itself or via another mechanism such as disassembly from and reassembly to the structure. For example, fluorescence loss in photobleaching (FLIP) experiments can be performed to test the potential flux of a cytokinesis protein between the cytosol and the bud neck. This involves bleaching an area of the cytosol multiple times, and then monitoring the change of fluorescence intensity at the bud neck. If the neck intensity stays the same, cytosol–neck flux does not occur; if the neck intensity drops quickly, the flux occurs for the protein.

3.3 DATA ANALYSIS

1. Load the image sequence in ImageJ creating a stack of images using maximum or sum projections (preferred).

   Note:

   - To analyze the acquired image data we use ImageJ (NIH) equipped with the MetaMorph import plugin (nd Stack Builder) (http://rsb.info.nih.gov/ij/plugins/track/builder.html). Different image acquisition software perhaps necessitates other plugins; however, in principle any image series in nonproprietary format (eg, mostly TIFF) can be opened in ImageJ directly using the Import → *Image Sequence function.*

2. Draw a polygon around the ROI (in ImageJ) where the bleach occurred. Make sure that the polygon contains the ROI during the whole acquisition by scrolling through the stack of images. If the cell moves too much outside of the boundary of the polygon, the cell should not be used for purposes of quantification. Alternatively, the stack can be aligned using the StackReg (Thevenaz, Ruttimann, & Unser, 1998) plugin for ImageJ.

3. Measure the stack (Plugin—Stacks—Measure Stack). The integrated density can be used for quantification of the bleached area. As a control, an unbleached area of equal size should be measured as well to correct for general photobleaching.
4. Plot the changes of fluorescence over time with Excel or statistical programs such as GraphPad or Origin.

Here, we show two examples to demonstrate the power of FRAP analysis in understanding the mechanisms of septin reorganization and cytokinesis during the cell cycle.

To gain insights into the mechanism underlying the septin hourglass-to-double ring conversion that occurs at the onset of cytokinesis, first, the total fluorescence intensity of septin (Cdc3-GFP) at the bud neck was monitored by 5D live-cell imaging from anaphase to cell separation (using the nuclear pore complex protein Nup57-mCherry as the cell cycle marker). The total intensity displayed a stereotypical about 25–30% decrease during the hourglass-to-double ring conversion (Fig. 5A) (Dobbelaere et al., 2003; Wloka et al., 2011), suggesting that septin filament disassembly occurs during this structural transition.

Next, a portion of the septin hourglass in cells of the same yeast strain was photobleached, and the fluorescence changes in the bleached and unbleached areas were followed over time (Fig. 5B; only Cdc3-GFP signal shown in black/white). The fluorescent molecules in the unbleached area apparently disappeared in situ (without diffusing into the bleached area), and new fluorescent molecules appeared at the ends of the hourglass. These data are complementary to those obtained with photoactivation and photoconversion studies (Fig. 3). In the case of FRAP, the behaviors of unbleached septin molecules are monitored; in other two approaches, the behaviors of activated or converted septin molecules are monitored. Together, these three types of analyses have led to the notion that the hourglass-to-double ring transition involves filament disassembly in the hourglass and filament reassembly at the ends of the hourglass.

To gain insights into the mechanism of myosin-II function during cytokinesis, we first performed 5D imaging on cells carrying Myo1-GFP (myosin-II) and Cdc3-mCherry (septin; as a cell cycle marker here) and found that nearly all the Myo1 molecules are localized to the bud neck ~30 min after bud emergence, and remain there until the onset of cytokinesis (indicated by the septin hourglass-to-double ring conversion) when Myo1 ring undergoes constriction coupled with disassembly (Wloka et al., 2013). FLIP analysis indicates that Myo1, once localized at the bud neck, shows little or no cytosol-neck fluxing (Wloka et al., 2013). Strikingly, FRAP analysis demonstrates that Myo1 is mobile at the bud neck at the small-budded stage (before cytokinesis) and becomes completely immobile during cytokinesis (Wloka et al., 2013). This immobility depends on the "putative assembly domain" located at its C-terminal end (Wloka et al., 2013), suggesting that Myo1 undergoes cell cycle–triggered filament assembly.

To determine how PS formation occurs in relationship to the AMR, we first performed 5D live imaging on cells carrying Chs2-GFP (chitin synthase-II essential

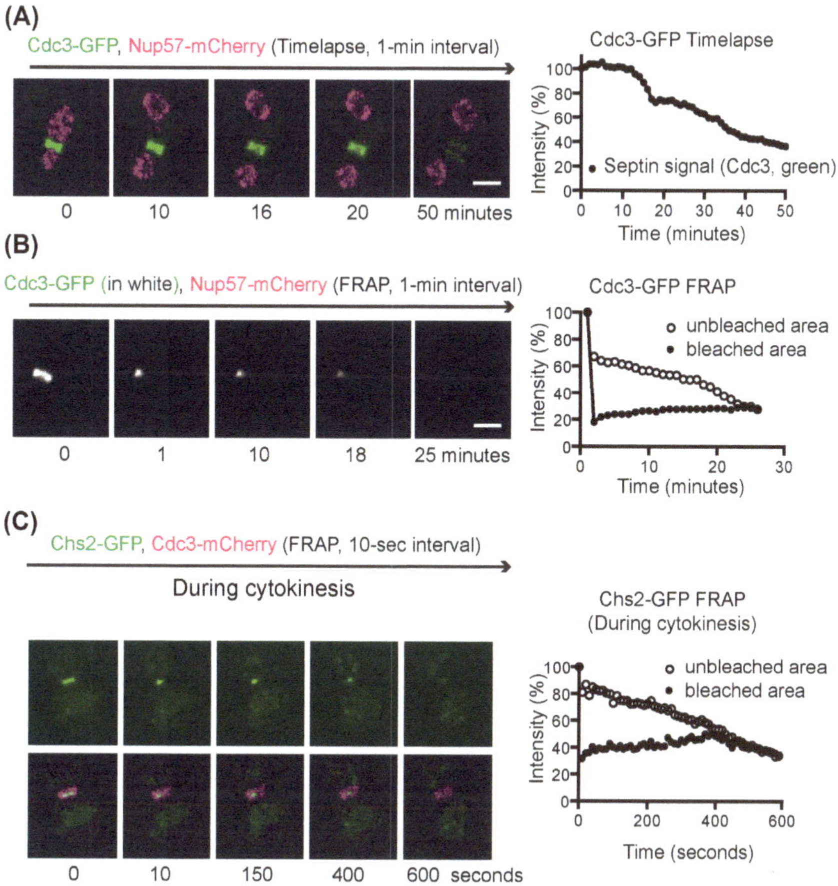

FIGURE 5

Fluorescence recovery after photobleaching (FRAP) analyses of the septin hourglass-to-double ring transition and Chs2 dynamics. (A) The budding yeast cells harboring Cdc3-GFP (septin) and Nup57-mCherry (nuclear envelope) were imaged using the slide method (left). The total fluorescence intensity of Cdc3-GFP at the bud neck was quantified and plotted over time (right). (B) The same yeast strain as in A was subjected to FRAP analysis (left), and the fluorescence intensities of Cdc3-GFP in the bleached and unbleached areas were quantified and plotted over time (right). The image sequences were constructed by maximum projection (left) as well as average projection (right) of 12 z-sections acquired in 0.4-μm increments. Scale bar = 2 μm. (C) The budding yeast cells harboring Chs2-GFP (chitin synthase-II) and Cdc3-mCherry (septin) were subjected to FRAP analysis during cytokinesis (left), and the fluorescence intensities of Chs2-GFP in the bleached and unbleached areas were quantified and plotted over time (right). The image sequences were constructed by maximum projection (left) as well as average projection (right) of 12 z-sections acquired in 0.4-μm increments.

for PS formation) and Cdc3-mCherry (septin; as a cell cycle marker here) and found that all Chs2 molecules are localized from the ER, presumably via the Golgi, to the bud neck within ~3 min at the onset of cytokinesis (Wloka et al., 2013). Chs2-GFP then undergoes "constriction" coupled with endocytosis (as indicated by the "puncta" in the cytosol) (Fig. 5C). Surprisingly, FRAP analysis indicates that this secretory cargo and a transmembrane protein is immobile during cytokinesis (Fig. 5C), and this immobility depends on Myo1, but not vice versa (Wloka et al., 2013). These data suggest that Myo1 acts as a scaffold for Chs2 organization to ensure that AMR constriction is spatiotemporally coupled with PS formation during cytokinesis. Thus, FRAP analysis of protein dynamics in wild-type and mutants during the cell cycle could provide significant insights into the functional mechanisms of core proteins involved in cytokinesis.

ACKNOWLEDGMENTS

We thank Yogini Bhavsar-Jog and Hiroki Okada for critically reading the manuscript, and Andrea Stout for technical advice. This work was supported by NIH grant GM115420 (to E.B.), and a predoctoral fellowship from the Boehringer Ingelheim Fonds (to C.W.).

REFERENCES

Balasubramanian, M. K., Bi, E., & Glotzer, M. (2004). Comparative analysis of cytokinesis in budding yeast, fission yeast and animal cells. *Current Biology: CB, 14*(18), R806–R818.

Beach, J. R., Shao, L., Remmert, K., Li, D., Betzig, E., & Hammer, J. A., 3rd (2014). Nonmuscle myosin II isoforms coassemble in living cells. *Current Biology: CB, 24*(10), 1160–1166. http://dx.doi.org/10.1016/j.cub.2014.03.071. pii:S0960-9822(14)00393-5.

Bi, E., Maddox, P., Lew, D. J., Salmon, E. D., McMillan, J. N., & Yeh, E. (1998). Involvement of an actomyosin contractile ring in *Saccharomyces cerevisiae* cytokinesis. *The Journal of Cell Biology, 142*(5), 1301–1312.

Bi, E., & Park, H. O. (2012). Cell polarization and cytokinesis in budding yeast. *Genetics, 191*(2), 347–387. http://dx.doi.org/10.1534/genetics.111.132886. pii:191/2/347.

Bidone, T. C., Tang, H., & Vavylonis, D. (2014). Dynamic network morphology and tension buildup in a 3D model of cytokinetic ring assembly. *Biophysical Journal, 107*(11), 2618–2628. http://dx.doi.org/10.1016/j.bpj.2014.10.034. pii:S0006-3495(14)01117-5.

Bretscher, A. (2003). Polarized growth and organelle segregation in yeast: the tracks, motors, and receptors. *The Journal of Cell Biology, 160*(6), 811–816.

Buttery, S. M., Yoshida, S., & Pellman, D. (2007). Yeast formins Bni1 and Bnr1 utilize different modes of cortical interaction during the assembly of actin cables. *Molecular Biology of the Cell, 18*(5), 1826–1838.

Chang, F., Woollard, A., & Nurse, P. (1996). Isolation and characterization of fission yeast mutants defective in the assembly and placement of the contractile actin ring. *Journal of Cell Science, 109*(Pt 1), 131–142.

Dobbelaere, J., Gentry, M. S., Hallberg, R. L., & Barral, Y. (2003). Phosphorylation-dependent regulation of septin dynamics during the cell cycle. *Developmental Cell, 4*(3), 345–357.

Echard, A., Hickson, G. R., Foley, E., & O'Farrell, P. H. (2004). Terminal cytokinesis events uncovered after an RNAi screen. *Current Biology: CB, 14*(18), 1685–1693.

Epp, J. A., & Chant, J. (1997). An IQGAP-related protein controls actin-ring formation and cytokinesis in yeast. *Current Biology: CB, 7*(12), 921–929.

Etienne-Manneville, S. (2004). Cdc42–the centre of polarity. *Journal of Cell Science, 117*(Pt 8), 1291–1300.

Fang, X., Luo, J., Nishihama, R., Wloka, C., Dravis, C., & Travaglia, M. (2010). Biphasic targeting and cleavage furrow ingression directed by the tail of a myosin-II. *The Journal of Cell Biology, 191*, 1333–1350.

Feng, Z., Okada, S., Cai, G., Zhou, B., & Bi, E. (2015). MyosinII heavy chain and formin mediate the targeting of myosin essential light chain to the division site before and during cytokinesis. *Molecular Biology of the Cell, 26*(7), 1211–1224. http://dx.doi.org/10.1091/mbc.E14-09-1363. pii:mbc.E14-09-1363.

Finger, F. P., Hughes, T. E., & Novick, P. (1998). Sec3p is a spatial landmark for polarized secretion in budding yeast. *Cell, 92*(4), 559–571.

Giansanti, M. G., Bonaccorsi, S., Williams, B., Williams, E. V., Santolamazza, C., & Goldberg, M. L. (1998). Cooperative interactions between the central spindle and the contractile ring during *Drosophila* cytokinesis. *Genes & Development, 12*(3), 396–410.

Gould, K. L., & Simanis, V. (1997). The control of septum formation in fission yeast. *Genes & Development, 11*, 2939–2951.

Guizetti, J., Schermelleh, L., Mantler, J., Maar, S., Poser, I., & Leonhardt, H. (2011). Cortical constriction during abscission involves helices of ESCRT-III-dependent filaments. *Science, 331*(6024), 1616–1620. http://dx.doi.org/10.1126/science.1201847. pii:science.1201847.

Guthrie, C., & Fink, G. R. (1991). Guide to yeast genetics and molecular biology. *Methods in Enzymology, 194*, 933 pp.

Holden, S. J., Pengo, T., Meibom, K. L., Fernandez Fernandez, C., Collier, J., & Manley, S. (2014). High throughput 3D super-resolution microscopy reveals *Caulobacter crescentus* in vivo Z-ring organization. *Proceedings of the National Academy of Sciences of the United States of America, 111*(12), 4566–4571. http://dx.doi.org/10.1073/pnas.1313368111. pii:1313368111.

Huang, J., Mishra, M., Palani, S., Chew, T. G., & Balasubramanian, M. K. (2016). Isolation of cytokinetic actomyosin rings from *Saccharomyces cerevisiae* and *Schizosaccharomyces pombe*. *Methods in Molecular Biology (Clifton, NJ), 1369*, 125–136. http://dx.doi.org/10.1007/978-1-4939-3145-3_10.

Laplante, C., Huang, F., Bewersdorf, J., & Pollard, T. D. (2016). High-speed super-resolution imaging of live fission yeast cells. *Methods in Molecular Biology (Clifton, NJ), 1369*, 45–57. http://dx.doi.org/10.1007/978-1-4939-3145-3_4.

Lillie, S. H., & Pringle, J. R. (1980). Reserve carbohydrate metabolism in *Saccharomyces cerevisiae*: responses to nutrient limitation. *Journal of Bacteriology, 143*, 1384–1394.

Lippincott, J., & Li, R. (1998). Sequential assembly of myosin II, an IQGAP-like protein, and filamentous actin to a ring structure involved in budding yeast cytokinesis. *The Journal of Cell Biology, 140*, 355–366.

Lippincott, J., Shannon, K. B., Shou, W., Deshaies, R. J., & Li, R. (2001). The Tem1 small GTPase controls actomyosin and septin dynamics during cytokinesis. *Journal of Cell Science, 114*(Pt 7), 1379–1386.

Lord, M., Laves, E., & Pollard, T. D. (2005). Cytokinesis depends on the motor domains of myosin-II in fission yeast but not in budding yeast. *Molecular Biology of the Cell, 16*(11), 5346–5355.

Maddox, A. S., Lewellyn, L., Desai, A., & Oegema, K. (2007). Anillin and the septins promote asymmetric ingression of the cytokinetic furrow. *Developmental Cell, 12*(5), 827–835.

Maupin, P., & Pollard, T. D. (1986). Arrangement of actin filaments and myosin-like filaments in the contractile ring and of actin-like filaments in the mitotic spindle of dividing HeLa cells. *Journal of Ultrastructure and Molecular Structure Research, 94*(1), 92–103.

Mishra, M., Kashiwazaki, J., Takagi, T., Srinivasan, R., Huang, Y., & Balasubramanian, M. K. (2013). In vitro contraction of cytokinetic ring depends on myosin II but not on actin dynamics. *Nature Cell Biology, 15*(7), 853–859. http://dx.doi.org/10.1038/ncb2781. pii: ncb2781.

Nguyen, P. A., Groen, A. C., Loose, M., Ishihara, K., Wuhr, M., & Field, C. M. (2014). Spatial organization of cytokinesis signaling reconstituted in a cell-free system. *Science, 346*(6206), 244–247. http://dx.doi.org/10.1126/science.1256773. pii:346/6206/244.

Oh, Y., Chang, K.-J., Orlean, P., Wloka, C., Deshaies, R., & Bi, E. (2012). Mitotic exit kinase Dbf2 directly phosphorylates chitin synthase Chs2 to regulate cytokinesis in budding yeast. *Molecular Biology of the Cell, 23*(13), 2445–2456.

Okada, S., Leda, M., Hanna, J., Savage, N. S., Bi, E., & Goryachev, A. B. (2013). Daughter cell identity emerges from the interplay of Cdc42, septins, and exocytosis. *Developmental Cell, 26*(2), 148–161. http://dx.doi.org/10.1016/j.devcel.2013.06.015. pii:S1534-5807(13)00354-7.

Ong, K., Wloka, C., Okada, S., Svitkina, T., & Bi, E. (2014). Architecture and dynamic remodelling of the septin cytoskeleton during the cell cycle. *Nature Communications, 5*, 5698. http://dx.doi.org/10.1038/ncomms6698. pii:ncomms6698.

Onishi, M., Ko, N., Nishihama, R., & Pringle, J. R. (2013). Distinct roles of Rho1, Cdc42, and Cyk3 in septum formation and abscission during yeast cytokinesis. *The Journal of Cell Biology, 202*(2), 311–329. http://dx.doi.org/10.1083/jcb.201302001. pii:jcb.201302001.

Park, H. O., & Bi, E. (2007). Central roles of small GTPases in the development of cell polarity in yeast and beyond. *Microbiology and Molecular Biology Reviews: MMBR, 71*, 48–96.

Patterson, G. H., & Lippincott-Schwartz, J. (2002). A photoactivatable GFP for selective photolabeling of proteins and cells. *Science, 297*(5588), 1873–1877.

Pollard, T. D. (2010). Mechanics of cytokinesis in eukaryotes. *Current Opinion in Cell Biology, 22*(1), 50–56. http://dx.doi.org/10.1016/j.ceb.2009.11.010. pii:S0955-0674(09)00216-6.

Pruyne, D., Gao, L., Bi, E., & Bretscher, A. (2004). Stable and dynamic axes of polarity use distinct formin isoforms in budding yeast. *Molecular Biology of the Cell, 15*(11), 4971–4989.

Ries, J., Kaplan, C., Platonova, E., Eghlidi, H., & Ewers, H. (2012). A simple, versatile method for GFP-based super-resolution microscopy via nanobodies. *Nature Methods, 9*(6), 582–584. http://dx.doi.org/10.1038/nmeth.1991. pii:nmeth.1991.

Sanger, J. M., & Sanger, J. W. (1980). Banding and polarity of actin filaments in interphase and cleaving cells. *The Journal of Cell Biology, 86*(2), 568–575.

Sburlati, A., & Cabib, E. (1986). Chitin synthetase 2, a presumptive participant in septum formation in *Saccharomyces cerevisiae*. *The Journal of Biological Chemistry, 261*(32), 15147–15152.

Schroeder, T. E. (1972). The contractile ring II: determining its brief existence, volumetric changes, and vital role in cleaving arbacia eggs. *The Journal of Cell Biology, 53*, 419–434.

Skop, A. R., Liu, H., Yates, J., 3rd, Meyer, B. J., & Heald, R. (2004). Dissection of the mammalian midbody proteome reveals conserved cytokinesis mechanisms. *Science, 305*(5680), 61–66.

Strauss, M. P., Liew, A. T., Turnbull, L., Whitchurch, C. B., Monahan, L. G., & Harry, E. J. (2012). 3D-SIM super resolution microscopy reveals a bead-like arrangement for FtsZ and the division machinery: implications for triggering cytokinesis. *PLoS Biology, 10*(9), e1001389. http://dx.doi.org/10.1371/journal.pbio.1001389. pii:PBIOLOGY-D-12-01887.

Strickland, L. I., & Burgess, D. R. (2004). Pathways for membrane trafficking during cytokinesis. *Trends in Cell Biology, 14*(3), 115–118.

Thevenaz, P., Ruttimann, U. E., & Unser, M. (1998). A pyramid approach to subpixel registration based on intensity. *IEEE Transactions on Image Processing: A Publication of the IEEE Signal Processing Society, 7*(1), 27–41. http://dx.doi.org/10.1109/83.650848.

Tolliday, N., VerPlank, L., & Li, R. (2002). Rho1 directs formin-mediated actin ring assembly during budding yeast cytokinesis. *Current Biology: CB, 12*(21), 1864–1870.

Vallen, E. A., Caviston, J., & Bi, E. (2000). Roles of Hof1p, Bni1p, Bnr1p, and Myo1p in cytokinesis in *Saccharomyces cerevisiae. Molecular Biology of the Cell, 11*(2), 593–611.

Vavylonis, D., Wu, J. Q., Hao, S., O'Shaughnessy, B., & Pollard, T. D. (2008). Assembly mechanism of the contractile ring for cytokinesis by fission yeast. *Science, 319*(5859), 97–100. http://dx.doi.org/10.1126/science.1151086. pii:1151086.

Weiss, E. L. (2012). Mitotic exit and separation of mother and daughter cells. *Genetics, 192*(4), 1165–1202. http://dx.doi.org/10.1534/genetics.112.145516. pii:192/4/1165.

Wloka, C., & Bi, E. (2012). Mechanisms of cytokinesis in budding yeast. *Cytoskeleton (Hoboken, NJ), 69*, 710–726. http://dx.doi.org/10.1002/cm.21046.

Wloka, C., Nishihama, R., Onishi, M., Oh, Y., Hanna, J., & Pringle, J. R. (2011). Evidence that a septin diffusion barrier is dispensable for cytokinesis in budding yeast. *Biological Chemistry, 392*(8–9), 813–829. http://dx.doi.org/10.1515/BC.2011.083.

Wloka, C., Vallen, E. A., Thé, L., Fang, X., Oh, Y., & Bi, E. (2013). Immobile myosin-II plays a scaffolding role during cytokinesis in budding yeast. *The Journal of Cell Biology, 200*(3), 271–286.

Yeong, F. M. (2005). Severing all ties between mother and daughter: cell separation in budding yeast. *Molecular Microbiology, 55*(5), 1325–1331.

Young, B. A., Buser, C., & Drubin, D. G. (2010). Isolation and partial purification of the *Saccharomyces cerevisiae* cytokinetic apparatus. *Cytoskeleton (Hoboken, NJ), 67*(1), 13–22. http://dx.doi.org/10.1002/cm.20412.

Zhang, M., Chang, H., Zhang, Y., Yu, J., Wu, L., & Ji, W. (2012). Rational design of true monomeric and bright photoactivatable fluorescent proteins. *Nature Methods, 9*(7), 727–729. http://dx.doi.org/10.1038/nmeth.2021. pii:nmeth.2021.

Zhao, J., & Wang, Q. (2016). Modeling cytokinesis of eukaryotic cells driven by the actomyosin contractile ring. *International Journal for Numerical Methods in Biomedical Engineering*. http://dx.doi.org/10.1002/cnm.2774.

CHAPTER

Imaging cytokinesis of *Drosophila* S2 cells

4

A. Kechad*,§, **G.R.X. Hickson***,§,[1]

*Sainte-Justine Hospital Research Center, Montreal, QC, Canada

§Université de Montréal, Montréal, QC, Canada

[1]Corresponding author: E-mail: gilles.hickson@umontreal.ca

CHAPTER OUTLINE

Abstract

Animal cell cytokinesis proceeds through three successive stages: a contractile ring stage, an intercellular bridge stage, and an abscission stage. Many studies have identified a complex network of key proteins required for successful cytokinesis. While each component interacts with, and depends on, several other components, our understanding of how these proteins cooperate in space and time to ensure faithful progression through the stages of cytokinesis remains incomplete. A full understanding of the complexity of the process and its underlying machinery necessitates experimental systems that allow both genetic manipulation and real-time visualization of the various components throughout the successive stages of cytokinesis. Cultured *Drosophila* S2 cells provide such a system. They are genetically tractable thanks to their exquisite sensitivity to RNA

Methods in Cell Biology, Volume 137, ISSN 0091-679X, http://dx.doi.org/10.1016/bs.mcb.2016.03.034

interference mediated by double-stranded RNAs, which can be generated with ease in the laboratory. Furthermore, S2 cells grow well under normal atmospheric conditions, and stable lines expressing fluorescently tagged proteins can be readily generated, making them ideal for long-term live-cell fluorescence microscopy.

Here we describe methodology for exploiting S2 cells for the study of cytokinesis, with an emphasis on live-cell imaging. We describe a variety of fluorescent markers available and their utility for highlighting different structures at different stages of cytokinesis. We describe our experimental setup that forms the basis for live-cell analysis of loss-of-function RNAi experiments, rescue experiments, and structure-function analyses of key regulators of cytokinesis. Finally, we describe the types of phenotypes that one can observe at the different stages of *Drosophila* S2 cell cytokinesis.

INTRODUCTION

Cytokinesis is the evolutionarily conserved process by which a cell physically separates into two, ensuring the proper partitioning of nuclei and cytoplasmic contents at the end of cell division. It is a highly regulated process that, in animal cells, occurs in three distinct and successive stages: a contractile ring stage; an intercellular bridge stage; and an abscission stage (reviewed in Barr & Gruneberg, 2007; Glotzer, 2009; Gould, 2016; Green, Paluch, & Oegema, 2012; Steigemann & Gerlich, 2009).

Cytokinesis employs robust molecular machinery that assembles with high spatiotemporal precision. Classical genetic studies and genome-wide RNAi screens identified a plethora of evolutionarily conserved proteins necessary for cytokinesis, including cell cycle regulators, cytoskeletal proteins, and membrane trafficking and remodeling proteins. The challenge remains to understand precisely how these proteins cooperate in space and time, how they interact with membranes, how the ensemble of the machinery evolves as cytokinesis progresses, and how that machinery adapts to suit differing developmental contexts.

Traditionally, genes were ascribed cytokinesis functions based on scoring fixed cells for the most obvious outcome for cytokinesis failure following successful mitosis: binucleate cells. However, defective cytokinesis does not always result in binucleation. For example, a failure in abscission will, strictly speaking, generate cells that remain connected by stable intercellular bridges. Furthermore, fixed cells represent snapshots in time. Although informative, they lack temporal information and are inadequate for the proper appreciation and characterization of dynamic phenotypes such as oscillating cleavage furrows, shedding of material from the intercellular bridge, and premature abscission, all of which have been documented through live imaging (El Amine, Kechad, Jananji, & Hickson, 2013; Goldbach et al., 2010; Hickson & O'Farrell, 2008; Kechad, Jananji, Ruella, & Hickson, 2012; Piekny & Glotzer, 2008; Renshaw, Liu, Lavoie, & Wilde, 2014; Straight, Field, & Mitchison, 2005). Here, we describe a simple yet powerful model for studying animal cell cytokinesis using live-cell imaging of *Drosophila* Schneider's line S2 tissue culture cells.

Many features make *Drosophila* tissue culture cells a powerful model system. First, the *Drosophila* genome exhibits less functional redundancy compared to mammalian genomes. This facilitates loss-of-function studies and interpretation of observed phenotypes. Second, most of the genes involved in cytokinesis and their functions are conserved from flies to mammals (Cabernard, 2012). Crucially, *Drosophila* cells are genetically tractable, thanks to exquisite sensitivity to RNAi elicited by long double-stranded RNAs (dsRNAs) (Clemens et al., 2000), which are both efficient and simple to generate in the laboratory. For these reasons, *Drosophila* tissue culture cells emerged as a system of choice for RNAi screening efforts (Dean, Rogers, Stuurman, Vale, & Spudich, 2005; Echard, Hickson, Foley, & O'Farrell, 2004; Foley & O'Farrell, 2004; Goshima et al., 2007; Kiger et al., 2003; Somma, Fasulo, Cenci, Cundari, & Gatti, 2002), many of which helped define the repertoire of essential cytokinesis genes. Finally, *Drosophila* cells are simple to work with. They are readily maintained at ambient temperature (20–25°C) and normal atmospheric conditions (no CO_2), which makes them ideal for live-cell microscopy, as does their amenity for transfection and the availability of genetically encoded fluorescent proteins such as GFP.

Together, these attributes provide a powerful model system in which expression of exogenous fluorescent proteins and depletion of endogenous proteins can be combined to allow reduction of function analyses, rescue experiments, and structure-function analyses to be performed in living cells. Here we describe the methodology in practice in our laboratory using S2 cells to study the functions and interdependencies of key regulators of cytokinetic progression in both space and time.

1. METHODS

1.1 MATERIALS AND SAMPLE PREPARATION

1.1.1 S2 cells and culture

Schneider's S2 cells derive from a primary culture of 1-day-old Oregon R *Drosophila* embryos (Schneider, 1972). They are spherical, 10–15 μm in diameter, and exhibit hemocyte/macrophage-like cell behavior and gene expression. We use S2 cells obtained from the University of California, San Francisco, cell culture facility, but S2 cells can be purchased from *Drosophila* Genome Resource Center (DGRC, Indiana) or Invitrogen/Thermo Fisher Scientific. Methods for culturing S2 cells have been described elsewhere in detail (Bettencourt-Dias & Goshima, 2009; Buster, Nye, Klebba, & Rogers, 2010; Goshima, 2010; Pereira, Matos, Lince-Faria, & Maiato, 2009; Rogers & Rogers, 2008). Briefly, we routinely proceed as follows:

1. Culture S2 cells in Schneider's *Drosophila* medium (Life Technologies/Thermo Fisher Scientific, cat#21720-001), supplemented with 10% heat-inactivated fetal calf serum (Invitrogen/Thermo Fisher Scientific, US origin cat#16140-071) and penicillin/streptomycin (Invitrogen/Thermo Fisher Scientific,

cat#15140-122). S2 cells are easily maintained under normal ambient temperature (20–25°C) and atmospheric conditions.

2. S2 cells have an approximate 22 h doubling time and should be passaged by simple pipetting every 4–7 days at a medium to high density (1:3 to 1:5 dilution) to maintain a good mitotic index and avoid cell death (Goshima, 2010).

Notes

- Special attention should be given when changing batch lots of cell media or serum as we have observed variation in growth rate and/or dsRNA treatment efficacy. New lots of media or serum should be systematically tested. Cells should be grown in the new media in parallel to the old media for several passages to assess for general cell health, growth rate, and RNAi sensitivity. It is recommended that batches that give good results should be purchased in bulk and stockpiled for future use.
- Temperature variability can also cause changes in cell growth rates and the kinetics of RNAi depletions and thus the onset of resulting phenotypes. Temperature should be monitored and kept stable, ideally between 22°C and 25°C.

1.1.2 Generation of stable cell lines expressing fluorescently tagged proteins

S2 cells readily take up DNA and have high transfection rates. It is possible to generate stable cell lines expressing multiple fluorescently tagged proteins of interest (or fragments thereof) and follow their behaviors during cytokinesis in live or fixed cells. Cloning and transfection protocols have been previously described in detail (Bettencourt-Dias & Goshima, 2009; Pereira et al., 2009). Briefly, the following workflow is employed in our laboratory for generating new fluorescent protein (FP)-expressing cell lines:

1. Amplify desired coding sequences of a gene of interest by PCR from cDNA clones obtained, for example, from the DGRC, adding CACC to the 5′ end of an initiating ATG.
2. TOPO-clone these into pENTR-D-TOPO Gateway entry plasmid (Thermo Fisher Scientific, cat# k2400-20) and sequence-verify.
3. Recombine coding sequences into expression vectors to form N- or C-terminal fusions with FPs such as EGFP, mCherry, Venus, Cerulean, Dendra2 using the *Drosophila* Gateway vector collection (https://emb.carnegiescience.edu/drosophila-gateway-vector-collection). We use destination plasmids that employ either the *actin05C* promoter (pAc) for constitutive expression or the *metallothionein* promoter (pMT) for $CuSO_4$-inducible expression. It is good practice to separately place FP tags at both the N- and C-termini, with and without the native stop codon as appropriate, in case the presence of the tag at either end interferes with protein function.
4. Transfect desired expression plasmids into S2 cells in serum-free media using Cellfectin II reagent (Thermo Fisher Scientific, cat#10362-100), together with

pCoHygro (Thermo Fisher Scientific) and select stable cell lines using 400 μg/mL hygromycin B (Thermo Fisher Scientific, cat#10687-010), supplemented to the growth media for 4 weeks. In our hands, up to three transgenes can be expressed simultaneously in the resulting cell lines.

Notes

- Transient transfection and selection present challenges to the cells that, from our experience, can compromise normal cell behaviors including division. We therefore do not perform experiments on transiently transfected cells, or until after hygromycin selection has been maintained for 4 weeks and then been discontinued for at least one passage.
- In some cases, we find that transgene expression may then be progressively lost after serial passaging. We therefore routinely cryopreserve aliquots of our newly made cell lines for later thawing, or simply remake lines periodically.

1.1.3 Choice of cytokinetic markers

To achieve an in depth characterization of cytokinesis dynamics and regulation, one would ideally like to visualize the various components of the cytokinetic machinery that comprise cortical, spindle, and membrane-associated proteins at different stages of cytokinesis. The choice of fluorescent marker(s) is important and may vary depending on the experiment in question. As discussed in the following section, there are a number of validated markers that are useful for examining cytokinesis in this system. These are summarized in Table 1, and some are also shown in Fig. 1.

1.1.3.1 General considerations

- When assessing RNAi depletion phenotypes, one should ideally examine a variety of cell lines expressing different FP-tagged cortical and spindle-associated proteins to maximize the chances of observing relevant phenotypes. An example from our own observations is that the myosin regulatory light chain (MRLC), Spaghetti squash (Sqh-FP), and Anillin-FP are both good markers for the contractile ring (CR) and midbody ring (MR), yet only Anillin-FP exhibits shedding from the nascent MR, even when both markers are coexpressed (El Amine et al., 2013). Examining multiple markers can also help mitigate concerns that relevant phenotypes are being masked or enhanced by (over-) expression of a given marker. Alterations in the extent and/or timing of the localization patterns of different markers following specific experimental manipulations can provide insight into the normal progression of cytokinesis.
- When studying the localization dynamics of a GFP-tagged protein of interest, it is often useful to coexpress it with a spindle marker fused to a different FP, eg, Tubulin-mCherry, or a chromosomal marker, eg, Histone H2B-mCherry (Hickson, Echard, & O'Farrell, 2006). This facilitates the identification of cells in early mitosis, ie, before the onset of cytokinesis. As both Tubulin-mCherry and Histone H2B-mCherry can be expressed constitutively from the *actin05C* promoter (pAC plasmid), including one of these markers allows the cells to be

Table 1 Summary of Localization Patterns of a Selection of Validated Markers for Examining S2 Cell Cytokinesis

Stages / FP-Markers	Easy ID of Cells in Early Mitosis?	CS/CR 	Midbody/ Nascent MR 	Shedding 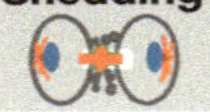	Mature MR 	MR Remnant 	Comments	References
Histone H2B	✓	✗	✗	✗	✗	✗	•Does not localize directly to division site. •Reliable chromosomal marker to identify different steps of mitosis.	Hickson et al. (2006)
Tubulin	✓	✓	✓	✓	✗	✗	•Reliable marker for mitosis, the central spindle and midbody. •Unreliable for late MRs and abscission.	Rogers et al. (2002) Vale, Spudich, & Griffis, (2009)
Fascetto (PRC1)	✗	✓	✓	✓	✗	✗	•Reliable central spindle and early midbody marker. •Unreliable for late MRs and abscission.	Hickson lab (unpublished)
IAL (Aurora B)	✓	✓	✓	✓	✗	✗	•Localizes to inner centromeres in early mitosis. •Reliable central spindle and early midbody marker. •Unreliable for late MRs and abscission.	Vale et al. (2009) Smurnyy et al. (2010) El Amine et al. (2013)
Polo	✓	✓	✓	✓	✗	✗	•Localizes to kinetochores and spindle poles in early mitosis.	D'Avino et al. (2007) Kachaner et al. (2014)

							•Reliable central spindle and midbody marker. •Unreliable for late MRs and abscission.	
Pavarotti (Kinesin-6)	✓	✓	✓	✓	✗	✗	•Reliable central spindle and midbody marker. •Unreliable for late MRs and abscission.	Echard et al. (2004) Goshima and Vale (2005)
Tum (RacGAP50C)	✗	✓	✓	✓	✓	✓	•Reliable cytokinesis marker, labels the central spindle early. •Relocalizes to ring structures later in cytokinesis.	D'Avino, Savoian, Capalbo, & Glover, (2006) Hickson lab (unpublished)
LifeAct	✗	✓	✓	✗	✗	✗	•Hard to distinguish mitosis from interphase. •Labels the cortex and early ring structures. •Unreliable for late MRs and abscission.	Riedl et al. (2008) El Amine et al. (2013)
Diaphanous	✗	✓	✓	✗	✗	✗	•Labels the cortex and ring structures. •Unreliable for late MRs and abscission.	Vale et al. (2009) Hickson lab (unpublished)
Anillin	✓	✓	✓	✓	✓	✓	•Reliable cytokinesis marker. Allows identification of mitotic cells upon nuclear envelope breakdown. •Labels ring structures throughout cytokinesis.	Hickson and O'Farrell (2008) Kechad et al. (2012) El Amine et al. (2013)

Continued

Table 1 Summary of Localization Patterns of a Selection of Validated Markers for Examining S2 Cell Cytokinesis—cont'd

Stages / FP-Markers	Easy ID of Cells in Early Mitosis?	CS/CR 	Midbody/ Nascent MR 	Shedding 	Mature MR 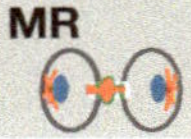	MR Remnant 	Comments	References
Nebbish (KIF14)	✗	✓	✓	✓	✓	✓	•Reliable cytokinesis marker. •Labels ring structures throughout cytokinesis.	Bassi et al. (2013) Hickson lab (unpublished)
Sticky (citron kinase)	✗	✓	✓	✓	✓	✓	•Labels ring structures throughout cytokinesis. •Other puncta also present that can be hard to distinguish from late MR.	Bassi et al. (2013) El Amine et al. (2013)
Septin2	✗	✓	✓	✓	✗	✗	•Labels the CR and nascent MR •Unreliable for late MRs and abscission.	Hickson lab (unpublished)
Sqh (MRLC)	✗	✓	✓	✗	✓	✓	•Reliable cytokinesis marker. •Labels ring structures throughout cytokinesis.	Rogers et al. (2004) Dean et al. (2005) Hickson et al. (2006) Kechad et al. (2012)

Localization patterns of fluorescently tagged cytokinetic markers expressed in Drosophila *S2 cells. Each marker is assessed for its reliability in identifying cells in early mitosis (ie, before anaphase onset) and labeling the different microtubules and/or cortical structures throughout cytokinesis: central spindle (CS), contractile ring (CR), midbody, midbody ring (MR), shedding events and remnants that persist after abscission.* ✓ *or* ✗ *marks indicate reliability or unreliability, respectively, of each fluorescently tagged marker for labeling the indicated structure.* MRLC, *myosin regulatory light chain.*

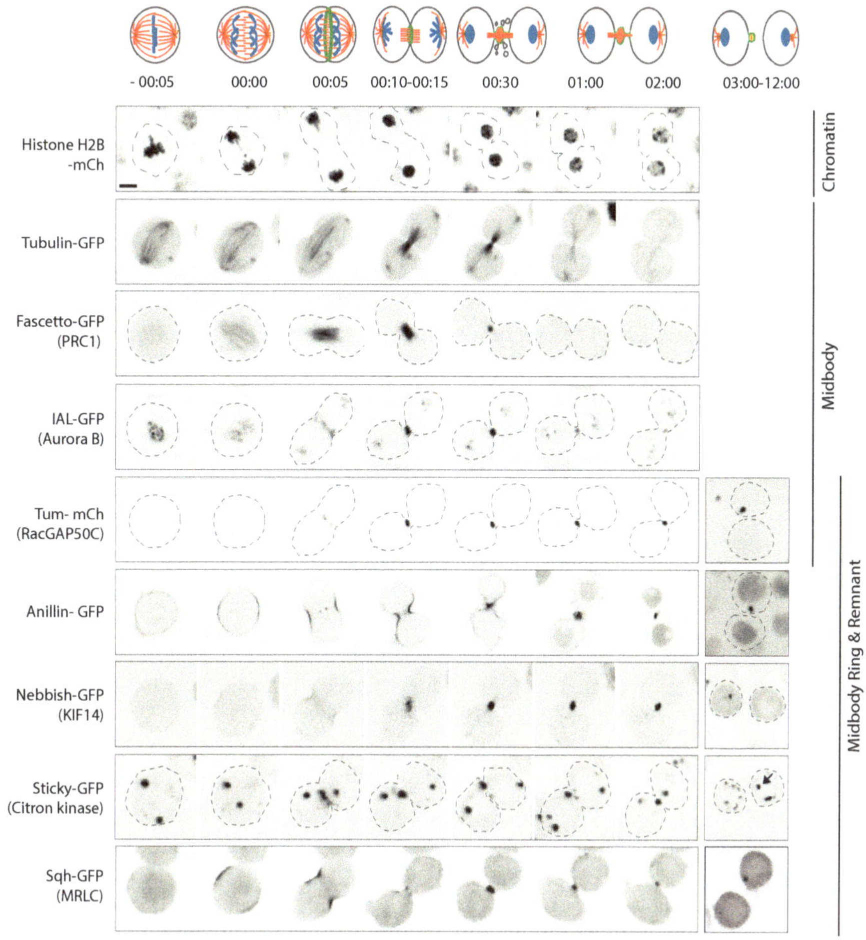

FIGURE 1 Localization patterns of specific cytokinetic markers.

Time-lapse sequences of a selection of S2 cell lines, expressing different FP-tagged markers that exhibit distinct localization patterns at the different stages of cytokinesis. All were expressed from the copper-inducible pMT promoter except Histone H2B-mCherry (mCh) and tubulin-GFP, which were constitutively expressed from the pAc promoter. *Dotted lines* represent cell outlines. The arrow in the sticky-GFP panel marks the midbody ring remnant after abscission. Spinning disk confocal images were acquired using a 63× 1.4 NA plan-apochromat oil-immersion objective. Single Z confocal images, with look up tables inverted using Adobe Photoshop CS6, are shown. Times are in h:min, scale bar is 5 μm. *MRLC*, myosin regulatory light chain.

imaged with and without induction of the second transgene, expressed from the pMT plasmid. This is particularly useful in the context of rescue experiments (see the later discussion). These markers are also likely to be the least perturbing, especially Histone H2B-FP, which does not even localize to the cytokinesis machinery.

1.1.3.2 Specific markers

- Tubulin-FP is an ideal marker for rapid detection of mitotic cells (Goshima & Vale, 2003) and, because it is excluded from the nucleus, binucleate cells. Importantly, it allows visualization of the central spindle, its compaction during furrowing and formation of the midbody, usually identifiable by a characteristic drop in tubulin fluorescence at the center of the midbody (Fig. 1). However, tubulin-FP becomes a poor marker for scoring abscission because the microtubules of the intercellular bridge are progressively lost and disappear before abscission (Fig. 1).
- FP-tagged versions of microtubule-associated proteins, such as the PRC1 orthologue, Fascetto, and the Aurora B kinase, IAL (Smurnyy, Toms, Hickson, Eck, & Eggert, 2010; Vale et al., 2009), localize to the central spindle and the early midbody and then rapidly disappear (Fig. 1).
- GFP-Polo kinase, also localizes to the central spindle and midbody during cytokinesis (Fig. 1), although it separates into two pools: a MAP205-dependent pool that localizes to the flanking regions of the midbody and an Aurora B kinase-dependent pool that localizes more transiently to the central midbody early during its formation (Kachaner et al., 2014).
- Other Microtubule-associated proteins, such as kinesin-6/Pavarotti-FP (Echard et al., 2004; Goshima & Vale, 2005) and its centralspindlin partner, RacGAP50C/Tumbleweed-FP (D'Avino et al., 2006; Hickson lab, unpublished), are reliable markers to visualize the maturation of the central spindle to the midbody and both of these persist in the late MR and MR remnants after abscission (Fig. 1). However, it is not possible to identify early mitotic cells with these markers alone.
- Anillin-FP is an ideal marker because of its localization patterns throughout cytokinesis (Field & Alberts, 1995; Hickson & O'Farrell, 2008). It is nuclear in interphase and becomes cortical upon nuclear envelope breakdown in prophase, allowing easy detection of cells in early mitosis. In anaphase, it becomes enriched at the furrow. Then, it localizes to the CR and MR in telophase and persists in the MR remnant after abscission (Fig. 1).
- LifeAct-FP (El Amine et al., 2013; Riedl et al., 2008) and the formin Diaphanous-FP (Vale et al., 2009) localize to the CR but become undetectable in the late MR. They are also unreliable markers for identifying early mitotic cells.
- Other cortical proteins such as Sqh-FP (MRLC) (Dean et al., 2005; Dean & Spudich, 2006; Hickson et al., 2006; Kechad et al., 2012; Rogers, Wiedemann, Hacker, Turck, & Vale, 2004) and Nebbish-FP (KIF-14) (Bassi, Audusseau, Riparbelli, Callaini, & D'Avino, 2013) and are reliable markers of the CR that persist in the MR and the MR remnant after abscission (Fig. 1). The citron kinase, sticky-FP, also localizes to the MR and MR remnants (Bassi et al., 2013;

El Amine et al., 2013), but in our hands, the presence of additional cytoplasmic puncta complicates the use of this as a marker (Fig. 1).

1.1.4 Double-stranded RNA preparation for RNAi experiments

Exquisite sensitivity to RNAi fueled the emergence of *Drosophila* S2 cells as a system of choice for genome-wide RNAi screens and it remains one of their major advantages for studying cytokinesis. Unlike mammalian cells that require transfection of carefully designed, expensive small interference RNAs, S2 cells readily take up long dsRNAs, which are easily prepared in the laboratory. Because *Drosophila* lacks an adaptive immune response, long dsRNAs do not trigger an interferon response, but rather the cells take them up and dice them into a pool of siRNAs that lead to efficient degradation of corresponding mRNAs. Detailed, authoritative protocols have been published elsewhere (Bettencourt-Dias & Goshima, 2009; Buster et al., 2010; Rogers & Rogers, 2008), but briefly we

1. amplify target cDNA regions (300–700 bp in length) by PCR, incorporating the T7 promoter sequence at both ends;
2. use PCR products as templates to synthesize complementary RNAs by in vitro transcription using T7-ribomax Express RNAi System (Promega, cat#P1300). Digest template DNA with DNAse I and precipitate RNAs with sodium acetate, wash with ethanol, air-dry, and resuspend the resulting RNA pellet in water, following the kit instructions. Anneal RNAs in a water bath heated to 90°C and allowed to cool to room temperature. Verify and quantify dsRNAs by agarose gel electrophoresis and adjust concentrations to 1 μg/μL.

Notes

- When designing primers for template production, the resulting amplimers should be verified for potential off-target hits, by performing BLAST searches with the parameters set to identify very short sequences of homology in the *Drosophila melanogaster* genome. We aim to avoid stretches of identity of more than 15 bp present in other genes.
- It is also good practice to generate multiple, nonoverlapping dsRNAs against the same mRNA target, to provide confidence that any arising phenotypes are indeed specific and not due to off-target effects.
- If antibodies to the target protein are available, the kinetics and efficacy of depletion of a given dsRNA should be monitored by western blot, ideally performed in parallel to live-cell imaging. In this way, estimates of depletion efficacy can accompany the phenotypic analysis. We suggest lysing cells at different times post dsRNA treatment and performing SDS-PAGE and immunoblot analysis. Comparing depleted lysates to a serially diluted control lysate will allow estimation of the level of protein depletion over time, and definition of the limits of detection of the antibody. Quantification of protein depletion is particularly important in the absence of observable phenotypes, as few conclusions can be drawn without it.

1.2 LIVE-CELL IMAGING

1.2.1 Sample preparation and experimental setup

1. Under aseptic conditions, seed desired S2 cells at medium density (2×10^6 mL^{-1}, approx. 50% confluency when cells have settled) in 200 μL complete media in 96-well plastic flat-bottomed plates (eg, Corning, cat#CS003595). Seed enough wells for duplicates and controls.
   Note: Limiting cell density is important for RNAi efficiency and to maintain a high mitotic index. For more stable (and/or abundant) targets that require longer incubation with dsRNA for phenotypes to become apparent, cells should be passaged after 4 days with fresh dsRNA added.
2. Where appropriate, add 1 μg of dsRNA (1 μg/μL) per well, mix by pipetting and incubate from 2 to 7 days, depending on the target gene in question. If incubation beyond 4 days is required, split cells 1:2 on day 4 and add 0.5 μg fresh dsRNA.
   Note: The incubation time required for the onset of phenotypes (eg, cytokinesis failure) should be determined empirically for each dsRNA and will depend on factors such as the stability of the target protein and any threshold(s) for its requirement.
3. Include negative and positive dsRNA controls:
   a. For a negative control, we typically use dsRNAs designed to target the *E. coli LacI* gene, but any other sequence that is not present in the cell line could be used (eg, kanamycin resistance gene). We also routinely include a control without any dsRNA just to monitor the general density and health of the culture.
   b. It is important to include positive controls for every experiment, because of potential variability of the kinetics and/or efficacy of RNAi due to factors such as media, temperature, cell density, and cell passage number. We routinely use dsRNA to deplete the RhoGEF, Pebble, or the kinesin-6, Pavarotti, since either of these ordinarily results in rapid and penetrant cytokinesis failure, producing binucleate cells from 2 to 3 days of treatment.
4. When appropriate, induce expression of pMT-driven transgenes with 250–500 μM $CuSO_4$ (2.5–5 μL of a 100 mM sterile-filtered stock solution per milliliter of growth media).
   Note: If performing a simple RNAi loss-of-function experiment, induction of expression of a relevant marker can be done 12–24 h prior to imaging. If performing rescue experiments, induction should be performed at the same time as dsRNA is added and maintained throughout any subsequent passages, if applicable.
5. 12–24 h prior to imaging, gently resuspend dsRNA-treated cells by micropipetting and transfer to 8-well chambered coverglass dishes (Lab-tek II, Nunc, cat#CA43300774) or 96-well glass bottomed plates (Whatman, cat#306180-030) and allow to settle for at least 30 min.
   Notes: Optimal cell density is important for imaging and should be around 2×10^6 mL^{-1} in a final volume of 400 μL (ie, pool duplicate wells, or split cells

1:2 into fresh media) for 8 wells dishes and 200 μL in 96 wells dishes. Note also that pipetting may rupture late intercellular bridges, which will need to be considered if abscission events are to be examined. For this reason, we recommend plating the cells 12–24 h prior to observing abscission events.

Since S2 cells are only semiadherent, they can move around and some will leave the field of view during long time-course imaging. Plating S2 cells on the lectin, concanavalin A (ConA), causes them to adhere and flatten out beautifully and this has been exploited in numerous important studies (Rogers, Rogers, Sharp, & Vale, 2002; Rogers, Wiedemann, Stuurman, & Vale, 2003; Vale et al., 2009). However plating on ConA also blocks the completion of cytokinesis and is thus not suitable for studying many stages of cytokinesis. S2R+ cells are a more adherent variant line that could alternatively be used.

1.2.1.1 Variation: codepletion experiments

Another powerful advantage of this system is that two target proteins can be reliably codepleted in the same cells, allowing genetic interactions to be uncovered (for examples, see (El Amine et al., 2013; Hickson et al., 2006)). However, such experiments require some additional considerations and controls. First, the incubation times required for each dsRNA need to be determined. If both dsRNAs require the same incubation time, they can be added to the cells together. However, if the times differ, the dsRNAs should be added sequentially according to their specific schedules. Second, individual depletions must also be performed in the presence of control (*lacI*) dsRNAs. This is important since depletion of a given target may be less efficient in the presence of a second dsRNA due to competition effects for components of the RNAi machinery.

An example of a codepletion experiment is given later (and in Fig. 2B), where dsRNAs targeting the septin gene, *peanut*, which require 7–8 day incubations before cytokinesis failures are observed, are to be combined with dsRNAs targeting Anillin, which induce cytokinesis failures by 3 days (Kechad et al., 2012).

1. 7 days prior to imaging, seed the cells and add *peanut* dsRNAs to two wells (one will be for a *peanut* only control). In parallel, set up two additional wells incubated only with *lacI* dsRNAs (for *lacI* and *anillin* only controls).
2. 3–4 days prior to imaging, split the cells and add *anillin* or *lacI* dsRNAs together with fresh *peanut* or *lacI* dsRNAs, as depicted in Fig. 2B. Incubate for a further 3 days.
3. 12–24 h prior to imaging, split the cells again and readd both *peanut* and *anillin* dsRNAs before transferring to imaging dishes and proceeding to the microscope system.

1.2.1.2 Variation: rescue experiments

Rescue experiments are invaluable controls for specificity of dsRNA-induced phenotypes and for testing the functionality of FP-tagged proteins. The idea is to specifically deplete an endogenous protein and replace it with an exogenous

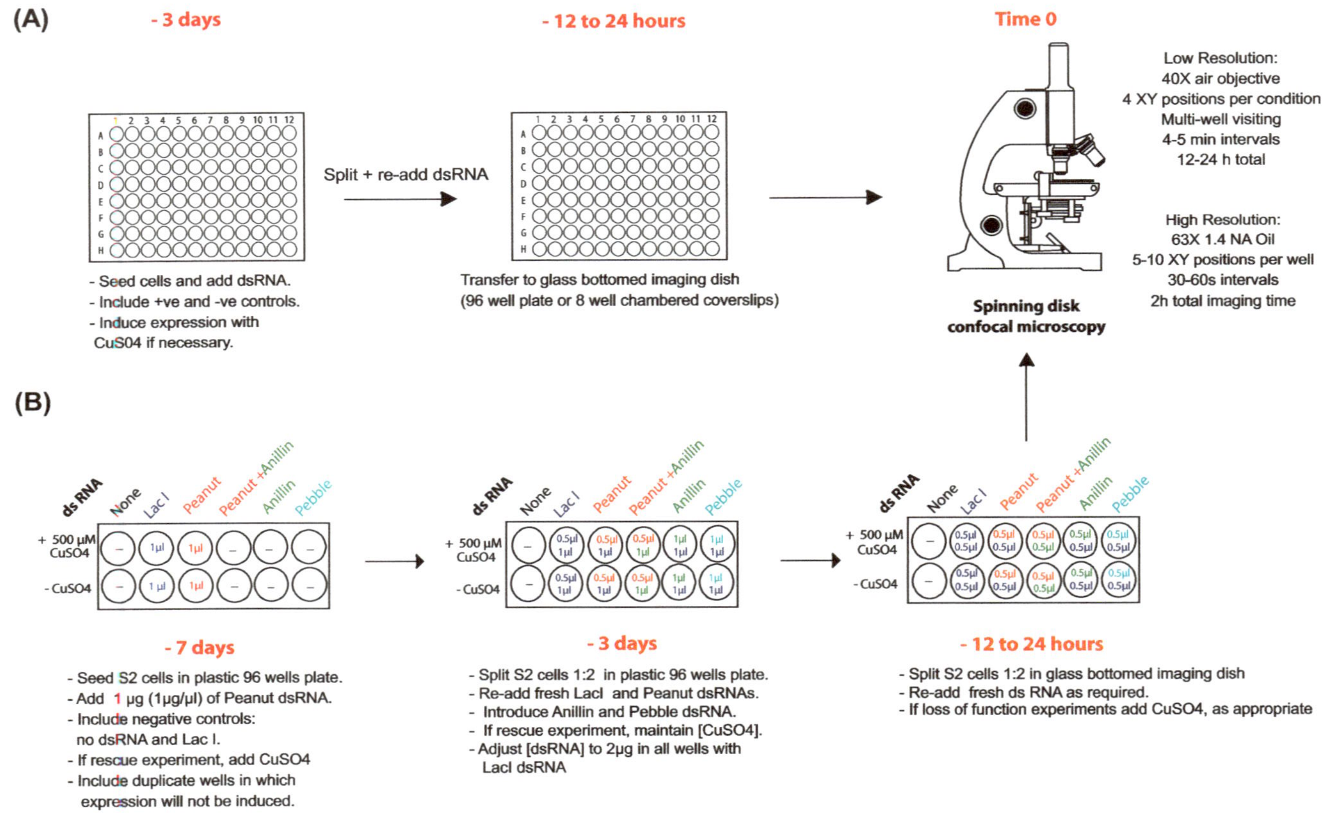

FIGURE 2 Typical experimental setup and workflow.

(A) General overview of a simple RNAi loss-of-function experiment: 3 days prior to imaging, plate the cells in a 96-well plastic plate and incubate with dsRNA along with positive (typically *pebble* dsRNA) and negative (typically *lacI dsRNA* and no dsRNA) controls. If a rescue experiment, induce expression with $CuSO_4$. 12–24 h prior to imaging, split cells, readd fresh dsRNA and transfer to glass bottomed imaging dish. If a loss-of-function experiment, induce expression with $CuSO_4$. At day 0, proceed to your imaging system. (B) A specific example of a codepletion experiment using dsRNAs that require different incubation times. First add dsRNAs that need the longest incubation time; here *peanut* dsRNA (7 days incubation), along with negative controls (*lacI dsRNA* and no dsRNA). Then split the cells and add dsRNAs that require a shorter incubation time; here *anillin* dsRNA (3 days incubation) along with positive control, *pebble* dsRNA (3 days incubation). 12–24 h prior to imaging, split cells, readd fresh dsRNA and transfer to a glass bottomed imaging dish. At day 0, proceed to your imaging system. Adjust dsRNA concentration in each well as indicated. Include a noninduced duplicate experiment, where appropriate, to confirm RNAi efficacy.

dsRNA-resistant, FP-tagged version. If the exogenous protein can rescue the defects caused by the dsRNA, it indicates that the RNAi phenotype is specific and that the exogenous protein (ie, FP-fusion) is functional.

The first task is to generate a dsRNA that will effectively target the endogenous mRNA but not the exogenous one. The simplest method is to generate dsRNAs against the 5′ and/or 3′ UTRs of the endogenous mRNA. We have successfully performed this for several genes, including Anillin (Hickson & O'Farrell, 2008).

However, the UTR sequences can sometimes be too short (resulting in weakly penetrant RNAi phenotypes) or their sequences not specific enough for the gene in question. In this case, an alternative strategy is to use a dsRNA that targets the coding sequence and express an engineered mRNA in which codon usage has been changed for the region targeted by the dsRNA. As a result, the expressed transgene still encodes the wild-type protein but is not recognized by the dsRNA. We are successfully using this approach for a number of targets, as have others such as for Sqh/MRLC (Dean & Spudich, 2006).

For rescue experiments, we generate cell lines that constitutively express Tubulin-mCherry (pAc promoter) and inducibly express the rescue construct (pMT). This then allows the rescue experiments to be imaged live both with and without induction. If the cells successfully progress through cytokinesis only when the transgene is induced, this provides strong confirmation that the RNAi has been effective and that the transgene is indeed rescuing. The pMT promoter can be somewhat leaky, potentially resulting in partial or even complete rescue in the absence of induction. However, this leaky expression will be readily detected on the microscope, aided by the fact that cytokinesis regulators tend to become highly enriched at the division site, and the corresponding levels can be correlated to the phenotype observed. An additional internal control can include subpopulations of cells within the culture that are not detectably expressing the rescuing transgene, even after induction.

1.2.2 Time-lapse image acquisition

1. Proceed to your imaging system.
   We use an Ultraview Vox spinning disk confocal system (Perkin−Elmer), equipped with a CSU-X1 scanning unit (Yokogawa) fitted to a fully motorized DM16000B inverted microscope (Leica Microsystems). It includes a motorized XY stage with linear encoders for greater positional accuracy and a piezo-electric Z top-plate for fast acquisition of Z-stacks (Applied Scientific Instrumentation). An Orca-R2 charged-coupled device camera (Hamamatsu) is used for detection. The system is controlled by Volocity 6.3 acquisition software (Improvision/Perkin Elmer).
2. Ensure your cells are protected against dehydration from evaporation for the duration of your experiment.
   Since S2 cells are healthy at room temperature and under normal atmospheric conditions, no special conditions or equipment are needed when the cells are on the microscope, provided there are no major fluctuations of ambient

temperature. However, you may wish to consider humidity control to counter evaporation of media when imaging over extended periods. Our system is equipped with a stage-top incubator (Pathology Devices) that includes humidity control, although we have found this unnecessary for most of our imaging needs, ie, up to 24 h imaging, with 400 μL media per well in 8-well chambered coverslip dishes (Nunc Lab-tek II).

3. Select an appropriate objective for *low-resolution imaging* (eg, 40× 0.85 NA air).
   When imaging a new cell line or assessing the effect of a dsRNA for the first time, we typically perform low-resolution time-lapse imaging for 24 h–48 h, starting 2–3 days after dsRNA treatment. This allows cells to be followed through mitosis, cytokinesis, abscission, and in many cases to the next mitosis. It also allows us to monitor the onset of any phenotypes and their potential progression as the RNAi takes full effect. A nonimmersion, air objective (40× 0.85 NA) is ideal for visiting all wells of an 8-well dish, allowing controls to be imaged in parallel.
4. Set laser power and camera exposure times.
   We routinely use two excitation laser lines: 488 nm for GFP and 561 nm for mCherry. To reduce photobleaching and phototoxicity, we limit laser power and exposure times as much as possible, although these will vary depending on protein expression levels and the specific aims of the experiment. We also routinely set camera binning to 2 × 2 to increase the signal to noise ratio, at the expense of resolution.
5. Select and save three to four fields of cells per condition and initiate acquisition of two-channel Z-stacks at 4–5 min intervals for 12–24 h.
   When saving the XYZ points, we use the microscope stand's motorized focus drive for setting the Z position. We then typically acquire, at each XYZ position, a Z-stack of 10 μm thickness with a 1 μm Z step size, using the piezoelectric stage. If you are capturing images from two channels consecutively with only one camera (as we are), special consideration should be given to the order of Z-stack versus channel acquisition. If the resulting data are to be analyzed for colocalization, for example, camera exposure time should be minimized and both channels should be captured at each Z position before moving to the next Z position. If colocalization in not an issue, the system can be set to collect the full stack images for one channel, before proceeding to the second channel. If emission filters are being employed, this latter configuration will allow for faster acquisition.
6. Perform *high-resolution imaging*, eg, using a 63× 1.4 NA oil-immersion objective.
   Once we have a general characterization of phenotypes from the low-resolution imaging, the time of their onset, success/failure rates, etc., we refine our analysis by repeating the experiment using high-resolution (spatial and temporal) time-lapse imaging conditions.
7. Identify and save the positions of cells in prophase/prometaphase.

Inclusion of markers such as tubulin-FP, Histone H2B-FP, or Anillin-FP facilitates this step. Work quickly at this step to be able to initiate the acquisition before cells enter anaphase. We typically select up to 10 XYZ positions (using the microscope focus drive for the Z position), all in the same well. We thus image different conditions consecutively rather than simultaneously.

8. Initiate the acquisition of two-channel Z-stacks at 30–60 s intervals for up to 2 h. At high resolution, we typically set our Z-stacks with a 0.25–0.5 μm Z step size.
9. Analyze collected data sets using Volocity Analysis software or third-party software (eg, Imaris, Bitplane).

Notes

- Pharmacological inhibitors that disrupt the mitotic spindle (eg, colchicine, taxol), actin dynamics (eg, Latrunculin A, jasplakinolide), or cell cycle progression (eg, MG132, RO3306) can be used to alter cytokinesis dynamics. It is possible to add these directly to the culture medium during image acquisition with a micropipette, although this takes some practice and luck not to disturb the poorly adherent S2 cells. Adherent S2R+ cells may be better suited for these kinds of experiments. It is also worth noting that the myosin-2 inhibitor, blebbistatin, which blocks cytokinesis in a variety of cell types, is ineffective in *Drosophila* cells (Straight et al., 2003).

1.2.3 Data analysis and expected phenotypes

1.2.3.1 Data analysis

1.2.3.1.1 Low-resolution imaging

From the long-term low-resolution imaging, we routinely analyze at least 50 division attempts from different fields for each condition. For each division attempt, we routinely score the times of anaphase onset, completion of CR closure, mature MR formation, and abscission or regression, as appropriate. This provides a range of temporal information and is very useful for determining the time of phenotypic onset for specific dsRNAs and potentially uncovering different aspects of the phenotype that manifest over time. For example, long time-course imaging revealed that the earliest observable phenotype upon sticky/citron kinase depletion is actually premature abscission, which occurs after only 2 days of dsRNA treatment, and which precedes the more obvious cytokinesis failures that occur after 3–4 days of dsRNA (El Amine et al., 2013). In a second example, low-resolution live imaging of Anillin-depleted cells helped define two populations of cells exhibiting slightly different phenotypes. In one population, the CR forms, but then rapidly becomes unstable, and oscillates back and forth across the cell equator before regressing 30 min after anaphase onset. In the second population, the CR does not oscillate but still fails to ingress to completion and regresses 1 h after anaphase onset (Kechad et al., 2012).

For each experiment, data are compared to both negative and positive controls, performed in parallel under identical conditions. Negative controls (eg, *lacI* RNAi) are necessary to exclude false positives while positive controls (eg, *pebble*/RhoGEF RNAi) exclude false negatives by providing confirmation that the cells remain responsive to RNAi. This can be particularly important for rescue experiments.

Additional controls that we often include in the context of rescue experiments are no induction of the rescuing transgene; and a dsRNA targeting both the endogenous and exogenous forms. If all conditions except the rescue condition show the expected loss-of-function phenotype, one can then be confident of rescue. Of course, rescue may be partial, showing incomplete penetrance if observed in only a subset of cells. It is also possible that cytokinetic progression be partially rescued, ie, cytokinesis could still fail, but in a different way or at a later stage. Such rescue experiments using separation-of-function alleles are a powerful means to dissect the requirements for each component of the cytokinetic machinery throughout cytokinetic progression.

1.2.3.1.2 High-resolution imaging

Once the long-term low-resolution imaging has defined the optimal experimental conditions (ie, duration of RNAi) and provided an overview of the experimental phenotypes, imaging is performed at higher spatial and temporal resolution to better capture and assess those phenotypes. The resulting data sets are ideal for more quantitative analyses such as intensity measurements, protein colocalization measurements, line profiles, 3D reconstructions to measure the volume of the different structures, all throughout the progression of cytokinesis.

1.2.3.2 Expected phenotypes

Using this live imaging approach, we have identified some common phenotypes that can arise from failure at each stage of progression through cytokinesis and we briefly describe them here (Fig. 3). One limitation of S2 cells is that they do not always divide in a perfectly stereotyped and reproducible way. As noted by others, they often have supernumerary centrosomes, which usually cluster together to form a bipolar spindle, but can also form multipolar spindles (Goshima, 2010). During the furrowing stage of cytokinesis, the cells have a tendency to bleb at the poles and can exhibit variability in furrow morphology. For this reason it is important to examine sufficient numbers of cells to have confidence in a given phenotype.

1.2.3.2.1 Early cytokinesis phenotypes

The first step of cytokinesis is formation of the central spindle in early anaphase. Anaphase onset can readily be scored by observing markers such as Histone H2B-FP and tubulin-FP (shortening of kinetochore fibers). The exclusion of cytoplasmic signal that the chromosomes induce can also be used to track the chromosomes with other fluorescent markers (eg, Anillin-FP), if expression levels are high enough. Failure to form the central spindle results in collapse of the microtubules causing rapid failure of furrowing within 15–20 min after anaphase.

The anaphase spindle delivers positive and negative signals that stimulate local formation and contraction of the cleavage furrow. The limiting event for furrow initiation is the activation of the small GTPase RhoA to a narrow zone that specifies furrow position. Rho activation stimulates local actin filament assembly and myosin recruitment and activation, triggering the formation of the CR. Rho also promotes

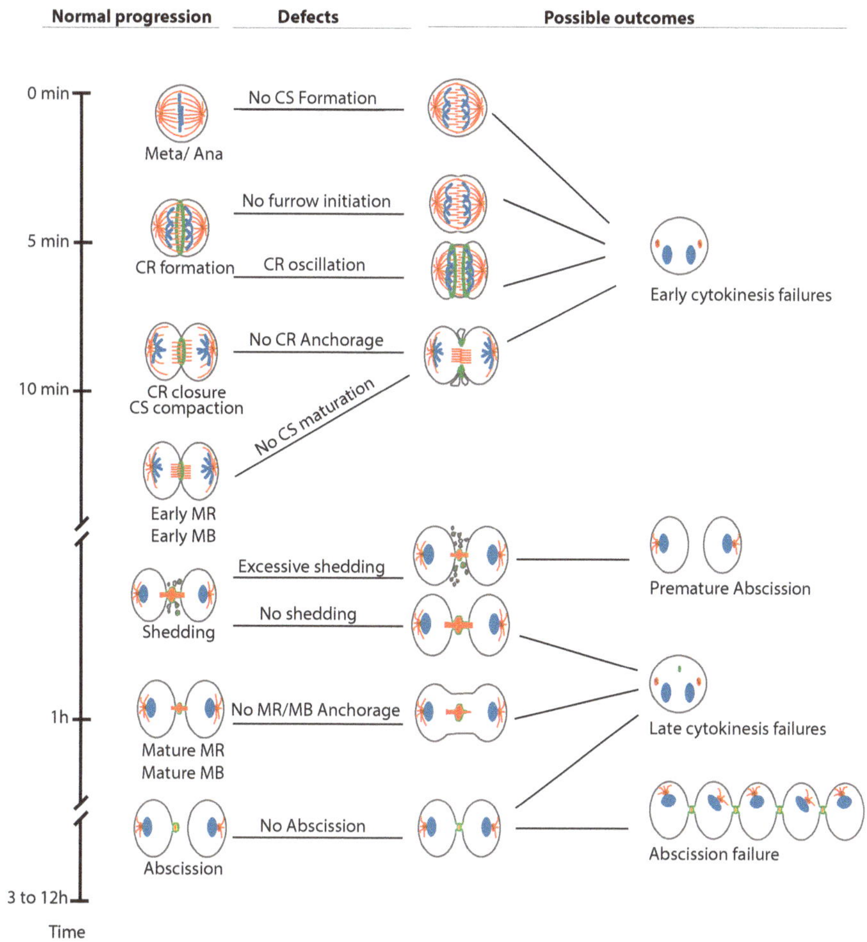

FIGURE 3 The progression of cytokinesis in S2 cells and possible phenotypic outcomes.

Cartoons describing the key steps during the normal progression of cytokinesis. These steps are highly regulated and prone to failure under certain experimental conditions. Here, we summarize some of the possible defects at each step and how they result in different phenotypes of cytokinesis failure. *CR*, contractile ring; *CS*, central spindle; *MB*, midbody; *MR*, midbody ring.

recruitment and activation, of scaffolding proteins such as Anillin and septins, which help organize and stabilize the CR. Failure to localize Rho or its activators, leads to failure in furrow ingression and rapid regression yielding a binucleated cell.

As the CR ingresses, it brings the equatorial membrane with it. Different scaffolding proteins mediate this process and defects in membrane anchorage result in uncoupling between the plasma membrane and the cell cortex, often leading to an excessive blebbing phenotype. During CR closure, several microtubule-associated proteins will move along the central spindle and accumulate in and around a central region of densely packed interdigitated microtubules that we refer to as the midbody, and which is characterized by a local drop in the tubulin-FP signal (Fig. 1). Cells that fail to form a midbody will often display a thicker central spindle that lacks the central drop in tubulin intensity. They will initiate furrowing, however, the CR will ingress with a delay to ultimately regress resulting in binucleated cells.

Cytokinesis failure during CR closure can also result from aberrant membrane organization. For example, depletion of the inositol 5-phosphtase, dOCRL, results in the ectopic localization of PtdIns(4,5)P(2) and CR components to endomembranes and leads to furrow regression (Ben El Kadhi, Roubinet, Solinet, Emery, & Carreno, 2011).

1.2.3.2.2 Late cytokinesis phenotypes

In S2 cells, the CR closes until it reaches a diameter of 1–2 μm, then it begins to transition to a stable MR. We have identified a 1-h time window, during which this early MR will undergo further maturation in parallel to midbody maturation that together form the intercellular bridge that connects the two sister cells until abscission. This maturation process is detectable by a gradual thinning of the MR structure accompanied by extrusions and internalization of different ring components such as Anillin, septins, Rho, sticky kinase, as well as components of the midbody such as Tumbleweed, Pavarotti, and Aurora B. Perturbations at this step can result in excess shedding events and can ultimately cause MR disintegration, causing premature abscission or rapid regression of the membrane (eg, citron kinase/*sticky* RNAi (El Amine et al., 2013)). Conversely, this process can also be blocked resulting in larger MRs that are unable to complete cytokinesis (eg, *peanut* RNAi (El Amine et al., 2013)).

Throughout this maturation process, the midbody and MR remain anchored to the plasma membrane. Defects in anchorage will cause rapid detachment of the plasma membrane and formation of binucleated cells with internal MR-like structures that can persist for hours (Kechad et al., 2012). Interestingly, many of the late cytokinetic failures that we observe happen soon after this 1-h time frame, suggesting that this is a particularly sensitive step during S2 cell cytokinesis.

1.2.3.2.3 Abscission phenotypes

Once the intercellular bridge is mature and well anchored to the membrane, the last stage of cytokinesis can proceed: abscission. Abscission is the longest and most variable stage of S2 cell cytokinesis. It can complete from 3 to >12 h after furrowing. Prior to this stage, the microtubules at the intercellular bridge gradually thin until they completely disappear. Sister cells remain linked by the late MR that is now <1 μm in diameter. A defect in abscission can yield sister cells that remain connected even as they undergo the next mitosis (eg, Shrub/CHMP4 RNAi (El Amine et al., 2013)).

1.2.3.2.4 Postabscission MR remnants

Upon normal completion of abscission of S2 cells, the MR remnant remains associated with one of the two sister cells. Although we have not systematically tracked the fate of these MR remnants, it is clear that they can persist for some time. MR remnants are readily observed using markers such as Anillin-GFP and they often persist in the cytoplasm of the cells until at least the following mitosis. This cytoplasmic localization suggests that they are engulfed, as has been shown in mammalian cells (Crowell, Gaffuri, Gayraud-Morel, Tajbakhsh, & Echard, 2014).

1.3 FIXED-CELL IMAGING

Although the real power of this system is for live-cell imaging, imaging of fixed cells remains a useful complementary approach. Importantly, immunofluorescence of fixed cells is the only way to localize endogenous proteins implicated in cytokinesis. It is important to confirm that tagged, and often overexpressed, versions of a protein of interest faithfully replicate the localization of the endogenous protein. However, it should be noted that cytokinetic structures such as the midbody and associated MR are densely packed structures and that certain epitopes may be locally masked. For example, it is well known that tubulin antibodies do not label the central region of the midbody (the part that is surrounded by the MR), even though electron micrographs and FP-tubulin fluorescence demonstrate that microtubules extend through this region. Thus a lack of an immunofluorescence signal at a particular location may not necessarily indicate a lack of protein localization there.

Here is the protocol that we use for indirect immunofluorescence:

1. Seed the cells at high confluence (6×10^6 mL^{-1}) in an 8-well chambered coverglass (Lab-tek II, cat#CA43300774) or 96-well glass bottomed dishes (Whatman, cat#306180-030). Allow the cells to settle overnight. We recommend setting more than one well for each condition.
   Critical: S2 cells are only loosely attached to the glass and many will be lost after the numerous washes. The following steps must be done very carefully to minimize cell loss. We perform all steps (addition and removal) with a standard 200 μL micropipette. Using this approach, we find that sufficient cells remain, especially near the edges of the wells. An alternative approach to increase cell numbers may include plating cells for approximately 5–10 min on ConA-coated glass prior to fixation. The cells will begin to adhere, but the pipetting step may disrupt late intercellular bridges.
2. Remove the medium and fix for 10 min at room temperature in 200 μL fixation buffer: phosphate buffered saline (PBS) containing 3.7% formaldehyde (Sigma, cat#f1635) and (optional) 0.1% glutaraldehyde (Sigma cat# g7651). The inclusion of glutaraldehyde better preserves microtubule morphology. Other fixatives such as methanol or trichloroacetic acid can be used. We recommend testing different fixatives to determine which one is most suitable to preserve epitopes and features of interest.
3. Wash once and permeabilize with PTX buffer: PBS containing 0.1% Triton-X100 (Fisher, cat# bp151500).
4. Block for 1 h with blocking buffer: PTX containing 5% normal goat serum (Jackson Immuno Research, cat#005-000-121).
5. Incubate with primary antibodies diluted in 100 μL blocking buffer. We generally perform this step overnight at 4°C.

6. Remove antibodies and carefully wash several times over a 30 min period in blocking buffer.
7. Incubate with appropriate goat secondary antibodies, such as ones coupled to Alexa Fluor (Molecular Probes/Thermo Fisher Scientific), diluted 1:1000 in blocking buffer for 1 h at room temperature. Dyes such as Alexa Fluor-phalloidin (Molecular Probes/Thermo Fisher Scientific) and Hoechst 33258 (Molecular Probes/Thermo Fisher Scientific, cat#H356) can also be included in this step (1:1000 dilution).
8. Remove antibody and gently wash several times over a 30-min period in PTX buffer.
9. Replace with 100 μL Fluoromount-G mounting medium (Southern Biotech, cat#0100-01) and proceed to your imaging system as soon as possible. Store at 4°C in the dark.

CONCLUSION AND PERSPECTIVES

Drosophila S2 cells provide a powerful system for dissecting the cell-intrinsic mechanisms underlying the different stages of animal cell cytokinesis: formation of the CR; formation of the intercellular bridge; and abscission. Growth at ambient temperature and atmospheric conditions makes S2 cells ideal for long time-course microscopy, and this allows the tracking of individual cells as they progress through cytokinesis. This live-cell imaging approach is essential for properly evaluating phenotypes and for uncovering alterations to the timing of events. S2 cells can be readily manipulated through RNAi-mediated protein depletion and expression of fluorescently tagged proteins, making them ideally suited for loss-of-function analyses and structure-function analyses of important cytokinesis regulators.

The genetic tractability of *Drosophila* tissue culture cells has recently been extended to include CRISPR-mediated genome editing (Bassett, Tibbit, Ponting, & Liu, 2014; Bottcher et al., 2014), which can be combined with RNAi (Housden et al., 2015). Further refinements include the engineering of related cell lines for integrase-mediated targeted insertions to improve the homogeneity of transgenic lines (Cherbas et al., 2015), and the development of lines expressing a fly-FUCCI system for analysis of cell cycle dynamics (Zielke, van Straaten, Bohlen, & Edgar, 2016). Finally, *Drosophila* tissue culture cells also offer a valuable resource for the biochemical purification and identification of protein complexes important during mitosis and cytokinesis (D'Avino et al., 2009). We fully expect this versatile system to continue making important contributions to our understanding of the molecular mechanisms governing cytokinetic progression.

ACKNOWLEDGMENTS

We thank members of the Hickson laboratory for discussions and helpful comments, and especially Mélanie Diaz for help with graphical design. We thank the Fonds de Recherche Québec

Santé (FRQS) for a doctoral studentship to A.K. and a Junior 2 scholarship to G.H. Research in the laboratory is supported by an infrastructure award from the Canada Foundation for Innovation and operating grants from the Canadian Institutes for Health Research (MOP97788), the Natural Sciences and Engineering Research Council, and the Cole Foundation.

REFERENCES

Barr, F. A., & Gruneberg, U. (2007). Cytokinesis: placing and making the final cut. *Cell, 131*(5), 847–860. http://dx.doi.org/10.1016/j.cell.2007.11.011. pii:S0092-8674(07)01463-8.

Bassett, A. R., Tibbit, C., Ponting, C. P., & Liu, J. L. (2014). Mutagenesis and homologous recombination in *Drosophila* cell lines using CRISPR/Cas9. *Biology Open, 3*(1), 42–49. http://dx.doi.org/10.1242/bio.20137120. pii:bio.20137120.

Bassi, Z. I., Audusseau, M., Riparbelli, M. G., Callaini, G., & D'Avino, P. P. (2013). Citron kinase controls a molecular network required for midbody formation in cytokinesis. *Proceedings of the National Academy of Sciences of the United States of America, 110*(24), 9782–9787. http://dx.doi.org/10.1073/pnas.1301328110. pii:1301328110.

Ben El Kadhi, K., Roubinet, C., Solinet, S., Emery, G., & Carreno, S. (2011). The inositol 5-phosphatase dOCRL controls PI(4,5)P2 homeostasis and is necessary for cytokinesis. *Current Biology: CB, 21*(12), 1074–1079. http://dx.doi.org/10.1016/j.cub.2011.05.030. pii:S0960-9822(11)00557-4.

Bettencourt-Dias, M., & Goshima, G. (2009). RNAi in *Drosophila* S2 cells as a tool for studying cell cycle progression. *Methods in Molecular Biology (Clifton, NJ), 545*, 39–62. http://dx.doi.org/10.1007/978-1-60327-993-2_3.

Bottcher, R., Hollmann, M., Merk, K., Nitschko, V., Obermaier, C., Philippou-Massier, J., … Forstemann, K. (2014). Efficient chromosomal gene modification with CRISPR/cas9 and PCR-based homologous recombination donors in cultured *Drosophila* cells. *Nucleic Acids Research, 42*(11), e89. http://dx.doi.org/10.1093/nar/gku289. pii:gku289.

Buster, D. W., Nye, J., Klebba, J. E., & Rogers, G. C. (2010). Preparation of *Drosophila* S2 cells for light microscopy. *Journal of Visualized Experiments: JoVE, 40*. http://dx.doi.org/10.3791/1982. pii:1982.

Cabernard, C. (2012). Cytokinesis in *Drosophila melanogaster. Cytoskeleton (Hoboken, NJ), 69*(10), 791–809. http://dx.doi.org/10.1002/cm.21060.

Cherbas, L., Hackney, J., Gong, L., Salzer, C., Mauser, E., Zhang, D., & Cherbas, P. (2015). Tools for targeted genome engineering of established *Drosophila* cell lines. *Genetics, 201*(4), 1307–1318. http://dx.doi.org/10.1534/genetics.115.181610. pii:genetics.115.181610.

Clemens, J. C., Worby, C. A., Simonson-Leff, N., Muda, M., Maehama, T., Hemmings, B. A., & Dixon, J. E. (2000). Use of double-stranded RNA interference in *Drosophila* cell lines to dissect signal transduction pathways. *Proceedings of the National Academy of Sciences of the United States of America, 97*(12), 6499–6503. http://dx.doi.org/10.1073/pnas.110149597. pii:110149597.

Crowell, E. F., Gaffuri, A. L., Gayraud-Morel, B., Tajbakhsh, S., & Echard, A. (2014). Engulfment of the midbody remnant after cytokinesis in mammalian cells. *Journal of Cell Science, 127*(Pt 17), 3840–3851. http://dx.doi.org/10.1242/jcs.154732. pii:jcs.154732.

D'Avino, P. P., Archambault, V., Przewloka, M. R., Zhang, W., Lilley, K. S., Laue, E. D., & Glover, D. M. (2007). Recruitment of Polo kinase to the spindle midzone during cytokinesis requires the Feo/Klp3A complex. *PLoS One, 2*(6), E572.

D'Avino, P. P., Savoian, M. S., Capalbo, L., & Glover, D. M. (2006). RacGAP50C is sufficient to signal cleavage furrow formation during cytokinesis. *Journal of Cell Science, 119*, 4402–4408.

D'Avino, P. P., Archambault, V., Przewloka, M. R., Zhang, W., Laue, E. D., & Glover, D. M. (2009). Isolation of protein complexes involved in mitosis and cytokinesis from *Drosophila* cultured cells. *Methods in Molecular Biology (Clifton, NJ), 545*, 99–112. http://dx.doi.org/10.1007/978-1-60327-993-2_6.

Dean, S. O., Rogers, S. L., Stuurman, N., Vale, R. D., & Spudich, J. A. (2005). Distinct pathways control recruitment and maintenance of myosin II at the cleavage furrow during cytokinesis. *Proceedings of the National Academy of Sciences of the United States of America, 102*(38), 13473–13478.

Dean, S. O., & Spudich, J. A. (2006). Rho kinase's role in myosin recruitment to the equatorial cortex of mitotic *Drosophila* S2 cells is for myosin regulatory light chain phosphorylation. *PLoS One, 1*, e131.

Echard, A., Hickson, G. R., Foley, E., & O'Farrell, P. H. (2004). Terminal cytokinesis events uncovered after an RNAi screen. *Current Biology: CB, 14*(18), 1685–1693.

El Amine, N., Kechad, A., Jananji, S., & Hickson, G. R. (2013). Opposing actions of septins and sticky on anillin promote the transition from contractile to midbody ring. *The Journal of Cell Biology, 203*(3), 487–504. http://dx.doi.org/10.1083/jcb.201305053. pii:jcb.201305053.

Field, C. M., & Alberts, B. M. (1995). Anillin, a contractile ring protein that cycles from the nucleus to the cell cortex. *The Journal of Cell Biology, 131*(1), 165–178.

Foley, E., & O'Farrell, P. H. (2004). Functional dissection of an innate immune response by a genome-wide RNAi screen. *PLoS Biology, 2*(8), E203.

Glotzer, M. (2009). The 3Ms of central spindle assembly: microtubules, motors and MAPs. *Nature Reviews. Molecular Cell Biology, 10*(1), 9–20. http://dx.doi.org/10.1038/nrm2609. pii:nrm2609.

Goldbach, P., Wong, R., Beise, N., Sarpal, R., Trimble, W. S., & Brill, J. A. (2010). Stabilization of the actomyosin ring enables spermatocyte cytokinesis in *Drosophila*. *Molecular Biology of the Cell, 21*(9), 1482–1493. http://dx.doi.org/10.1091/mbc.E09-08-0714. pii:E09-08-0714.

Goshima, G. (2010). Assessment of mitotic spindle phenotypes in *Drosophila* S2 cells. *Methods in Cell Biology, 97*, 259–275. http://dx.doi.org/10.1016/S0091-679X(10)97015-5. pii:S0091-679X(10)97015-5.

Goshima, G., & Vale, R. D. (2003). The roles of microtubule-based motor proteins in mitosis: comprehensive RNAi analysis in the *Drosophila* S2 cell line. *The Journal of Cell Biology, 162*(6), 1003–1016.

Goshima, G., & Vale, R. D. (2005). Cell Cycle-dependent Dynamics and Regulation of Mitotic Kinesins in Drosophila S2 Cells. *Molecular Biology of the Cell., 16*(8), 3896–3907.

Goshima, G., Wollman, R., Goodwin, S. S., Zhang, N., Scholey, J. M., Vale, R. D., & Stuurman, N. (2007). Genes required for mitotic spindle assembly in *Drosophila* S2 cells. *Science, 316*(5823), 417–421. http://dx.doi.org/10.1126/science.1141314. pii:1141314.

Gould, G. W. (2016). Animal cell cytokinesis: the role of dynamic changes in the plasma membrane proteome and lipidome. *Seminars in Cell & Developmental Biology*. http://dx.doi.org/10.1016/j.semcdb.2015.12.012. pii:S1084-9521(15)30025-2.

Green, R. A., Paluch, E., & Oegema, K. (2012). Cytokinesis in animal cells. *Annual Review of Cell and Developmental Biology*. http://dx.doi.org/10.1146/annurev-cellbio-101011-155718.

Hickson, G. R., Echard, A., & O'Farrell, P. H. (2006). Rho-kinase controls cell shape changes during cytokinesis. *Current Biology: CB, 16*(4), 359–370.

Hickson, G. R., & O'Farrell, P. H. (2008). Rho-dependent control of anillin behavior during cytokinesis. *The Journal of Cell Biology, 180*(2), 285–294. http://dx.doi.org/10.1083/jcb.200709005. pii:jcb.200709005.

Housden, B. E., Valvezan, A. J., Kelley, C., Sopko, R., Hu, Y., Roesel, C., … Perrimon, N. (2015). Identification of potential drug targets for tuberous sclerosis complex by synthetic screens combining CRISPR-based knockouts with RNAi. *Science Signaling, 8*(393), rs9. http://dx.doi.org/10.1126/scisignal.aab3729. pii:8/393/rs9.

Kachaner, D., Pinson, X., El Kadhi, K. B., Normandin, K., Talje, L., Lavoie, H., … Archambault, V. (2014). Interdomain allosteric regulation of Polo kinase by Aurora B and Map205 is required for cytokinesis. *The Journal of Cell Biology, 207*(2), 201–211. http://dx.doi.org/10.1083/jcb.201408081. pii:jcb.201408081.

Kechad, A., Jananji, S., Ruella, Y., & Hickson, G. R. (2012). Anillin acts as a bifunctional linker coordinating midbody ring biogenesis during cytokinesis. *Current Biology: CB, 22*(3), 197–203. http://dx.doi.org/10.1016/j.cub.2011.11.062. pii:S0960-9822(11)01395-9.

Kiger, A. A., Baum, B., Jones, S., Jones, M. R., Coulson, A., Echeverri, C., & Perrimon, N. (2003). A functional genomic analysis of cell morphology using RNA interference. *Journal of Biology, 2*(4), 27. http://dx.doi.org/10.1186/1475-4924-2-27.

Pereira, A. J., Matos, I., Lince-Faria, M., & Maiato, H. (2009). Dissecting mitosis with laser microsurgery and RNAi in *Drosophila* cells. *Methods in Molecular Biology (Clifton, NJ), 545*, 145–164. http://dx.doi.org/10.1007/978-1-60327-993-2_9.

Piekny, A. J., & Glotzer, M. (2008). Anillin is a scaffold protein that links RhoA, actin, and myosin during cytokinesis. *Current Biology: CB, 18*(1), 30–36. http://dx.doi.org/10.1016/j.cub.2007.11.068. pii:S0960-9822(07)02414-1.

Renshaw, M. J., Liu, J., Lavoie, B. D., & Wilde, A. (2014). Anillin-dependent organization of septin filaments promotes intercellular bridge elongation and Chmp4B targeting to the abscission site. *Open Biology, 4*, 130190. http://dx.doi.org/10.1098/rsob.130190. pii:rsob.130190.

Riedl, J., Crevenna, A. H., Kessenbrock, K., Yu, J. H., Neukirchen, D., Bista, M., … Wedlich-Soldner, R. (2008). Lifeact: a versatile marker to visualize F-actin. *Nature Methods, 5*(7), 605–607. http://dx.doi.org/10.1038/nmeth.1220. pii:nmeth.1220.

Rogers, S. L., & Rogers, G. C. (2008). Culture of *Drosophila* S2 cells and their use for RNAi-mediated loss-of-function studies and immunofluorescence microscopy. *Nature Protocols, 3*(4), 606–611. http://dx.doi.org/10.1038/nprot.2008.18. pii:nprot.2008.18.

Rogers, S. L., Rogers, G. C., Sharp, D. J., & Vale, R. D. (2002). *Drosophila* EB1 is important for proper assembly, dynamics, and positioning of the mitotic spindle. *The Journal of Cell Biology, 158*(5), 873–884.

Rogers, S. L., Wiedemann, U., Hacker, U., Turck, C., & Vale, R. D. (2004). *Drosophila* RhoGEF2 associates with microtubule plus ends in an EB1-dependent manner. *Current Biology: CB, 14*(20), 1827–1833.

Rogers, S. L., Wiedemann, U., Stuurman, N., & Vale, R. D. (2003). Molecular requirements for actin-based lamella formation in *Drosophila* S2 cells. *The Journal of Cell Biology, 162*(6), 1079–1088.

Schneider, I. (1972). Cell lines derived from late embryonic stages of *Drosophila melanogaster*. *Journal of Embryology and Experimental Morphology, 27*(2), 353–365.

Smurnyy, Y., Toms, A. V., Hickson, G. R., Eck, M. J., & Eggert, U. S. (2010). Binucleine 2, an isoform-specific inhibitor of *Drosophila* Aurora B kinase, provides insights into the mechanism of cytokinesis. *ACS Chemical Biology, 5*(11), 1015–1020. http://dx.doi.org/10.1021/cb1001685.

Somma, M. P., Fasulo, B., Cenci, G., Cundari, E., & Gatti, M. (2002). Molecular dissection of cytokinesis by RNA interference in *Drosophila* cultured cells. *Molecular Biology of the Cell, 13*(7), 2448–2460.

Steigemann, P., & Gerlich, D. W. (2009). Cytokinetic abscission: cellular dynamics at the midbody. *Trends in Cell Biology, 19*(11), 606–616. http://dx.doi.org/10.1016/j.tcb.2009.07.008. pii:S0962-8924(09)00158-5.

Straight, A. F., Cheung, A., Limouze, J., Chen, I., Westwood, N. J., Sellers, J. R., & Mitchison, T. J. (2003). Dissecting temporal and spatial control of cytokinesis with a myosin II inhibitor. *Science, 299*(5613), 1743–1747.

Straight, A. F., Field, C. M., & Mitchison, T. J. (2005). Anillin binds nonmuscle myosin II and regulates the contractile ring. *Molecular Biology of the Cell, 16*(1), 193–201.

Vale, R. D., Spudich, J. A., & Griffis, E. R. (2009). Dynamics of myosin, microtubules, and Kinesin-6 at the cortex during cytokinesis in *Drosophila* S2 cells. *The Journal of Cell Biology, 186*(5), 727–738. http://dx.doi.org/10.1083/jcb.200902083. pii:jcb.200902083.

Zielke, N., van Straaten, M., Bohlen, J., & Edgar, B. A. (2016). Using the fly-FUCCI system for the live analysis of cell cycle dynamics in cultured *Drosophila* cells. *Methods in Molecular Biology (Clifton, NJ), 1342*, 305–320. http://dx.doi.org/10.1007/978-1-4939-2957-3_19.

CHAPTER

Studying cytokinesis in *Drosophila* epithelial tissues

5

D. Pinheiro[*,§,¶], **Y. Bellaïche**[*,§,1]

Institut Curie, PSL Research University, Paris, France

[§]*Sorbonne Universités, Paris, France*

[¶]*Abel Salazar Biomedical Sciences Institute, University of Porto, Porto, Portugal*

[1]*Corresponding author: E-mail: yohanns.bellaiche@curie.fr*

CHAPTER OUTLINE

Abstract

Epithelial tissue cohesiveness is ensured through cell–cell junctions that maintain both adhesion and mechanical coupling between neighboring cells. During development, epithelial tissues undergo intensive cell proliferation. Cell division, and particularly cytokinesis, is coupled to the formation of new adhesive contacts, thereby preserving tissue integrity and propagating cell polarity. Remarkably, the geometry of the new interfaces is determined by the combined action of the dividing cell and its neighbors. To

Methods in Cell Biology, Volume 137, ISSN 0091-679X, http://dx.doi.org/10.1016/bs.mcb.2016.04.013

further understand the interplay between the dividing cell and its neighbors, as well as the role of cell division for tissue morphogenesis, it is important to analyze cytokinesis in vivo. Here we present methods to perform live imaging of cell division in *Drosophila* epithelial tissues and discuss some aspects of image processing and analysis.

INTRODUCTION

Epithelial tissues have as main function to physically separate body compartments, thereby allowing the coexistence of distinct biochemical and mechanical microenvironments. Epithelial architecture is dictated by the assembly of specialized junctions along the apical–basal axis of cells, namely the tight junctions (TJ) and the *adherens* junctions (AJ). In flies, however, the septate junctions are the functional equivalent to the TJs and are localized basally regarding the AJs. Such adhesive complexes are connected to the underlying actin–myosin cytoskeleton, to ensure not only tissue cohesiveness, but also mechanical coupling between neighboring cells (Brieher & Yap, 2013). During development, epithelial tissues, both in vertebrates and invertebrates, undergo intensive cell proliferation, cell–cell rearrangements, cell shape changes, cell delamination events and, even, large tissue-scale elongation, invagination and folding (for review see Heisenberg & Bellaïche, 2013; Munjal & Lecuit, 2014). Thus, it is critical to understand how epithelia can act as a barrier, yet exhibit such plasticity.

In this chapter, we will focus on how to study cytokinesis in vivo by combining fly genetics and advanced microscopy techniques. Cell division involves profound cell shape changes, particularly during cytokinesis, the process whereby the dividing cell cytoplasm is partitioned into the future daughter cells. Cytokinesis involves the formation and constriction of an actin–myosin ring at the cell equator. As constriction proceeds, the connection between the future daughters is progressively closed, until abscission individualizes the two cells (for review see Green, Paluch, & Oegema, 2012). Recent work, in both vertebrates and invertebrates, showed how epithelial cytokinesis is coupled to the formation of new adhesive contacts (Firmino, Rocancourt, Saadaoui, Moreau, & Gros, 2016; Founounou, Loyer, & Le Borgne, 2013; Guillot & Lecuit, 2013; Herszterg, Leibfried, Bosveld, Martin, & Bellaïche, 2013; Morais-de-Sá & Sunkel, 2013). Such mechanism should allow the maintenance of polarity and adhesion during tissue proliferation, an essential aspect of epithelia integrity. In the *Drosophila* pupal notum and pupal wing, it was shown that the cells neighboring the dividing cell participate in setting the length of the newly formed contact, supporting that, in a tissue, division is a multicellular process. Altogether studying cell division in vivo is likely to be a rather critical step towards a better understanding of the mechanisms regulating epithelial dynamics, as well as tissue cohesiveness.

1. METHODS AND EXPERIMENTAL SETUP

In this section, we will provide a detailed protocol to image cytokinesis during morphogenesis of *Drosophila* pupal epithelial tissues, namely the dorsal thorax (notum) and the pupal wing, which have been extensively used to decipher the mechanisms of epithelial polarity and dynamics (Bardet et al., 2013; Bosveld et al., 2012; Etournay et al., 2015; Georgiou, Marinari, Burden, & Baum, 2008; Guirao et al., 2015; Herszterg et al., 2013; Langevin et al., 2005; Leibfried, Fricke, Morgan, Bogdan, & Bellaiche, 2008; Levayer, Dupont, Levayer, Dupont, & Moreno, 2016; Marinari et al., 2012). For that, we show as an example a time-lapse movie of a pupa expressing E-Cadherin tagged with GFP (E-Cad::GFP (Huang, Zhou, Dong, Watson, & Hong, 2009)), which labels the AJs and MyoII fused with mChFP (MyoII::mChFP (Martin, Kaschube, & Wieschaus, 2009)), which labels the contractile ring (Fig. 1B).

1.1 MOUNTING THE *DROSOPHILA* PUPAE FOR TIME-LAPSE IMAGING OF EPITHELIAL TISSUES

To optimize the imaging session, pupae can be timed; the pupal notum displays two peaks of cell proliferation at 15–18 hours after puparium formation (hAPF) and at 21–24hAPF, while the pupal wing displays a single peak of cell divisions around 18hAPF (Etournay et al., 2015; Guirao et al., 2015). Therefore, white pupae should be collected at puparium formation and kept in petri dishes, lined with moist paper, until mounting. Alternatively, pupae can be collected later and the timing can then be set by head eversion.

The material required for the mounting is detailed below:

1. 25 × 75 × 1.0 mm slides;
2. 18 × 18 mm cover slips;
3. 24 × 40 mm cover slips;
4. double-sided tape—Scotch 3M;
5. glass pasteur pipettes 150 mm;
6. Voltalef oil 10S (VWR Prolabo);
7. quick-dry nail polish;
8. Kleenex paper towel; and
9. a set of forceps (Dumont #5—fine forceps).

In Fig. 1A,A' we provide a schematic representation of the mounting chamber (top and side views).

1. Carefully dry the pupae in a paper towel with a brush, to ensure that it will adhere to the double-sided tape.
2. Glue the ventral side of the pupae to the slide. To avoid hurting the pupae, you can easily hold it by the spiracles.
3. Using the forceps carefully pull the spiracles to detach the operculum.
4. Insert the forceps (alternatively you can use fine scissors) through the opercula opening, parallel to the pupae and tear the pupal case from head to abdomen.
5. Gently pull the pupal case at each side of the incision and glue it on the slide.
6. Glue a stripe of wet paper on the slide to maintain humidity in the mounting chamber.

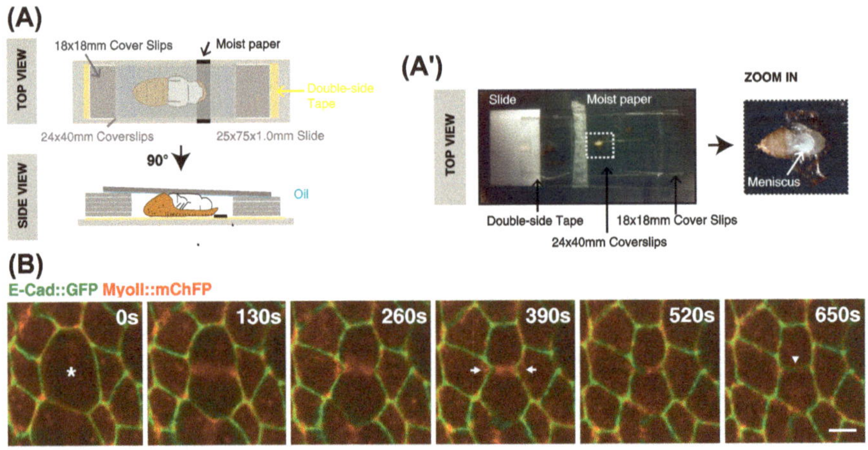

FIGURE 1 A simple method to perform live imaging of the *Drosophila* pupal notum.

(A,A') Schematic representation of the mounting chamber (side and top views) and an example of a pupal notum prepared for live imaging (top view). (B) Apical view of a dividing cell (*asterisk*) and its neighbors expressing E-Cad::GFP and MyoII::mChFP. *Arrows*: MyoII accumulation in the neighboring cells at the edges of the furrow. *Arrowhead*: new *adherens* junction formation between the daughter cells. Scale bar: 5 μm (B).

7. Carefully place a coverslip previously coated with a thin layer of oil on top of the pillars at each side of the pupae. Stabilize the coverslip with nail polish and make sure a meniscus is formed between the pupal notum and the coverslip. Avoid pressing too hard with the coverslip.

The protocol to image the *Drosophila* pupal wing is very similar with a key difference regarding the positioning of the pupae on the slide—which should be placed laterally, rather than ventrally, so that the pupal wing will be facing the coverslip.

1.2 IMAGE ACQUISITION

Samples can be imaged with a confocal spinning disk microscope (CSU-X1, Yokogawa spinning disc head mounted on an inverted microscope and piloted by Metamorph, Roper Scientific) using either a 60× or an 100× objective. To image at 100×, a confocal spinning disk wide is a great alternative, as the field of imaging is 4× larger than the classical confocal spinning disk microscopes, allowing you to record more events of cell division in a single time-lapse. A laser-scanning confocal microscope can also be used; however, in our experience, photobleaching is higher. The microscope should be enclosed in a temperature box, which can be stably tuned at 25°C or 29°C, depending on the experiment's requirements; in this example, we set the microscope at 25°C (Fig. 1B).

High sensitivity cameras are crucial for live imaging experiments, in order to increase the signal-to-noise ratio and allow long-term imaging with reduced photobleaching. In the lab, we use either a CoolSNAP HQ2 (Photometrics) or a CMOS (Hamamtsu) camera. Alternatively, an EMCCD camera can be used to further increase sensitivity; however, the pixel size is bigger when compared to the CMOS camera (0.13 × 0.13 μm vs 0.065 × 0.065 μm, at 100×, Binning 1 × 1). The time of exposure and the laser power should be optimized to avoid saturation, this is especially important when you wish to perform intensity quantifications; in this case, we used 20% laser power, for both 491 and 561 nm lasers (confocal spinning disk microscope—raw laser power at 491 nm is 50 mW and 100 mW at 561 nm), and 200 ms of exposure (Fig. 1B).

Since the pupae epithelium can move, it is preferable to acquire a z-stack (with a step size of 0.2–0.5 μm) along the apical–basal axis of the tissue at each time point. Depending on the size of the stack acquired, the time resolution can increase up to 1 frame/5 s. For example, to image contractile ring constriction, which takes on average 320 s in the pupal notum, we set a 20-s interval sampling frequency to acquire a 7–8 μm z-stack (0.5 μm step). To image cell division at the tissue scale, time resolution drops to 1–5 min/stack, depending on how many positions are acquired. To generate such movies we use the autofocus option of the Metamorph software, to avoid drift along the z-axis.

1.3 LASER ABLATION AND PHOTOBLEACHING

To probe the contractile ring pulling forces, for example, it is possible to couple live imaging with laser ablation experiments. For that, we use a laser-scanning confocal microscope (LSM710 NLO, Carl Zeiss) equipped with the Ti:Sapphire laser (Mai Tai DeepSee, Spectra Physics) at 800 nm with <100 fs pulses with a 80 MHz repetition rate, typically set at 20–25% power. Following laser ablation, confocal images or a confocal stack can be acquired (see Herszterg et al., 2013 for an example). Upon laser ablation of the contractile ring, the measured recoil velocity is indicative of both the ring tension and the pulling forces generated by the contractile ring (Mayer et al., 2010). It can be useful to perform such experiments in different mutant conditions to understand how a certain gene affects force production within the contractile ring (assuming similar friction in *wild-type* and mutant conditions).

To analyze the dynamics of a protein of interest, for example, MyoII within the contractile ring, we can perform FRAP experiments (fluorescence recovery after photobleaching). For that, we use a confocal spinning disk equipped with a FRAP module (iLas^2 system). The number of iterations and the laser power used to photobleach a region of a given size has to be optimized; for the example mentioned above, we used the 491 nm laser, 100% laser power, with 40–50 iterations. Importantly, the size of the bleached region should be carefully considered and kept constant for all experiments (for further details see Fritzsche & Charras, 2015). Following photobleaching, a time-lapse movie can be acquired. Photoconversion and photoactivation experiments are also possible in the described setup and can be very instructive and complementary to the photobleaching experiments described above.

2. DATA PROCESSING AND ANALYSIS

In this section, we provide an overview of methods suitable to project and analyze time-lapse movies in Fiji (http://fiji.sc/), focused in the context of cytokinesis. To open the time-lapses, we use the Bio-Formats importer (File > Import > Bio-Formats), which is suitable for many life sciences file formats. Furthermore, it extracts basic metadata (such as spatial calibration), if they are available in the file.

2.1 PROJECTING CONFOCAL Z-STACKS

As mentioned above, the pupae epithelium can move along the z-axis during the acquisition of the time-lapse, therefore the first challenge when analyzing this type of data is the z-stack projection. Rather than projecting the entire stack, which greatly increases out-of-focus fluorescence, the projection should be restricted to the relevant stack positions, which will probably change over time; in this example, we restricted the projection to the level of the AJs (top view) (Fig. 1B). For that, we generated a set of custom-made macros for Fiji (available upon request). After running the macro, you should manually select, by a single click, the z position of interest at each frame; in this case, it corresponds to the maximum E-Cad::GFP intensity (Fig. 1B). The macro automatically registers the coordinates in a results table, which is then used as an input for the extraction of the z positions throughout the time-lapse. As a single z may not be sufficient to fully visualize the structure of interest—AJs span 1.5–2 μms, which corresponds to 3–4 z positions (0.5 μm step)—you can extract z positions above and below the one you have selected, at each frame, thereby generating a new stack centered at the levels of the AJs. This stack can then be easily projected using the Z Project option in Fiji (Image > Stacks > Z Project) (Fig. 1B).

To use the time-lapses for intensity measurements, it is preferable to generate a sum projection and to apply a bleach correction before the quantification (Image > Adjust > Bleach Correction > Histogram Matching). Such method can be useful, for example, to quantify the recovery of a protein of interest upon photobleaching experiments. For that, the mean intensity at the bleached region, and corresponding background, can be measured manually in Fiji, at each frame. After subtracting the background, you can plot the mean intensity at the bleached region over time. Using the best-fit to the experimental data, you can extract several parameters describing the dynamics of your protein of interest, namely the turnover time, as well as the mobile and immobile fraction.

2.2 GENERATING KYMOGRAPHS

A very useful way to visualize the dynamics of the AJ or of MyoII after projection is to generate kymographs. Several methods exist for this purpose; a simple option is to use the Re-slice tool in Fiji (Image > Stacks > Re-slice; before using the Re-slice, you need to change the image properties (Image > Properties—exchange the time for z). To improve the reslicing, you can adjust the thickness of the line and/or apply a Stack Reg on your time-lapse, to align the stack images (Plugin > Stacks > Stack Reg; you may have to download this plugin online). Although, this Re-slice option can be very useful, when working with in vivo data it often happens that the alignment even after applying the Stack Reg is not ideal, thus we developed a Macro in Fiji that allows you define, at each frame, the landmarks for your kymograph (available upon request). After running the macro, you can manually select, by click, the two positions defining the extraction box at each frame. The length and width of the extraction box are adjustable. Upon extraction of the box, it is automatically rotated to align the extracted regions over time, thus generating a kymograph. For examples, see Herszterg et al. (2013).

2.3 MEASURING THE RATE OF CONTRACTILE RING CONSTRICTION

To monitor the constriction rate, we developed a Macro for Fiji that allows the user to quickly extract the contractile ring length over time (available upon request). Then, the rate of constriction is determined as the slope of the linear fit of the contractile ring length, normalized to the length at the onset of cytokinesis, as a function of time (see Herszterg et al., 2013). For dividing cells that constrict very slowly, only the linear part of the curves can be fitted to determine the constriction rate.

2.4 MEASURING THE RECOIL VELOCITY UPON CONTRACTILE RING LASER ABLATION

To measure the recoil velocity upon contractile ring laser ablation, you can generate a kymograph along the contractile ring encompassing both the dividing cell and its neighbors. Using the kymograph, the dividing cell diameter over time is measured using a custom-made MATLAB code (available upon request). For this specific example, the recoil velocity is then measured between t_0—time of laser ablation, and t_{20}—20 s after laser ablation (averaging the two time points closest to t_{20}). The amount of constriction prior to contractile ring laser ablation can be measured as the ratio of the difference in cell diameter prior to contractile ring ablation and upon full cell relaxation. A similar strategy can be employed to measure the recoil velocity in other contexts, for example, upon AJ laser ablation.

2.5 ANALYZING THE CONTRIBUTION OF CELL PROCESSES TO TISSUE-SCALE MORPHOGENESIS

More automated tools for image analysis adapted to epithelial tissues allow selective z-stack projection of the time-lapse, followed by a pipeline to segment and track cells at the tissue scale, over time. Cell features, such as area, topology, divisions, delaminations or cell–cell rearrangements, are extracted in a semiautomated fashion. In the lab, we developed a custom-made pipeline optimized to segment and track the cell contours throughout morphogenesis of the *Drosophila* pupal notum (Guirao et al.,

2015). Such pipeline can also be applied to other epithelial tissues with minimal adjustments. Alternatively, you can perform a similar analysis using the Fiji plugins for segmentation (Plugins > Segmentation). Recently, a software package—EpiTools (Heller et al., 2016)—was developed for biologists with little computer expertise and, it is thus accompanied by a detailed website, which includes tutorials.

Finally, to understand how cell shape changes, cell divisions, rearrangements and delaminations actually contribute for tissue-scale morphogenesis, an approach unifying the quantitative characterization of all those processes is required to be able to relate them and bridge the gap between the cell and the tissue scale. In the lab, such formalism was recently developed and tested on computer simulations, in the *Drosophila* pupal notum and in the pupal wing (Guirao et al., 2015). The authors collected and averaged several animals to build an archetypal *wild-type* tissue and determine the biological variability, thereby allowing a robust and meaningful comparison with mutant conditions. Using this approach, it became clear that the interplay between cell processes is much more complex than originally anticipated; for example, cell division can contribute either positively or negatively to tissue deformation depending on the region of the notum and on the developmental time point. Furthermore, the cell processes are not strictly independent of each other; for example, blocking cell division in the pupal notum, in addition to suppressing the contribution of division, also affects the amount and the anisotropy of cell shape changes and cell rearrangements, as well as their contribution for tissue deformation (Guirao et al., 2015). Alternatively, a different framework was developed and applied to the *Drosophila* pupal wing (Etournay et al., 2015).

3. MARKERS AND GENETIC TOOLS TO STUDY EPITHELIAL CYTOKINESIS IN VIVO

To perform live imaging, it is critical to use GFP/mChFP reporter lines, thus we compiled in Table 1 a list of useful fly stocks to study cytokinesis in vivo. As previously discussed, in an epithelial tissue, the cells neighboring the dividing cell also participate in the formation of the new adhesive contact (Firmino et al., 2016; Founounou et al., 2013; Herszterg et al., 2013). These findings highlighted the importance of studying how cell division, and in particular cytokinesis, impacts the surrounding tissue. Thus, the ability to generate adjacent patches of genetically distinct cells is essential to correctly address this question. In flies, two techniques have been widely used to generate clones, namely mitotic recombination, mediated by FRT sites and the flp-out clone technique (for review see del Valle Rodríguez, Didiano, & Desplan, 2011). Combining these alleles with reporters and/or mutant lines allows you to discriminate the localization and function of a particular gene exclusively in the dividing cell or its neighbors. Using FRT-mediated recombination, we generated adjacent patches of cells expressing MyoII::GFP or MyoII::RFP, allowing us to determine that MyoII accumulation upon contractile ring constriction takes place in the neighboring cells, rather than in the dividing cell itself (Herszterg

Table 1 Alleles and Transgenes Useful to Study Cytokinesis In Vivo

***Drosophila* Stock**	**Reference or Source**
ubi-*H2B:RFP*	Bosveld et al. (2012)
ubi-E-Cad::GFP	Oda, Tsukita, and Takeichi (1998)
E-Cad::GFP	Huang et al. (2009)
E-Cad::3XGFP	Unpublished line generated by the Bellaiche Lab
E-Cad::3XmTagRFP	Unpublished line generated by the Bellaiche Lab
E-Cad::3XmKate2	Unpublished line generated by the Bellaiche Lab
MyoII::mChFP	Martin et al. (2009)
MyoII::3XmKate2	Unpublished line generated by the Bellaiche Lab
MyoII::3XGFP	Unpublished line generated by the Bellaiche Lab
UAS-CAAX::mOrg	Kanca, Caussinus, Denes, Percival-Smith, and Affolter (2014)
ubi-PLCγPH:GFP	Herszterg et al. (2013)
UAS-PH::ChFP	Herszterg et al. (2013)
Utr(ABD)::GFP	Rauzi, Lenne, and Lecuit (2010)
Lifeact::GFP	Bloomington Stock Center
Ubla-Ani(RBD)::GFP	Munjal, Philippe, Munro, and Lecuit (2015)
ubi-Rok::GFP	Bloomington Stock Center
ubi-*α-Tub::GFP*	Grieder, de Cuevas, and Spradling (2000)
ubi-*RFP:α-Tub*	Basto et al. (2008)
Act > y^+ > Gal4	Bloomington Stock Center
Tub > GFP > Gal4	Bloomington Stock Center
FRT40A, MyoII::GFP	Herszterg et al. (2013)
FRT40A, MyoII::RFP	Herszterg et al. (2013)

et al., 2013). A similar approach was used to demonstrate that, upon contractile ring constriction in the dividing cell, the neighboring cells become interposed in between the dividing cell membranes until midbody formation (Founounou et al., 2013; Herszterg et al., 2013).

CONCLUSIONS AND FUTURE PERSPECTIVES

Here we presented a simple protocol to image and analyze cytokinesis in vivo in two *Drosophila* epithelial tissues—the pupal notum and pupal wing. Combining this experimental setup with advanced microscopy techniques and the power of *Drosophila* genetics allows researchers to study cell division in a tissue context.

A very important aspect for future work would be to regulate actin–myosin dynamics at the subcellular scale. With this in mind, an optogenetic tool to locally inhibit cell contractility was recently developed and tested in the *Drosophila* ectoderm (Guglielmi, Barry, Huber, & De Renzis, 2015). The authors adapted the CRY2-CIBN protein dimerization system to recruit, upon light exposure, the catalytic domain of the inositol polyphosphate 5-phosphatase OCRL to the plasma membrane, where it modulates the levels of both phosphoinositides and actin. As a proof-of-principle, the authors showed that activation of CRY2-OCRL in the whole embryo, or even in patches of ventral cells, is able to arrest ventral furrow formation or even revert it, if invagination had already started at the time of CRY2 activation.

A critical difficulty of studying cell division in vivo is the intrinsically complex 3D geometry of an epithelial tissue, with two sets of junctions assembled apically and basolaterally, as well as the contacts established between epithelial cells and the underlying extracellular matrix. To understand cell division in 3D, correlative electron microscopy would be a useful complementary approach to the live imaging described above. Alternatively, super resolution microscopy can also be employed to increase the spatial resolution. Such methods may allow researchers to further understand, for example, the organization of the actin–myosin cytoskeleton at the subcellular scale.

ACKNOWLEDGMENTS

We thank Boris Guirao for comments on the manuscript. Diana Pinheiro acknowledges fellowships from FCT (SFRH/BD/51700/2011) and FRM (FDT20150531972) for funding. Work in the Bellaïche lab is supported by ANR-MaxForce, ANR-MorphoDro, ERC Advanced (TiMoprh, 340784), ARC (SL220130607097), ANR Labex DEEP 11-LBX-0044, ANR-10-IDEX-0001-02 PSL grants.

REFERENCES

Bardet, P. L., Guirao, B., Paoletti, C., Serman, F., Léopold, V., Bosveld, F., … Bellaïche, Y. (2013). PTEN controls junction lengthening and stability during cell rearrangement in epithelial tissue. *Developmental Cell, 25*(5), 534–546. http://dx.doi.org/10.1016/j.devcel.2013.04.020.

Basto, R., Brunk, K., Vinadogrova, T., Peel, N., Franz, A., Khodjakov, A., & Raff, J. W. (2008). Centrosome amplification can initiate tumorigenesis in flies. *Cell, 133*(6), 1032–1042. http://dx.doi.org/10.1016/j.cell.2008.05.039.

Bosveld, F., Bonnet, I., Guirao, B., Tlili, S., Zhimin, W., Petitalot, A., … Bellaïche, Y. (2012). Mechanical control of morphogenesis by fat/dachsous/four-jointed planar cell polarity pathway. *Science (New York, NY), 336*, 724–727. http://dx.doi.org/10.1126/science.1221071.

Brieher, W. M., & Yap, A. S. (2013). Cadherin junctions and their cytoskeleton(s). *Current Opinion in Cell Biology, 25*(1), 1–8. http://dx.doi.org/10.1016/j.ceb.2012.10.010.

Etournay, R., Popović, M., Merkel, M., Nandi, A., Blasse, C., Aigouy, B., … Eaton, S. (2015). Interplay of cell dynamics and epithelial tension during morphogenesis of the *Drosophila* pupal wing. *eLife, 4*, e07090. http://dx.doi.org/10.7554/eLife.07090.

Firmino, J., Rocancourt, D., Saadaoui, M., Moreau, C., & Gros, J. (2016). Cell division drives epithelial cell rearrangements during gastrulation in chick. *Developmental Cell, 36*(3), 249–261. http://dx.doi.org/10.1016/j.devcel.2016.01.007.

Founounou, N., Loyer, N., & Le Borgne, R. (2013). Septins regulate the contractility of the actomyosin ring to enable adherens junction remodeling during cytokinesis of epithelial cells. *Developmental Cell, 24*(3), 242–255. http://dx.doi.org/10.1016/j.devcel.2013.01.008.

Fritzsche, M., & Charras, G. (2015). Dissecting protein reaction dynamics in living cells by fluorescence recovery after photobleaching. *Nature Protocols, 10*(5), 660–680. http://dx.doi.org/10.1038/nprot.2015.042.

Georgiou, M., Marinari, E., Burden, J., & Baum, B. (2008). Cdc42, Par6, and aPKC regulate Arp2/3-mediated endocytosis to control local adherens junction stability. *Current Biology, 18*(21), 1631–1638. http://dx.doi.org/10.1016/j.cub.2008.09.029.

Green, R. A., Paluch, E. K., & Oegema, K. (2012). Cytokinesis in animal cells. *Annual Review of Cell and Developmental Biology, 31*, 169–213. http://dx.doi.org/10.1146/annurev-cellbio-101011-155718.

Grieder, N. C., de Cuevas, M., & Spradling, A. C. (2000). The fusome organizes the microtubule network during oocyte differentiation in *Drosophila*. *Development (Cambridge, England), 127*(19), 4253–4264.

Guglielmi, G., Barry, J. D., Huber, W., & De Renzis, S. (2015). An optogenetic method to modulate cell contractility during tissue morphogenesis. *Developmental Cell, 35*(35), 1–15. http://dx.doi.org/10.1016/j.devcel.2015.10.020.

Guillot, C., & Lecuit, T. (2013). Adhesion disengagement uncouples intrinsic and extrinsic forces to drive cytokinesis in epithelial tissues. *Developmental Cell, 24*(3), 227–241. http://dx.doi.org/10.1016/j.devcel.2013.01.010.

Guirao, B., Rigaud, S. U., Bosveld, F., Bailles, A., Lopez-Gay, J., Ishihara, S., … Bellaïche, Y. (2015). Unified quantitative characterization of epithelial tissue development. *eLife, 4*, e08519. http://dx.doi.org/10.7554/eLife.08519.

Heisenberg, C.-P., & Bellaïche, Y. (2013). Forces in tissue morphogenesis and patterning. *Cell, 153*(5), 948–962. http://dx.doi.org/10.1016/j.cell.2013.05.008.

Heller, D., Hoppe, A., Restrepo, S., Gatti, L., Tournier, A. L., Tapon, N., … Mao, Y. (2016). EpiTools: an open-source image analysis toolkit for quantifying epithelial growth dynamics. *Developmental Cell, 36*(1), 103–116. http://dx.doi.org/10.1016/j.devcel.2015.12.012.

Herszterg, S., Leibfried, A., Bosveld, F., Martin, C., & Bellaïche, Y. (2013). Interplay between the dividing cell and its neighbors regulates adherens junction formation during cytokinesis in epithelial tissue. *Developmental Cell, 24*(3), 256–270. http://dx.doi.org/10.1016/j.devcel.2012.11.019.

Huang, J., Zhou, W., Dong, W., Watson, A. M., & Hong, Y. (2009). Directed, efficient, and versatile modifications of the *Drosophila* genome by genomic engineering. *Proceedings of the National Academy of Sciences of the United States of America, 106*(20), 8284–8289. http://dx.doi.org/10.1073/pnas.0900641106.

Kanca, O., Caussinus, E., Denes, A. S., Percival-Smith, A., & Affolter, M. (2014). Raeppli: a whole-tissue labeling tool for live imaging of *Drosophila* development. *Development (Cambridge, England), 141*(2), 472–480. http://dx.doi.org/10.1242/dev.102913.

Langevin, J., Morgan, M. J., Rossé, C., Racine, V., Sibarita, J. B., Aresta, S., … Bellaïche, Y. (2005). *Drosophila* exocyst components sec5, sec6, and Sec15 regulate DE-Cadherin trafficking from recycling endosomes to the plasma membrane. *Developmental Cell, 9*(3), 365–376. http://dx.doi.org/10.1016/j.devcel.2005.07.013.

Leibfried, A., Fricke, R., Morgan, M. J., Bogdan, S., & Bellaiche, Y. (2008). *Drosophila* Cip4 and WASp define a branch of the Cdc42-Par6-aPKC pathway regulating e-cadherin endocytosis. *Current Biology, 18*(21), 1639–1648. http://dx.doi.org/10.1016/j.cub.2008.09.063.

Levayer, R., Dupont, C., Levayer, R., Dupont, C., & Moreno, E. (2016). Tissue crowding induces caspase-dependent competition for space. *Current Biology, 26*(5), 1–8. http://dx.doi.org/10.1016/j.cub.2015.12.072.

Marinari, E., Mehonic, A., Curran, S., Gale, J., Duke, T., & Baum, B. (2012). Live-cell delamination counterbalances epithelial growth to limit tissue overcrowding. *Nature, 484*(7395), 542–545. http://dx.doi.org/10.1038/nature10984.

Martin, A. C., Kaschube, M., & Wieschaus, E. F. (2009). Pulsed contractions of an actin-myosin network drive apical constriction. *Nature, 457*(7228), 495–499. http://dx.doi.org/10.1038/nature07522.

Mayer, M., Depken, M., Bois, J. S., Jülicher, F., Grill, S. W., Mayer, M., … Grill, S. W. (2010). Anisotropies in cortical tension reveal the physical basis of polarizing cortical flows. *Nature, 467*(7315), 617–621. http://dx.doi.org/10.1038/nature09376.

Morais-de-Sá, E., & Sunkel, C. (2013). Adherens junctions determine the apical position of the midbody during follicular epithelial cell division. *EMBO Reports, 14*(8), 696–703. http://dx.doi.org/10.1038/embor.2013.85.

Munjal, A., & Lecuit, T. (2014). Actomyosin networks and tissue morphogenesis. *Development (Cambridge, England), 141*(9), 1789–1793. http://dx.doi.org/10.1242/dev.091645.

Munjal, A., Philippe, J.-M., Munro, E., & Lecuit, T. (2015). A self-organized biomechanical network drives shape changes during tissue morphogenesis. *Nature, 524*(7565), 351–355. http://dx.doi.org/10.1038/nature14603.

Oda, H., Tsukita, S., & Takeichi, M. (1998). Dynamic behavior of the cadherin-based cell-cell adhesion system during *Drosophila* gastrulation. *Developmental Biology, 203*(2), 435–450. http://dx.doi.org/10.1006/dbio.1998.9047.

Rauzi, M., Lenne, P.-F., & Lecuit, T. (2010). Planar polarized actomyosin contractile flows control epithelial junction remodelling. *Nature, 468*(7327), 1110–1114. http://dx.doi.org/10.1038/nature09566.

del Valle Rodríguez, A., Didiano, D., & Desplan, C. (2011). Power tools for gene expression and clonal analysis in *Drosophila*. *Nature Methods, 9*(1), 47–55. http://dx.doi.org/10.1038/nmeth.1800.

CHAPTER

6

Visualization of cleavage furrow proteins in fixed dividing spermatocytes

A. Frappaolo, S. Sechi, G. Belloni, R. Piergentili, M.G. Giansanti[1]

Istituto di Biologia e Patologia Molecolari del CNR, Università di Roma Sapienza, Roma, Italy

[1]*Corresponding author: E-mail: mariagrazia.giansanti@uniroma1.it*

CHAPTER OUTLINE

Methods in Cell Biology, Volume 137, ISSN 0091-679X, http://dx.doi.org/10.1016/bs.mcb.2016.03.035

Abstract

Cytokinesis separates the cytoplasmic organelles and the duplicated genome into two daughter cells at the end of cell division. In animal cell cytokinesis, assembly and constriction of the contractile apparatus must be finely coordinated with plasma membrane remodeling and vesicle trafficking at the cleavage furrow. Accurate control of these events during cell cleavage is a fundamental task in all organisms and is also essential for maintaining ploidy and preventing neoplastic transformation. *Drosophila* male meiosis provides a well-suited cell system for exploring the molecular mechanisms underlying cytokinesis, combining the powerful tools of *Drosophila* genetics with unique cytological characteristics. Remarkably the large size of male meiotic cells highly facilitates cytological analysis of cytokinesis. Here we describe the main procedures that we use for fixing and visualizing cleavage furrow proteins in male meiotic cells. Moreover, we detail our protocol to detect protein interactions in fixed dividing spermatocytes by applying in situ proximity ligation assay.

INTRODUCTION: OVERVIEW OF CYTOKINESIS IN *DROSOPHILA* SPERMATOCYTES

Cytokinesis, the final event of cell division, partitions the cytoplasmic organelles and the genomic content of each dividing cell into two daughter cells (D'Avino, Giansanti, & Petronczki, 2015; Green, Paluch, & Oegema, 2012). As such, this process is fundamental for proper growth in all living organisms (D'Avino et al., 2015). Moreover, cytokinesis is crucial for maintaining ploidy during metazoan development and in adult tissues (D'Avino et al., 2015). Cytokinesis errors have been associated with several diseases, including blood disorders, age-related macular degeneration, Lowe syndrome, female infertility, and cancer (Lacroix & Maddox, 2012). Cytokinesis failures promote tumorigenesis by generating genetically unstable tetraploid cells (Lacroix & Maddox, 2012; Zack et al., 2013). Remarkably, computational analysis of somatic copy number alterations from 4934 human cancers revealed that 37% of human tumors have undergone at least one genome-doubling event during their progression (Zack et al., 2013).

Thus, the identification of new cytokinesis players and the characterization of their mechanism of action will help the design of new therapeutic strategies against cancer and other diseases.

In animal cells, cytokinesis is accomplished by constriction of a plasma membrane-anchored contractile ring, a transient organelle composed of F-actin filaments, and active Myosin II (D'Avino et al., 2015). Actomyosin ring constriction drives the plasma membrane and the underlying equatorial cortex inward, leading to formation of a cleavage furrow at the division site. Cleavage furrow ingresses until the two daughter cells remain connected by a thin cytoplasmic bridge. In most cell models, the narrow membrane bridge connecting the two daughters is eventually cut during the last stage of cytokinesis-dubbed abscission, which completely separates the nascent daughter cells (D'Avino et al., 2015; Mierzwa & Gerlich, 2014; Salbreux, Charras, & Paluch, 2012).

The scaffolding proteins, Septins and Anillin, ensure the tight anchoring of the actomyosin ring to the plasma membrane during furrow ingression and abscission (D'Avino, 2009; Mostowy & Cossart, 2012). Formation of the cytokinetic ring is triggered by accumulation of active Rho A GTPase at an equatorial position at the cortex that controls formin-mediated actin polymerization at the plasma membrane and indirectly promotes Myosin II activation (D'Avino, Savoian, & Glover, 2005; D'Avino et al., 2015). Central spindle microtubules dictate the division plate by delivering Rho A regulators to the equatorial cortex (D'Avino et al., 2005, 2015). Accumulation of active Rho A and its downstream effectors at the equatorial cortex requires a conserved guanine nucleotide exchange factor (GEF) which is targeted to the equatorial cortex by associating with the MgcRacGAP/RacGAP50C component of the well-conserved heterotetrameric, central spindlin complex (D'Avino et al., 2005; Piekny, Werner & Glotzer, 2005).

Studies in mammalian cells and in animal model systems indicate that targeted vesicle transport is necessary for cleavage furrow ingression as well as during abscission (Giansanti & Fuller, 2012; Neto, Collins, & Gould, 2011; Tang, 2012). Work from several groups has involved components of the secretory and endocytic/recycling trafficking pathways in cytokinesis (Montagnac, Echard, & Chavrier, 2008; Neto et al., 2011; Prekeris & Gould; 2008; Tang, 2012). Evidence also indicates that cytokinesis depends on a special lipid composition at the cleavage site (Echard, 2012). Phosphoinositide lipids, particularly phosphatidylinositol 4-phosphate [PI(4)P] and PI 4,5-biphosphate [PI(4,5)P2], provide critical membrane signals that control contractile ring assembly and dynamics, as well as new membrane addition (Brill, Wong, & Wilde, 2011; Cauvin & Echard, 2015; Echard, 2012).

Genetic dissection of cytokinesis in suitable model organisms, including *Drosophila melanogaster, Caenorhabditis elegans*, and *Schizosaccharomyces pombe*, has been successfully used to identify new cytokinesis players and to analyze the molecular mechanisms underlying this process (reviewed in Bathe & Chang, 2010; Eggert, Mitchison, & Field, 2006; Glotzer, 2005; Green et al., 2012; Pollard & Wu, 2010).

Drosophila male meiosis offers several advantages for combined genetic and cytological studies of the structural and morphological events that characterize cytokinesis (Giansanti & Fuller, 2012; Giansanti, Sechi, Frappaolo, Belloni, & Piergentili, 2012). Clusters of primary and secondary spermatocytes undergo meiotic divisions in synchrony (Fig. 1). Each primary spermatogonium undergoes four rounds of synchronous mitotic divisions giving rise to a cyst of 16 primary spermatocytes, which in turn undergo a premeiotic S phase. During the extended G2 phase of the cell cycle, each spermatocyte undergoes an impressive growth, which results in a 25-fold increase in volume (Lindsley & Tokuyasu, 1980). Thus, at the onset of the first meiotic division, approximately 90 h after premeiotic S phase, the 16 primary spermatocytes, contained in each cyst are considerably larger than most somatic cells. Primary spermatocytes synchronously divide meiotically after the G2/M transition, producing a cyst of 64 haploid round spermatids. Starting from late anaphase, dividing spermatocytes exhibit a prominent central spindle,

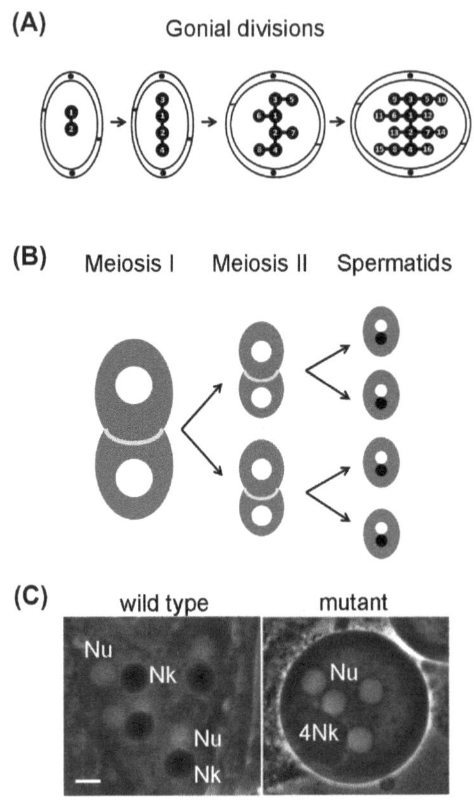

FIGURE 1 Spermatogenesis in *Drosophila melanogaster*.

(A) A single primary spermatogonium undergoes four mitotic divisions followed by incomplete cytokinesis, leading to the formation of 16 primary spermatocytes that remain connected by "ring canals." Two cyst cells engulf the progeny of the primordial primary spermatogonium throughout this process. (B) Each mature primary spermatocyte undergoes two meiotic divisions. At the end of meiosis II, each spermatid is characterized by the presence of a single, haploid nucleus, associated with a mitochondrial derivative of similar size, dubbed nebenkern. White circles, nuclei; black circles, nebenkern; light gray curved lines, contractile rings. (C) Onion-stage spermatids viewed by phase-contrast microscopy. Each wild-type spermatid contains a single light nucleus (Nu) associated with a dark nebenkern (Nk). A multinucleate spermatid from a mutant defective in male meiotic cytokinesis containing a large nebenkern (4Nk) associated with 4 nuclei of similar size (Nu). Scale Bar, 10 μm.

made up of bundles of interpolar microtubules that overlap between the two separating nuclei (Figs. 2–4). Central spindle microtubules are pinched in the middle by a clear actomyosin contractile ring during cytokinesis (Figs. 2 and 4). Immunofluorescence and in vivo analysis of contractile ring-associated proteins and other

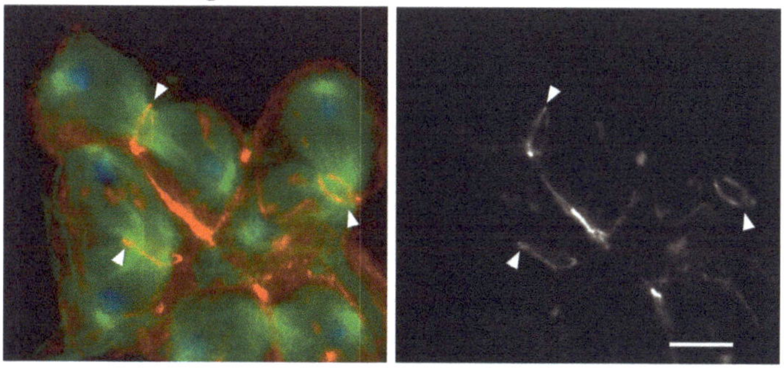

FIGURE 2 ***Drosophila*** **telophase spermatocytes display prominent central spindles encircled by F-actin rings.**

Dividing spermatocytes were simultaneously stained for F-actin (*red*), α-tubulin (*green*), and DNA (*blue*). *Arrowheads* point to F-actin rings. Scale bar, 10 μm.

cleavage furrow proteins is highly facilitated by the large size of these cells (Figs. 2–4).

During *Drosophila* spermatogonial and spermatocyte divisions, cytokinesis is incomplete; daughter cells remain interconnected by stable intercellular bridges called ring canals (Fig. 1A; Robinson & Cooley, 1996). Aside from this, the core molecular machineries and the molecular mechanisms required for cleavage furrow formation and ingression in mitotic cytokinesis are largely conserved in spermatocytes (Giansanti et al., 2012). Indeed, most of the gene products identified as players of mitotic cytokinesis by screening of late lethal mutants or through RNAi in cultured cells are also required for male meiotic cytokinesis (reviewed in Giansanti et al., 2012).

During metaphase of male meiosis, nuclei become spindle-shaped and the nuclear envelope that encircles meiotic chromosomes does not disassemble, but becomes fenestrated at the spindle poles. Dividing spermatocyte nuclei are also surrounded by five layers of double membranes called parafusorial membranes (Church & Lin, 1982; Tates, 1971). Alignment of mitochondria along the spindle microtubules, outside the nuclear membranes, allows equitable partitioning of these organelles to each daughter cell at the end of cytokinesis. Following the second meiotic division, all the mitochondria contained in each early round spermatid, aggregate to one side of the nucleus and fuse to form a complex, interlaced structure resembling an onion, dubbed the nebenkern (Fuller, 1993). When viewed by phase-contrast optics, each cyst of early round spermatids contains 64 spermatids, each containing a phase-dark nebenkern paired with a single phase-light haploid nucleus of similar size (Fig. 1B and C). Mutations affecting meiotic cytokinesis disrupt distribution of mitochondria between daughter cells resulting in spermatids containing abnormally enlarged mitochondrial derivatives associated with two or four

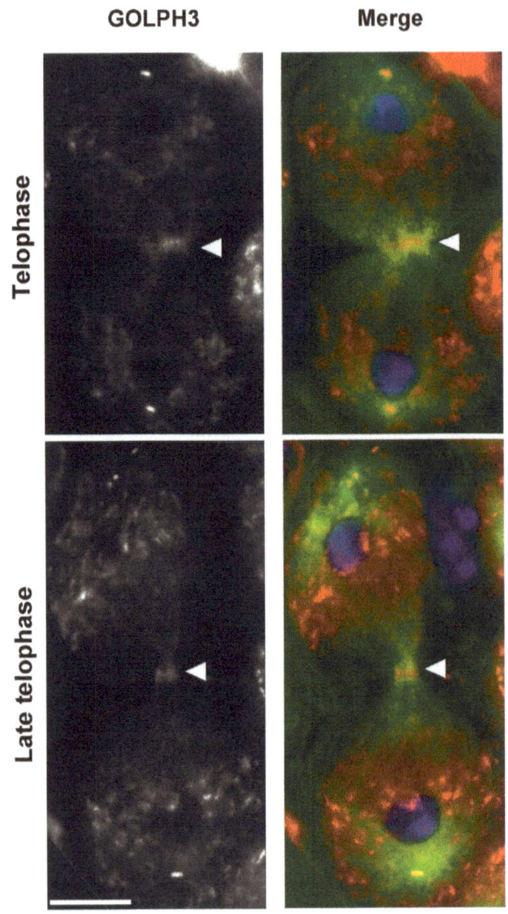

FIGURE 3 GOLPH3 protein accumulates at cleavage furrows of dividing spermatocytes.

Dividing spermatocytes were simultaneously stained for GOLPH3 (*red*), α-tubulin (*green*), and DNA (*blue*). In telophase cells, GOLPH3 is associated with vesicles in the polar region of the cells and accumulates at the cleavage furrows. *Arrowheads* point to cleavage furrows. Scale bar, 10 μm.

nuclei (Fig. 1C; Giansanti & Fuller, 2012). Thus the presence of multiple nuclei in early round spermatids allows a rapid scoring of mutants with cytokinesis alterations in *Drosophila* male meiotic divisions (Giansanti et al., 2012). When defective male meiotic cytokinesis is coupled with abnormal chromosome segregation, cysts of early round spermatids display large nebenkerne associated with multiple nuclei of variable size (Carmena et al., 1998; Gao et al., 2008).

In this context, another advantage offered by male meiosis when studying cytokinesis is the lack of a robust spindle assembly checkpoint (SAC) (Rebollo &

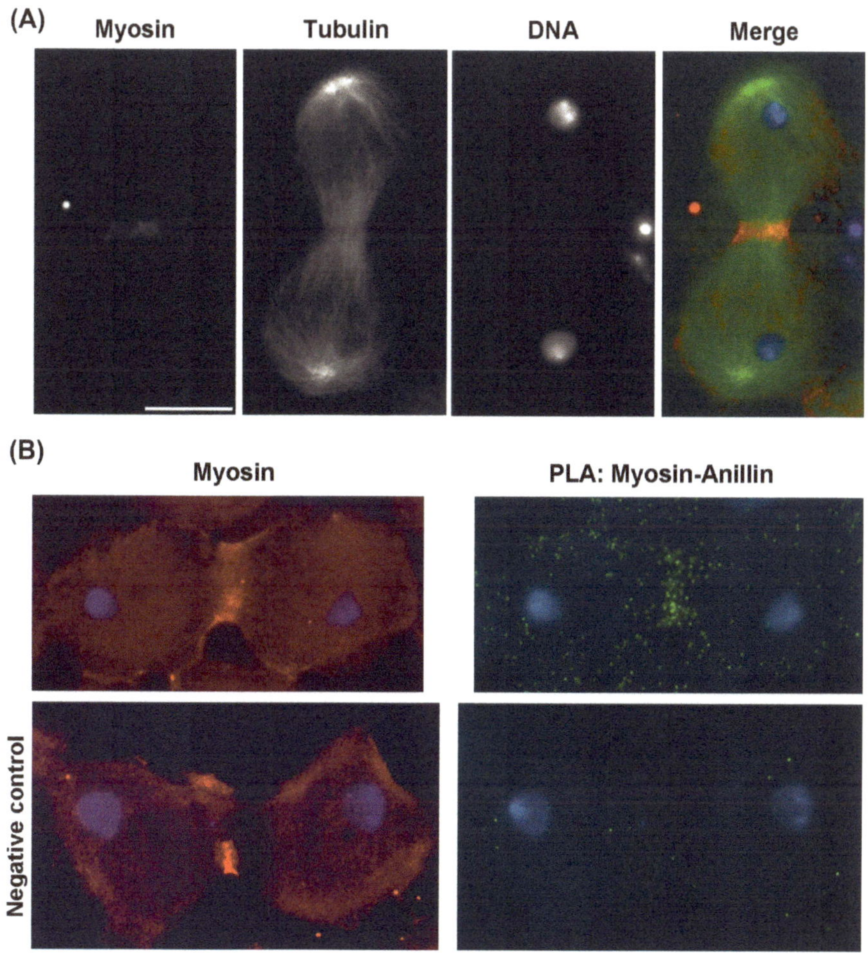

FIGURE 4 Immunolocalization of Myosin II at the cleavage furrow and visualization of its interaction with Anillin.

(A) Dividing spermatocytes were simultaneously stained for Myosin II (*red*), α-tubulin (*green*), and DNA (*blue*). (B) Proximity ligation assay (PLA) using rabbit anti-Myosin II and mouse anti-Anillin in telophase cells. The negative control was obtained by omitting anti-Anillin antibodies in the reaction. Scale bar, 10 μm.

González, 2000), ie, the signaling pathway which monitors proper kinetochore attachment to spindle microtubules (Musacchio & Salmon, 2007) and blocks cells in metaphase. In *Drosophila* spermatocytes, the SAC is weak resulting in a small delay in anaphase onset and meiotic progression in the presence of unattached chromosomes or malformed spindles (Gao et al., 2008; Rebollo & González, 2000; Szafer-Glusman, Fuller, & Giansanti, 2011). This spermatocyte peculiarity allows to investigate whether a given gene product required for chromosome segregation is also involved in cytokinesis. For example, mutations affecting the *Drosophila* homolog of human Polo like kinase cause a metaphase arrest in *Drosophila* neuroblasts due to the activation of the SAC checkpoint (Llamazares et al., 1991). Conversely, the spermatocyte system reveals the requirement for this multifunctional kinase for central spindle assembly and F-actin ring formation during later stages of meiotic divisions (Carmena et al., 1998; Herrmann, Amorim, & Sunkel, 1998).

Drosophila spermatocytes are also emerging as a well-suited model to dissect membrane trafficking events during cytokinesis in a tissue of a multicellular organism (Giansanti & Fuller, 2012). In this context, it is important to note that in *Drosophila* interphase spermatocytes, as well as in other *Drosophila* cells, the Golgi apparatus comprises approximately 20 separate stacks that can be visualized by staining for the golgin Lava lamp (Lva) (Farkas, Giansanti, Gatti, & Fuller, 2003; Robinett, Giansanti, Gatti, & Fuller, 2009; Sisson, Field, Ventura, Royou, & Sullivan, 2000). The peculiar organization in multiple separate stacks highly facilitates the analysis of any structural alterations in secretory mutants that might be not detectable in cell types containing a pericentriolar Golgi ribbon (Kondylis & Rabouille, 2009). In dividing spermatocytes, as well as in other dividing cells, the Golgi organelles disassemble in small vesicle-like structures (Fig. 3) that concentrate at the cell poles (Farkas et al., 2003; Giansanti et al., 2006; Giansanti, Belloni, & Gatti, 2007).

Cytokinesis of male meiotic cells appears particularly sensitive to mutations affecting vesicle trafficking pathways when compared to cytokinesis of somatic cells (Sechi, Frappaolo, Belloni, & Giansanti, 2015). The extra demand of spermatocytes on membrane trafficking have been related to the lack of plasma membrane protrusions and microvilli (Tates, 1971) that might be used as a plasma membrane reservoir to provide the additional surface area during cytokinesis (Giansanti & Fuller, 2012). Moreover, these cells are larger than most somatic cells and complete two successive divisions in less than 2 h with a short intervening interphase. Indeed, large screens for mutants affecting male meiotic cytokinesis allowed identifying many vesicle trafficking and membrane remodeling factors required for this process (Giansanti et al., 2012). The components of the secretory and endocytic/recycling pathways that have been involved in spermatocyte cytokinesis include the conserved oligomeric Golgi complex (COG) subunits Cog5 and Cog7, the Syntaxin 5 ER-to-Golgi vesicle-docking protein, the Rab11 GTPase, the Arf6 GTPase, the TRAPPII complex subunit Brunelleschi, and the exocyst complex proteins Sec8 and Exo84 (Belloni et al., 2012; Brill, Hime, Scharer-Schuksz, & Fuller, 2000; Dyer et al., 2007; Farkas et al., 2003; Gatt & Glover, 2006; Giansanti et al., 2007, 2015; Robinett et al., 2009; Sechi et al., 2014, 2015; Xu et al., 2002). Mutational analysis has also highlighted the requirement for special lipids in spermatocyte cytokinesis. Mutations in Elovl, a protein involved in biosynthesis of very long-chain fatty acids, impair both cleavage-furrow ingression and central spindle assembly in *Drosophila* male meiotic cells, suggesting an interplay between membrane lipids and the actomyosin apparatus (Szafer-Glusman et al., 2008) Proteins involved in the phosphoinositide cycle have been implicated in *Drosophila* furrowing (Brill et al., 2011). The *Drosophila* PI transfer protein (PITP) Giotto/Vibrator (Gio/Vib) concentrates at the endoplasmic reticulum membranes and at the cleavage furrow of dividing spermatocytes and is required for furrow ingression in both spermatocytes and larval neuroblasts (Gatt & Glover, 2006; Giansanti et al., 2006). The PI 4 kinase III β Fwd is also essential for male meiotic cytokinesis (Brill, Hime, Scharer-Schuksz, & Fuller, 2000). Wild-type function of Fwd is required to recruit Rab11 at the Golgi, for synthesis of PI 4-phosphate [PI(4)P] on Golgi membranes and for localization of Rab11- and PI(4)P-containing organelles at the cleavage site of dividing spermatocytes (Polevoy et al., 2009). We have recently demonstrated that GOLPH3, described as a PI(4)P effector at the Golgi (Sechi, Frappaolo, Belloni, Colotti, & Giansanti, 2015), accumulates at the cleavage furrow and is essential for cytokinesis in *Drosophila* spermatocyte and neuroblast dividing cells. GOLPH3 is essential for maintenance of central spindlin and Rho1 at cell equator and stabilization of Myosin II and Septin rings (Sechi et al., 2014). We have shown that GOLPH3 function during cytokinesis is dependent on its binding to PI(4)P. Mutations that affect interaction with PI(4)P disrupt localization of GOLPH3 at both the Golgi stacks and the cleavage furrow. Telophase spermatocytes from mutants with defective GOLPH3-PI(4)P interaction also fail to accumulate PI(4)P- and Rab11-associated secretory organelles at the cleavage site. Based on these data, GOLPH3 might act as a key molecule in coupling vesicle trafficking with actomyosin ring assembly and stability during cytokinesis.

Here we will detail the main procedures that we use for fixing and staining dividing spermatocytes to visualize contractile ring components and other cleavage furrow proteins. In addition, we will illustrate our protocol to visualize protein interactions in fixed dividing spermatocytes by applying in situ proximity ligation assay (PLA).

1. FIXATION PROTOCOLS

1.1 PROTOCOL 1: FIXATION WITH 4% METHANOL-FREE FORMALDEHYDE

This protocol describes a fixation procedure that preserves F-actin filaments and can be used to visualize the actomyosin ring with rhodamine-conjugated phalloidin (Fig. 2; Gao et al., 2008; Giansanti et al., 1998; Szafer-Glusman et al., 2008). In addition, this procedure is suitable to fix larval testes expressing GFP-tagged proteins (Szafer-Glusman et al., 2011), like for example, the *Drosophila* myosin regulatory light chain Spaghetti Squash (Sqh) tagged with GFP (Sqh-GFP, Royou, Sullivan, & Karess, 2002). However, this protocol is not as adequate as protocols 2 and 3 to preserve central spindle microtubules.

1. Dissect larval testes in TB buffer from either third-instar larvae, pupae, or young adult males.
2. Transfer testes into a 2–3 μL of TB, placed on a clean 20 × 20 mm coverslip.
3. Place a clean slide over the coverslip without pressing.
4. After inverting the sandwich, examine each preparation under a phase-contrast microscope, and select specimens that are enough flattened and contain telophase cells.
5. Quickly freeze slides in liquid nitrogen.
6. Remove the coverslips using a razor blade and immerse slides in chilled 95% ethanol (at −20°C).
7. Fix samples in 4% methanol-free formaldehyde (in 1X PBS) for 7 min.
8. Wash samples twice (2 × 5′) in 1X PBS at room temperature.
9. Immerse samples in PBT for 20–30 min.
10. Wash the samples twice (2 × 5′ in 1X PBS) before analyzing each preparation by viewing with a 40× objective using phase-contrast optics.
11. Discard the specimens in which the morphology of dividing cells appears damaged or grainy.
12. Proceed to visualize α-tubulin along with either F-actin or the GFP-tagged Sqh (from testes expressing this fluorescent protein).

1.1.1 Materials

1.1.1.1 Supplies and equipments

Dissecting forceps No. 5 (Dumont Cat. No. 11252-20)
20 × 20-mm coverslips (LLG Labware—Cat. No. 9.160 823)
Glass slides Superfrost PLUS 75 × 25 × 1-mm (Thermo Scientific, Menzel-Glaser Cat. No. J1800AMNZ)
Razor blade
Dissecting microscope
Upright phase-contrast microscope equipped with a 40× objective

1.1.1.2 Solutions and reagents

Testis buffer (TB): 183 mM KCl, 47 mM NaCl, 10 mM Tris–HCl, 1 mM EDTA
Liquid nitrogen
95% Ethanol
1X PBS (Dulbecco's phosphate buffered saline—Sigma-Aldrich D5652)
Triton X-100 (Sigma-Aldrich Cat. No. T8787)
PBT (1X PBS containing 0.1% Triton)
Methanol-free formaldehyde (Polysciences Inc., Cat. No. 18814-20)

1.2 PROTOCOL 2: FIXATION WITH METHANOL AND FORMALDEHYDE

This fixation protocol, which allows an optimal preservation of spindle microtubules, has been extensively used to immunolocalize cleavage furrow components at different stages of meiotic divisions (that can be clearly distinguished by analyzing the meiotic spindle morphologies). We recently used this protocol to simultaneously stain for GOLPH3 and tubulin (Sechi et al., 2014, Fig. 3). Since this procedure involves methanol fixation, it is not suitable for F-actin staining with rhodamine-conjugated phalloidin.

1.2.1 Method

1. Dissect testes from third instar larvae in saline buffer.
2. Transfer testes into a drop of PBS containing 0.5% Triton X-100 for 2 min to allow tissue permeabilization.
3. Transfer testes into a 4 μL drop of 3.7% formaldehyde (in 1X PBS) on a 20 × 20-mm coverslip fix and let them fix for 2 min at room temperature.
4. Place a glass slide over the coverslip without applying any pressure and invert the sandwich.
5. Examine preparations by viewing through a phase-contrast optical system using a 40× objective.
6. Select the specimens displaying meiotic divisions and discard the other ones.
7. Let the preparations fix for additional 8 min at room temperature before immersing the slides in liquid nitrogen.
8. Remove the coverslip using a razor blade and, quickly, immerse the slides in methanol at −20°C for 15 min.
9. Transfer the slides in PBT buffer for 10 min.
10. Transfer the slides in the blocking solution (PBT containing 3% BSA) for 25–30 min.
11. Rinse the slides in 1X PBS for 5 min, before incubating with primary antibodies.

1.2.2 Materials

1.2.2.1 Supplies and equipments

Dissecting forceps no. 5 (Dumont Cat. No. 11252-20)
20 × 20-mm coverslips (LLG Labware—Cat. No. 9.160 823)
Glass slides Superfrost PLUS 75x25x1-mm (Thermo Scientific—Menzel-Glaser Cat. No. J1800AMNZ)
Dissecting microscope
Upright phase-contrast microscope equipped with a 40× objective

1.2.2.2 Solutions and reagents

Saline buffer (0.7% NaCl in distilled water)
1X PBS (Dulbecco's phosphate buffered saline—Sigma-Aldrich D5652)
PBS containing 0.5% Triton X-100 (Sigma-Aldrich T8787)
Triton X-100 (Sigma-Aldrich Cat. No. T8787)
Bovine serum albumine (BSA, MP Biomedicals Cat. No. 0218057580)
PBT (1X PBS containing 0.1% Triton X-100)
PBT containing 3% BSA

Formaldehyde (3.7% in 1X PBS)
Liquid nitrogen
Methanol, −20°C

1.3 PROTOCOL 3: FORMALDEHYDE FIXATION

1. Dissect larval or pupal testes in saline buffer.
2. Transfer testes into a drop of 3.7% formaldehyde in PBS and let testes fix for 20 min at room temperature.
3. Transfer testes into a drop of 45% acetic acid for about 30 s.
4. Transfer testes into a drop (4−5 μL) of 60% acetic acid, placed on a 20 × 20-mm coverslip.
5. After 2−3 min, place a clean glass slide over the coverslip exerting a slight pressure.
6. Rapidly check the quality of each preparation by viewing under a phase-contrast microscope (discard the preparations that are devoid of meiotic divisions).
7. Immerse preparations in liquid nitrogen.
8. Leave the slides freezed in liquid nitrogen for at least 5 min.
9. Remove the coverslip using a razor blade, and quickly immerse slides in ethanol at −20°C for 10 min.
10. Transfer into PBT, wash twice for 10 min at room temperature.
11. Wash in 1X PBS for 5 min before proceeding with either immunofluorescence or PLA assay.

1.3.1 Materials

1.3.1.1 Supplies and equipments

Dissecting forceps no. 5 (Dumont Cat. No. 11252-20)
Coverslips 20 × 20-mm (LLG Labware—Cat No. 9.160 823)
Coverslips 22 × 22-mm (LLG Labware—Cat No. 9.160 824)
Glass Slides Superfrost 76 × 26 × 0.1-mm (Thermo Scientific—Menzel-Glaser Cat. No. ABAA00008)
Dissecting microscope
Upright phase-contrast microscope equipped with a 40× objective

1.3.1.2 Solutions and reagents

Saline buffer (0.7% NaCl in H_2O)
1X PBS (Dulbecco's phosphate buffered saline—Sigma-Aldrich D5652)
BSA (bovine serum albumine—MP Biomedicals Cat. No. 0218057580)
Triton X-100 (Sigma-Aldrich Cat. No. T8787)
PBT (1X PBS containing 0.1% Triton X-100-Sigma-Aldrich T8787)
PBT containing 3% BSA

Formaldehyde (3.7% in 1X PBS)
Acetic acid (45% and 60% in distilled H_2O)
Liquid nitrogen
Chilled ethanol

2. STAINING PROTOCOLS

2.1 PROTOCOL 4: VISUALIZATION OF SPINDLE MICROTUBULES WITH F-ACTIN IN FIXED SPERMATOCYTES

This protocol, that derives from the one described by Szafer-Glusman et al. (2008) with minor changes, proved to be very efficient in detecting F-actin rings.

2.1.1 Method

1. Fix testes according to protocol 1.
2. Dilute mouse anti α-tubulin antibody in 1X PBT to a suitable concentration.
3. Remove excess of PBS around each testis preparations and add 20 μL of α-tubulin solution.
4. Incubate in a humid chamber, overnight at 4°C or alternatively for 1 h at room temperature.
5. Dilute anti-Mouse IgG antibodies in 1X PBS to a suitable concentration.
6. After two washes in 1X PBS (2 × 5 min), remove excess of PBS around each testis preparations and add anti-Mouse IgG.
7. Incubate in the dark, a humid chamber, for 40 min at room temperature.
8. Dilute rhodamine-labeled Phalloidin 1:20 in 1X PBS.
9. After two washes in 1X PBS (2 × 5 minutes), remove excess of PBS around each testis preparations and add rhodamine-labeled phalloidin.
10. Incubate in the dark, a humid chamber, for 40 min at 37°C.
11. Wash two times in 1X PBS (2 × 5 minutes).
12. Let preparations air-dry in the dark.
13. To prevent photobleaching and to visualize DNA, dispense 30 μL of Vectashield mounting medium with DAPI onto each specimen and gently lower a coverslip over the drop.
14. Remove excess of Vectashield mounting medium around from the edges of the coverslip using a tissue paper.
15. Leave the slides at −20°C for at least 2 h before analyzing preparations using a fluorescence microscope.

2.1.2 Materials

2.1.2.1 Supplies and equipments

Coverslips 22x22-mm (LLG Labware—Cat No. 9.160 824)
Humid chamber
Upright fluorescence microscope

2.1.2.2 Solutions and reagents

1X PBS (Dulbecco's phosphate buffered saline—Sigma-Aldrich Cat. No. D5652)
Mouse anti-α-Tubulin (Sigma-Aldrich Cat. No. T6199)
Fluorescein (FITC) Affinity purified $F(ab')_2$ Fragment Goat anti-Mouse IgG I (Jackson ImmunoResearch Cat. No. 115-096-006)
Rhodamine phalloidin (Tebu Bio Cat. No. PHDR1)
Vectashield mounting Medium with DAPI (Vector Laboratories Cat. No. H-1200)

2.2 PROTOCOL 5: SIMULTANEOUS STAINING OF TUBULIN AND CLEAVAGE FURROW PROTEINS BY DOUBLE ANTIBODY STAINING

To immunolocalize cytokinesis proteins using Protocol 5, testis preparations can be either fixed using Protocol 2 or Protocol 3. Protocol 2 proved to be optimal for visualizing GOLPH3 protein at the cleavage site. Protocol 3 proved to be efficient to detect Myosin 2 (Fig. 4), Anillin Septins and Rab11 (Giansanti et al., 2007; Sechi et al., 2014).

2.2.1 Method

1. Dilute rabbit antibodies against the cleavage furrow protein (for example Rabbit anti-Myosin II, Fig. 4), together with mouse anti α tubulin, in 1X PBT containing 3% BSA to a suitable concentration.
2. Remove excess of PBS around each testis preparations and add 20 μL of the antibody solution.
3. Incubate in a humid chamber, overnight at 4°C.
4. Dilute anti-Mouse IgG antibodies in 1X PBS to a suitable concentration.
5. After two washes in 1X PBS (2 × 5 minutes), remove excess of PBS around each testis preparations and add anti-Mouse IgG.
6. Incubate in the dark, a humid chamber, for 40 min at room temperature.
7. Dilute Alexa 555-conjugated anti-rabbit IgG 1:300 in 1X PBT.
8. After two washes in 1X PBS (2 × 5 minutes), remove excess of PBS around each testis preparations and add Alexa 555-conjugated anti-rabbit IgG.
9. Incubate in the dark, a humid chamber, for 40 min at room temperature.
10. Perform steps 11−15 of Protocol 4.

2.2.2 Materials

2.2.2.1 Supplies and equipments

Coverslips 22 × 22-mm (LLG Labware—Cat. No. 9.160 824)
Humid chamber
Upright fluorescence microscope

2.2.2.2 Solutions and reagents

Saline buffer (0.7% NaCl in distilled H_2O)
1X PBS (Dulbecco's phosphate buffered saline—Sigma-Aldrich D5652)
BSA (MP Biomedicals Cat. No. 0218057580)
Triton X-100 (Sigma-Aldrich Cat. No. T8787)
PBT (1X PBS containing 0.1% Triton X-100)
PBT containing 3% BSA
Mouse anti-α-Tubulin (Sigma-Aldrich Cat. No. T6199)
Rabbit anti-Myosin II (Royou et al., 2002)
Fluorescein (FITC) Affinity purified F(ab')$_2$ Fragment Goat anti-Mouse IgG I (Jackson ImmunoResearch Cat. No. 115-096-006)
F(ab')$_2$-Goat Alexa555-conjugated anti rabbit IgG (Thermofisher Scientific Cat. No. A-21430)
Vectashield mounting medium with DAPI (Vector Laboratories Cat. No. H-1200)

2.3 PROTOCOL 6: PROXIMITY LIGATION ASSAY IN FIXED TESTES

Protocol 3 proved to be very efficient to fix testis preparations to be used for in situ PLA. The approach based on in situ PLA technology provides an excellent method to study protein interactions with spatiotemporal resolution in fixed tissues (Fredriksson et al., 2002). PLA requires the use of two primary antibodies raised in two different species. The secondary antibodies (PLA probes) are labeled with short DNA oligonucleotides acting as primers for rolling circle amplification. If the PLA probes are in close proximity (ie, distance between antigens of 10–30 nm), a positive PLA signal is generated that can be detected as an individual fluorescent signal. The PLA signals can be analyzed and mapped to a subcellular location. Thus, this assay can be used to test the interaction of two cytokinesis proteins at the cleavage furrow.

2.3.1 Method

1. Perform the blocking step either by immersing slides in PBT containing 3% BSA for 20 min or alternatively by immersing in the blocking solution contained in the kit (Duolink In Situ PLA Probes) and following the instructions provided by the purchaser.
2. Dilute the two primary antibodies (raised against each protein to be tested) to a suitable concentration in PBT containing 3% BSA or in the Duolink in Situ Antibody Diluent, equipped with the kit (Duolink In Situ PLA Probes).
3. Add the primary antibody solution to each preparation.
4. Incubate preparations with the primary antibody solution in a humid chamber for 2 h at room temperature or alternatively over night at 4°C.
5. Carry out the following steps, including the PLA probes incubation and the detection protocol in accordance with the procedures described in the Duolink In Situ-Fluorescence User Guide, using the Duolink In Situ PLA Probes and Duolink In Situ Detection Reagents.

6. Following the detection steps, mount the specimens Vectashield mounting medium containing with DAPI as described in Protocol 4.
7. Leave the slides at −20°C for at least 2 h before analyzing preparations using a fluorescence microscope.

2.3.2 Materials

2.3.2.1 Supplies and equipments

Coverslips 22 × 22-mm (LLG Labware—Cat. No. 9.160 824)
Humid chamber
Upright fluorescence microscope

2.3.2.2 Solutions and reagents

Saline buffer (0.7% NaCl in H_2O)
1X PBS (Dulbecco's phosphate buffered saline—Sigma-Aldrich D5652)
BSA (—MP Biomedicals 0218057580)
PBT (1X PBS containing 0.1% Triton X-100-Sigma-Aldrich T8787)
PBT containing 3% BSA
Rabbit anti-Myosin II (Royou et al., 2002)
Mouse anti-Anillin (Giansanti et al., 2015)
Fluorescein (FITC) Affinity purified $F(ab')_2$ Fragment Goat anti-Mouse IgG I (Jackson ImmunoResearch Cat. No. 115-096-006)
$F(ab')_2$-Goat Alexa555-conjugated anti rabbit IgG (Thermofisher Scientific Cat. No. A-21430)
Duolink In Situ reagents:
Duolink In Situ PLA Probes (Sigma-Aldrich: Anti-Mouse PLUS Cat. No. DUO92001; Anti-Rabbit MINUS Cat. No. DUO92005)
Duolink In Situ Detection Reagents (Sigma-Aldrich Cat. No. DUO92014)
Vectashield mounting medium with DAPI (Vector Laboratories Cat. No. H-1200)

ACKNOWLEDGMENTS

Our work was supported by a grant from Associazione Italiana per la Ricerca sul Cancro (AIRC) (Grant number IG14671) to M.G. Giansanti.

REFERENCES

Bathe, M., & Chang, F. (2010). Cytokinesis and the contractile ring in fission yeast: towards a systems-level understanding. *Trends in Microbiology, 18*, 38–45.

Belloni, G., Sechi, S., Riparbelli, M. G., Fuller, M. T., Callaini, G., & Giansanti, M. G. (2012). Mutations in Cog7 affect Golgi structure, meiotic cytokinesis and sperm development during *Drosophila* spermatogenesis. *Journal of Cell Science, 125*, 5441–5452.

Brill, J. A., Hime, G. R., Scharer-Schuksz, M., & Fuller, M. T. (2000). A phospholipid kinase regulates actin organization and intercellular bridge formation during germline cytokinesis. *Development, 127*, 3855–3864.

Brill, J. A., Wong, R., & Wilde, A. (2011). Phosphoinositide function in cytokinesis. *Current Biology, 21*, R930–R934.

Carmena, M., Riparbelli, M. G., Minestrini, G., Tavares, A. M., Adams, R., Callaini, G., & Glover, D. M. (1998). *Drosophila* polo kinase is required for cytokinesis. *The Journal of Cell Biology, 143*, 659–671.

Cauvin, C., & Echard, A. (2015). Phosphoinositides: lipids with informative heads and mastermind functions in cell division. *Biochimica et Biophysica Acta, 1851*, 832–843.

Church, K., & Lin, H. P. (1982). Meiosis in *Drosophila melanogaster.* II. The prometaphase-I kinetochore microtubule bundle and kinetochore orientation in males. *The Journal of Cell Biology, 93*, 365–373.

D'Avino, P. P., Savoian, M. S., & Glover, D. M. (2005). Cleavage furrow formation and ingression during animal cytokinesis: a microtubule legacy. *Journal of Cell Science, 118*, 1549–1558.

D'Avino, P. P., Giansanti, M. G., & Petronczki, M. (2015). Cytokinesis in animal cells. *Cold Spring Harbor Perspectives in Biology, 13*, 7:a015834.

D'Avino, P. P. (2009). How to scaffold the contractile ring for a safe cytokinesis – lessons from Anillin-related proteins. *Journal of Cell Science, 122*, 1071–1079.

Dyer, N., Rebollo, E., Domínguez, P., Elkhatib, N., Chavrier, P., Daviet, L., ... González-Gaitán, M. (2007). Spermatocyte cytokinesis requires rapid membrane addition mediated by ARF6 on central spindle recycling endosomes. *Development, 134*, 4437–4447.

Echard, A. (2012). Phosphoinositides and cytokinesis: the "PIP" of the iceberg. *Cytoskeleton, 69*, 893–912.

Eggert, U. S., Mitchison, T. J., & Field, C. M. (2006). Animal cytokinesis: from parts list to mechanisms. *Annual Review Biochemistry, 75*, 543–566.

Farkas, R. M., Giansanti, M. G., Gatti, M., & Fuller, M. T. (2003). .The *Drosophila* Cog5 homologue is required for cytokinesis, cell elongation, and assembly of specialized Golgi architecture during spermatogenesis. *Molecular Biology of the Cell, 14*, 190–200.

Fredriksson, S., Gullberg, M., Jarvius, J., Olsson, C., Pietras, K., Gustafsdottir, S. M., ... Landegren, U. (2002). Protein detection using proximity-dependent DNA ligation assays. *Nature Biotechnology, 20*, 473–477.

Fuller, M. T. (1993). Spermatogenesis. In M. Bate, & A. Martinez-Arias (Eds.), *The development of* Drosophila (pp. 71–147). Cold Spring Harbor, New York: Cold Spring Harbor Press.

Gao, S., Giansanti, M. G., Buttrick, G. J., Ramasubramanyan, S., Auton, A., Gatti, M., & Wakefield, J. G. (2008). Australin: a chromosomal passenger protein required specifically for *Drosophila melanogaster* male meiosis. *The Journal of Cell Biology, 180*, 521–535.

Gatt, M. K., & Glover, D. M. (2006). The *Drosophila* phosphatidylinositol transfer protein encoded by vibrator is essential to maintain cleavage-furrow ingression in cytokinesis. *Journal of Cell Science, 119*, 2225–2235.

Giansanti, M. G., & Fuller, M. T. (2012). What *Drosophila* spermatocytes tell us about the mechanisms underlying cytokinesis. *Cytoskeleton, 69*, 869–881.

Giansanti, M. G., Bonaccorsi, S., Williams, B., Williams, E. V., Santolamazza, C., Goldberg, M. L., & Gatti, M. (1998). Cooperative interactions between the central spindle and the contractile ring during *Drosophila* cytokinesis. *Genes & Development, 12*, 396–410.

Giansanti, M. G., Bonaccorsi, S., Kurek, R., Farkas, R. M., Dimitri, P., Fuller, M. T., & Gatti, M. (2006). The class I PITP giotto is required for *Drosophila* cytokinesis. *Current Biology, 16*, 195–201.

Giansanti, M. G., Belloni, G., & Gatti, M. (2007). Rab11 is required for membrane trafficking and actomyosin ring constriction in meiotic cytokinesis of *Drosophila* males. *Molecular Biology of the Cell, 18*, 5034–5047.

Giansanti, M. G., Sechi, S., Frappaolo, A., Belloni, G., & Piergentili, R. (2012). Cytokinesis in *Drosophila* male meiosis. *Spermatogenesis, 2*, 185–196.

Giansanti, M. G., Vanderleest, T. E., Jewett, C. E., Sechi, S., Frappaolo, A., Fabian, L., … Blankenship, J. T. (2015). Exocyst-dependent membrane addition is required for anaphase cell elongation and cytokinesis in *Drosophila*. *PLoS Genetics, 11*, e1005632.

Glotzer, M. (2005). The molecular requirements for cytokinesis. *Science, 307*, 1735–1739.

Green, R. A., Paluch, E., & Oegema, K. (2012). .Cytokinesis in animal cells. *Annual Review Cell and Developmental Biology, 28*, 29–58.

Herrmann, S., Amorim, I., & Sunkel, C. E. (1998). The POLO kinase is required at multiple stages during spermatogenesis in *Drosophila melanogaster*. *Chromosoma, 107*, 440–451.

Kondylis, V., & Rabouille, C. (2009). The Golgi apparatus: lessons from *Drosophila*. *FEBS Letters, 583*, 3827–3838.

Lacroix, B., & Maddox, A. S. (2012). Cytokinesis, ploidy and aneuploidy. *The Journal of Pathology, 226*, 338–351.

Lindsley, D. T., & Tokuyasu, K. T. (1980). Spermatogenesis. In M. Ashburner, & T. R. F. Wright (Eds.), *Genetics and Biology of* Drosophila (pp. 225–294). New York: Academic Press.

Llamazares, S., Moreira, A., Tavares, A., Girdham, C., Spruce, B. A., Gonzalez, C., … Sunkel, C. E. (1991). Polo encodes a protein kinase homolog required for mitosis in *Drosophila*. *Genes & Development, 5*, 2153–2165.

Mierzwa, B., & Gerlich, D. W. (2014). Cytokinetic abscission: molecular mechanisms and temporal control. *Developmental Cell, 31*, 525–538.

Montagnac, G., Echard, A., & Chavrier, P. (2008). Endocytic traffic in animal cell cytokinesis. *Current Opinion in Cell Biology, 20*, 454–461.

Mostowy, S., & Cossart, P. (2012). Septins: the fourth component of the cytoskeleton. *Nature Reviews Molecular Cell Biology, 13*, 183–194.

Musacchio, A., & Salmon, E. D. (2007). The spindle-assembly checkpoint in space and time. *Nature Reviews Molecular Cell Biology, 8*, 379–393.

Neto, H., Collins, L. L., & Gould, G. W. (2011). Vesicle trafficking and membrane remodelling in cytokinesis. *The Biochemical Journal, 437*, 13–24.

Piekny, A., Werner, M., & Glotzer, M. (2005). Cytokinesis: welcome to the Rho zone. *Trends in Cell Biology, 15*, 651–658.

Polevoy, G., Wei, H. C., Wong, R., Szentpetery, Z., Kim, Y. J., Goldbach, P., … Brill, J. A. (2009). Dual roles for the *Drosophila* PI 4-kinase four wheel drive in localizing Rab11 during cytokinesis. *The Journal of Cell Biology, 187*, 847–858.

Pollard, T. D., & Wu, J. Q. (2010). Understanding cytokinesis: lessons from fission yeast. *Nature Reviews Molecular Cell Biology, 11*, 149–155.

Prekeris, R., & Gould, G. W. (2008). Breaking up is hard to do – membrane traffic in cytokinesis. *Journal of Cell Science, 121*, 1569–1576.

Rebollo, E., & González, C. (2000). Visualizing the spindle checkpoint in *Drosophila* spermatocytes. *EMBO Reports, 1*, 65–70.

Robinett, C. C., Giansanti, M. G., Gatti, M., & Fuller, M. T. (2009). TRAPPII is required for cleavage furrow ingression and localization of Rab11 in dividing male meiotic cells of *Drosophila*. *Journal of Cell Science, 122*, 4526–4534.

Robinson, D. N., & Cooley, L. (1996). Stable intercellular bridges in development: the cytoskeleton lining the tunnel. *Trends in Cell Biology, 6*, 474–479.

Royou, A., Sullivan, W., & Karess, R. (2002). Cortical recruitment of nonmuscle myosin II in early syncytial *Drosophila* embryos: its role in nuclear axial expansion and its regulation by Cdc2 activity. *The Journal of Cell Biology, 158*, 127–137.

Salbreux, G., Charras, G., & Paluch, E. (2012). Actin cortex mechanics and cellular morphogenesis. *Trends in Cell Biology, 22*, 536–545.

Sechi, S., Colotti, G., Belloni, G., Mattei, V., Frappaolo, A., Raffa, G. D., … Giansanti, M. G. (2014). GOLPH3 is essential for contractile ring formation and Rab11 localization to the cleavage site during cytokinesis in *Drosophila melanogaster*. *PLoS Genetics, 10*, e1004305.

Sechi, S., Frappaolo, A., Belloni, G., & Giansanti, M. G. (2015a). The roles of the oncoprotein GOLPH3 in contractile ring formation and membrane trafficking during cytokinesis. *Biochemical Society Transactions, 43*, 117–121.

Sechi, S., Frappaolo, A., Belloni, G., Colotti, G., & Giansanti, M. G. (2015b). The multiple cellular functions of the oncoprotein Golgi phosphoprotein 3. *Oncotarget, 6*, 3493–3506.

Sisson, J. C., Field, C., Ventura, R., Royou, A., & Sullivan, W. (2000). Lava lamp, a novel peripheral golgi protein, is required for *Drosophila melanogaster* cellularization. *The Journal of Cell Biology, 151*, 905–918.

Szafer-Glusman, E., Giansanti, M. G., Nishihama, R., Bolival, B., Pringle, J., Gatti, M., & Fuller, M. T. (2008). A role for very-long-chain fatty acids in furrow ingression during cytokinesis in *Drosophila* spermatocytes. *Current Biology, 18*, 1426–1431.

Szafer-Glusman, E., Fuller, M. T., & Giansanti, M. G. (2011). Role of Survivin in cytokinesis revealed by a separation-of-function allele. *Molecular Biology of the Cell, 22*, 3779–3790.

Tang, B. L. (2012). Membrane trafficking components in cytokinesis. *Cellular Physiology and Biochemistry: International Journal of Experimental Cellular Physiology, Biochemistry, and Pharmacology, 30*, 1097–1108.

Tates, A. D. (1971). *Cytodifferentiation during spermatogenesis in* Drosophila melanogaster*: An electron microscope study* (Ph. D. thesis). Leiden: Rijksuniversiteit.

Xu, H., Brill, J. A., Hsien, J., McBride, R., Boulianne, G. L., & Trimble, W. S. (2002). Syntaxin 5 is required for cytokinesis and spermatid differentiation in *Drosophila*. *Developmental Biology, 251*, 294–306.

Zack, T. I., Schumacher, S. E., Carter, S. L., Cherniack, A. D., Saksena, G., Tabak, B., … Beroukhim, R. (2013). Pan-cancer patterns of somatic copy number alteration. *Nature Genetics, 45*, 1134–1140.

CHAPTER

7

Monitoring complete and incomplete abscission in the germ line stem cell lineage of *Drosophila* ovaries

J. Mathieu[*,§], **J.-R. Huynh**[*,§,1]

[*]Institut Curie, Paris, France

[§]CNRS UMR3215, Inserm U934, Paris, France

[1]Corresponding author: E-mail: Jean-Rene.Huynh@curie.fr

CHAPTER OUTLINE

Abstract

In most species, cytokinesis is blocked in germ cells during at least some stage of their development. Abscission is difficult to assess directly in germ cells which are located in internal organs. Here, we described several indirect and direct methods to monitor the completion of abscission in *Drosophila* germ line cells. These methods are based on the

Methods in Cell Biology, Volume 137, ISSN 0091-679X, http://dx.doi.org/10.1016/bs.mcb.2016.03.033

observation that cells still connected by some cytoplasm share some degree of synchronization of their cell cycle. This synchrony can be detected on fixed tissue (Section 1.1), including using EdU incorporation to label S-phase (Section 1.2). Mitotic synchrony can also be observed using short-term live imaging (Section 1.3). Finally, we describe how the completion of abscission can be monitored using photoactivatable markers diffusing or not between two cells (Section 1.4).

INTRODUCTION

Abscission is the last step of cytokinesis, which allows the physical separation of the two daughter cells. Its duration is variable in different cell types and can last from 2 to 4 h in vertebrate culture cells, to at least 12 h in *Drosophila* germ line stem cells (GSCs). Some cells do not even proceed to cytokinesis or abscission, as is the case for mammalian megakaryocytes, or germ line cysts. The formation of germ line cysts is an evolutionary conserved feature of germ line development, present from invertebrates to humans, in both males and females (Pepling, de Cuevas, & Spradling, 1999). It is part of the amplification step of germ line cell development, where progenitors divide several times with incomplete cytokinesis, forming cysts of interconnected and synchronous cells linked by cytoplasmic bridges called ring canals.

How the timing of abscission is regulated in different cell types and during development remains poorly understood. *Drosophila* female germ line cells provide a good genetic model system to study such regulation, as both complete and incomplete cytokinesis take place in neighboring cells in the germarium at the tip of each ovary (Huynh & St Johnston, 2004). Indeed, the GSC performs complete abscission, while its daughter cell, called cystoblast (CB), divides four times with incomplete cytokinesis, ultimately forming a cyst of 16 interconnected cells (Fig. 1A and B). The GSC cycle lasts 24 h, and after mitosis, GSCs remain connected to their daughter CB until the G2 phase of the next cycle (de Cuevas & Spradling, 1998). Monitoring exact abscission timing of the GSC remains technically challenging as the ovary must be imaged outside of the fly abdomen and kept alive during the whole abscission process, ie, for more than 10 h. Although methods to image live germaria for long periods are available, they do not allow proper optical resolution to follow subtle cellular events (Morris & Spradling, 2011). Therefore, only indirect methods are currently available to assess abscission duration.

An important characteristic of mitotic *Drosophila* germ cells is the presence of an organelle called the fusome. The fusome is made of endoplasmic reticulum-derived vesicles and required to orient germ cells divisions and to synchronize cell cycle in dividing cysts (Deng & Lin, 1997; Grieder, de Cuevas, & Spradling, 2000; Huynh, 2005; Snapp, Iida, Frescas, Lippincott-Schwartz, & Lilly, 2004). In GSCs, it adopts a characteristic shape for each stage of the cell cycle (Ables & Drummond-Barbosa, 2013; de Cuevas & Spradling, 1998). It has a spherical shape in late G2 and during mitosis. An additional plug of fusome forms within the ring canal that links the GSC and the CB shortly after mitotic exit in G1 phase. The anterior round fusome then starts to elongate toward the plug, which also adopts a bar

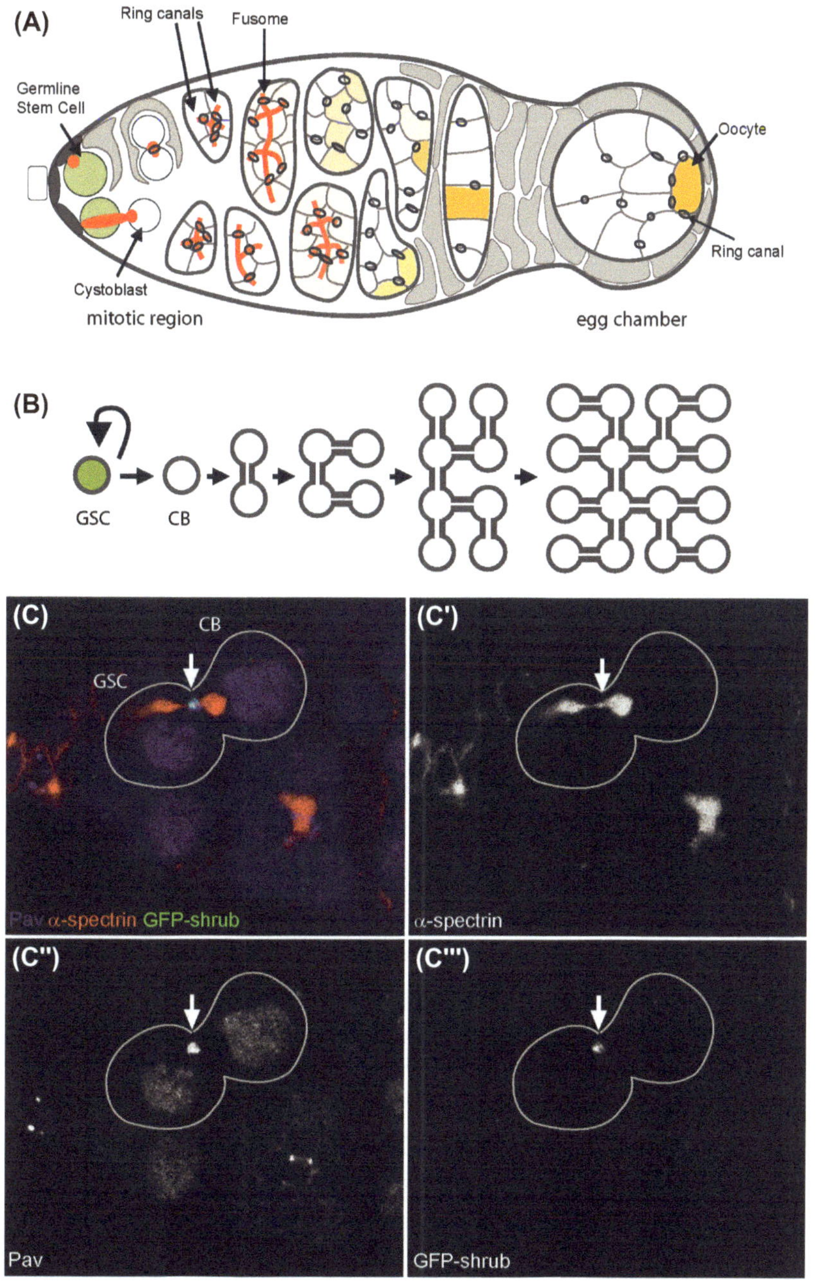

FIGURE 1

(A) Scheme of a germarium. Germ line stem cell (GSC)/cystoblast (CB) pair linked by a fusome (red) at the exclamation point stage is represented at the anterior. Ring canals (*gray ellipses*) linking cyst cells are maintained throughout oogenesis: they are found from the mitotic region to the encapsulated egg chambers. Anterior is on the left. (B) Schematic view of the pattern of divisions taking place in the mitotic region. A GSC (green) self-renews and generates one CB. The CB divides four times synchronously with incomplete cytokinesis, forming a 16-cell cyst. (C) GSC/CB pair linked by an exclamation point–shaped fusome, marked with α-spectrin (red in C, white in C′), Pavarotti (Pav, blue in C, white in C″) and GFP-Shrub (green in C, white in C‴). A bulge of the fusome is visible (*arrow*), which colocalizes with the midbody markers Pav and Shrub.

shape. During S-phase these two parts join, and ultimately fuse in early G2 phase. This unique fusome then adopts an exclamation point–like shape which is maintained until abscission occurs in early/mid G2. Then, each individualized GSCs and CBs inherit a spherical fusome (Fig. 2). The midbody, on the side of which abscission occurs, is formed at the end of cytokinesis between the daughter cells. In *Drosophila* females, its formation in GSC/CB pair has not been precisely

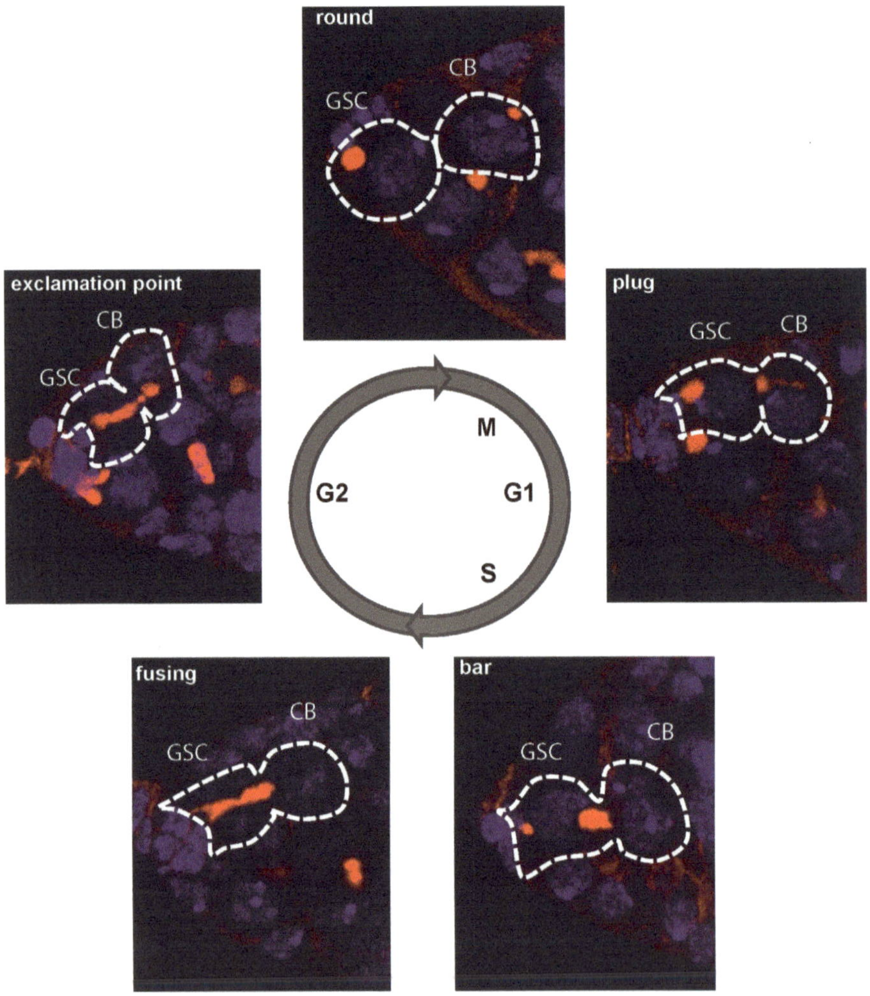

FIGURE 2

Correlation between the shape of the fusome and cell cycle stages in germ line stem cell (GSC)/cystoblast (CB) pairs (surrounded by *dotted lines*). The fusome is stained with α-spectrin (red) and the DNA with Hoechst (blue). Cell cycle stages are indicated in the *gray circle*.

described so far. However, markers of the midbody are detected on a tiny bulge on the fusome at the late exclamation point stage, before the fusome is split and shared by the two individualized cells (Eikenes et al., 2015; Matias, Mathieu, & Huynh, 2015). At this stage, the ring canal between the GSC and the CB has constricted and is not detected anymore (Fig. 1C). In dividing cysts, the fusome branches and passes though ring canals. After mitosis, a plug forms in each new ring canal and rotates to fuse with the original fusome (Huynh & St Johnston, 2004). No midbody forms in dividing cysts.

In vertebrate cells, membrane scission takes place first on one side of the midbody (Mierzwa & Gerlich, 2014). In mammalian cell culture, the process of abscission can be observed either directly, as a rupture of the cytoplasmic bridge linking the daughter cells by light videomicroscopy, or by the appearance of a gap in the microtubule bundle within the bridge by fluorescence live imaging (Dambournet et al., 2011; Steigemann et al., 2009). Abscission timing is generally quantified as the time spent from mitotic exit to abscission. As mentioned previously, monitoring abscission duration of the *Drosophila* GSC remains challenging as abscission takes place around 10 h after anaphase. Culture conditions used so far do not allow long-term live imaging and good optical resolution required to precisely quantify abscission timing. However, it is possible to image for shorter periods of time and observe, at low frequency, the rupture of the fusome between the GSC and the CB.

We and others have shown that abscission delay in *Drosophila* female GSC leads to the formation of stem cysts, where two or more GSC-like cells remain connected by a fusome, share the same cytoplasm, and cycle synchronously (Fig. 3A–C) (Eikenes et al., 2015; Mathieu et al., 2013; Matias et al., 2015; Sanchez et al., 2016). Conversely, precocious abscission in GSC/CB pairs can be detected as a loss of synchrony of their S-phase (Fig. 3D). In cysts, ectopic abscission can be visualized by a rupture of the fusome (Fig. 3E) (Mathieu et al., 2013).

Here we present methods and protocols that we and others have used to monitor abscission in *Drosophila* germ cells. The fusome is visualized in fixed or live samples, with or without cell cycle markers. This allows the detection of synchronously cycling cells within stem cysts, precocious abscission in GSC/CB pairs, or breaking fusomes if ectopic abscission takes place in cysts. Moreover, photoactivation experiments are described, which allow testing diffusion of a photoactivatable tubulin-PA-GFP between connected cells, hence the abscission status at a given time point.

1. METHODS

1.1 IMMUNOSTAINING OF *DROSOPHILA* GERMARIA

1.1.1 Dissection of ovarioles

Dissection is done using a dissection scope (medium magnification (×20)), with a black background, and illumination from the side to enhance contrast.

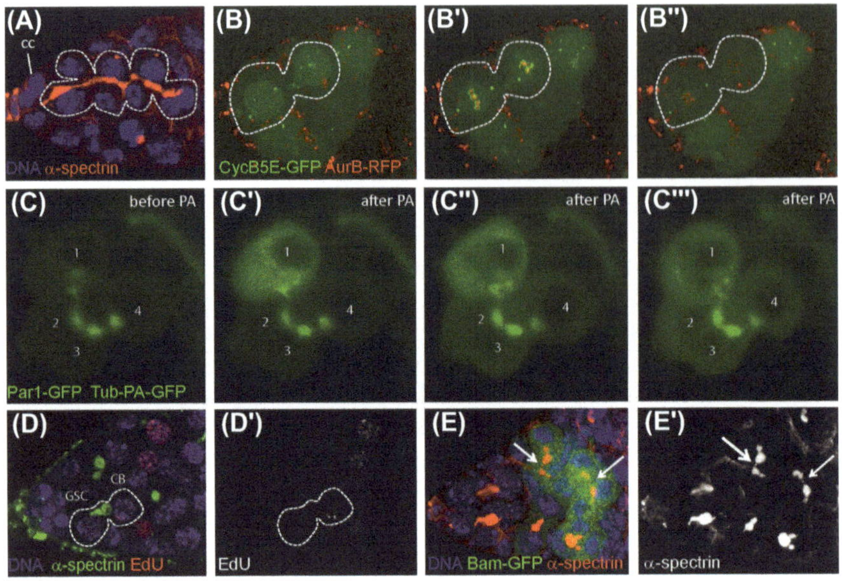

FIGURE 3

Examples of abscission defects observed in germ line stem cells (GSCs) (A—D) and in cysts (E). (A—C) Delayed abscission in GSCs is characterized by the formation of stem cysts (surrounded by a *dotted line* in A—B). Stem cysts are composed of connected cells anchored to the caps cells (cc), linked by a fusome (stained in red by α-spectrin in A). Stem cysts cycle synchronously (as seen by CyclinB5E-GFP, green, and Aurora-B-RFP, red, in three different time points of mitosis in B, B′, and B″) and share the same cytoplasm (as seen by the diffusion of Tub-PA-GFP from cell 1 to cell 4 after photoactivation of Tub-PA-GFP in C). In C, the fusome is marked by Par1-GFP. Photoactivation is done in cell 1 (C′), the activated Tub-PA-GFP diffuses throughout the cells of the stem cyst (C″), until the GFP level become similar in all cells (C‴). (D) Precocious abscission in the GSC/CB pair can be detected by an asymmetric incorporation of EdU (EdU red in D, white in D′; α-spectrin green in D, white in D′). In this example, EdU is only detected in the CB and not in the GSC. (E) Precocious abscission in cysts can be detected by a breaking fusome (stained by α-spectrin, red in E, white in E′) and appearance of a bulge of fusome at the breaking point (*arrows*). Here the cysts are marked by Bam-GFP, showing that the abscission takes place in properly determined and differentiating cysts.

1. Collect newly hatched females of the selected genotype and feed them overnight on fresh yeast paste.[1]
2. Submerge one female in PBS solution in a dissection well, and remove the ovaries from its abdomen using a pair of fine forceps such as Dumont 5 or 55.

[1]Phenotype penetrance may vary with the age of the flies.

3. Transfer the pair of ovaries into a depression slide (Dutscher # 020302).
4. Repeat this first round of dissection for 10 flies. Do not dissect more than 10–12 flies per staining, otherwise dissection lasts too long and ovaries become damaged.
5. Proceed to the fine dissection step of ovarioles using a pair of sharpened dissection needles: hold the posterior side of the ovary with one needle, while pick into an egg chamber at the vitellogenic stage with the other needle. Tease the ovarioles apart one by one by pulling them toward the anterior, until the muscle sheet appears broken, but be careful not to separate them completely from the ovaries. They should remain attached to the ovary by the terminal filaments. Otherwise, they will float and get lost during the staining and washing procedure (Fig. 4B, steps 1 and 2). Breaking the muscle sheet allows a better fixation and penetration of staining reagents.
6. Transfer ovarioles to a 0.5 mL tube using a glass Pasteur pipette. Do not use a plastic tip as the ovarioles tend to stick to it.

1.1.2 Immunostaining of ovarioles

7. Remove most of the PBS (do not let ovarioles dry out) and fix 17 min with 300–500 μL of 4% PFA (made from a 16% EM-grade solution and diluted in PBS) at room temperature (RT). Put the 0.5 mL tube on a benchrocker.
8. Place the dissection tube back on a rack and gently tip the tube once or twice to let the ovarioles sediment to the bottom of the tube for 3 more min (in total, the ovarioles are fixed for 20 min[2]).
9. Remove the PFA, and wash once with PBS-0.2% Triton, until ovaries sediment to the bottom of the tube.
10. Remove PBS-0.2% Triton, and perform three additional washes in PBS-0.2% Triton of 15 min each on the benchrocker.
11. Optional: Remove the PBS-0.2% Triton and block in PBS-0.2% Triton-5% BSA for 1 h.
12. Remove the PBS-0.2% Triton and incubate overnight at RT or 4°C with primary antibodies diluted in PBS-0.2% Triton on a benchrocker. To stain the fusome, use a mouse anti-α-spectrin monoclonal antibody at 1:1000 (DSHB). To detect mitotic cells, use a rabbit anti-Phospho-S10-Histone3 polyclonal antibody at 1:1000 (Upstate # 06-570). Other primary antibodies can be used as appropriate.
13. Put the tube back on a rack and gently tip the tube once or twice to let ovaries sediment to the bottom of the tube.
14. Remove the antibody dilution. Antibody dilution can be reused several times if kept at 4°C.
15. Wash three times 20 min with PBS-0.2% Triton on a benchrocker.

[2]Fixation timing is especially important when using flies expressing GFP-tagged protein. Longer fixation timing decreases GFP fluorescence.

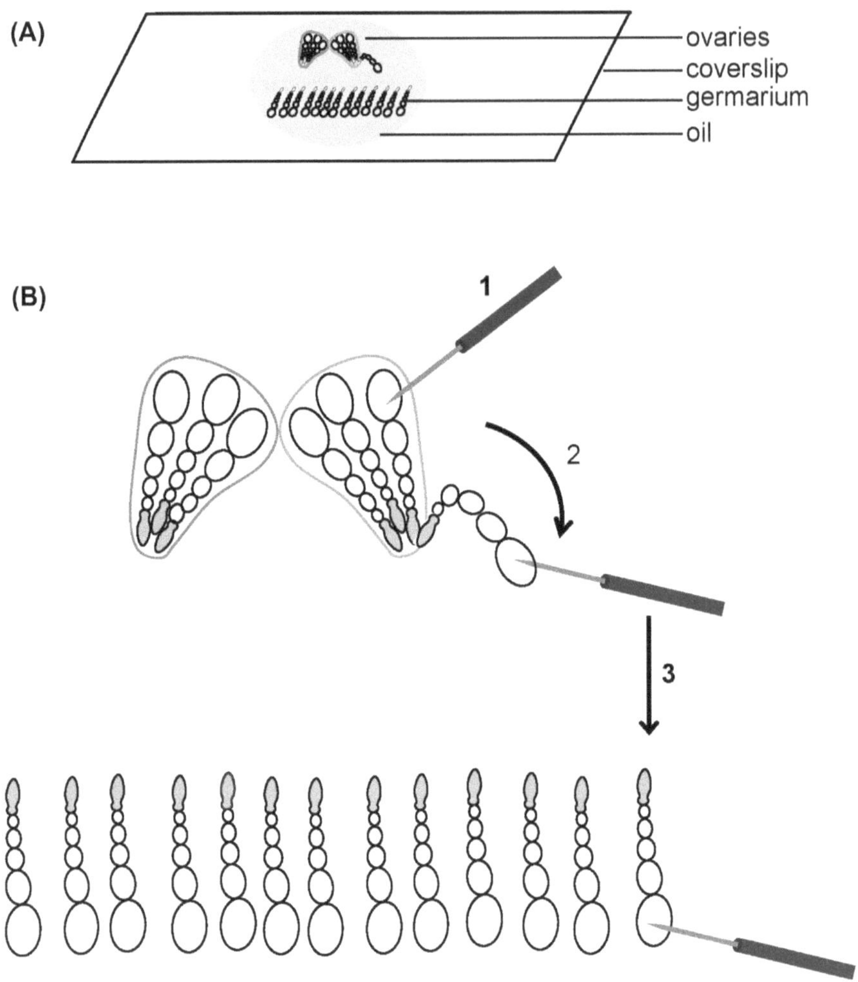

FIGURE 4

(A) Scheme representing the ovaries mounted for short-term live imaging. (B) Close up on the fine dissection steps: (1) Pick an egg chamber at the previtellogenic stage with a dissection needle (2) Tease the ovarioles apart by pulling them until they detaches from the ovary. For regular immunostaining, tease apart ovarioles but do not detach the ovarioles from ovaries. (3) Pull the ovarioles on the glass surface and align them for imaging.

16. Incubate 2 h at RT on a benchrocker with a dilution containing the secondary antibodies. Add Hoechst to visualize DNA in the secondary antibody dilution.
17. Let ovarioles sediment to the bottom of the tube and remove the antibody dilution.
18. Wash three times 20 min with PBS-0.2% Triton on a benchrocker.

19. Let ovarioles sediment to the bottom of the tube; remove the PBS-0.2% Triton.
20. Add a drop of mounting solution (50% glycerol in PBS, from Citifluor Ltd #AF1-25) and let ovarioles equilibrate overnight at 4°C or RT. During equilibration, ovaries move to the middle-bottom section of tubes.

1.1.3 Mounting ovarioles on a microscope slide

The mounting step is done using a dissection scope (high magnification (×50)), with a black background, and illumination from the side to enhance contrast.

21. Collect ovarioles in the mounting medium with a cut plastic tip, and spread them on a slide. Do not aspirate the ovarioles too high in the tip, otherwise it will be more difficult to deposit on a slide.
22. Put a drop (around 4 μL) of fresh mounting medium on a new slide.
23. Detach each ovariole from each ovary with a dissection needle. Cut each ovariole using two dissection needles as scissors, so that only young egg chambers (before any yolk is visible in the oocyte) remain attached to the germaria.
24. Transfer one by one the ovarioles from the first slide into the drop of mounting medium of the second slide using a dissection needle. A homemade hook at the tip of the needle may help getting the ovariole out of the mounting medium.
25. Once all ovarioles have been transferred, place a coverslip on the drop.
26. Wait until the medium has stopped spreading and seal with nail polish.
27. Image using a confocal microscope (Fig. 3A and E).

1.2 EdU INCORPORATION AND REVELATION

EdU is a nucleoside analog of thymidine and is incorporated into DNA during DNA synthesis. It can be detected with fluorescent markers and allow marking S-phase. If two cells have incorporated similar amount of EdU during a pulse, it indicates that their previous S-phase were synchronous.

The Click-iT Edu Imaging kit from Invitrogen is used for this experiment, reagents are from the kit (#C10084).

1.2.1 Edu incorporation

1. Perform a fine dissection of ovarioles as described in Section 1.1.1, steps 1–5 except that Schneider medium (Life Technologies # 21720024) supplemented with 10% fetal bovine serum (FBS) is used instead of PBS.
2. Transfer ovarioles into a 1.5 mL tube using a Pasteur pipette. Do not use a plastic tip as the ovarioles tend to stick to it.
3. Place the tube on a rack and gently tip the tube once or twice and let the ovarioles sediment to the bottom of the tube.
4. Remove as much medium as possible, being careful not to let the ovarioles dry.

5. Add 500 μL of the EdU labeling medium (10 μM of EdU in Schneider medium-10% FBS).
6. Incubate for 25 min with gentle agitation of a benchrocker at RT.
7. Put the tube on a rack and gently tip the tube once or twice and let the ovarioles sediment to the bottom of the tube for 5 min (the incorporation lasts 30 min in total).

1.2.2 Fixation and revelation

8. Remove as much EdU labeling solution as possible.
9. Add 500 μL of 4% PFA diluted in PBS and incubate for 12 min on a bench-rocker, and 3 min on a rack with ovarioles sedimenting (fixation lasts 15 min in total).
10. Remove the fixative solution and wash with 500 μL of PBS.
11. Remove the PBS and incubate 20 min with PBS-0.5% Triton.
11. Proceed to the revelation reaction exactly as described by the manufacturer's instructions. Use 500 μL for each reaction.
12. Wash the reaction solution twice in PBS.
13. Proceed to immunostaining as in Section 1.1.2, steps 12–20. Note that some primary antibodies, such as anti-α-spectrin, do not work as well after EdU revelation, and requires an increased concentration compared to simple immunostaining (Fig. 3D).

1.3 PREPARING OVARIOLES FOR SHORT-TERM LIVE IMAGING OF GERMARIA

The dissection is performed using a dissection scope (high magnification (×50)), with a black background, and illumination from the side to enhance contrast. Images can then be acquired with a regular confocal microscope, a spinning disk confocal or a widefield microscope with deconvolution. One female is dissected for each imaging session.

1. Collect newly hatched females of the selected genotype[3] and feed them overnight on fresh yeast paste. The phenotype penetrance may vary with the age of the flies.
2. Submerge a female in halocarbon oil (VWR International # VWRC24627.188) in a dissection well, and remove the ovaries from its abdomen using a pair of fine forceps such as Dumont 5 or 55. The oil in the dissection cup can be kept for several days for the first step of ovary dissection.
3. Put a drop of fresh halocarbon oil on a coverslip adapted to the stage of a spinning disc microscope.

[3]To label the fusome, Par1-GFP is expressed under the control of nanos-Gal4 driver and gives a strong signal, but be aware that Par1-GFP disappears from the fusome during mitosis.

4. Transfer the pair of ovaries into this drop.
5. Proceed to the fine dissection of the ovarioles using a pair of sharpened dissection needles: hold the posterior side of the ovary with one needle, while pick into an egg chamber with the other needle. Choose a previtellogenic egg chamber stage, because the yolk will spread and be fluorescent during acquisition. Tease the ovarioles apart one by one by pulling them, letting them adhere to the glass surface of the coverslip, which is necessary for good image quality. Align the germaria side by side, this will be helpful to find their positions once at the microscope (Fig. 4).
6. Place the coverslip on a controlled stage of an inverted spinning disc confocal microscope.
7. Move the objective to the bright remnants of ovaries, and make the focus on its side; then follow autofluorescent traces of the ovarioles until you find a first germarium. The others will be located adjacent to it if you have placed germaria as suggested during the fine dissection step.
8. Save the positions of the germaria you plan to image (multiposition mode of acquisition software).
9. Start imaging the germaria. Z-stacks of 1 μm give enough spatial resolution to capture branched or breaking fusome. Acquire images every 1 min, which allows the germaria to survive for more than 1 h (Fig. 3B).

1.4 PHOTOACTIVATION OF DIFFUSIBLE PROTEINS TO ASSESS ABSCISSION STATUS IN GERM CELLS

We have used photoactivatable GFP (PA-GFP) tag and photoconvertible Dendra2 tag (from GFP to RFP fluorescence). We have used a Zeiss 780 microscope equipped with a two-photon MaiTai (Spectra Physics) laser scalable to 609−1040 nm. In our hand, photoactivation of PA-GFP worked both with single photon and 2-photon activation. In contrast, Dendra2 photoconversion was only successful with single photon. We thus used two-photon activation for a better accuracy with PA-GFP. As rough examples, we used the following settings (which could vary a lot depending on laser usage): (1) single-photon activation: 405 nm, laser power 6%, scan speed 6, four iterations, pixel dwell time 6.3 μs; (2) two-photon activation: 820 nm, laser power 6%, scan speed 6, three iterations.

As a diffusible protein, we used tubulin fused to PA-GFP (Murray & Saint, 2007). Other proteins fused to Kaede (a GFP to RFP photoswitchable tag) can also be used and are available at Bloomington Stock center (#26161 and #38622).

1. Collect newly hatched females of selected genotypes and feed them overnight on fresh yeast paste. Tubulin-PA-GFP is used as a diffusible protein, which can pass from one cell to the other before abscission has occurred. The fusome is labeled by Par1-GFP. The expression of both transgenes in the germ cells is driven by nanos-Gal4.

2. Proceed to the fine dissection step of ovarioles in halocarbon oil as described in Section 1.3, steps 2–5.
3. Place the coverslip on the stage of an inverted confocal microscope equipped with a 405-nm (or 820 nm) laser.
4. Move the objective to the bright remnants of the ovaries, and make the focus on its side; then follow autofluorescent traces of ovarioles until you find a first germaria. Other germaria will be located adjacent to it. Move to the one located at the extreme left.
5. Select a germaria and a cell in which you want to perform the photoactivation to calibrate the experiment. Ideally, select a two-cell cyst where tubulin-PA-GFP is supposed to diffuse freely from one cell to the other.
6. Select a 7-μm-diameter circle within one of the two cells as the region of photoactivation.
7. Proceed to photoactivation of the tubulin-PA-GFP. The power of the laser should be strong enough to get a strong activation signal, but not too strong to avoid burning the cell[4]. As an indication, we used the following settings: 405 nm, laser 6%, four iterations, scan speed 6, pixel dwell time 6.3 μs.
8. After photoactivation is performed, single sections at the same focal plane are acquired every 10 s to follow the diffusion (or not) of the tubulin-PA-GFP. If the cytoplasms of the two cells are still connected, the cytoplasmic fluorescence intensity decreases in the photoactivated cell and increases in the sister cell. Conversely, if abscission has occurred between the cells, no diffusion will be detected. Fluorescence intensity is measured postacquisition with the Fiji software (Fig. 3C).

CONCLUSION

Here, we presented simple experiments that allow indirect determination of the abscission status between GSC and CB or within cyst cells during *Drosophila* female germ line development. Long-term live imaging of germaria at a good optical resolution would provide the ultimate tool to directly quantify abscission timing in female germ line cells; overcoming this challenge will be the next breakthrough in the field.

ACKNOWLEDGMENTS

We thank the Department of Genetics and Developmental Biology Imaging facility for technical support. This work is supported by the CNRS, Institut Curie, and grants from the ARC (PJA 20141202045) and ANR (AbsCyStem).

[4]The strength of the Gal4 driver may vary from cell to cell.

REFERENCES

Ables, E. T., & Drummond-Barbosa, D. (2013). Cyclin E controls *Drosophila* female germline stem cell maintenance independently of its role in proliferation by modulating responsiveness to niche signals. *Development, 140*(3), 530–540. http://dx.doi.org/10.1242/dev.088583.

Dambournet, D., Machicoane, M., Chesneau, L., Sachse, M., Rocancourt, M., El Marjou, A., … Echard, A. (2011). Rab35 GTPase and OCRL phosphatase remodel lipids and F-actin for successful cytokinesis. *Nature Cell Biology, 13*(8), 981–988. http://dx.doi.org/10.1038/ncb2279.

de Cuevas, M., & Spradling, A. (1998). Morphogenesis of the *Drosophila* fusome and its implications for oocyte specification. *Development. [development (cambridge, England)], 125*, 2781–2789.

Deng, W., & Lin, H. (1997). Spectrosomes and fusomes anchor mitotic spindles during asymmetric germ cell divisions and facilitate the formation of a polarized microtubule array for oocyte specification in *Drosophila. Developmental Biology, 189*, 79–94.

Eikenes, Å. H., Malerød, L., Christensen, A. L., Steen, C. B., Mathieu, J., Nezis, I. P., … Haglund, K. (2015). ALIX and ESCRT-III coordinately control cytokineticl abscission during germline stem cell division in vivo. *PLoS Genetics, 11*(1), e1004904. http://dx.doi.org/10.1371/journal.pgen.1004904.

Grieder, N. C., de Cuevas, M., & Spradling, A. (2000). The fusome organizes the microtubule network during oocyte differentiation in *Drosophila. Development, 127*, 4253–4264.

Huynh, J. R., & St Johnston, D. (2004). The origin of asymmetry: early polarisation of the *Drosophila* germline cyst and oocyte. *Current Biology, 14*(11), R438–R449. http://dx.doi.org/10.1016/j.cub.2004.05.040.

Huynh, J. R. (2005). The fusome as cell-cell channel of *Drosophila* ovarian cyst. In F. Baluska, D. Volkmann, & P. W. Barlow (Eds.), *Cell–cell channels* (pp. 217–235). Springer-Landes Biosciences.

Mathieu, J., Cauvin, C., Moch, C., Radford, S. J., Sampaio, P., Perdigoto, C. N., … Huynh, J. R. (2013). Aurora B and cyclin B have opposite effects on the timing of cytokinesis abscission in *Drosophila* germ cells and in vertebrate somatic cells. *Developmental Cell, 26*(3), 250–265. http://dx.doi.org/10.1016/j.devcel.2013.07.005.

Matias, N. R., Mathieu, J., & Huynh, J. R. (2015). Abscission is regulated by the ESCRT-III protein shrub in *Drosophila* germline stem cells. *PLoS Genetics, 11*(2), e1004653. http://dx.doi.org/10.1371/journal.pgen.1004653.

Mierzwa, B., & Gerlich, D. W. (2014). Cytokinetic abscission: molecular mechanisms and temporal control. *Developmental Cell, 31*(5), 525–538. http://dx.doi.org/10.1016/j.devcel.2014.11.006.

Morris, L. X., & Spradling, A. C. (2011). Long-term live imaging provides new insight into stem cell regulation and germline-soma coordination in the *Drosophila* ovary. *Development, 138*(11), 2207–2215. http://dx.doi.org/10.1242/dev.065508.

Murray, M. J., & Saint, R. (2007). Photoactivatable GFP resolves *Drosophila* mesoderm migration behaviour. *Development, 134*(22), 3975–3983. http://dx.doi.org/10.1242/dev.005389.

Pepling, M. E., de Cuevas, M., & Spradling, A. (1999). Germline cysts: a conserved phase of germ cell development? *Trends in Cell Biology, 9*, 259.

Sanchez, C. G., Teixeira, F. K., Czech, B., Preall, J. B., Zamparini, A. L., Seifert, J. R., … Lehmann, R. (2016). Regulation of ribosome biogenesis and protein synthesis controls germline stem cell differentiation. *Cell Stem Cell, 18*(2), 276–290. http://dx.doi.org/10.1016/j.stem.2015.11.004.

Snapp, E. L., Iida, T., Frescas, D., Lippincott-Schwartz, J., & Lilly, M. A. (2004). The fusome mediates intercellular endoplasmic reticulum connectivity in *Drosophila* ovarian cysts. *Molecular Biology of the Cell, 15*(10), 4512–4521. http://dx.doi.org/10.1091/mbc.E04-06-0475.

Steigemann, P., Wurzenberger, C., Schmitz, M. H., Held, M., Guizetti, J., Maar, S., & Gerlich, D. W. (2009). Aurora B-mediated abscission checkpoint protects against tetraploidization. *Cell, 136*(3), 473–484. http://dx.doi.org/10.1016/j.cell.2008.12.020.

CHAPTER

Incomplete cytokinesis/binucleation in mammals: the powerful system of hepatocytes

8

M. Fortier*[,§,¶], S. Celton-Morizur*[,§,¶,1,a], C. Desdouets*[,§,¶,1,a]

*Inserm, U1016, Paris, France

[§]CNRS, UMR 8104, Paris, France

[¶]Paris Descartes University, Paris, France

[1]Corresponding authors: E-mail: severine.morizur@inserm.fr; chantal.desdouets@inserm.fr

CHAPTER OUTLINE

[a]These authors contribute equally.

Methods in Cell Biology, Volume 137, ISSN 0091-679X, http://dx.doi.org/10.1016/bs.mcb.2016.04.006

Abstract

Polyploidy, the state of having greater than a diploid DNA content (tetraploid, octoploid, etc.) is a characteristic feature of mammalian hepatocytes and accompanies late fetal development and postnatal maturation of the liver. During the weaning period, diploid hepatocytes can engage either into normal cell division cycle giving rise to two diploid hepatocytes or follow a scheduled division program characterized by incomplete cytokinesis. In that case, diploid hepatocytes undergo mitosis, but do not form a contractile ring. Indeed, cleavage-plane specification is never established, because of the deficiencies of actin cytoskeleton reorganization. Furthermore, microtubules fail both to contact the cortex and to deliver their molecular signal, preventing localization and activation of RhoA. Therefore, cytokinesis aborts and a binucleate tetraploid liver cell is generated, which subsequently plays a pivotal role in liver progressive polyploidization. In this chapter, we describe detailed protocols to monitor hepatocyte proliferation and cytokinesis process by in situ and dynamic ex vivo approaches.

INTRODUCTION

Polyploidy, also known as whole-genome amplification refers to organisms containing more than two sets of chromosomes (eg, 4n, 8n) (Gentric & Desdouets, 2014; Pandit, Westendorp, & de Bruin, 2013). This phenomenon is commonly observed in plants, fungi, insects, fishes, and amphibians (Otto & Whitton, 2000). The additional set (or sets) of chromosomes may originate from the same individual ("autopolyploid") or from the hybridization of two different species ("allopolyploid"). Although diploid is the normal status for mammalian cells, various studies have demonstrated during the past decades a major implication of "diploid—polyploid conversion" during physiopathological processes in different tissues. Thus, cardiomyocytes, muscle cells, platelet progenitor megakaryocytes, and hepatocytes develop a certain degree of polyploidy during their normal life cycle. Importantly, polyploid cells can be also formed in response to stress and injury [eg, uterine smooth muscle cells during pregnancy (van der Heijden & James, 1975) and vascular smooth muscle cells during hypertension (Hixon et al., 2000)] and during the tumorigenesis process [eg, colon, esophagus, cervical cancer (Ganem, Storchova, & Pellman, 2007; Storchova & Pellman, 2004)]. The biological properties of polyploid cells remain obscure but, in certain tissues, it has been already discovered that polyploidization leads to genetic diversity and also confers specific cell function (Duncan et al., 2012; Gentric & Desdouets, 2014).

As previously described, polyploidy is a characteristic feature of mammalian hepatocytes. Polyploidy characterizes up to 90% of adult hepatocytes in mice (Duncan et al., 2010) and around 40% in humans (Toyoda et al., 2005). Polyploid hepatocytes are defined by the number of nuclei per cell (cellular ploidy) as well as the DNA content of each nucleus (nuclear ploidy) (Fig. 1) (Gentric & Desdouets, 2015). Polyploid hepatocytes can be tetraploid [eg, binucleate with two diploid (2n) nuclei or mononucleate with a single tetraploid (4n) nucleus] or octoploid [eg, binucleate with two tetraploid (4n) nuclei or mononucleate with a single octoploid (8n) nucleus]. Hepatocyte polyploidization is an age-dependent process. Hepatocytes are almost

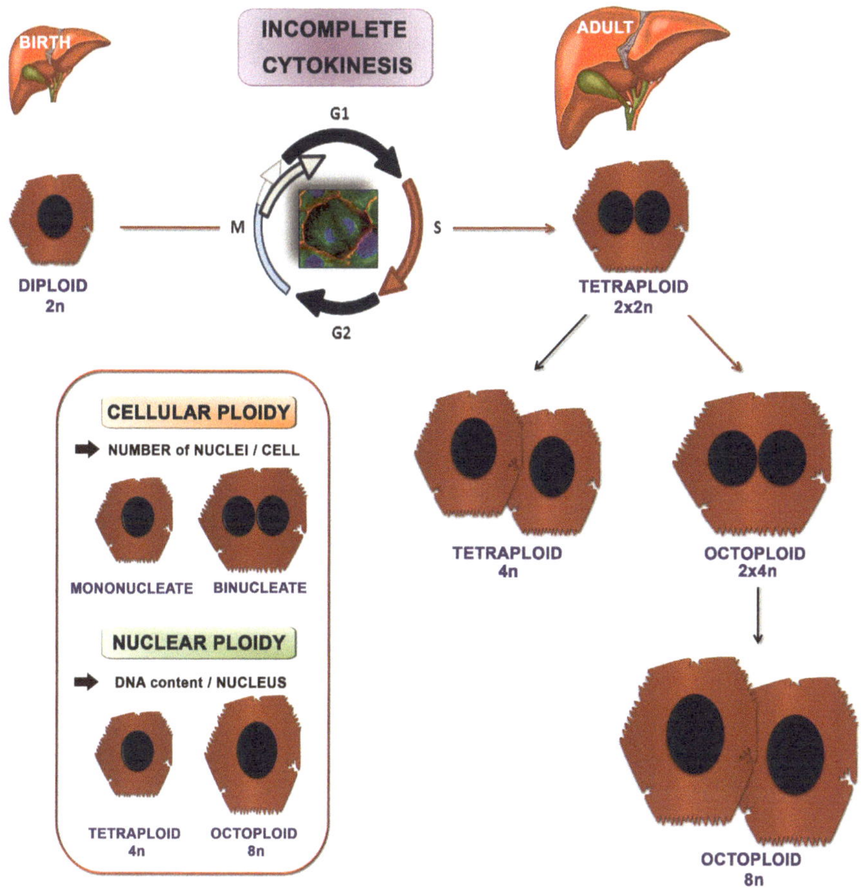

FIGURE 1 Polyploidization during postnatal liver growth.

At birth, hepatocytes are exclusively diploid (mononucleate 2n). During postnatal liver development after weaning, mononucleate 2n hepatocytes can engage either into normal cell division cycle giving rise to two mononucleate 2n hepatocytes or follow an adaptive cell cycle with cytokinesis failure (*red arrow*) giving rise to one binucleate tetraploid (2 × 2n) hepatocyte. By this process, progressive polyploidization takes place in the liver parenchyma and tetraploid and octoploid cell classes with one or two nuclei are formed. Polyploid hepatocytes are then characterized by the number of nuclei per cell (cellular ploidy) as well as the DNA content of each nucleus (nuclear ploidy).

diploid in young animals and polyploid contingents increase with age (Guidotti et al., 2003; Kudryavtsev, Kudryavtseva, Sakuta, & Stein, 1993). The primary mechanism for hepatocyte polyploidization is incomplete cytokinesis (Guidotti et al., 2003; Margall-Ducos, Morizur-Celton, Couton, Bregerie, & Desdouets, 2007). First, diploid hepatocytes (2n) accomplish adequate karyokinesis but fail to complete cytokinesis, generating binucleate tetraploid daughter cells (2 × 2n). If binucleate tetraploid hepatocyte reenters the cell cycle, this cell will either undergo cytokinesis, leading to the genesis of two mononucleate tetraploid hepatocytes (4n), or fail again, leading to the genesis of a binucleate octoploid hepatocyte (2 × 4n) (Fig. 1). We demonstrated in 2009 in rodents that the initiation of incomplete cytokinesis events and the subsequent appearance of binucleated tetraploid cells in the liver are strictly controlled by the suckling-to-weaning transition (Celton-Morizur, Merlen, Couton, Margall-Ducos, & Desdouets, 2009). Furthermore, we showed that the disruption of cytokinesis program in hepatocytes is due to an absence of cytoskeleton reorganization: actin cytoskeleton is not reorganized into the division plane in anaphase–telophase transition, then impairing cell elongation. In concert, microtubules fail to contact the cortex and therefore, molecular signals, essential for furrow induction (eg, Aurora B and polo-like kinase 1), are not correctly delivered. Consequently, activation of the RhoA GTPase (the major cytokinesis orchestrator) to the central cortex is impaired, leading to the genesis of binucleate progeny (Margall-Ducos et al., 2007).

Here, we present different methods to assess hepatocyte proliferation and cytokinesis process both in vivo by in situ immunostaining and ex vivo by using primary culture of hepatocytes, which is a very powerful tool to investigate ex vivo the process of tetraploidization.

1. EXPERIMENTAL PROCEDURES

1.1 IN SITU IMMUNOSTAINING

1.1.1 Liver harvesting

- *Materials*
  - Rodents (C57BL/6NRj mice—RjHan:Wistar rats from Janvier Labs)
  - 1X phosphate buffered saline (PBS)
  - Formalin (Carlo Erba reagents, 415691)
  - 70% Ethanol
  - Tissue-Tek VIP (Sakura)

Mice or rats are housed under standard light/dark conditions and receive pelleted food (weaning) and water ad libitum. Animals are weaned at 19 days after birth.

Animals are killed by cervical dislocation (mice) or by lethal injection of anesthetic (rats). Their livers are harvested and incubated in 4% phosphate-buffered formalin during 12 h. They are washed once in 70% ethanol before to be embedded in paraffin via Tissue -Tek VIP.

1.1.2 Liver section

- *Materials*
  - Microtome
  - Superfrost Ultra Plus Adhesion Slides (ThermoScientific)

- Water bath
- 55°C Incubator

- *Methods*

1. Section paraffin blocks at a thickness of 10 μm on a microtome and float on a 37°C water bath containing distilled water.
2. Transfer the sections onto a Superfrost Ultra Plus Adhesion slide. Slides dry overnight in a 55°C incubator. They can then be used immediately or stored at room temperature for future experiments. *However, we recommend not to use the slides more than 1 week after the sectioning.*
3. Deparaffinize sections in xylene 3 × 5 min.
4. Hydrate with 100% ethanol (2 × 5 minutes) and transfer once through 90% (1 × 5 min) and then 80% alcohol (1 × 5 minutes). Finally, rinse in distilled water for 5−10 min. Keep the slides in the water until ready to perform antigen retrieval. *From this step, slides should not be allowed to dry as it could induce nonspecific antibody binding and high background staining.*

1.1.3 Cellular ploidy analysis (Fig. 2)

To investigate hepatocyte cellular ploidy, colabeling of nuclear (Hoechst) and plasma membrane (β-catenin) which allows to discriminate mononucleate and binucleate hepatocytes are performed.

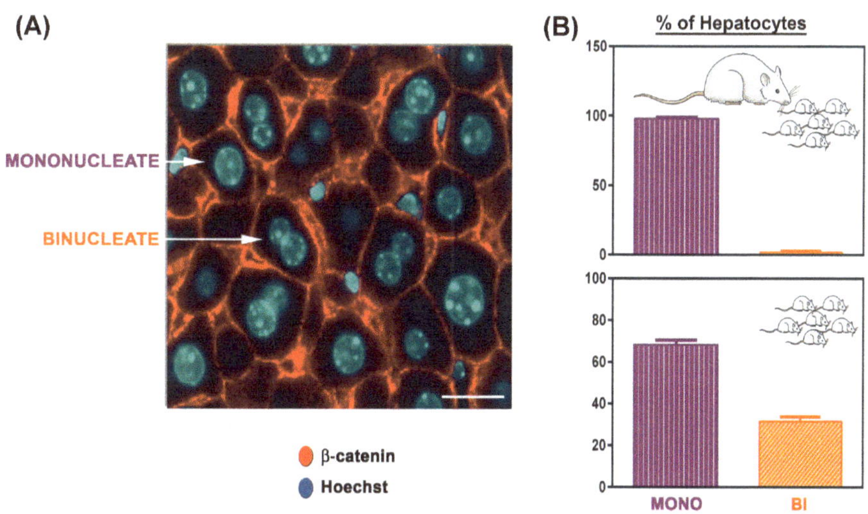

FIGURE 2 Cellular ploidy analysis.

(A) Images of liver sections after double staining with β-catenin (plasma membrane labeling, *red*) and Hoechst (nucleus, *blue*) to discriminate between mononucleate and binucleate hepatocytes. Scale bar: 10 μm. (B) Illustration of the percentage of mononucleate and binucleate hepatocytes observed in the liver 19 days suckling rats (upper panel) and 23 days weaned rats (lower panel).

- *Materials*
  - Pressure cooker (Eurobio)
  - Epitope retrieval solution: Tris 0.1M/0.1% Tween-20, pH 9.4
  - Hydrophobic pen
  - Antibody diluent, Dako REAL (Dako, S2022)
  - 1X PBS
  - 1X phosphate buffered saline tween (PBS-T): PBS with 0.1% Tween-20
  - Mouse anti−β-catenin antibody (BD Transduction Laboratories, 610154)
  - Goat anti−mouse IgG (H + L) secondary antibody, Alexa Fluor 594 conjugate (Molecular Probes, A11032)
  - Hoechst 33258 at1 μg/mL (Sigma, B2883)
  - Fluorescent mounting medium (Dako, S3023).
- *Methods*

Labeling	β-Catenin
Blocking solution	2% Goat serum In Dako REAL 1 h at room temperature
Primary antibody incubation	Anti−β-catenin antibody 1/200 in Dako REAL 1 h at 37°C
Secondary antibody incubation	Anti−mouse Alexa Fluor 594 1/500 in Dako REAL 1 h at 37°C

1. Transfer the slides in epitope retrieval solution and place them in a pressure cooker for 15 min at 95°C. At the end of the run, wait 20 min and then transfer slides in a bath of distilled water, during 10 min.
2. Place the slides in a bath of 1X PBS solution for 2 min.
3. Delimitate an interest zone using hydrophobic pen on the slides.
4. Add blocking solution on the delimited zone during the recommended time.
5. Aspirate the blocking solution and add β-catenin primary antibody solution.
6. Rinse with 1X PBS-T thrice.
7. Add secondary antibody solution.
8. Rinse with 1X PBS-T thrice.
9. Incubate in Hoechst solution (diluted at 1/500 in 1X PBS-T) during 20 min to counterstain nuclei.
10. One wash with 1X PBS-T.
11. Add fluorescent mounting medium and a glass coverslip. Seal with nail polish.
    - *Data analysis*

Z-axis stacks were collected using a piezoelectric device mounted at the base of a 20× magnification, on a Olympus BX63F microscope and an ORCA-Flash4.0 LT C11440-42U Hamamatsu camera controlled by Metamorph software (Molecular Device). A total of 40 planes (0.2-μm slice) are captured and compiled as single two-dimensional projections using ImageJ software. All images are imported into Adobe Photoshop CS for contrast manipulation and figure assembly.

For diploid and tetraploid hepatocyte counting experiments, at least 15 randomly chosen fields of liver tissue should be imaged (more than 2500 cells analyzed).

Fig. 2 shows an example of in situ β-catenin/Hoechst staining (Panel A) and the result of cellular ploidy analysis between suckling and weaned rats (Panel B).

1.1.4 Cytokinesis process analysis (Fig. 3)

- *Materials*
  - Pressure cooker
  - Epitope retrieval solution: Tris 0.1 M/0.1% Tween-20, pH 9.4
  - Hydrophobic pen
  - Antibody diluent, Dako REAL
  - 1X PBS
  - 1X PBS-T: PBS with 0.1% Tween-20
  - Mouse anti−β-tubulin Tub 2.1 antibody (Sigma, T4026)
  - Goat anti−rabbit IgG (H + L) secondary antibody, Alexa Fluor 488 conjugate (Molecular Probes, A11034)
  - Hoechst 33258 at 1 μg mL
  - Fluorescent mounting medium
- *Methods*

Labeling	β-Tubulin
Blocking solution	2% Goat serum In Dako REAL 1 h at room temperature
Primary antibody incubation	Anti−β-tubulin antibody 1/200 in Dako REAL 1 h at 37°C
Secondary antibody incubation	Anti−mouse Alexa Fluor 488 1/500 in Dako REAL 1 h at 37°C

Proceed in the same manner as described in Section 1.1.3.

Notes

This staining could also be coupled with β-catenin staining. In that case, we recommend to first use anti−β-tubulin antibody and its specific secondary antibody and then anti−β-catenin antibody and its secondary antibody.

- *Data analysis*

Images are taken using a Nikon Statif Eclipse E600 microscope with 40× magnification, 1.4-0.7 NA PL-APO objectives, a DXM1200 cooled charge-coupled-device (CCD) camera (Nikon), and ACT-1 (Universal Imaging). At least 50 cytokinesis events per animal should be randomly imaged.

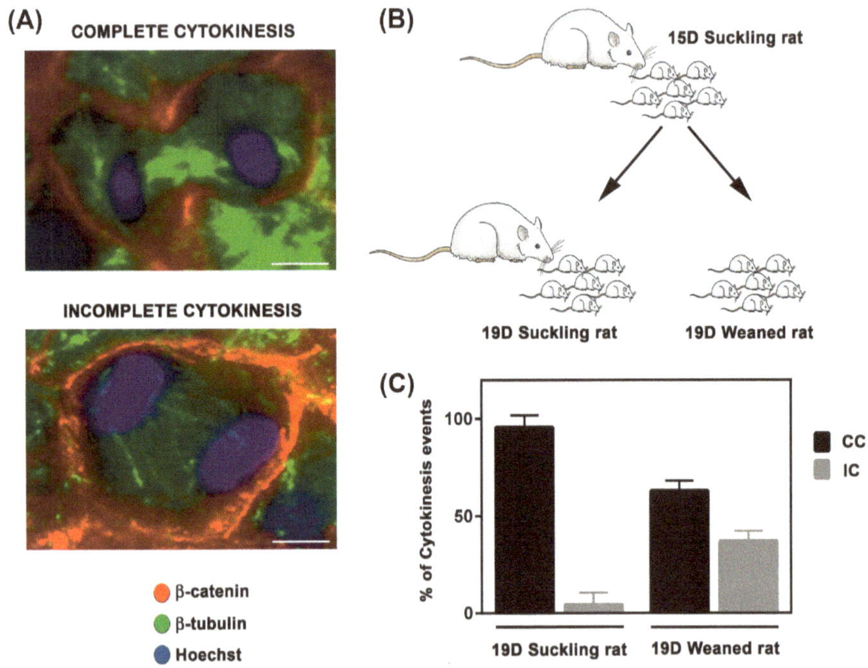

FIGURE 3 Analyzing cytokinesis failure events during postnatal liver growth.

(A) Immunostaining of liver sections to detect cytokinesis figures: β-catenin (plasma membrane labeling, *red*), β-tubulin (microtubule labeling, *green*) and condensed chromatin staining (Hoechst, *blue*). Upper panel illustrates normal cytokinesis process with ingression of cellular membrane and formation of a central midbody. Lower panel illustrates incomplete cytokinesis process with the absence of furrow ingression and a round cellular shape. Scale bar: 5 μm. (B and C) Illustration of the percentage of complete cytokinesis (CC) and incomplete cytokinesis (IC) observed in suckling (aged 19 days) and in weaned rats (aged 19 days, weaned at 15 days).

Adapted from Celton-Morizur, S., Merlen, G., Couton, D., Margall-Ducos, G., & Desdouets, C. (2009). The insulin/Akt pathway controls a specific cell division program that leads to generation of binucleated tetraploid liver cells in rodents. The Journal of Clinical Investigation, 119*(7), 1880–1887.*

Fig. 3 shows examples of in situ β-tubulin/β-catenin/Hoechst staining (Fig. 3A) and the percentage of complete/incomplete cytokinesis events (Fig. 3B) between suckling and weaned rat.

1.2 PRIMARY CULTURE OF HEPATOCYTES

Primary culture, usually isolated from whole liver (animal/human) by a two-step process involving collagenase digestion, is commonly used to investigate proliferative status and cell cycle progression of hepatocytes. This ex vivo system provides a great tool to better assess the role of hepatocytes in pathophysiological processes of the liver.

1.2.1 Solution preparation

Solution	Product	Reference	Final Concentration
Coating solution	Collagen from calf skin 1 mg/mL	Sigma C8919	0.4 mg/mL
	Glacial acetic acid 100%	–	0.1%
	H_2O	Deionized/ sterilized water	
Perfusion solution	NaCl	Sigma S5886	137 mM
	KCl	Sigma P5405	2.68 mM
	Na_2HPO_4	Sigma S5136	232 μM
	4-(2-Hydroxyethyl)-1-piperazinee-thanesulfonic acid	Sigma H6147	10 mM
	Ethylenedia-minetetraacetic acid	Sigma E3889	0.5 mM
	H_20	Deionized/ sterilized water	
	pH	–	7.65
Liver digest solution	Collagenase from *Clostridium histolyticum*	Sigma C5138	0.05%
	$CaCl_2.2H_2O$	Sigma C7902	10 mM
	Perfusion solution pH 7.65	-	
Hepatocyte culture medium	William's E medium, GlutaMAX supplement	Gibco 32551-020	
	Penicilin/streptomycin	Gibco 15140-122	100 U/mL
	Fungizone antimycotic B	Gibco 15290-026	100 μg/mL
	Insulin	Sigma I0516	4 μg/mL
	Bovin serum albumin 7.5%	Sigma A8412	0.1%
	Dexamethasone	Sigma D2915	25 nM
Mitogenic medium	Epidermal growth factor	Sigma E4127	50 ng/mL
	Pyruvate sodium	Sigma P4562	20 mM
	Hepatocyte culture medium	-	

1.2.2 Liver perfusion (Fig. 4)

- *Materials*
  - Anesthetic
  - Ethanol 70%
  - Peristaltic pump
  - Water bath at 37°C
  - Dissecting instruments: scissors, tweezers

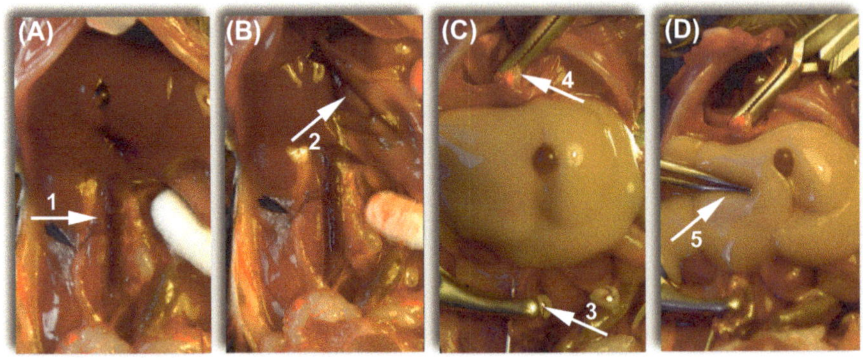

FIGURE 4 Key steps of liver perfusion.

(A) Opening of the abdomen and cleaning of cava vein (*arrow 1*). (B) Organ localization before insertion of the needle with the highlighting of portal vein (*arrow 2*). (C) White and inflated liver during washing step. Two clamps are placed, one at needle and cava vein (*arrow 3*) and the second at the sus-hepatic vein (*arrow 4*). (D) Digestion step: tweezer leaving a mark on liver (*arrow 5*).

- Clamps (2)
- Surgical infusor (B.BRAUN Intrafix Primeline I.S. Spinlock, 4064009)
- Needle (B.BRAUN Epicranien Venofix AAIG 0,5 G25, 4056370)
- Compresses and Q-tip
- Perfusion solution
- Digestion solution
- Hepatocyte culture medium
- 10-cm dish

- *Methods*

Per animal, 50 mL of 4-(2-hydroxyethyl)-1-piperazineethanesulfonic acid (HEPES) solution, 25 mL of digestion solution, and 50 mL of hepatocyte culture medium at 37°C are required.

1. Install perfusion system with surgical infusor, needle, and peristaltic pump. Prime the circuit with HEPES solution at 3.5 mL/min flow rate (mouse) or 5 mL/min (rat).
2. Anesthetize the animal.
3. Open the animal from the lower abdomen to the rib cage.
4. Shift the intestine on the right of the animal to highlight the inferior cava vein and portal vein.
5. Remove gently the conjunctive tissue of the inferior cava vein with Q-tip.
6. Turn off the peristaltic pump.
7. Insert needle into the inferior cava vein and clamp the vein and the needle.
8. Open the rib cage and clamp the sus-hepatic vein.

9. Cut the portal vein.
10. Turn on the peristaltic pump to deliver the perfusion solution inside the liver. The blood leaves progressively the hepatic tissue and the chelating agent ethylenediaminetetraacetic acid in the perfusion solution progressively removes hepatic calcium disrupting cell−cell tight junctions.
11. Wash the liver for 10 min with the perfusion solution (the liver should be white and inflated).
12. Change the perfusion solution with the digestion solution. Collagenase will help to separate the hepatocytes from the liver stroma by digestion of the extracellular matrix. The liver continues to inflate, and then it spreads.
13. Stop the digestion (around 4 min). To appreciate liver digestion efficiency, press the liver with a tweezer, this should induce a mark like a scar.
14. Take the liver and place it in a 10 cm-diameter culture dish filled with the hepatocyte culture medium (prewarmed at 37°C).
15. Open the Glisson capsule with a tweezer and release hepatocytes in the medium by shaking the liver.

Notes

The digestion efficiency depends on the quality of the collagenase. So, before experiments, you need to evaluate the collagenase activity (digestion and cellular viability) by testing different batches.

1.2.3 Primary hepatocyte isolation

- *Materials*
  - Centrifuge
  - Sterile cell strainers: 70 and 40 μm diameter (BD Falcon)
  - Water bath at 37°C
  - Hepatocyte culture medium
  - Fetal calf serum (FCS, Gibco)
- *Methods*

Before manipulation, prewarm the hepatocyte culture medium and FCS in the water bath to obtain 37°C solutions.

1. Recover cells from the dish to a 50-mL tube by filtering them through a 70-μm cell strainer. This step helps to remove undigested tissue and connective tissue.
2. Centrifuge the 50-mL tube at 400 rpm during 3 min at room temperature.
3. Discard the supernatant (containing nonparenchymal cells) and gently add 10 mL of the hepatocyte culture medium to the pellet. Rotate slowly the tube manually to gently resuspend the pellet. Complete with 40 mL of hepatocyte culture medium.
4. Repeat steps 2 and 3 twice.
5. Remove the supernatant, and add between 10 and 20 mL (according to the size of the pellet) of hepatocyte culture medium supplemented with 10% FCS.
6. Filter the suspended cells through a 40-μm cell strainer.

7. Evaluate the cellular yield and viability with trypan blue exclusion assay.
   Notes

   In order to analyze cell cycle parameters, hepatocyte viability has to be ≥85%.

1.2.4 Cell plating and culture conditions

- *Materials*
  - 3.5 and 6 cm dishes (BD Falcon)
  - Round glass coverslip
  - Coating solution (collagen from calf skin, Sigma C8919)
  - 1X PBS
  - Water bath at 37°C
  - Hepatocyte culture medium at 37°C
  - FCS at 37°C
  - Epidermal growth factor (EGF) (Sigma, E4127), 50 ng/mL
  - Sodium pyruvate (Sigma, P4562), 20 mM
  - Trypan blue
  - Incubator at 37°C and 5% CO_2 atmosphere
- *Methods*

Dishes should first be coated with an extracellular matrix (collagen-coating solution) at 37°C for at least 30 min, washed with 1X PBS twice, and then dried. For the 3.5-cm dishes, place a round glass coverslip inside before coating. This coverslip will be used for videomicroscopy experiment (see Section 1.2.6.2).

Coating step will provide hepatocyte attachment and maintenance of the hepatic phenotype. Consequently, hepatocyte primary culture in that context can be considered as a 2D culture.

1. Display:
   a. 600,000 (mouse) to 800,000 (rat) isolated hepatocytes in 2.5 mL of hepatocyte culture medium +10% FCS, in 6-cm dishes.
   b. 250,000 (mouse) to 300,000 (rat) isolated hepatocytes in 1.5 mL of hepatocyte culture medium +10% FCS, in 3.5-cm dishes.

   To evaluate plating quality, mortality and cell aggregates should be taken into account.
2. Place plates in 37°C incubator with 5% CO_2.
3. After complete adhesion of the hepatocytes (about 4 h), wash the dishes with fresh serum-free hepatocyte culture medium twice, to remove dead cells.
4. Finally, place plates in 37°C incubator with 5% CO_2 in culture medium overnight.
5. Change medium every morning (*First point: T12*) with mitogenic culture medium (supplemented with EGF and pyruvate).

1.2.5 Analysis of proliferative status (Fig. 5)

After collagenase perfusion, hepatocytes exit quiescence and enter into the G1 phase to the late G1 restriction checkpoint (Loyer et al., 1996). Mitogenic treatment with EGF and pyruvate allows hepatocytes to bypass this checkpoint and to progress synchronously into S phase. At 36–48 h of plating, a peak in phase S is reached and a maximal mitotic index is observed at 60 h (See Fig. 5A).

1.2.5.1 Cell fixation

- *Materials*
  - 1X PBS
  - -20°C Methanol
  - -20°C Ethanol 75%–acetic acid 25%
- *Methods*

Labeling	Fixation Solution	Timing for Fixation
Bromodeoxyuridine	Ethanol 75%–acetic acid 25%	20 min/–20°C
Phospho histone H3	Methanol	6 min/–20°C

1. Place dishes on ice.
2. Aspirate the medium and add 1X PBS.
3. Rinse a second time with 1X PBS.
4. Add 2 mL of cold fixer according to the envisaged labeling.
5. Place dishes at –20°C during indicated time.
6. Discard the fixator and rinse thrice with 1X PBS, and finally display 2 mL of 1X PBS. Fixated dishes could be used immediately or be stored at 4°C for 1 week.

1.2.5.2 Immunostaining

- *Materials*
  - Hydrophobic pen
  - 1X PBS
  - 1X PBS-T
  - 5-Bromo-2′-deoxy-uridine labeling and detection kit I (Roche, 11296736001) containing:
    - Bromodeoxyuridine (BrdU) labeling medium (1000X)
    - Anti-BrdU antibody
    - Anti–mouse-Ig-fluorescein antibody
    - Buffer 3
  - Antibody diluent, Dako REAL
  - Rabbit anti-phospho histone H3 (anti-PHH3) antibody (Millipore, 06-570)

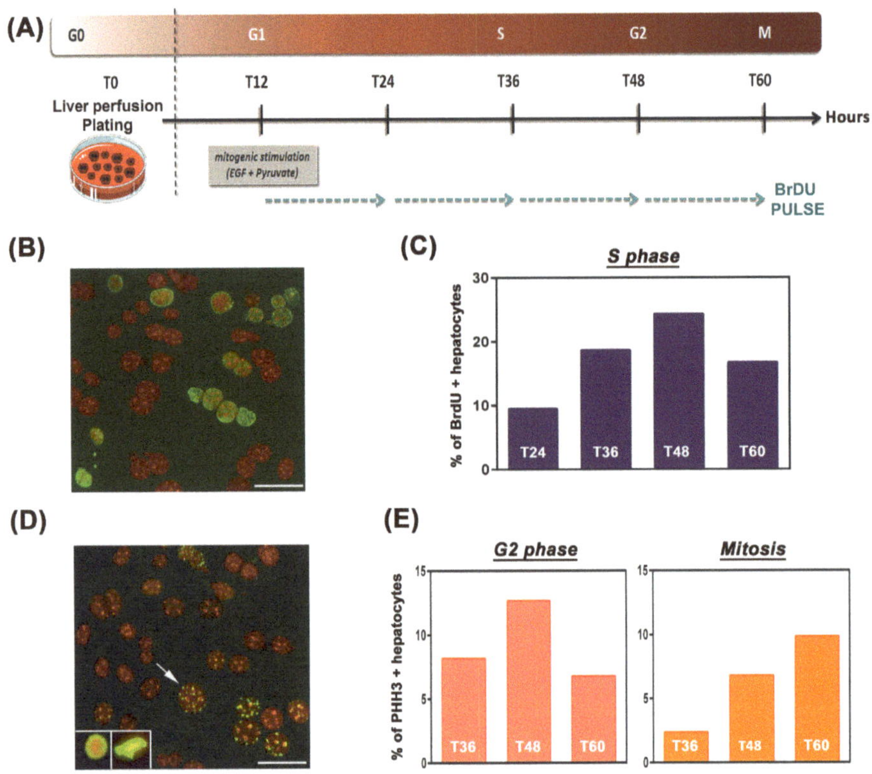

FIGURE 5 Primary culture and hepatocyte proliferation analysis.

(A) Schematic protocol of primary hepatocyte culture. Bromodeoxyuridine (BrdU) is incorporated in culture medium during time course kinetic (at T12 h, T24 h, T36 h, T48 h) for a duration of 12 h. T0: Liver perfusion and hepatocytes plating; T12 h: 12 h after plating. (B) Immunostaining of primary hepatocytes with anti-BrdU (*green*) and Hoechst (*red*) 36 h after plating. Scale bar: 15 μm. (C) Quantitative representation of BrdU-positive hepatocytes during the time course of the primary culture (T24 h, T36 h, T48 h, T60 h). The proliferative profile of hepatocytes describes a Gaussian curve. (D) Immunostaining of primary hepatocytes with anti-PHH3 (*green*) and Hoechst (*red*) 48 h after plating. Hepatocytes in G2 phase present dot labeling (*arrows*) and hepatocytes in mitosis have a dense staining (windows). Scale bar: 15 μm. (E) Quantitative representation of PHH3-positive G2 (left panel) and mitotic (right panel) hepatocytes during the time course of the primary culture (T36 h, T48 h, T60 h).

- Goat anti–rabbit IgG (H + L) secondary antibody, Alexa Fluor 488 conjugate (Molecular Probes, A11034)
- Hoechst 33258 (1 μg/mL)
- Fluorescent mounting medium

• *Methods*

To detect BrdU incorporation in hepatocyte nuclei, 1X BrdU labeling medium is added during culture at key timing (T12, T24, T36, and T48 h after plating; Fig. 5A) for a duration of 12 h.

Labeling	Bromodeoxyuridine (BrdU) (S Phase)	Phospho Histone H3 (PHH3) (G2/M Phases)
Blocking solution	2% Goat serum +10% bovine serum albumin Diluted in 1X TBS–0.1% Tween-20 30 min at room temperature	2% Goat serum Diluted in Dako REAL 1 h at room temperature
Primary antibody incubation	Anti-BrdU antibody Diluted at 1/50 in Buffer 3 30 min at 37°C	Anti-PHH3 antibody Diluted at 1/200 in Dako REAL Overnight at 4°C
Secondary antibody incubation	Anti–mouse-Ig-fluorescein antibody Diluted at 1/50 in Buffer 3 30 min at 37°C	Anti–rabbit Alexa Fluor 488 Diluted at 1/500 in Dako REAL 1 h at 37°C

1. Delimit a rounded area in the fixated dish using a hydrophobic pen.
2. Add blocking solution on the delimited zone during the recommended time.
3. Aspirate the blocking solution and add appropriate primary antibody solution.
4. Rinse with 1X PBS-T thrice.
5. Add appropriate secondary antibody solution.
6. Rinse with 1X PBS-T thrice.
7. Incubate in Hoechst solution (diluted at 1/500 in 1X PBS-T) to counterstain nuclei.
8. One wash with 1X PBS-T.
9. Add fluorescent mounting medium and a glass coverslip. Seal with nail polish.

- *Data Analysis*

Fluorescent images are taken as described in Section 1.1.4. To appreciate hepatocyte proliferation, BrdU- and PPH3-positive hepatocytes are counted during the time course kinetic.

Fig. 5 illustrates an example of the global proliferative status of hepatocytes isolated from the liver of 1-month-old mouse.

1.2.6 Analysis of mitosis and cytokinesis progression (Figs. 6 and 7)

In mammals, cytokinesis is a fundamental process that requires complex changes in cell shape and a profound reorganization of the cellular cytoskeleton (D'Avino, Giansanti, & Petronczki, 2015). Astral microtubules emanating from the spindle poles and central spindle microtubules between separating chromosomes are essential for the activation of equatorial cortical regions (D'Avino, Savoian, & Glover, 2005). Indeed, numerous signaling proteins (MKLP1, MgcRacGAP, the

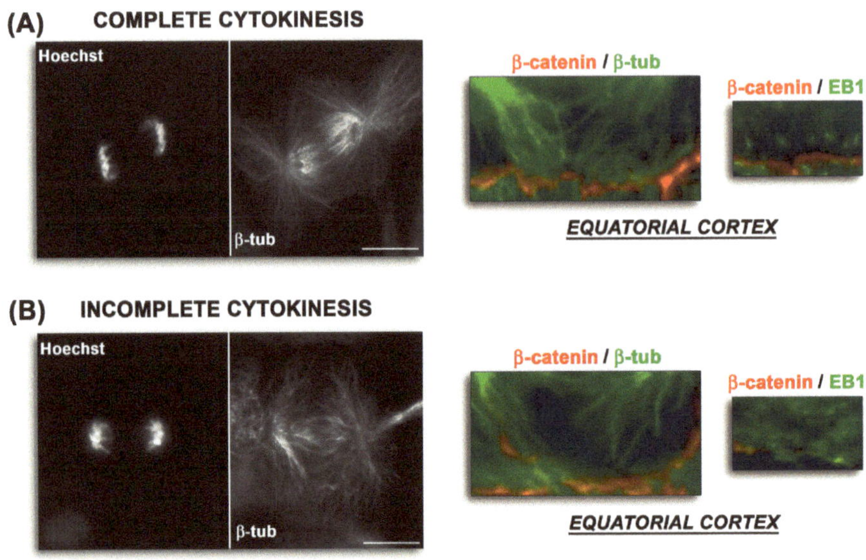

FIGURE 6 Analysis of microtubules dynamic in primary hepatocytes.

(A) During complete cytokinesis event, immunostaining for β-tubulin indicates the presence of antiparallel microtubules (the central spindle) and microtubules toward the cell cortex in the furrow and polar regions, during anaphase. At the central cortex, astral microtubules are correctly anchored and EB1 localization to the microtubules tips is weakly observed. Scale bar: 10 μm. (B) During incomplete cytokinesis event, hepatocytes present a disrupted central spindle and astral equatorial microtubules, as well as reduced astral polar microtubules. As astral microtubules fail to contact the equatorial cortex, EB1 is still present on microtubule tips. Scale bar: 10 μm.

Adapted from Margall-Ducos, G., Morizur-Celton, S., Couton, D., Bregerie, O., & Desdouets, C. (2007). Liver tetraploidization is controlled by a new process of incomplete cytokinesis. Journal of Cell Science, 120*(20), 3633–3639.*

passenger protein Aurora B, and the Rho GTPase exchange factor ECT2) are involved in relaying information from the spindle to the cortex and ultimately anchor RhoA, the master regulator of contractile ring formation (Eggert, Mitchison, & Field, 2006). Activation of RhoA GTPase at the central cellular cortex promotes rapid remodeling of the cortical actomyosin cytoskeleton (Bement, Benink, & von Dassow, 2005) (Yuce, Piekny, & Glotzer, 2005) (Kamijo et al., 2006) (Mierzwa & Gerlich, 2014) (Jordan & Canman, 2012).

To define more precisely molecular mechanisms regulating incomplete cytokinesis process in hepatocytes, immunostaining of these different actors can be performed in primary cultured hepatocytes.

1.2.6.1 Analysis of microtubules dynamic

Astral and central spindle microtubules are essential to deliver spatially restricted signals to the cortex and consequently to promote cleavage plane specification

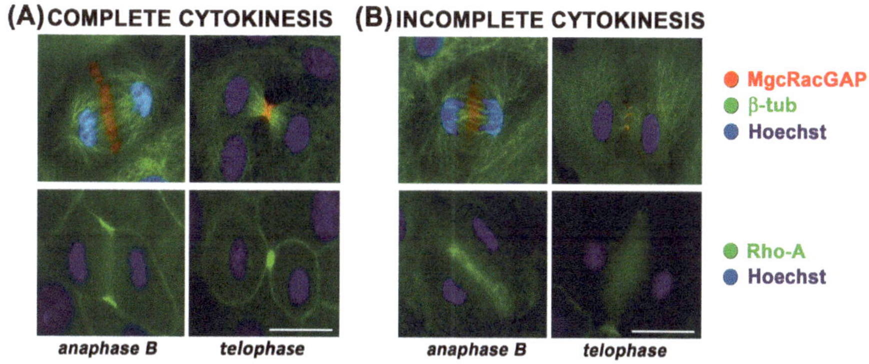

FIGURE 7 Immunostaining of cytokinesis players in primary hepatocytes.

(A) During complete cytokinesis process (left panel), MgcRacGAP (*red*, upper panel) is localized during anaphase both on the central spindle and on astral equatorial microtubules. During telophase, MgcRacGAP is concentrated at the midbody. RhoA (*green*, lower panel) accumulates at the equatorial cortex in early telophase in hepatocytes that complete cytokinesis; then it concentrates at the cleavage furrow and finally at the midbody. Scale: 10 μm. (B) During incomplete cytokinesis process (right panel), MgcRacGAP (*red*, upper panel) is observed on the remaining interdigitating microtubules in anaphase B and telophase but is never localized on unattached astral equatorial microtubules. Furthermore, RhoA (*green*, lower panel) does not correctly localize at the equatorial cortex, in which astral microtubules concentration is decreased. Scale: 10 μm.

Adapted from Margall-Ducos, G., Morizur-Celton, S., Couton, D., Bregerie, O., & Desdouets, C. (2007). Liver tetraploidization is controlled by a new process of incomplete cytokinesis. Journal of Cell Science, 120*(20), 3633–3639.*

(D'Avino et al., 2005). During normal cytokinesis, astral microtubules become stabilized upon contacting the cortex (Burgess & Chang, 2005). In that case, EB1, a microtubule tips protein only associated with elongating microtubules (Strickland, Wen, Gundersen, & Burgess, 2005), is rarely observed in this cortical region.

Then, detailed analysis of the microtubule network represents a very dynamical way to investigate the cytokinesis process in hepatocytes.

- *Material*
  - Hydrophobic pen
  - Antibody diluent, Dako REAL
  - 1X PBS
  - 1X PBS-T: PBS with 0.1% Tween-20
  - Mouse anti–β-catenin antibody (BD Transduction Laboratories, 610154)
  - Mouse anti–β-tubulin antibody (Sigma, T4026)
  - Mouse anti–EB1 antibody (BD Transduction Laboratories, 610535)

- Goat anti–mouse IgG (H + L) secondary antibody, Alexa Fluor 594 conjugate (Molecular Probes, A11032)
- Goat anti–rabbit IgG (H + L) secondary antibody, Alexa Fluor 488 conjugate (Molecular Probes, A11034)
- Hoechst 33258 (1 μg/mL)
- Fluorescent mounting medium

- *Methods*

Labeling	Fixation Solution	Timing for Fixation	Primary Antibody Dilution	Secondary Antibody Dilution
B-Catenin	Methanol	5 min/−20°C	1/200	Mouse Alexa Fluor 594 1/500
B-Tubulin	Methanol	5 min/−20°C	1/100	Mouse Alexa Fluor 488 1/500
EB1	Methanol	5 min/−20°C	1/100	Mouse Alexa Fluor 488 1/500

1. Fix cells according to the used primary antibody.
2. Remove fixator and add 1X PBS. At this step, dishes can be kept at 4°C until 1 week.
3. Remove 1X PBS and replace it by 1X PBS-T for 5 min.
4. Remove 1X PBS-T and delimitate an interest zone using Dako pen.
5. Add blocking solution (as described in Section 1.2.5.2) on the delimited zone during the recommended time.
6. Aspirate the blocking solution and add primary antibody solution.
7. Rinse with 1X PBS-T thrice.
8. Add secondary antibody solution.
9. Rinse with 1X PBS-T thrice.
10. Incubate in Hoechst solution (diluted at 1/500 in 1X PBS-T) for 20 min to counterstain nuclei.
11. Wash with 1X PBS-T.
12. Add fluorescent mounting medium and a glass coverslip. Seal with nail polish.

- *Data Analysis*

Single-plane fluorescent images are taken as described in Section 1.1.4.

Fig. 6 illustrates the differences observed in the microtubules network in hepatocytes entering normal cytokinesis and hepatocytes which fails cytokinesis.

Indeed, in hepatocytes that do not complete cytokinesis, microtubules network behavior is normal until early anaphase; thereafter, astral microtubules fail to contact the equatorial cortex and EB1 staining is still present on microtubule tips, illustrating the absence of cortical anchorage.

1.2.6.2 Immunostaining of cytokinesis key players

- *Material*
  - Hydrophobic pen

- Antibody diluent, Dako REAL
- 1X PBS
- 1X PBS-T
- Mouse anti–AIM-1 antibody (BD Transduction Laboratories, 611083)
- Mouse anti–MgcRacGAP antibody (from T. Kitamura, Institute of Medical Research Science, Tokyo, Japan)
- Mouse anti–P-myosin regulatory light chain antibody (Cell Signaling, 3675)
- Mouse anti-polo-like-kinase 1 antibody (Merck-Millipore, 05,844)
- Mouse anti-RhoA antibody (Santa Cruz, SC418)
- Goat anti–mouse IgG (H + L) secondary antibody, Alexa Fluor 594 conjugate
- Goat anti–rabbit IgG (H + L) secondary antibody, Alexa Fluor 488 conjugate
- Hoechst 33258
- Fluorescent mounting medium

• *Methods*

Labeling	Fixation Solution	Timing for Fixation	Primary Antibody Dilution	Secondary Antibody Dilution
AIM-1	Methanol	5 min/ −20°C	1/100	Mouse Alexa Fluor 594 1/500
MgcRacGAP	Methanol	5 min/ −20°C	1/100	Rabbit Alexa Fluor 594 1/500
P-myosin regulatory light chain	PFA 4%	15 min/4°C	1/50	Mouse Alexa Fluor 594 1/500
Polo-like-kinase 1	Methanol	5 min/ −20°C	1/200	Mouse Alexa Fluor 594 1/500
RhoA	Trichloroacetic acid 10%	15 min/4°C	1/100	Mouse Alexa Fluor 488 1/500

For the immunostaining of these different proteins, proceed as described in Section 1.2.6.1. Double immunostaining with β-tubulin antibody can also be performed to better characterize the behavior of these proteins.

• *Data Analysis*

Single-plane fluorescent images are taken as described in Section 1.1.4. For RhoA staining, *z*-axis stacks were collected using a piezoelectric device mounted at the base of a 63× magnification, on a Olympus BX63F microscope and an ORCA-Flash4.0 LT C11440-42U Hamamatsu camera controlled by Metamorph software (Molecular Device). A total of 20 planes (0.2 μm-thick slices) are captured for each cell and compiled as single two-dimensional projections using ImageJ software.

Fig. 7 illustrates that microtubule-dependent molecular signal such as MgcRac-GAP is not correctly delivered to the equatorial cortex; consequently, RhoA is not correctly anchored at the equatorial cortex, as normally observed during complete cytokinesis process.

1.2.6.3 Time-lapse videomicroscopy (Fig. 8)

Live-cell video microscopy of cultured primary hepatocytes represents another powerful tool to investigate cell behavior during cell cycle progression and in particular during mitosis/cytokinesis process. It has to be noticed that during primary culture of hepatocytes in our conditions, hepatocytes divided just once and a maximum mitotic index of 10% is reached at 60 h of culture (Guidotti et al., 2003).

- *Material*
    - 3.5-cm dishes (Falcon)
    - Coating solution
    - 1X PBS
- *Methods*

1. Glass coverslips with stimulated primary hepatocytes are mounted on the microscope after 48 h of culture after mitogenic treatment.

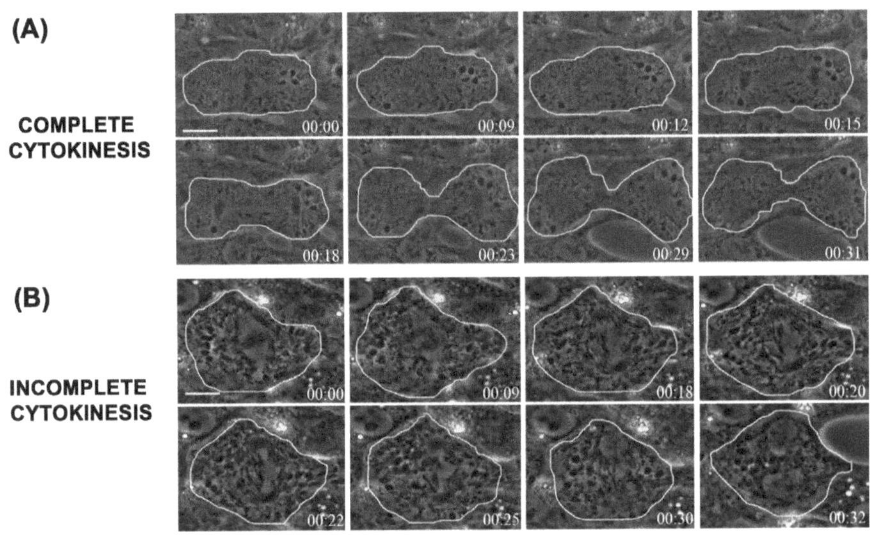

FIGURE 8 Monitoring cytokinesis process by time-lapse videomicroscopy.

Primary hepatocytes are isolated from the liver before (15-day-old rat) (A) and after weaning (25-day-old rat) (B) and cultured for 48 h in mitogenic (EGF/pyruvate) conditions. (A) During complete cytokinesis process, hepatocytes progress normally through anaphase: cells elongated preceding furrow formation and ingression. (B) During incomplete cytokinesis process, neither dynamic shape changes (cell elongation) nor furrow ingression is observed. Images are shown at selected time points (minutes). The outlines show cell shape.

Adapted from Margall-Ducos, G., Morizur-Celton, S., Couton, D., Bregerie, O., & Desdouets, C. (2007). Liver tetraploidization is controlled by a new process of incomplete cytokinesis. Journal of Cell Science, 120*(20), 3633–3639.*

2. Before imaging, culture medium is replaced by a fresh mitogenic medium to remove dead cells which can perturb videomicroscopy.
3. During imaging, hepatocytes were on a stage heated at 37°C under a 5% CO_2 atmosphere.
4. Cells are filmed every 90 s with a Leica DMIRBE using a 63× lens (numerical aperture, 1.4), a condenser (working distance, 23 mm; numerical aperture, 0.53), and a Pentamax cooled CCD camera (Popper Scientific) coupled to an electronic shutter.
5. Metamorph 7.1 is used for computer-based image acquisition and analysis of live-cell data.

- *Data analysis*

Time-lapse observations in Fig. 8 illustrates the differences observed between primary hepatocytes before (complete cytokinesis/Fig. 8A; Movie 1) and after weaning (incomplete cytokinesis/Fig. 8B; Movie 2).

CONCLUSION

The liver is a quite fascinating organ. In addition to its crucial role in a wide array of metabolic processes, it also harbors regenerative capabilities that favor its response to several types of damage [partial hepatectomy (Gerlyng et al., 1993; Sigal et al., 1999), metabolic overload (Madra, Styles, & Smith, 1995; Yamada et al., 1998), chemical overload (Martin, McGregor, Sansom, Gould, & Harrison, 2001; Melchiorri et al., 1993), oxidative damage (Gorla, Malhi, & Gupta, 2001; Nakatani, Inouye, & Mirochnitchenko, 1997), and viral infection (Toyoda et al., 2005; Toyoda, Kumada, Bregerie, Brechot, & Desdouets, 2006)]. In these circumstances, the liver polyploidy profile can also be modified. However, the cellular and molecular mechanisms regulating polyploidy in these different situations are not fully understood. The combination of in situ and dynamic ex vivo approaches could then provide great tools to better decipher hepatocyte polyploidization and cell cycle regulation during liver physiopathological process.

ACKNOWLEDGMENTS

The studies were supported by grants from the Institut National de la Santé et de la Recherche Médicale (INSERM), the Agence Nationale de la Recherche (ANR 2010 BLANC 1123 02), the Fondation ARC pour la recherche contre le cancer (Fondation ARC no. SFI20111203568), the INCa-CHC (2009-1-CHC-03-INSERM), the Ligue Comité de Paris (RS06/75-57), and the CORDDIM Région Ile de France.

SUPPLEMENTARY DATA

Supplementary data related to this article can be found online at http://dx.doi.org/10.1016/bs.mcb.2016.04.006.

REFERENCES

Bement, W. M., Benink, H. A., & von Dassow, G. (2005). A microtubule-dependent zone of active RhoA during cleavage plane specification. *Journal of Cell Biology, 170*(1), 91–101.

Burgess, D. R., & Chang, F. (2005). Site selection for the cleavage furrow at cytokinesis. *Trends in Cell Biology, 15*(3), 156–162.

Celton-Morizur, S., Merlen, G., Couton, D., Margall-Ducos, G., & Desdouets, C. (2009). The insulin/Akt pathway controls a specific cell division program that leads to generation of binucleated tetraploid liver cells in rodents. *The Journal of Clinical Investigation, 119*(7), 1880–1887.

D'Avino, P. P., Giansanti, M. G., & Petronczki, M. (2015). Cytokinesis in animal cells. *Cold Spring Harbor Perspectives in Biology, 7*(4), a015834.

D'Avino, P. P., Savoian, M. S., & Glover, D. M. (2005). Cleavage furrow formation and ingression during animal cytokinesis: a microtubule legacy. *Journal of Cell Science, 118*(Pt 8), 1549–1558.

Duncan, A. W., Hanlon Newell, A. E., Smith, L., Wilson, E. M., Olson, S. B., Thayer, M. J., … Grompe, M. (2012). Frequent aneuploidy among normal human hepatocytes. *Gastroenterology, 142*(1), 25–28.

Duncan, A. W., Taylor, M. H., Hickey, R. D., Hanlon Newell, A. E., Lenzi, M. L., Olson, S. B., … Grompe, M. (2010). The ploidy conveyor of mature hepatocytes as a source of genetic variation. *Nature, 467*(7316), 707–710.

Eggert, U. S., Mitchison, T. J., & Field, C. M. (2006). Animal cytokinesis: from parts list to mechanisms. *Annual Review of Biochemistry, 75*, 543–566.

Ganem, N. J., Storchova, Z., & Pellman, D. (2007). Tetraploidy, aneuploidy and cancer. *Current Opinion in Genetics & Development, 17*(2), 157–162.

Gentric, G., & Desdouets, C. (2014). Polyploidization in liver tissue. *The American Journal of Pathology, 184*(2), 322–331.

Gentric, G., & Desdouets, C. (2015). Liver polyploidy: Dr Jekyll or Mr Hide? *Oncotarget, 6*(11), 8430–8431.

Gerlyng, P., Abyholm, A., Grotmol, T., Erikstein, B., Huitfeldt, H. S., Stokke, T., & Seglen, P. O. (1993). Binucleation and polyploidization patterns in developmental and regenerative rat liver growth. *Cell Proliferation, 26*(6), 557–565.

Gorla, G. R., Malhi, H., & Gupta, S. (2001). Polyploidy associated with oxidative injury attenuates proliferative potential of cells. *Journal of Cell Science, 114*(Pt 16), 2943–2951.

Guidotti, J. E., Bregerie, O., Robert, A., Debey, P., Brechot, C., & Desdouets, C. (2003). Liver cell polyploidization: a pivotal role for binuclear hepatocytes. *The Journal of Biological Chemistry, 278*(21), 19095–19101.

Hixon, M. L., Obejero-Paz, C., Muro-Cacho, C., Wagner, M. W., Millie, E., Nagy, J., … Gualberto, A. (2000). Cks1 mediates vascular smooth muscle cell polyploidization. *The Journal of Biological Chemistry, 275*(51), 40434–40442.

Jordan, S. N., & Canman, J. C. (2012). Rho GTPases in animal cell cytokinesis: an occupation by the one percent. *Cytoskeleton (Hoboken), 69*(11), 919–930.

Kamijo, K., Ohara, N., Abe, M., Uchimura, T., Hosoya, H., Lee, J. S., & Miki, T. (2006). Dissecting the role of Rho-mediated signaling in contractile ring formation. *Molecular Biology of the Cell, 17*(1), 43–55.

Kudryavtsev, B. N., Kudryavtseva, M. V., Sakuta, G. A., & Stein, G. I. (1993). Human hepatocyte polyploidization kinetics in the course of life cycle. *Virchows Archiv B Cell Pathology Including Molecular Pathology, 64*(6), 387–393.

Loyer, P., Cariou, S., Glaise, D., Bilodeau, M., Baffet, G., & Guguen-Guillouzo, C. (1996). Growth factor dependence of progression through G1 and S phases of adult rat hepatocytes in vitro. Evidence of a mitogen restriction point in mid-late G1. *The Journal of Biological Chemistry, 271*(19), 11484–11492.

Madra, S., Styles, J., & Smith, A. G. (1995). Perturbation of hepatocyte nuclear populations induced by iron and polychlorinated biphenyls in C57BL/10ScSn mice during carcinogenesis. *Carcinogenesis, 16*(4), 719–727.

Margall-Ducos, G., Morizur-Celton, S., Couton, D., Bregerie, O., & Desdouets, C. (2007). Liver tetraploidization is controlled by a new process of incomplete cytokinesis. *Journal of Cell Science, 120*(20), 3633–3639.

Martin, N. C., McGregor, A. H., Sansom, N., Gould, S., & Harrison, D. J. (2001). Phenobarbitone-induced ploidy changes in liver occur independently of p53. *Toxicology Letters, 119*(2), 109–115.

Melchiorri, C., Chieco, P., Zedda, A. I., Coni, P., Ledda-Columbano, G. M., & Columbano, A. (1993). Ploidy and nuclearity of rat hepatocytes after compensatory regeneration or mitogen-induced liver growth. *Carcinogenesis, 14*(9), 1825–1830.

Mierzwa, B., & Gerlich, D. W. (2014). Cytokinetic abscission: molecular mechanisms and temporal control. *Developmental Cell, 31*(5), 525–538.

Nakatani, T., Inouye, M., & Mirochnitchenko, O. (1997). Overexpression of antioxidant enzymes in transgenic mice decreases cellular ploidy during liver regeneration. *Experimental Cell Research, 236*(1), 137–146.

Otto, S. P., & Whitton, J. (2000). Polyploid incidence and evolution. *Annual Review of Genetics, 34*, 401–437.

Pandit, S. K., Westendorp, B., & de Bruin, A. (2013). Physiological significance of polyploidization in mammalian cells. *Trends in Cell Biology, 23*(11), 556–566.

Sigal, S. H., Rajvanshi, P., Gorla, G. R., Sokhi, R. P., Saxena, R., Gebhard, D. R., Jr., … Gupta, S. (1999). Partial hepatectomy-induced polyploidy attenuates hepatocyte replication and activates cell aging events. *American Journal of Physiology, 276*(5 Pt 1), G1260–G1272.

Storchova, Z., & Pellman, D. (2004). From polyploidy to aneuploidy, genome instability and cancer. *Nature Reviews Molecular Cell Biology, 5*(1), 45–54.

Strickland, L. I., Wen, Y., Gundersen, G. G., & Burgess, D. R. (2005). Interaction between EB1 and p150glued is required for anaphase astral microtubule elongation and stimulation of cytokinesis. *Current Biology, 15*(24), 2249–2255.

Toyoda, H., Bregerie, O., Vallet, A., Nalpas, B., Pivert, G., Brechot, C., & Desdouets, C. (2005). Changes to hepatocyte ploidy and binuclearity profiles during human chronic viral hepatitis. *Gut, 54*(2), 297–302.

Toyoda, H., Kumada, T., Bregerie, O., Brechot, C., & Desdouets, C. (2006). Conserved balance of hepatocyte nuclear DNA content in mononuclear and binuclear hepatocyte populations during the course of chronic viral hepatitis. *World Journal of Gastroenterology, 12*(28), 4546–4548.

Van der Heijden, F. L., & James, J. (1975). Polyploidy in the human myometrium. *Zeitschrift für Mikroskopisch-Anatomische Forschung, 89*(1), 18–26.

Yamada, T., Sogawa, K., Kim, J. K., Izumi, K., Suzuki, Y., Muramatsu, Y., … Matsumoto, K. (1998). Increased polyploidy, delayed mitosis and reduced protein phosphatase-1 activity associated with excess copper in the Long Evans Cinnamon rat. *Research Communications in Molecular Pathology and Pharmacology, 99*(3), 283–304.

Yuce, O., Piekny, A., & Glotzer, M. (2005). An ECT2-centralspindlin complex regulates the localization and function of RhoA. *The Journal of Cell Biology, 170*(4), 571–582.

CHAPTER

9 Specific membrane dynamics during neural stem cell division

A. Ettinger*, Y. Kosodo[§], W.B. Huttner[¶,1]

*Institute of Epigenetics and Stem Cells, Munich, Germany

[§]Korea Brain Research Institute, Daegu, Korea

[¶]Max Planck Institute of Molecular Cell Biology and Genetics, Dresden, Germany

[1]Corresponding author: E-mail: huttner@mpi-cbg.de

CHAPTER OUTLINE

Methods in Cell Biology, Volume 137, ISSN 0091-679X, http://dx.doi.org/10.1016/bs.mcb.2016.05.007

Abstract

Neural stem and progenitor cells in the developing cerebral cortex, but also when grown in culture, display a range of distinct phenomena during cytokinesis. Cleavage furrow ingression in neural progenitor cells can bisect their basal processes and, later on, result in midbody formation at the apical surface. After abscission, these midbodies are released as membrane-bound particles into the extracellular space, in contrast to uptake and degradation of postabscission midbodies in other cell types. Whether these cellular dynamics are unique to neural stem cells, or more ubiquitously found, and what biological significance these processes have for cell differentiation or cell–cell communication, are open questions that require a combination of approaches. Here, we discuss techniques to study the specific membrane dynamics underlying the basal process splitting and postabscission midbody release in neural stem cells. We provide some basic concepts and protocols to isolate, enrich and stain released midbodies, and follow midbody dynamics over time. Moreover, we discuss techniques to prepare cortical sections for high-voltage electron microscopy to visualize the fine basal processes of progenitor cells.

INTRODUCTION

In final steps of telophase, phosphoregulation events along the prospective cleavage plane trigger constriction of the cell cortex which results in cleavage furrow ingression and midbody formation. Vesicle dynamics and membrane remodeling are key to cleavage furrow constriction, midbody formation, and necessary for abscission and daughter cell separation (Steigemann & Gerlich, 2009).

Cleavage furrow ingression and postabscission midbody fate may display unexpected dynamics: Neuroepithelial (NE) progenitor cells are highly polarized, with an apical membrane facing the lumen of the neural tube and a process extending to the basal lamina. The nuclei of these progenitor cells are translocated along

the apical–basal cell axis during the cell cycle. This movement of progenitor cell nuclei, called interkinetic nuclear migration, results in the multilayered, or pseudostratified, appearance of the neuroepithelium (Taverna & Huttner, 2010). Even though nuclei are transported basally in the G1-phase, mitosis occurs at the apical side. During neuronal progenitor cytokinesis, the cleavage furrow ingresses asymmetrically from the distant basal side. Especially during the early neurogenic stage in mammalian and fish brains, cleavage furrow ingression may bisect the basal process and eventually leave both daughter cells with an extension to the basal side (Kosodo et al., 2008). The asymmetric constriction also leads to midbodies being formed at the apical surface of NE cells. These midbodies can subsequently be released into the extracellular space after abscission is completed and accumulate in the lumen of the neural tube (Dubreuil, Marzesco, Corbeil, Huttner, & Wilsch-Brauninger, 2007; Marzesco et al., 2005). This release of postabscission midbodies has already been suggested from early electron microscopy studies on cultured bone marrow-derived cells (Mullins & Biesele, 1977) and also been observed in *Xenopus* neural plate divisions (Kieserman, Glotzer, & Wallingford, 2008) and in neural stem cells in culture (Ettinger et al., 2011) and may thus represent a second route of midbody elimination next to midbody uptake and degradation by autophagy (Pohl & Jentsch, 2009). As a nonexclusive alternative, midbody release may precede uptake and autophagy, as observations in some cell lines (Crowell, Gaffuri, Gayraud-Morel, Tajbakhsh, & Echard, 2014) and in *Caenorhabditis elegans* (Chai et al., 2012) suggest.

Apical–basal polarity and the outcome of neural progenitor cytokinesis is a critical determinant of cell fate specification in neuronal progenitor cells (Florio & Huttner, 2014). However, how midbody fate might direct neural differentiation is presently not clear, but studies in *C. elegans* and *Drosophila* at least indicate that midbody remnants may associate with specific cell types during differentiation in a stereotypical manner (Ou, Gentili, & Gönczy, 2014; Salzmann et al., 2014). Moreover, even though the underlying mechanisms are not well understood, positioning of midbody remnants in *C. elegans* embryos impacts axis formation (Singh & Pohl, 2014), and midbody fate may play a crucial role for cellular reprogramming and tumorigenicity (Kuo et al., 2011).

A combination of biochemical, immunofluorescence, live-cell time-lapse microscopy, and ultrastructural approaches is required to determine how the specific membrane dynamics associated with these distinct cell types control cell behavior, differentiation, and tissue-level organization. We discuss several methods that have been used by us and other labs to identify midbodies and follow their fate and discuss how techniques for isolation and purification of released midbodies can be upscaled for "-omics" type of analyses. Further, we provide some basic concepts for setting up long-term time-lapse microscopy to monitor membrane and midbody dynamics over time. Finally, we introduce contrasting techniques for analysis of basal processes of neuronal progenitor cells with high-voltage electron microscopy.

1. GENERAL CONSIDERATIONS

1.1 MARKERS AND PROBES

The first challenge to investigate membrane dynamics associated with cytokinesis is to identify the relevant structures and associated cytokinesis events. We and others have used a number of factors associated with central spindle microtubules, the cell cortex, intracellular vesicles, and the plasma membrane to identify the midzone, cleavage furrow, and midbody to determine dynamics and mechanisms of cleavage furrow ingression, midbody formation, abscission, and postabscission midbody fate. Many of the following proteins have important roles in cytokinesis and/or abscission, so care has to be taken when designing experiments that rely on (over) expression of a single or multiple factors and should, as a minimum, include controls for cell viability and cytokinesis progression. Lentivirus-mediated low-copy number integration (Stehbens, Pemble, Murrow, & Wittmann, 2012), CRISPR/Cas9 genome engineering techniques (Cong et al., 2013; Jinek et al., 2012), or bacterial artificial chromosome (BAC) vectors (Poser et al., 2008) may help to achieve expression levels that better reflect physiological levels and should be considered in the planning phase of live-cell studies. Potential markers for relevant structures are summarized in Table 1.

In addition to the use of markers to design immunolabeling or live cell probes, hallmarks of late cytokinesis may also be identified by structural or ultrastructural features that may aid in the characterization of spatial and temporal relationships. For example, the midbody can be identified in phase contrast transmitted light microscopy without further contrasting techniques and the morphology of spindle microtubules, cleavage furrow and vesicles at the midzone can aid to classify early, mid, and late phases of cytokinesis progression.

1.1.1 Central spindle and midzone

In telophase, two antiparallel microtubule arrays form the central spindle. The density of microtubules in the midzone may itself be used as a marker for the prospective cleavage plane and should be clearly distinguishable in immunofluorescence for tubulin and in electron microscopy. Of note, when using tubulin immunostaining to highlight midbodies, one typically observes a "gap" in staining at the zone of overlapping microtubules, which is thought to result from low accessibility in the dense and quasi crystalline arrangement of microtubules. In electron microscopy, however, the antiparallel microtubule bundles are seen to occupy the center of the midbody (Ettinger et al., 2011). Since filamentous structures like microtubules have birefringent properties, an alternative modality for visualization is polarized light microscopy. This method has successfully been applied to visualize the mitotic spindle in live cells without further contrasting techniques (Oldenbourg, Salmon, & Tran, 1998) and may be an attractive alternative in combination with multiple fluorophores or on its own in live-cell studies.

Table 1 Selection of Markers Used for Cytokinesis Dynamics and Midbody Formation

Structure	Marker/ Probe	Application	Example References
Central spindle and midzone	Tubulin	Immunoblotting, immunofluorescence	Carlton and Martin-Serrano (2007), Ettinger et al. (2011), Gai et al. (2011), Kieserman et al. (2008)
	Kif23	Immunoblotting, immunofluorescence, live-cell marker	Ettinger et al. (2011), Gromley et al. (2005), White and Glotzer (2012)
	MgcRacGAP	Immunofluorescence	Hirose et al. (2001)
	Cep55	Live-cell marker	Carlton and Martin-Serrano (2007), Lee et al. (2008)
Cortex and cleavage furrow	CRIK	Immunoblotting, immunofluorescence	Ettinger et al. (2011), Gai et al. (2011)
	Anillin	Immunoblotting, live-cell marker	Dubreuil et al. (2007), Gai et al. (2011)
	RhoA	Immunofluorescence	Gai et al. (2011)
Vesicles	ALIX	Immunofluorescence, live-cell marker	Carlton and Martin-Serrano (2007), Guizetti et al. (2011)
	CHMP-4b	Immunofluorescence	Guizetti et al. (2011)
	Tsg101	Live-cell marker	Carlton and Martin-Serrano (2007)
Plasma membrane	DiI	Immunofluorescence, live-cell probe	Kosodo et al. (2008)
	Annexin V	Immunofluorescence, live-cell probe	Arai et al. (2015), Chai et al. (2012)
	Prominin-1	Immunofluorescence, live-cell marker	Dubreuil et al. (2007), Marzesco et al. (2005)

Already prior to cleavage furrow ingression, a number of microtubule-associated proteins localize to the central spindle midzone. Among them, components of the centralspindlin complex have been widely used to determine midzone and identify the midbody after cleavage furrow ingression (Gromley et al., 2005; White & Glotzer, 2012). Centralspindlin consists of a heterotetramer of the proteins KIF23 (MKLP1, CHO1) (Kuriyama, Gustus, Terada, Uetake, & Matuliene, 2002) and RACGAP1 (MgcRacGAP1) (Hirose, Kawashima, Iwamoto, Nosaka, & Kitamura, 2001). We have used antibodies against KIF23 successfully in biochemical assays and immunofluorescence to detect inherited and released midbodies in cultured cells and also KIF23 as fusion protein with EGFP in live-cell

time-lapse microscopy. Of note, key for generation of stably expressing KIF23-EGFP neural stem cell lines was the use of a KIF23-EGFP BAC (Ettinger et al., 2011; Poser et al., 2008) with intact intronic and upstream regulatory regions to ensure proper expression levels and reduce adverse effects on cell divisions.

1.1.2 Cell cortex and cleavage furrow ingression

Activity of a number of kinases and guanine nucleotide exchange factors regulates contraction of the actomyosin cortex at the prospective cleavage plane. In several instances, kinases and regulatory factors are transported to the cleavage furrow through interactions with central spindle microtubules. To identify midbodies, we used Citron Rho-interacting kinase (CIT, CRIK (Di Cunto et al., 1998)), which localizes to the contractile ring, concentrates at the midbody, and remains associated with postabscission midbodies (Ettinger et al., 2011).

In addition to CRIK, actin-interacting proteins like Anillin (ANLN) and the small GTPase RhoA drive cleavage furrow ingression (Gai et al., 2011). They localize to the contractile ring and are attractive markers to highlight cleavage plane and furrow progression. As EGFP fusion protein, or as a marker in immunofluorescence, we used Anillin to highlight the ingression of the contractile ring through the basal process of mouse NE cells (Kosodo et al., 2008) and monitored release of postabscission midbodies from chicken spinal cord slice cultures (Dubreuil et al., 2007).

1.1.3 Vesicle fusion and abscission

Vesicle transport and membrane dynamics at the midbody are thought to trigger plasma membrane fission during abscission. Notably, members of the ESCRT (endosomal sorting required for transport) family localize to the midbody, potentially through interaction with the centrosome protein 55 (Cep55) (Carlton & Martin-Serrano, 2007; Lee, Elia, Ghirlando, Lippincott-Schwartz, & Hurley, 2008) where they are required for completion of cytokinesis and may also recruit microtubule-severing enzymes (Yang et al., 2008). The ESCRT-III family proteins ALIX (ALG-interacting protein X; also known as PDCD6IP, programmed cell death 6 interacting protein), Tsg101 (Tumor susceptibility gene 101), and CHMP-4B have been used in live-cell microscopy and immunofluorescence to determine mechanisms of cortical constriction and abscission at the midbody (Carlton & Martin-Serrano, 2007; Guizetti et al., 2011) and could be used to identify stages of midbody formation. Of note, they may also distinguish distinct zones of the midbody, in particular the central part versus the more extended "stalk" regions (Guizetti et al., 2011).

1.1.4 Plasma membrane and cytoplasm

In addition to regulatory and structural proteins, membrane components can lend themselves to follow cytokinesis progression. In its simplest form, labeling of the plasma membrane by lipophilic probes such as DiI allows to determine membrane

dynamics associated with cytokinesis. Using sparse labeling of NE cells in organotypical slice cultures, we were successful in live-cell recording of basal-to-apical cleavage furrow progression resulting in splitting of basal processes of NE cells (Kosodo et al., 2008).

In NE progenitor cells in the developing mouse embryo, the transmembrane protein Prominin-1 (CD-133) localizes to defined domains of the apical membrane. Midbodies formed at the apical side in final phases of progenitor divisions are thus decorated with Prominin-1, a feature that led to the discovery of midbody release from progenitor cells (Dubreuil et al., 2007; Marzesco et al., 2005).

Moreover, changes in the plasma membrane can be utilized to detect midbodies. Around the time of abscission, the inner-leaflet lipid phosphatidylserine (PS) gets concentrated at the midbody and switches to the outer leaflet where it can be detected with Annexin V-fluorescent probes by immunofluorescence or in live-cell microscopy (Arai et al., 2015; Chai et al., 2012). It will be interesting to explore how this feature of postabscission midbodies can be utilized to determine the timing of abscission and investigate how changes in membrane components facilitate membrane fusion during abscission.

1.2 ASSAY DESIGN

To determine structural and functional aspects of NE cytokinesis dynamics, midbody formation, and midbody release, we employed a combination of techniques to benefit from combining data from multiple angles. Each of the approaches comes with their own advantages and disadvantages.

1.2.1 Biochemical assays

For biochemical assays on released midbodies, especially in combination with mass spectrometry, getting enough starting material can be challenging. While some protocols can be upscaled easily, this is difficult to accomplish when sample material is scarce and hard to obtain, for instance in the case of collecting ventricular fluid from E11.5-E13.5 mouse embryos. Moreover, one has to keep in mind that these type of assays can only yield data about populations and can, for example, not differentiate between freshly released and "aged" midbodies.

For studying midbody release from cells in culture, we faced the difficulty how to compare differential centrifugation and immunoblotting data from different cell lines with different cell division parameters. In the end, we decided to plate the same number of cells and collect medium according to equivalent rounds of cell doubling rather than after a fixed time (Ettinger et al., 2011). The reasoning behind this was that the speed of cell division will be the main parameter influencing the generation, and potentially the release, of midbodies into the extracellular medium. There have been efforts toward modeling the relationship between midbody generation and removal (Crowell, Tinevez, & Echard, 2013), and it will be interesting to incorporate such models into future assay designs.

1.2.2 Immunofluorescence and live-cell microscopy

Immunofluorescence staining of fixed cells can help to obtain insight into the spatial relationships and mechanisms of cytokinesis and midbody release. A number of antibodies and fluorescent probes work well on paraformaldehyde- or methanol-fixed cells. Care has to be taken, however, to optimize fixation and staining conditions. For example, to investigate PS accumulation at the midbody, it is necessary to incubate live samples with an Annexin V-probe to label PS, fix with paraformaldehyde, and only use digitonin as a very mild permeabilization agent to not extract lipids in the staining procedure (Arai et al., 2015).

In combination with drug treatments and/or inhibition of gene expression by RNA interference (RNAi), immunofluorescence is also a powerful approach to investigate structural and functional connections. In RNAi experiments, it is key to verify gene depletion with (co)immunostaining, immunoblotting, or with quantitative PCR (qPCR), and include nontargeting controls to be able to exclude unspecific or off-target effects. Analysis of individual cells aids to determine cell-to-cell variations, but collection and analysis of a high number of microscopy images can become quite time consuming. For routine analysis, it may be worthwhile to develop automated image analysis approaches. Good open source software packages (for example, Cell Profiler (Carpenter et al., 2006), Fiji (Schindelin et al., 2012), or Icy (de Chaumont et al., 2012)) are available that allow to set up analysis workflows for segmentation and quantification of cells and subcellular structures. For automated analysis of inherited midbodies, we established a Cell Profiler analysis pipeline to reliably detect the number of cells and midbodies (Ettinger et al., 2011).

A big advantage of immunofluorescence is the detection of endogenous protein levels. However, even though one may be able to select individual time points for an immunofluorescence experiment, it is difficult to assess dynamic processes from individually fixed specimens. Also, fixation may destroy or change structures in unpredicted ways. Live-cell microscopy is a powerful approach to investigate spatial and temporal dynamics of cellular structures. In the last decade, a number of novel light microscopy methods expanded both the spatial and time scales at which biological processes can be observed. In addition, better detectors and the ever increasing palette of fluorescent proteins allow quantitative analysis of multiple fluorophores. Finally, computational approaches can help to collect more data from microscopy images in unbiased ways and develop better models of cell dynamics.

In live-cell microscopy, proteins of interest are typically expressed as fusion constructs with a fluorescent protein, for instance, EGFP. However, both the addition of fluorescent protein tags as well as changed expression levels by widely used viral promoters may interfere with function of the endogenous protein or have unpredictable effects on cell function. Genomic tagging with CRISPR/Cas9, stable shRNA depletion of endogenous transcripts, or low-expression promoters should be considered if endogenous level of the protein of interest is low. An important consideration for live-cell experiments is also the choice of microscope modality and optimization of acquisition parameters to minimize phototoxicity in a time-lapse experiment (see Section 4). In combination with automated microscopy and high-content analysis, both immunofluorescence and live-cell microscopy also become powerful techniques in high-throughput screens (Kittler et al., 2004; Neumann et al., 2010).

1.2.3 Electron microscopy analysis

Even though superresolution techniques are pushing the resolution of light microscopy beyond the diffraction limit (Huang, Bates, & Zhuang, 2009), electron microscopy is frequently still the method of choice to determine cellular ultrastructure. With new fixation techniques and better image reconstruction algorithms, it is possible to get to the three-dimensional structure of macromolecular assemblies inside cells. In particular, high pressure freezing and electron tomography techniques can resolve functional macromolecular complexes in unprecedented detail, and improved techniques have revolutionized structural biology (Nogales & Scheres, 2015).

2. ISOLATION AND DETECTION OF RELEASED POSTABSCISSION MIDBODIES

To enrich released postabscission midbodies, we have adapted centrifugation protocols that were originally devised for fractionation and purification of cellular organelles. These protocols have successfully been applied to enrich midbodies for immunoblotting and immunofluorescence from mouse embryonic ventricular fluid (Marzesco et al., 2005), cell culture supernatants (Ettinger et al., 2011), and for lipidome analysis of released midbodies (Arai et al., 2015). In these protocols, we apply either differential centrifugation, in which centrifugation time and speed are increased step by step to sequentially collect fractions of particles of decreasing size or density gradient centrifugation to separate particles by their size/density in continuous or discontinuous sucrose gradients.

2.1 CENTRIFUGATION BASICS

Separation of structures of specific size by centrifugation depends on the following parameters: Rotor geometry, revolution speed, medium density, and centrifugation time. Rotor geometry and revolution speed determine the relative centrifugal force (RCF) acting on the sample. Medium density is especially relevant in density gradient centrifugation protocols. To allow particles to move through a given unit of volume, it is moreover crucial to adjust centrifugation time. The connection between medium viscosity, particle size, and sedimentation velocity is essentially given by Stokes' law (Eq. (1)) (adapted from Graham & Rickwood, 1999):

$$F = 3\pi\mu dv = \frac{1}{6}\pi d^3 (\rho_p - \rho_f) g \tag{1}$$

where F is the frictional force, μ is viscosity coefficient of the liquid medium, d is particle diameter, v the velocity of sedimentation, ρ_p is the particle density, ρ_f is the density of the liquid medium, and g is the gravitational force. By rearranging this equation, the velocity of sedimentation can be stated as (Eq. (2)):

$$v = \frac{d^2(\rho_p - \rho_f)g}{18\mu} \tag{2}$$

From this equation, it is evident that particle diameter and gravitational force are the main parameters that determine sedimentation velocity when the medium is of uniform density. If particle density is smaller than the density of the surrounding medium, particles will float up. In equilibrium centrifugation protocols, this property is exploited to separate particles into zones of equivalent density by centrifugation over extended periods of time in a dense medium or gradient.

During centrifugation, the relevant force is the centrifugal force, which is expressed in relation to earth's gravity as RCF (Eq. (3)):

$$\mathrm{RCF} = \frac{r\omega^2}{g} = 11.18\, r\left(\frac{N}{1000}\right)^2 \tag{3}$$

where r is the radius of the rotor, ω is the angular velocity in radians, g is earth's gravitational acceleration, and N is the revolutions per minute (rpm). In practical terms, it is therefore always important to consider rotor geometry when designing or adapting centrifugation protocols. What if a specific rotor used in a published protocol is not available in the lab? A useful metric to estimate how to adjust centrifugation parameters is the k-factor. k-Factors are typically provided by centrifuge manufacturers for different types of centrifuge/rotor combinations and are calculated from the rotor geometry (Eq. (4)):

$$k = \frac{2.53 \cdot 10^5 \ln(r_{max}/r_{min})}{(N/1000)^2} \tag{4}$$

where r_{max}, r_{min} are maximum and minimum diameter of the rotor, and N is the centrifugation speed in rpm. k-Factors can be used to convert the time required to sediment particles in one type of rotor to another type of rotor (Eq. (5)).

$$\frac{T_1}{K_1} = \frac{T_2}{K_2} \tag{5}$$

where T_1 is the time in hours to pellet a particle in rotor 1 with k-factor K_1, and T_2 is the time to pellet a particle in rotor 2 with K_2. k-Factors are usually given for the maximum speed of the rotor but can be adjusted for lower speeds using the formula (Eq. (6)):

$$k_{adj} = \left(\frac{N_{max}}{N_{adj}}\right)^2 \tag{6}$$

Table 2 Rotors Used for Isolation of Released Midbodies

Rotor	Manufacturer	*k*-factor	N_{max} (rpm)	Centrifuge
F-45-24-11 (or similar)	Eppendorf	360[a]	13,200	Eppendorf 5427 R (or similar)
TLA-55	Beckman	66	55,000	Optima MAX ultracentrifuge
SW40 Ti	Beckman	137	40,000	Optima ultracentrifuge

[a] *Estimated from rotor diameter.*

where N_{max} is the maximum speed the rotor in rpm and N_{adj} is the lower speed in rpm at which the rotor is run. The rotors we used for isolation of released midbodies are summarized in Table 2.

2.2 DIFFERENTIAL AND DENSITY GRADIENT CENTRIFUGATION

2.2.1 Protocol A: differential centrifugation

For differential centrifugation of midbodies released from mouse embryonic NE cells or neural stem cells in culture, we started from a volume of 1.5 mL. For mouse embryos, we collected ventricular fluid with a glass capillary into phosphate buffered saline (PBS) containing protease inhibitors. For cultured neural stem cells, we collected 1.5 mL supernatant of cells cultured in a total of 2 mL medium in 6-well format. After centrifugation, pellets can be resuspended in buffer for subsequent immunofluorescence staining or lysis buffer for Western blotting. Collection of ventricular fluid of ~20 E13.5 embryos should yield enough material for downstream assays like immunoblotting or immunofluorescence but may require upscaling for "-omics" type of analyses. For cultured cells, we typically started with 100,000 cells in 6-well format in a total of 2 mL medium and collected supernatants after two rounds of cell doubling (47 h in the case of neural stem cells). The centrifugation scheme is summarized in Fig. 1A.

Materials:

- Dilution buffer: PBS, with protease inhibitors (Roche complete mini)
- Desktop centrifuge (Eppendorf)
- TLA-55 fixed angle rotor (Beckman Coulter)
- Ultracentrifuge (Optima MAX table top ultracentrifuge, Beckman Coulter)

All centrifugation steps are carried out at 4°C, and samples are kept on ice.

1. Transfer 1.5 mL of diluted ventricular fluid or 1.5 mL of conditioned cell culture medium into 1.5 mL centrifuge tubes. Spin at 300 g for 5 min to sediment cells and large particles (P1 pellet).

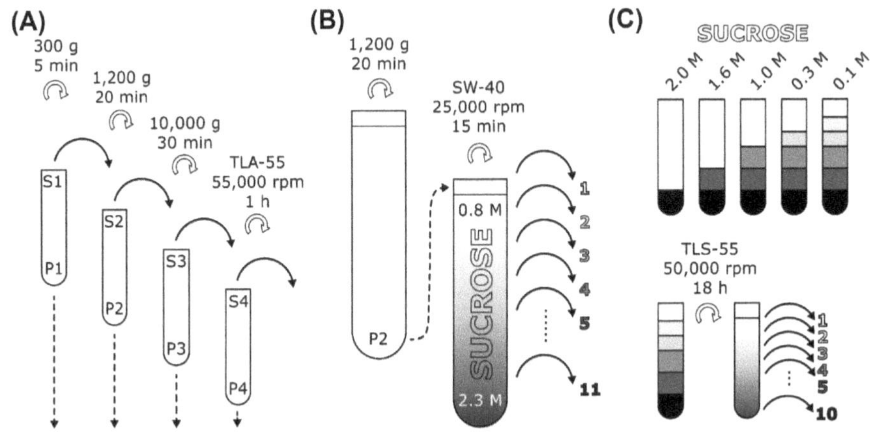

FIGURE 1 Schematics of centrifugation protocols.

(A) Differential centrifugation protocol. P1-4, pellets, S1-4, supernatant. (B) Sucrose gradient centrifugation procedure. A first spin yields a "P2" pellet which is resuspended and overlaid on a sucrose gradient. 1–11, (exemplary) fractions collected from gradient after centrifugation. (C) Schematic of floatation gradient. Sample is overlaid with sucrose solution of decreasing concentration. After centrifugation, fractions (eg, 1–10) are collected from gradient.

2. Transfer the supernatant (S1) into fresh 1.5 mL tubes and spin at 1200 g for 20 min. The pellet of this step should contain intermediate size particles including midbodies (P2 pellet).
3. Transfer the supernatant (S2) into fresh 1.5 mL tubes and spin at 10,000 g for 30 min (P3 pellet).
4. Transfer the supernatant (S3) into ultracentrifuge tubes in a TLA-55 rotor and spin at 55,000 rpm for 1 h to yield the final P4 pellet and S4 supernatant.
5. Resuspend pellets depending on downstream assay.

2.2.2 Protocol B: continuous density gradient centrifugation

For continuous velocity gradient centrifugation of cell culture supernatants, we started from collecting conditioned medium of one 6-well plate (2 mL per well, or 12 mL in total). In a first centrifugation step, particles are sedimented, then resuspended and overlaid onto a sucrose density gradient. After centrifugation, fractions are collected top to bottom. The centrifugation scheme is outlined in Fig. 1B. One can use a refractometer to determine the percentage of sucrose from a small aliquot (10–20 μL) of the collected fraction. Because it is difficult to prepare gradients and collect fractions in exactly the same way, this is an important control to be able to compare different gradient centrifugation experiments.

Materials:

- Gradient maker (Auto Densi Flow, Labconco, United States)
- Gradient buffer: 10 mM HEPES-NaOH pH 7.3, 5 mM EDTA
- 0.8 M sucrose in gradient buffer
- 2.3 M sucrose in gradient buffer
- SW-40 Ti swing-out rotor (Beckman Coulter)
- Ultracentrifuge (Optima, Beckman Coulter)

All centrifugation steps are carried out at 4°C, and samples are kept on ice.

1. Prepare c. 11 mL of a continuous sucrose density gradient using a gradient maker. Keep at 4°C.
2. Collect supernatant from one 6-well plate (c. 12 mL). Centrifuge at 1200 g for 20 min in a standard cell culture centrifuge with a swing-out rotor.
3. Resuspend pellet in 250 μL gradient buffer.
4. Carefully overlay samples onto sucrose density gradient. Centrifuge in SW-40 swing-out rotor at 25,000 rpm for 15 min (after reaching the final speed) with slow acceleration and slow brake profiles.
5. Collect 1-mL fractions from top to bottom.
6. Optional: Determine sucrose concentration in each fraction with refractometer.
7. Proceed with downstream assay.

2.2.3 Protocol C: discontinuous density gradient centrifugation

An alternative approach to enrich midbodies has been developed by (Arai et al., 2015). Here, equilibrium density gradient centrifugation is applied to separate particles by density in a discontinuous floatation gradient. This protocol has been optimized for isolation of midbodies from E11.5 mouse embryos for lipidomics analysis and is sketched out in Fig. 1C.

Materials:

- Acetate buffer: 150 mM ammonium acetate
- 100× Protease inhibitors (Roche cOmplete mini in water)
- Sucrose at 2.5, 1.6, 1.0, 0.3, 0.1 M
- TLS-55 swing-out rotor (Beckman Coulter)
- Ultracentrifuge (Optima MAX table top ultracentrifuge)

All centrifugation steps are carried out at 4°C, and samples are kept on ice.

1. Dissect embryos in ice-cold acetate buffer.
2. Collect and pool ventricular fluid to a total of 76 μL. Add 4 μL protease inhibitors and 320 μL 2.5 M sucrose. Transfer to bottom of ultracentrifuge tube.
3. Carefully overlay with 400 μL 1.6 M sucrose, 400 μL 1.0 M sucrose, 300 μL 0.3 M sucrose, and 300 μL 0.1 M sucrose.
4. Spin at 50,000 rpm in a TLS-55 rotor for 18 h with no break at the end of the run.
5. Collect 180 μL fractions into 1.5 mL tubes.
6. Optional: Use 10 μL per fraction to determine sucrose concentration in a refractometer.
7. Proceed with downstream assay.

2.3 DOWNSTREAM ASSAYS

Depending on whether pellets or fractions are used for biochemistry, immunofluorescence, or electron microscopy, samples are processed in different ways. It is also possible to combine different downstream assays, for example, by using a small amount of a fraction for immunofluorescence or EM and subjecting the remaining sample to protein precipitation for immunoblotting.

2.3.1 Protocol D: preparation of samples for immunoblotting

Materials:

- 2× concentrated SDS sample buffer (Tris–HCl 100 mM pH7.4, 10% glycerol, 4% SDS, 200 mM DTT, 0.02% bromphenol blue)
- Optional (for protein precipitation): Methanol, chloroform
- Heating block set at 95°C
- 4–12% precast gels (NuPAGE Novex, Bis-Tris, Thermo Fisher Scientific)

For immunoblotting of isolated midbodies from 6-well plates:

1. Add 20 μL of 2× concentrated SDS sample buffer to P1–P4 pellets. To obtain cell lysate, directly add 200 μL of 2× SDS sample buffer per well of a 6-well plate.
2. Boil samples at 95°C for 5 min.
3. Keep at −20°C or run on SDS gel.

To separate proteins of high and low molecular weight on one gel, we routinely use 4–12% gradient gels, followed by blotting onto PVDF membrane. To compare immunoblotting results in P1–P4 and cells, we load equivalent volumes for SDS gel electrophoresis. Half of the final volume in P1–P4 fraction and 1/20 of cell samples should be sufficient.

For immunoblotting of fractions from gradients, we use a protein precipitation step:

1. Transfer sample to a 1.5 mL centrifuge tube. If sample volume is low, dilute with water to at least 300 μL. Add one volume 100% methanol and 1/4 volume of chloroform.
2. Vortex gently. Centrifuge at 13,000 rpm in a table top centrifuge at room temperature for 5 min.
3. Transfer aqueous phase into fresh 1.5 mL tube. Add 3/4 volumes of 100% methanol.
4. Vortex gently. Centrifuge at 13,000 rpm for 5 min at room temperature to pellet protein.
5. Remove supernatant. Dissolve protein pellets in desired volume of 2× SDS sample buffer. Boil at 95°C for 5 min. Keep at −20°C or proceed with gel electrophoresis.

2.3.2 Protocol E: preparation of isolated midbodies for immunofluorescence

To fix and stain isolated midbodies in P1 and P2 fractions, we adapted protocols developed for spin-down of mitotic spindles (Maresca & Heald, 2006). Magnesium in BRB80 buffer helps maintaining microtubule integrity, and precipitation with methanol is efficient in immobilizing particles onto glass coverslips. Centrifugation speed and time were adjusted for a swing-out rotor to spin isolated midbodies onto coverslips placed on the bottom of a multiwell plate.

Materials:

- Clean #1.5 round coverslips, 15 mm diameter
- 24-well plate
- BRB80 buffer (80 mM PIPES-KOH pH 6.8, 1 mM $MgCl_2$, 1 mM EGTA) with 30% glycerol and 0.5% Triton X-100.
- BRB80 with 40% glycerol.
- Ice-cold 100% methanol (water-free).
- Centrifuge with swing-out rotor for multiwell plates.

1. Place coverslips into wells of 24-well plate. Overlay with BRB80 with 40% glycerol.
2. Resuspend pellets in 400 μL BRB80 with 30% glycerol and 0.5% Triton X-100. Carefully overlay samples onto 40% glycerol cushion.
3. Spin in swing-out rotor for 40 min at 2250 g.
4. Remove coverslips and immediately immerse in ice-cold methanol. Precipitate in methanol for 5 min at −20°C.
   Note: The 15-mm coverslips nicely cover the whole surface of the wells of a 24-well plate but may be hard to remove. A fine syringe needle, bent with forceps to a 90 degree angle may help to lift coverslips from the plate.
5. Rehydrate with PBS and proceed with immunostaining protocol.

3. LIVE-CELL IMAGING

High-resolution live-cell time-lapse microscopy can provide quantitative spatial and temporal data of cellular processes. We applied long-term time-lapse epifluorescence and confocal microcopy to study basal process splitting in mouse NE progenitor cells (Kosodo et al., 2008) and to determine dynamics of midbody release in cultured cells (Arai et al., 2015; Ettinger et al., 2011). In all epifluorescence microscopy modalities, including confocal and multiphoton microscopy, one cannot avoid making compromises between resolution, acquisition speed, imaging time, signal-to-noise ratio (SNR), and photobleaching, all of which are intricately linked. There are several good resources covering various aspects of basic and advanced microscopy techniques with emphasis on quantitative imaging (eg, Waters & Wittmann, 2014b).

Nevertheless, we highlight and summarize some crucial aspects below that should be considered for live-cell imaging of cytokinesis dynamics and midbody fate.

In the following, we focus on the use of live-cell microscopy to observe midbody release in cultured cells. Nevertheless, the principles of basic live-cell microscopy practice discussed later can certainly also be applied for imaging of cell divisions in organotypical slice cultures. This poses, however, additional requirements to sample preparation and careful optimization of environmental conditions to ensure survival of tissue slices during imaging (Attardo, Calegari, Haubensak, Wilsch-Bräuninger, & Huttner, 2008; Dubreuil et al., 2007; Miyata, Kawaguchi, Okano, & Ogawa, 2001).

3.1 TIME-LAPSE MICROSCOPY BASICS

3.1.1 Resolution and sampling

Abbe's law states that the resolution of an objective in the XY plane, the minimal distance d_{xy} at which 2 points can still be resolved, depends on the wavelength λ and the numerical aperture (NA) of the objective (Eq. (7)).

$$d_{xy} = \frac{\lambda}{2\,\mathrm{NA}} \tag{7}$$

In the third dimension, along the axis of the objective, the resolution d_Z can be calculated as follows (Eq. (8)):

$$d_z = \frac{2\lambda}{\mathrm{NA}^2} \tag{8}$$

From Abbe's law, it is evident that the NA is really the determining factor for the possible resolution of the microscope, independent of microscope modality. It is a common misconception that confocal microscopes provide a much better resolution than epifluorescence microscopes. Rather, confocal systems reduce the contribution of out-of-focus light to the image and can thus increase image contrast.

The second factor for obtaining the best possible resolution of the microscope is appropriate sampling on the detection side. Most commonly, charge-coupled device (CCD) cameras or scientific complementary metal-oxide-semiconductor (sCMOS) cameras are used nowadays in epifluorescence microscopy. They have the advantage of very high quantum efficiencies, the ratio by which incoming photons are converted into an electric signal and speed. In line-scan confocal systems, photomultipliers (PMTs) and, since recently, hybrid detectors (a combination of photocathodes and avalanche photodiodes) are used to detect photons. In live-cell microscopy, where sample intensities are usually low, cameras have the advantage because of their higher quantum efficiencies, and their speed is a plus when it comes to imaging of fast biological processes.

In modern digital microscopy, images are treated essentially as signals that are sampled from an analogous source (light from the sample) to a discrete unit, the pixels of the digital image. The question is, how small must a single pixel be so that one can make use of the possible resolution of the objective? The Nyquist–Shannon sampling theorem is cited frequently in digital image acquisition to address this question. It establishes the minimal conditions that have to be met so that an underlying ground truth (the analogous signal) can still be faithfully interpolated from discrete data. From the theorem, it follows that the equivalent pixel size of the image has to be significantly smaller than half of the minimal resolution of the objective. For example, for a 1.2 NA objective and 520 nm, the possible resolution is

$$d_{xy} = \frac{520\ \mathrm{nm}}{2 \cdot 1.2} \approx 216\ \mathrm{nm}$$

Hence, the desired pixel size of the image should be less that 216/2 = 108 nm.

When CCD or sCMOS cameras are used, the size of the individual photodiodes on the chip determines the pixel size of the image and thus the sampling frequency. The only way to sample at higher resolution (and decrease the pixel size of the image) is to either increase magnification, by using an objective with higher magnification, introducing an additional magnifying lens in front of the camera, or to use a camera with a chip that has smaller physical pixels. In line-scan confocal microscopes, the sampling rate can be adjusted more flexibly by adjusting the scan parameters. However, decreasing the pixel size to increase the sampling rate on line-scan confocal systems means that the time the excitation laser beam is moved over the sample is increased. This comes at the expense of increased photobleaching (discussed in greater detail later).

3.1.2 Signal and noise

There are different sources of noise in digital imaging that will affect how precisely the sample signal can be recorded. For cameras, the main sources of noise are camera read noise, fixed pattern noise, dark noise and Poisson, or "shot" noise (Lambert & Waters, 2014). Read noise originates from imprecisions in how photoelectrons on the chip are converted into electrical signals. Read noise is the dominating form of noise in quantitative microscopy and can be assessed by taking a "dark image" where no light is delivered onto the chip. Fixed pattern noise comes from small differences in sensitivity between photodiodes on the chip. It is relatively easy to identify because the differences will be present in every image taken with the same camera. Luckily, fixed pattern noise can easily be removed or dealt with by postacquisition image correction methods. Dark noise, or thermal noise, results from electrons that are released by heat rather than incoming photons on the chip. In most modern cameras, dark noise is suppressed by cooling of the chip and plays only a negligible role. Poisson noise originates from the fact that recording a microscopy image is essentially nothing else than counting the number of photons arriving on the chip in a given amount of time. The number of photons collected per unit of time is, however, not constant, but will follow counting statistics, and thus have a Poisson distribution. There is no way to eliminate Poisson noise. One can, however, minimize the effects of Poisson noise by collecting more photons, for example, by increasing the exposure time.

Especially in low-intensity applications like live-cell microscopy, SNR becomes critical to detect structures of interest. If the signal of the structure of interest is too low, it may be lost in the noise of the detector. Because the magnitude of read noise is constant depending on the readout mode of the camera, collecting more photons helps to reduce the contribution of camera noise to the image. Collecting more photons also leads to better SNR by reducing effects of Poisson noise. More photons can be collected by either increasing the exposure time or reducing the scan speed in line-scan confocal microscopes. However, this may increase photobleaching. Alternatively, CCD cameras allow to bin pixels. This allows to combine an $N \times N$ array of photodiodes on the chip to one unit that can collect more photons but will be read out as a single pixel. Thus, read noise will be added only once to the signal instead to each individual pixel. Binning can significantly improve SNR but will likely lead to a loss in resolution because the image is now sampled at larger intervals (see also earlier discussion).

3.1.3 Photobleaching, phototoxicity, and optimizing imaging conditions

Photobleaching is the process by which fluorophores are irreversibly destroyed. It affects both chemical dyes and fluorescent proteins. A combination of factors may affect photobleaching rates, but use of excess excitation light is certainly the main cause of photobleaching. In time-lapse microscopy, it is crucial to reduce the excitation light needed to obtain a signal to be able to image the same sample again and again over extended periods of time. Therefore, it is key that the microscope setup is optimized to collect the maximum number of photons. Matching filter sets to the spectra of fluorophores can lead to a significant boost in signal, and testing of fluorescent protein tags or probes becomes crucial when proteins of interest are expressed at low levels. Moreover, factors that contribute to unspecific background signals should be minimized. Sources of unwanted background could be fluorescent components in cell culture media (eg, phenol red), unspecific labeling of fluorescent dyes like Hoechst, or ambient room light. Further, use of hardware-triggered shutters can reduce software overheads during image acquisition and thus decrease light exposure of the sample. If highest resolution is not required, for example, for tracking whole nuclei, camera binning can help to collect more light and thus reduce exposure time. Also, lower magnification objectives with comparable NA will collect more light. Last but not least, good microscopy practice helps to reduce the time a sample is exposed to light: Taking snapshot images for setting up imaging experiments, rather than excessive use of "live" view modes, is mandatory for live specimens.

It also has to be kept in mind that photobleaching and phototoxicity are linked. Too much light may kill your sample, and imaging of dying samples will make experiments impossible to interpret. Therefore, one should control cell health at the end of a live-cell experiment in transmitted light (see also Fig. 2) and plan ahead how to reduce the total amount of light. Key questions to reduce light exposure are: Are filters and fluorophores matched? What is the shortest exposure time to still

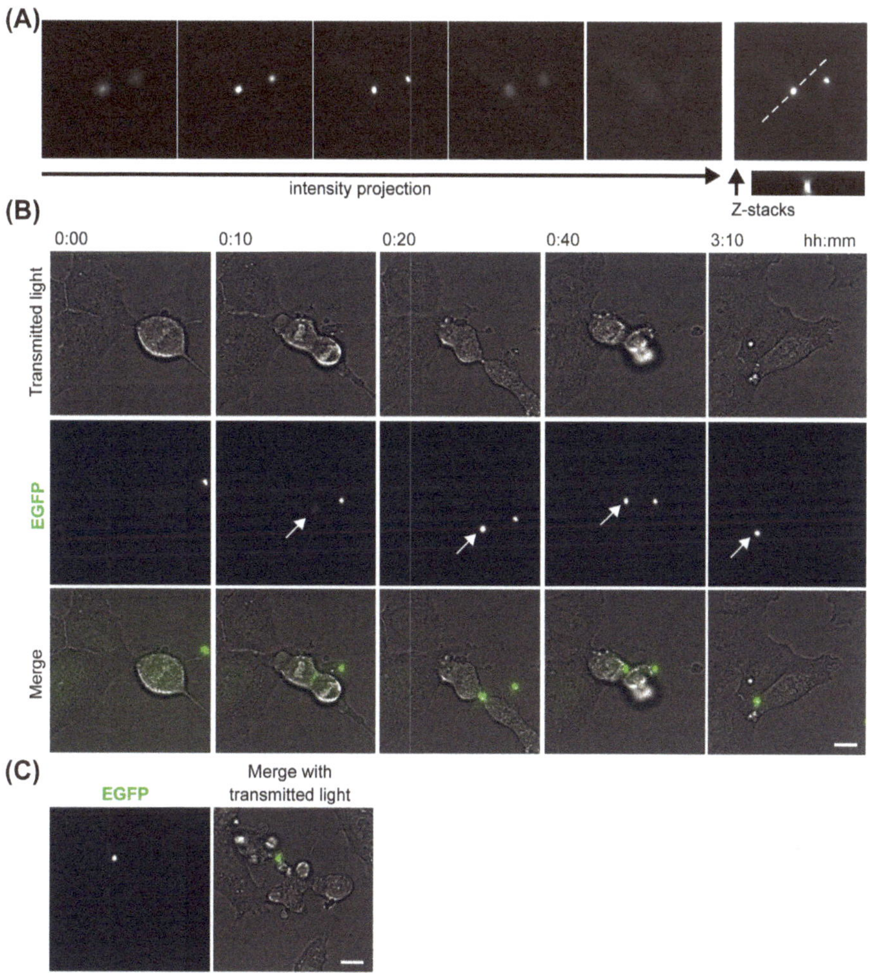

FIGURE 2 Live-cell time-lapse microscopy of midbody dynamics.

NIH-3T3 cells stably expressing a Kif23-EGFP BAC construct were observed with a 1.4 NA water immersion objective on a Zeiss Axiovert 200 M inverted microscope under 5% CO_2 supply and at 37°C (for details, see text). (A) Left panels, optical sections (Z-stacks) collected at 1-μm intervals. Shown are five example sections of a 10-image stack. Right-most panel, maximum intensity projection. Below, orthogonal view of the Z-stacks along the dotted line through a newly formed midbody. (B) Example images of a 32-h time lapse of one cell undergoing cytokinesis. Midbody formation can be followed by Kif23-EGFP localizing first to midzone and then to the midbody (*arrows*). Time points are indicated. (C) Example of a dying cell during the same time-lapse experiment. Note that extensive blebbing is only visible in transmitted light. Scale bars (B, C), 10 μm.

get a reasonable signal? Is it necessary to take a high number of optical sections (Z-stacks) per time point? Can a lower magnification objective or camera binning be used? How many acquisitions per minute/hour are really necessary for appropriate temporal sampling?

3.1.4 Environmental control and live-imaging hardware

To keep samples alive and happy under the microscope, key environmental parameters have to be maintained (Ettinger & Wittmann, 2014). For mammalian cells, temperature and CO_2 are the main factors that have to be controlled. There are a variety of commercial solutions for microscope incubators with different degrees of sophistication. For a dedicated live-cell microscope, however, we recommend a full enclosure incubator around stage, microscope stand, and objective revolver. This has the big advantage that stage and objectives are maintained at 37°C and do not act as heat sinks, especially in the case of immersion objectives. Objective heaters in form of attachable collars do exist, but subjecting this key optical component to cycles of heating and cooling during live-cell imaging experiments seems like a recipe to shorten its lifetime. CO_2 can be supplied in form of premixed gas (eg, 5% CO_2/air mixture) or from a 100% bottle or CO_2 line and mixed to the desired concentration by adjusting the flow rate or by a feedback-loop controlled CO_2 mixer. To not dry the sample during the course of a long-term time-lapse experiment, humidification of the incubator or at least of the gas supply is necessary. Air bubblers or semipermeable tubing run through water are good options. In the case of bubble bottles, one has to pay attention, however, that the bottle is placed such that the bubbling gas does not introduce vibrations into the imaging system.

For live-cell imaging, it is recommended to have an inverted automated microscope where filters, shutters, and stage can be controlled by the acquisition software. This greatly increases the flexibility and speed of experiments and allows the acquisition of multiple stage positions in one experiment, a crucial factor for obtaining data of multiple cells in a single long-term time-lapse experiment. Moreover, several microscope vendors now offer hardware-controlled focus systems. These systems are much more reliable than autofocus algorithms. Hardware focus typically employs an angled far-red low-power laser that is reflected by the coverglass/medium interface onto a detection photodiode when the sample is in focus. By a feedback loop, the system can then regulate the z-movement of objective or stage to maintain focus even if the sample is displaced over time. Another advantage is that the far-red laser of the focus system usually does not interfere with imaging and no additional image acquisition (and thus light exposure) of the sample is necessary as is the case for software autofocus.

Finally, the choice of microscopy modality is important: For imaging of cultured cells with fluorescent protein tags that have good signal-to-background properties, wide-field epifluorescence microscopy is probably the best choice. For specimens where out-of-focus fluorescence contributes to the signal so that it becomes difficult to detect or quantify structures of interest, confocal microscopes help to reduce out-of-focus light. For live-cell confocal microscopy, spinning disk systems provide a good balance between confocality, sensitivity, and speed but are less light efficient than wide-field systems. Line-scan confocals can provide even better confocality than spinning disk microscopes, but they are two orders of magnitude worse than wide-field microscopes when it comes to collecting and detecting fluorescence (Waters & Wittmann, 2014a).

3.2 LIVE-CELL MICROSCOPY OF MIDBODY RELEASE

For live-cell time-lapse microscopy of midbody release in cultured cells, we used the following microscope setup:

- Zeiss Axiovert 200 M inverted wide-field microscope equipped with motorized filter wheels, automated shutters, and motorized stage.
- Environmental control with incubator built around the microscope and CO_2 mixer (Visitron Systems, Puchheim, Germany), set at 37°C and 5% CO_2/95% air.
- Bubble humidifier and stage-top mounted small chamber for CO_2 supply.
- X-Cite metal-halide epifluorescence light source, attenuated to 25–50% intensity.
- Water Immersion 40× C-Apochromat NA 1.4 (Zeiss, Germany).
- CoolSnap ES cooled CCD camera with Sony Interline ICX285 CCD chip with 1392 × 1040 pixels (physical pixel size 6.45 × 6.45 μm).

The 1.4 NA water immersion objective provides excellent combination of high resolution, large field of view, increased working distance, and very good light efficiency. To ensure cell viability over dozens of hours of imaging, it is crucial to reduce epifluorescence light by decreasing the intensity at the epifluorescence lamp and optimize imaging parameters. We routinely used the following settings:

- Bright field/transmitted light: 2 × 2 binning, 20 ms exposure, single plane
- Fluorescence/EGFP: 2 × 2 binning, 15–20 ms exposure, 8–12 z-planes at 0.8–1.5 μm spacing
- no software autofocus

Under these conditions, we were successful in acquiring image stacks covering 6–15 μm at 5–15 min time intervals up to 48 h. Depending on cell division speed and cell size, this was necessary to capture three-dimensional movements of residual and released midbodies over more than one round of cell division and determine the timing of midbody release. Due to binning, the final dimension of the images was 696 × 520 pixels with ~310 nm pixel size. However, as midbodies are typically >500 nm, this was sufficient to resolve two or more neighboring midbodies. Since the microscope was not equipped with a hardware focusing system, we kept a range of z-sections for all stage positions, which was more reliable than the MetaMorph autofocus option. If available, we highly recommend using a hardware focus system to ensure focus stability during long-term imaging experiments. For all live-cell time-lapse microscopy experiments, we established stable cell lines expressing a Kif23-EGFP BAC fusion construct, which ensured very low cell-to-cell and experiment-to-experiment variation with no apparent effects on cytokinesis.

Materials:

- Labtek #1.5 chamber coverglass or Matek #1.5 glass bottom 35 mm dishes
- Optional: 0.1% gelatine (Sigma) in PBS or 0.01% poly-L-lysine (Sigma) for coating
- Optional: DMEM without phenol red, supplied with glutamine, 10% fetal bovine serum, and penicillin/streptomycin
- Optional: annexin V-Cy5 solution (BioVision)

For some cell lines, coating coverslips improved cell attachment and viability.

1. Optional: (1) Add gelatine onto coverslips. Incubate at room temperature for few minutes or until needed. Remove gelatine just before cell plating. (2) Cover chambers or glass-bottom dishes with poly-L-lysine. Incubate at room temperature for 15 min. Remove poly-L-lysine and wash 3× with PBS or water before plating.
2. Plate cells into imaging chamber. For long-term time-lapse imaging, plate at densities that allow for sufficient space for expansion. We found 5000–10,000 cells/cm^2 a good starting point. Allow cells to settle and attach overnight.
3. Optional: (1) Change to imaging medium without phenol red prior imaging. (2) Change to medium without phenol red containing annexin V-Cy5 at 1:5000 dilution.
4. Set up microscope: Find cells in transmitted or epifluorescence light. If experiment and microscope setup is permitting, choose multiple stage positions. One way to do this is to first find cells with the eye piece and mark positions. Mitotic rounded-up cells can be easily identified in bright field mode. Then, switch to camera mode and readjust focus and X-Y positions if necessary in the microscope software.
5. Postprocess images for analysis.

To analyze midbody release from neural stem cells and other cell lines, we collapsed z-stacks (maximum intensity projection), adjusted brightness and contrast, and generated merged bright field/epifluorescence datasets for manual tracking of midbodies (Fig. 2A and B). The open software image analysis tool Fiji (ImageJ) (Schindelin et al., 2012) comes with a powerful macro language that allows to automate such image processing tasks. It is highly recommended to check bright field images to control cell health, since it is not necessarily obvious from fluorescence channels if cells died during the imaging experiment (Fig. 2C).

4. HIGH-VOLTAGE ELECTRON MICROSCOPY

4.1 INTRODUCTION

High-voltage electron microscopy (HVEM) allows the observation of relatively thick (>5 μm) specimens because of the increased penetration power of electrons at high accelerating voltage (more than 1000 kV). HVEM is especially valuable when studying the structures underlying complex processes of neuronal and glial cells in the mammalian central nervous system, because they are too small to be studied by conventional light microscopy and too extensive in 3D to be easily analyzed by thin-section electron microscopy (Hama, Arii, & Kosaka, 1994). Thanks to the high penetrating power of the electron and the large depth of focus with the aid of stereoscopy, HVEM has also provided valuable information on the 3D organization of ultrafine structures (Arii & Hama, 1987). We utilized the unique strength of HVEM to visualize the fine structure of the basal process of mammalian NE cells in embryonic brain (Kosodo et al., 2008), a cellular structure which is now considered to be a key to understand the fate determination of NE cells (Florio & Huttner, 2014; Kosodo & Huttner, 2009; Matsuzaki & Shitamukai, 2015). It should be noted that superresolution light microscopy and serial block face scanning electron microscopy can also be applied to visualize the fine structure of the basal process. It will be important to compare these methodologies in future studies. In this section, we will describe the procedures we used to visualize the basal process structure at M-phase using HVEM (see Kosodo et al., 2008).

4.2 CHOICE OF THE STAINING METHOD

When performing HVEM studies, contrasting of the object of interest is the first consideration. The high penetrating power of electrons upon HVEM does not yield images of fine cellular structures such as the basal process of M-phase NE cells unless the target structure is sufficiently stained. We initially tried preembedding immunostaining using a specific marker of mitosis (MPM-2, Upstate) to visualize the basal process structure at M-phase by HVEM. After applying the primary antibody (MPM-2), biotin-conjugated secondary antibody and streptavidin-FluoroNanogold were utilized. Following the 3-3′-diaminobenzidine tetrahydrochloride (DAB) reaction, silver enhancement was applied using HQ-Silver (Nanoprobe) to enhance the contrast for HVEM. It turned out that contrast and resolution of the sample were not sufficient to visualize the thin basal process(es), even though the cell bodies at M-phase were adequately stained. To reveal the fine structure of the basal process of M-phase NE cells by HVEM, we found that Golgi silver impregnation is superior to MPM-2 immunoperoxidase with regard to contrast and resolution. However, in the case of Golgi impregnation, there is no specificity of the stained cell in terms of certain cell cycle phases. Therefore, appropriate criteria must be applied to identify an M-phase NE cell in tissue (described in Section 4.3.2, Point 5). Below, we will describe the detailed procedure of Golgi impregnation for HVEM analyses, which is based on a previous study (Hama, Arii, & Kosaka, 1989).

4.3 EXPERIMENTAL PROCEDURES

4.3.1 *Golgi impregnation*

Materials:

- Paraformaldehyde, 4% in 0.1 M phosphate buffer pH 7.4
- Glutaraldehyde, 0.1% in 0.1 M phosphate buffer pH 7.4

- FD Rapid GolgiStain Kit (FD NeuroTechnologies)
- Ethanol series (50%, 70%, 90%, 100%)
- Forceps for mouse dissection and tissue handling

1. Dissect embryonic mouse brain (typically E10.5 to visualize the basal process of M-phase NE cells at early neurogenic stages in the cerebral cortex).
2. Fix overnight at 4°C in 4% paraformaldehyde and 0.1% glutaraldehyde in 0.1 M phosphate buffer.
   Note: It is indicated in the FD Rapid GolgiStain Kit protocol that using fixative during tissue preparation should be avoided unless absolutely necessary. However, we found that the staining efficiency is highest when using the above conditions of fixation; to this end, we compared "no fixative," "4% paraformaldehyde only," and "4% paraformaldehyde with 0%, 0.002%, 0.01%, 0.05%, and 0.1% glutaraldehyde." It should be noted that the background staining increases with 0.1% glutaraldehyde (which, however, is required for proper fixation).
3. Rinse tissue briefly with 0.1 M phosphate buffer (pH 7.4)
4. Put tissue into solutions **A** + **B**, and store at room temperature in the dark (1–2 weeks). Replace solutions once after 12 h. Prepare solutions at least 24 h prior to use.
   Note that solutions **A** and **B** are toxic as they contain mercuric chloride, potassium dichromate, and potassium chromate. For details, see the Kit instruction.
5. Transfer tissue into solution **C** and store at room temperature for 3 days up to 1 week. Replace solution once after 12 h.
6. Rinse tissue in distilled water 2 times, 4 min each, in a plastic dish.
7. Place tissue into a mixture of one part solution **D**, one part solution **E**, and two parts of Milli-Q water for 10 min.
8. Rinse tissue in distilled water 2 times, 4 min each.
9. Dehydrate tissue, using stepwise 50%, 70%, 90%, 100%, 100%, 100% ethanol, 4 min each.

4.3.2 Embedding and sectioning

Materials:

- Rotary microtome (eg, RM2255, Leica)
- Ultramicrotome (eg, EM UC7, Leica)
- Light microscope (eg, Olympus BX61)
- Oven
- Small iron to warm up Epon block
- Mesh grid (Square mesh oyster grid D75, Veco)
- Liquid Release Agent (Electron Microscopy Science, #70880)
- Propylene oxide (PO)
- Epon-Araldite (EA)

Araldite M	11.3 g (10 mL)
DDSA	24.0 g (24 mL)
Epon 812	12.1 g (10 mL)
DMP 30	0.66–0.88 mL

1. Infiltrate the dehydrated tissue with PO and EA resin as follows:
   PO:EA = 1:0; 10–20 min at room temperature twice,
   PO:EA = 1:1; overnight at room temperature with gentle shaking, and
   PO:EA = 0:1; 2 h at room temperature.
2. Embed in mold, and harden the Epon block in oven at 60°C overnight.
3. Cut Epon block containing brain tissue into 100 μm-thick sections by rotary microtome. Warm up Epon block briefly using a heated iron (50–60°C) to soften the block for smooth cutting by the rotary microtome.
4. Flat-embed on glass slides coated by Liquid Release Agent according to the manufacturer's instruction.
5. Identification of M-phase NE cells under a light microscope using a 40× objective, applying the following criteria:
   a. the shape of the cell body is nearly round;
   b. the location of the cell body is at the ventricular surface.
6. Sections containing appropriate cells should be selected. Small pieces of sections are cut with a razor blade from the glass slide and remounted on an Epon block.
7. Serial 5-μm sections are prepared using an ultramicrotome and mounted on 75-mesh grids (Fig. 3).

4.3.3 HVEM observation and analysis

Materials:

- High-voltage electron microscope (HVEM; eg, Hitachi H-1250 M at the National Institute for Physiological Sciences (NIPS), Okazaki, Japan)
- Side-entry rotation tilt stage with feedback control mechanism for the tilt angle for stereomicroscopy
- IMOD software (Colorado University)

1. Place the prepared specimen on the rotation tilt stage under a magnifying lens, and then mount to the HVEM.
2. Observation under the HVEM can be started with 750× magnification to identify target cells. To observe the fine structure of the basal process of an M-phase cell, 3000×–5000× magnification is optimal. Accelerating voltage should be 1000 kV.
3. Use of rotation tilt stage allows stereo observation. Stereo-paired photomicrographs are prepared by tilting the specimen stage ±8 degree (one-paired image should have 16° difference for stereo observation)

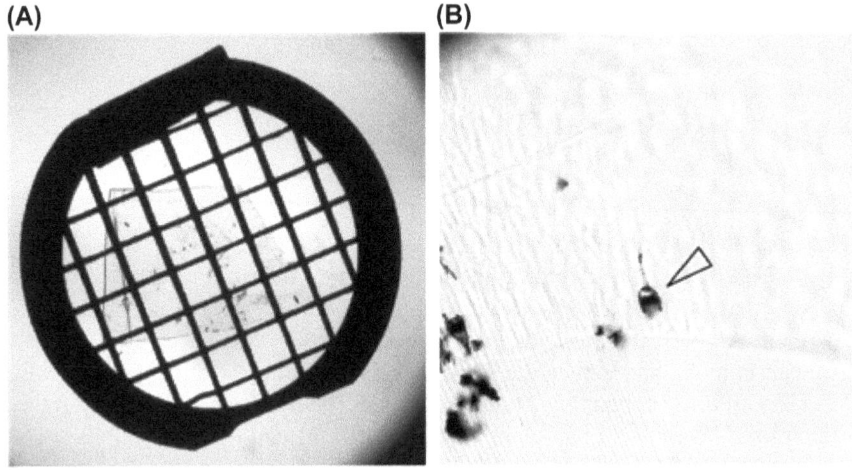

FIGURE 3 Light microscope images of sample preparation for high-voltage electron microscopy (HVEM) analysis.

(A) Golgi-stained embryonic mouse brain was embedded in Epon, and the section (5 μm) was mounted on 75-mesh grids. (B) Magnified view of M-phase neuroepithelial cell extending the process toward the basal side (*arrowhead*). Down, apical surface. Note that the apical side of the cell body is not visible in the particular section shown. The HVEM image of this sample is available in Fig. 2B of Kosodo et al., 2008.

4. For electron tomography analysis, record a series of tilted images from $-60°$ to $+60°$ with 2–5 increments along the y axis.
5. Recorded series of tilted images are reconstructed to obtain a three-dimensional view by using IMOD software (visit the website for details; http://bio3d.colorado.edu/imod/).

SUMMARY

To assess membrane dynamics associated with cytokinesis in specialized cell types such as neural stem cells, a combination of approaches is key. Isolation of released midbodies by centrifugation may provide enough material for downstream analyses with "-omics" approaches to generate hypotheses and reveal unprecedented mechanistic networks. Quantitative live-cell fluorescence microscopy allows to determine cellular and subcellular dynamics on spatial and temporal scales, dissect cell-to-cell variation, and result in surprising and novel observations. Ultrastructural analysis may require specialized electron microscopy protocols and approaches but provides structural details that can hardly be obtained with other methods.

Together, such multidisciplinary approaches of biochemical analyses and single-cell observations continue to unravel unexpected networks and cellular dynamics.

Comparison of isolated cells to tissue-level organization can clarify functional relationships and place observations into their physiological context. Importantly, live-cell microscopy allows to study cellular processes over time and thus adds the temporal dimension, which is particularly important for investigating dynamics of cytokinesis.

ACKNOWLEDGMENTS

We thank present and past members of the Huttner lab for discussions and contributions to methods described in this chapter; in particular, Anne-Marie Marzesco and Yoko Arai for developing centrifugation protocols for isolation of midbodies; Veronique Dubreuil and Judith Schenk for setting up live-cell microscopy. Jennifer Waters, Talley Lambert (Nikon Imaging Center, Harvard Medical School) and Torsten Wittmann (University of California San Francisco) for their efforts of teaching good microscopy practice and sharing their expertise in epi-fluorescence and confocal live-cell microscopy.

A part of this work was supported by the KBRI basic research program through the Korea Brain Research Institute funded by the Ministry of Science, ICT, and Future Planning (No. 2231-415) (Y.K). W.B.H. was supported by grants from the Deutsche Forschungsgemeinschaft (DFG) and the European Research Council (ERC), by the DFG-funded Center for Regenerative Therapies Dresden (CRTD), and by the Fonds der Chemischen Industrie.

REFERENCES

Arai, Y., Sampaio, J. L., Wilsch-Bräuninger, M., Ettinger, A. W., Haffner, C., & Huttner, W. B. (2015). Lipidome of midbody released from neural stem and progenitor cells during mammalian cortical neurogenesis. *Frontiers in Cellular Neuroscience, 9*, 325.

Arii, T., & Hama, K. (1987). Method of extracting three-dimensional information from HVTEM stereo images of biological materials. *Journal of Electron Microscopy, 36*, 177–195.

Attardo, A., Calegari, F., Haubensak, W., Wilsch-Bräuninger, M., & Huttner, W. B. (2008). Live imaging at the onset of cortical neurogenesis reveals differential appearance of the neuronal phenotype in apical versus basal progenitor progeny. *PLoS One, 3*, e2388.

Carlton, J. G., & Martin-Serrano, J. (2007). Parallels between cytokinesis and retroviral budding: a role for the ESCRT machinery. *Science, 316*, 1908–1912.

Carpenter, A. E., Jones, T. R., Lamprecht, M. R., Clarke, C., Kang, I. H., Friman, O., … Sabatini, D. M. (2006). CellProfiler: image analysis software for identifying and quantifying cell phenotypes. *Genome Biology, 7*, R100.

Chai, Y., Tian, D., Yang, Y., Feng, G., Cheng, Z., Li, W., & Ou, G. (2012). Apoptotic regulators promote cytokinetic midbody degradation in *C. elegans*. *Journal of Cell Biology, 199*, 1047–1055.

de Chaumont, F., Dallongeville, S., Chenouard, N., Hervé, N., Pop, S., Provoost, T., … Olivo-Marin, J. C. (2012). Icy: an open bioimage informatics platform for extended reproducible research. *Nature Methods, 9*, 690–696.

Cong, L., Ran, F. A., Cox, D., Lin, S., Barretto, R., Habib, N., ... Zhang, F. (2013). Multiplex genome engineering using CRISPR/Cas systems. *Science, 339*, 819–823.

Crowell, E. F., Gaffuri, A.-L., Gayraud-Morel, B., Tajbakhsh, S., & Echard, A. (2014). Engulfment of the midbody remnant after cytokinesis in mammalian cells. *Journal of Cell Science, 127*, 3840–3851.

Crowell, E. F., Tinevez, J.-Y., & Echard, A. (2013). A simple model for the fate of the cytokinesis midbody remnant: implications for remnant degradation by autophagy. *BioEssays: News and Reviews in Molecular, Cellular and Developmental Biology, 35*, 472–481.

Di Cunto, F., Calautti, E., Hsiao, J., Ong, L., Topley, G., Turco, E., & Dotto, G. P. (1998). Citron rho-interacting kinase, a novel tissue-specific Ser/Thr kinase encompassing the Rho-Rac-binding protein Citron. *Journal of Biological Chemistry, 273*, 29706–29711.

Dubreuil, V., Marzesco, A. M., Corbeil, D., Huttner, W. B., & Wilsch-Brauninger, M. (2007). Midbody and primary cilium of neural progenitors release extracellular membrane particles enriched in the stem cell marker Prominin-1. *Journal of Cell Biology, 176*, 483–495.

Ettinger, A., & Wittmann, T. (2014). Fluorescence live cell imaging. *Methods in Cell Biology, 123*, 77–94.

Ettinger, A. W., Wilsch-Bräuninger, M., Marzesco, A.-M., Bickle, M., Lohmann, A., Maliga, Z., ... Huttner, W. B. (2011). Proliferating versus differentiating stem and cancer cells exhibit distinct midbody-release behaviour. *Nature Communications, 2*, 503.

Florio, M., & Huttner, W. B. (2014). Neural progenitors, neurogenesis and the evolution of the neocortex. *Development, 141*, 2182–2194.

Gai, M., Camera, P., Dema, A., Bianchi, F., Berto, G., Scarpa, E., ... Di Cunto, F. (2011). Citron kinase controls abscission through RhoA and Anillin. *Molecular Biology of the Cell, 22*, 3768–3778.

Graham, J. M., & Rickwood, D. (1999). *Biological centrifugation*. Taylor & Francis.

Gromley, A., Yeaman, C., Rosa, J., Redick, S., Chen, C.-T., Mirabelle, S., ... Doxsey, S. J. (2005). Centriolin anchoring of exocyst and SNARE complexes at the midbody is required for secretory-vesicle-mediated abscission. *Cell, 123*, 75–87.

Guizetti, J., Schermelleh, L., Mäntler, J., Maar, S., Poser, I., Leonhardt, H., ... Gerlich, D. W. (2011). Cortical constriction during abscission involves helices of ESCRT-III-dependent filaments. *Science, 331*, 1616–1620.

Hama, K., Arii, T., & Kosaka, T. (1989). Three-dimensional morphometrical study of dendritic spines of the granule cell in the rat dentate gyrus with HVEM stereo images. *Journal of Electron Microscopy Technique, 12*, 80–87.

Hama, K., Arii, T., & Kosaka, T. (1994). Three-dimensional organization of neuronal and glial processes: high voltage electron microscopy. *Microscopy Research and Technique, 29*, 357–367.

Hirose, K., Kawashima, T., Iwamoto, I., Nosaka, T., & Kitamura, T. (2001). MgcRacGAP is involved in cytokinesis through associating with mitotic spindle and midbody. *Journal of Biological Chemistry, 276*, 5821–5828.

Huang, B., Bates, M., & Zhuang, X. (2009). Super-resolution fluorescence microscopy. *Annual Review of Biochemistry, 78*, 993–1016.

Jinek, M., Chylinski, K., Fonfara, I., Hauer, M., Doudna, J. A., & Charpentier, E. (2012). A programmable dual-RNA-guided DNA endonuclease in adaptive bacterial immunity. *Science, 337*, 816–821.

Kieserman, E. K., Glotzer, M., & Wallingford, J. B. (2008). Developmental regulation of central spindle assembly and cytokinesis during vertebrate embryogenesis. *Current Biology, 18*, 116–123.

Kittler, R., Putz, G., Pelletier, L., Poser, I., Heninger, A.-K., Drechsel, D., … Buchholz, F. (2004). An endoribonuclease-prepared siRNA screen in human cells identifies genes essential for cell division. *Nature, 432*, 1036–1040.

Kosodo, Y., & Huttner, W. B. (2009). Basal process and cell divisions of neural progenitors in the developing brain. *Development, Growth & Differentiation, 51*, 251–261.

Kosodo, Y., Toida, K., Dubreuil, V., Alexandre, P., Schenk, J., Kiyokage, E., … Huttner, W. B. (2008). Cytokinesis of neuroepithelial cells can divide their basal process before anaphase. *EMBO Journal, 27*, 3151–3163.

Kuo, T.-C., Chen, C.-T., Baron, D., Onder, T. T., Loewer, S., Almeida, S., … Doxsey, S. (2011). Midbody accumulation through evasion of autophagy contributes to cellular reprogramming and tumorigenicity. *Nature Cell Biology, 13*, 1214–1223.

Kuriyama, R., Gustus, C., Terada, Y., Uetake, Y., & Matuliene, J. (2002). CHO1, a mammalian kinesin-like protein, interacts with F-actin and is involved in the terminal phase of cytokinesis. *Journal of Cell Biology, 156*, 783–790.

Lambert, T. J., & Waters, J. C. (2014). Assessing camera performance for quantitative microscopy. In J. C. Waters, & T. Wittmann (Eds.), *Quantitative imaging in cell biology* (Vol. 123, pp. 35–53). Elsevier.

Lee, H. H., Elia, N., Ghirlando, R., Lippincott-Schwartz, J., & Hurley, J. H. (2008). Midbody targeting of the ESCRT machinery by a noncanonical coiled coil in CEP55. *Science, 322*, 576–580.

Maresca, T. J., & Heald, R. (2006). Methods for studying spindle assembly and chromosome condensation in *Xenopus* egg extracts. *Methods in Molecular Biology (Clifton, NJ), 322*, 459–474.

Marzesco, A.-M., Janich, P., Wilsch-Bräuninger, M., Dubreuil, V., Langenfeld, K., Corbeil, D., & Huttner, W. B. (2005). Release of extracellular membrane particles carrying the stem cell marker prominin-1 (CD133) from neural progenitors and other epithelial cells. *Journal of Cell Science, 118*, 2849–2858.

Matsuzaki, F., & Shitamukai, A. (2015). Cell division modes and cleavage planes of neural progenitors during mammalian cortical development. *Cold Spring Harbor Perspectives in Biology, 7*, a015719.

Miyata, T., Kawaguchi, A., Okano, H., & Ogawa, M. (2001). Asymmetric inheritance of radial glial fibers by cortical neurons. *Neuron, 31*, 727–741.

Mullins, J. M., & Biesele, J. J. (1977). Terminal phase of cytokinesis in D-98s cells. *Journal of Cell Biology, 73*, 672–684.

Neumann, B., Walter, T., Hériché, J.-K., Bulkescher, J., Erfle, H., Conrad, C., … Ellenberg, J. (2010). Phenotypic profiling of the human genome by time-lapse microscopy reveals cell division genes. *Nature, 464*, 721–727.

Nogales, E., & Scheres, S. H. W. (2015). Cryo-em: a unique tool for the visualization of macromolecular complexity. *Molecular Cell, 58*, 677–689.

Oldenbourg, R., Salmon, E. D., & Tran, P. T. (1998). Birefringence of single and bundled microtubules. *Biophysical Journal, 74*, 645–654.

Ou, G., Gentili, C., & Gönczy, P. (2014). Stereotyped distribution of midbody remnants in early *C. elegans* embryos requires cell death genes and is dispensable for development. *Cell Research, 24*, 251–253.

Pohl, C., & Jentsch, S. (2009). Midbody ring disposal by autophagy is a post-abscission event of cytokinesis. *Nature Cell Biology, 11*, 65–70.

Poser, I., Sarov, M., Hutchins, J. R. A., Hériché, J.-K., Toyoda, Y., Pozniakovsky, A., … Hyman, A. A. (2008). BAC TransgeneOmics: a high-throughput method for exploration of protein function in mammals. *Nature Methods, 5*, 409–415.

Salzmann, V., Chen, C., Chiang, C.-Y. A., Tiyaboonchai, A., Mayer, M., & Yamashita, Y. M. (2014). Centrosome-dependent asymmetric inheritance of the midbody ring in *Drosophila* germline stem cell division. *Molecular Biology of the Cell, 25*, 267–275.

Schindelin, J., Arganda-Carreras, I., Frise, E., Kaynig, V., Longair, M., Pietzsch, T., … Cardona, A. (2012). Fiji: an open-source platform for biological-image analysis. *Nature Methods, 9*, 676–682.

Singh, D., & Pohl, C. (2014). Coupling of rotational cortical flow, asymmetric midbody positioning, and spindle rotation mediates dorsoventral axis formation in *C. elegans*. *Developmental Cell, 28*, 253–267.

Stehbens, S., Pemble, H., Murrow, L., & Wittmann, T. (2012). Imaging intracellular protein dynamics by spinning disk confocal microscopy. *Methods in Enzymology, 504*, 293–313.

Steigemann, P., & Gerlich, D. W. (2009). Cytokinetic abscission: cellular dynamics at the midbody. *Trends in Cell Biology, 19*, 606–616.

Taverna, E., & Huttner, W. B. (2010). Neural progenitor nuclei IN motion. *Neuron, 67*, 906–914.

Waters, J. C., & Wittmann, T. (2014a). Concepts in quantitative fluorescence microscopy. *Methods in Cell Biology, 123*, 1–18.

Waters, J. C., & Wittmann, T. (Eds.). (2014b). Quantitative imaging in cell biology. *Methods in Cell Biology, 123*. Elsevier.

White, E. A., & Glotzer, M. (2012). Centralspindlin: at the heart of cytokinesis. *Cytoskeleton (Hoboken, NJ), 69*, 882–892.

Yang, D., Rismanchi, N., Renvoisé, B., Lippincott-Schwartz, J., Blackstone, C., & Hurley, J. H. (2008). Structural basis for midbody targeting of spastin by the ESCRT-III protein CHMP1B. *Nature Structural & Molecular Biology, 15*, 1278–1286.

CHAPTER

Understanding post-mitotic roles of the midbody during cell differentiation and polarization

10

E. Peterman, R. Prekeris[1]

University of Colorado Anschutz Medical Campus, Aurora, CO, United States

[1]Corresponding author: E-mail: rytis.prekeris@ucdenver.edu

CHAPTER OUTLINE

Abstract

The midbody (MB) is a microtubule-rich structure that forms between dividing cells during final stages of cytokinesis. Previously thought to be a transient structure, MBs are now suggested to have additional roles beyond regulating cytokinesis. While the role MBs play during abscission are now well established, their function in regulating polarity

Methods in Cell Biology, Volume 137, ISSN 0091-679X, http://dx.doi.org/10.1016/bs.mcb.2016.04.001

and cell signaling are only beginning to be understood. Due to the newly found interest in the structure and functions of MBs, new techniques must be developed for further studies of this once-thought transient structure. Here, we describe several approaches used to explore postmitotic roles of the MBs.

INTRODUCTION

The midbody (MB) is a subcellular structure that forms during cell division and was initially described over 100 years ago by Walther Flemming. The MB forms during the ingression of the cleavage furrow as the central spindle microtubules are compacted and cross-linked within a thin intracellular bridge connecting two daughter cells. It is now well established that the MBs play key roles in orchestrating cytokinesis by recruiting various mitotic kinases, such as Aurora B and Plk1, as well as Rab11/FIP3-containing furrow endosomes, membrane severing ESCRT complex, and the microtubule severing enzyme spastin, all of which are responsible for mediating abscission during late stages of cytokinesis (Elia, Sougrat, Spurlin, Hurley, & Lippincott-Schwartz, 2011; Schiel et al., 2012; Yang et al.,2008).

Recently, it has become apparent that in addition to regulating cytokinesis, MBs also have nonmitotic roles. Several studies have highlighted that the MB can be asymmetrically inherited during cell division, which may confer properties (such as a polarity cue or a signaling platform) to the cell which inherited it (Fig. 1A) (Ettinger et al., 2011; Kuo et al., 2011; Pollarolo, Schulz, Munck, & Dotti, 2011; Singh & Pohl, 2014; Wilcock, Swedlow, & Storey, 2007). Conversely, other studies have revealed that symmetric abscission leads to a release of the MB to the extracellular space (Fig. 1B) (Elia et al., 2011). Interestingly, some recent studies suggest that these extracellular MBs may be engulfed by surrounding cells, although the functional significance of this MB uptake remains unclear. It was proposed that MB uptake process relies on the enrichment of phosphatidylserines in the outer leaflet of the MB membrane (Chai et al., 2012), a process that appears to be similar to an actin-dependent phagocytosis-like uptake of apoptotic cells (Crowell, Gaffuri, Gayraud-Morel, & Tajbakhsh, 2014).

In the past few years, the MB has been shown to help define the polarity of daughter cells in vivo and in vitro. It was found that during spinal cord neurogenesis in chick embryos, the MB is inherited and displaced toward the apical surface of dividing neural progenitors. Additionally, upon relocation to the apical surface, the MB positions itself at the site of forming neurites (Wilcock et al., 2007). In the fly notum, polarized accumulation of postmitotic furrow markers at the MB (such as RhoA and Aurora A) also preceded neurite or "sprout" outgrowth in newly born neurons (Pollarolo et al., 2011). Finally, during the early development of *Caenorhabditis elegans*, the MB was shown to function as a polarity cue which is necessary for directing mitotic spindle orientation for the following cell division (Singh & Pohl, 2014). Similarly, using 3D mammalian epithelial cultures, we have recently demonstrated that the formation of the apical membrane initiation site (AMIS) is

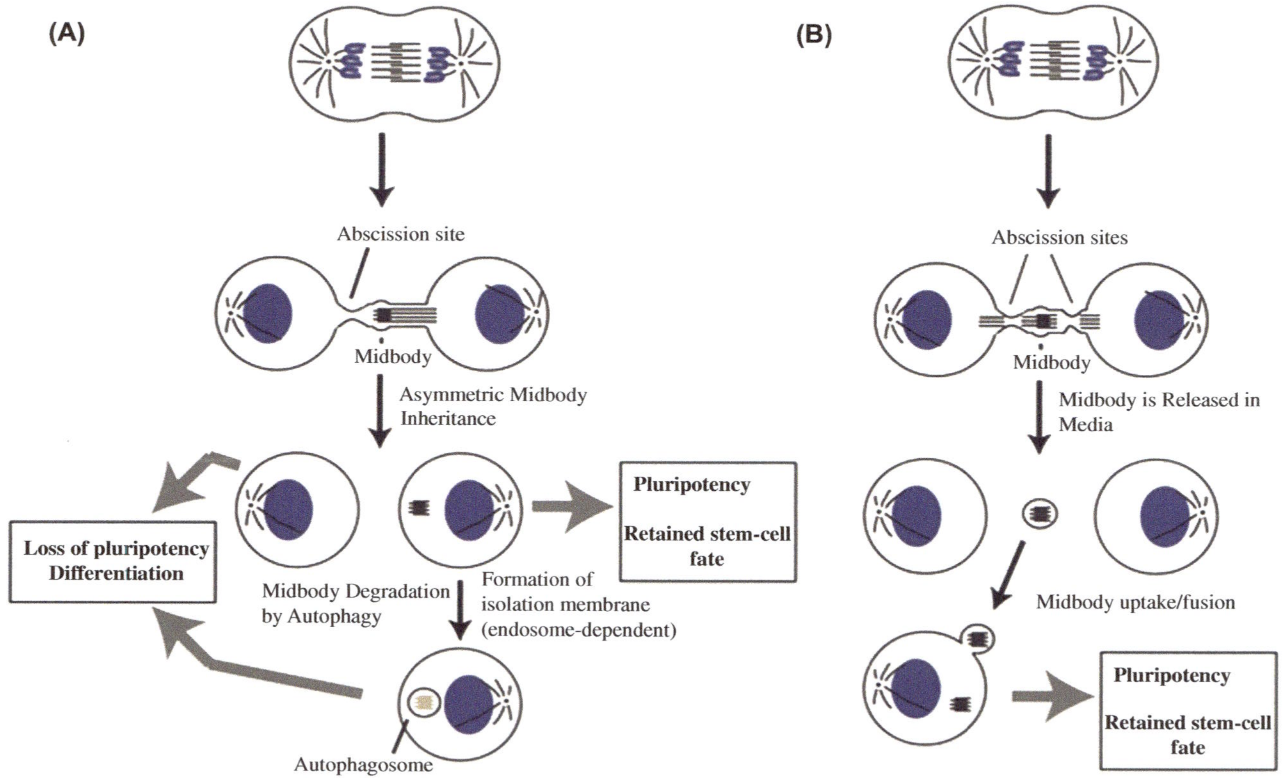

FIGURE 1 Midbodies can either be asymmetrically inherited or released into the extracellular space.

(A) Model detailing the possible fates of an asymmetrically inherited midbody (MB). Upon asymmetric inheritance, the MB can remain as a postmitotic MB in the cell and potentially influence pluripotency, or can be degraded via autophagy. (B) Model detailing the possible fate of an MB following symmetric division, where it is released into extracellular space and can be potentially uptaken by surrounding cells.

guided by the MB, thus allowing for Rab11/FIP5 endosomes to deliver apical cargo during lumenogenesis (Li, Mangan, Cicchini, Margolis, & Prekeris, 2014).

Numerous recent studies have now begun to recognize the potential roles for MBs during stem cell maintenance and or carcinogenesis (Dionne, Wang, & Prekeris, 2015). It has been shown that cells that have differentiated in vitro from human embryonic stem cells (hESCs) have fewer MBs as compared to hESCs (Kuo et al., 2011). Additionally, MBs accumulate in the basal compartment of seminiferous tubules in testes, where germline cells reside (Kuo et al., 2011). Somewhat controversial to this, another phenomenon that has been observed is the capacity for different cell types to release MBs at different frequencies. For example, proliferating cells (such as stem cells) tend to release MBs at a higher rate, while cancer cells appear to retain them (Ettinger et al., 2011). Furthermore, it is known that postmitotic intracellular MBs can be degraded via selective macroautophagy. This autophagic degradation of MBs is necessary to prevent their accumulation, and it was suggested that cancer cells may have a decreased capacity of inducing postmitotic MB degradation (Kuo et al., 2011).

Despite accumulating evidence that MBs influence cell polarity and differentiation, many questions remain. How cells regulate MB uptake and degradation is unknown. What determines whether cells undergo asymmetric cytokinesis and which cell inherits the MB also remains unclear. Finally, despite many correlative studies, there is no clearly defined signaling pathway(s) that is dependent on MB accumulation and can regulate cell "stemness" and proliferation. Here we describe a few newly developed techniques and approaches that lead to the beginning of defining the postmitotic MB functions and regulation.

1. MIDBODY AND EPITHELIAL CELL POLARITY

Epithelial tissues are composed of polarized cells, which function as selectively permeable barriers. The plasma membrane of epithelial cells is divided into apical and basolateral domains, and specialized protein complexes between adjacent cells, such as the tight junctions (TJs), maintain the separation of apical and basolateral plasma membrane. Additionally, epithelial cells coordinate their polarization with neighboring cells within 3D space to form an apical lumen, a key step in the establishment of renal and gut architecture, and thereby function (Blasky, Mangan, & Prekeris, 2015). Despite the importance of lumenogenesis for epithelia function, the mechanisms governing this process remain to be fully understood. One of the models of de novo lumen formation proposes that upon first cell division, Rab11/FIP5 protein complex-containing apical endosomes are transported to the site of the forming lumen, where they fuse with a specialized apical PM site known as the AMIS to initiate single lumen formation (Fig. 2A). These Rab11-endosomes were shown to contain gp135 (apical glycoprotein), Crumbs3 (apical CRB polarity complex), TUBA (GEF for Cdc42), myosin-Vb, Sec15 (Exocyst subunit), and

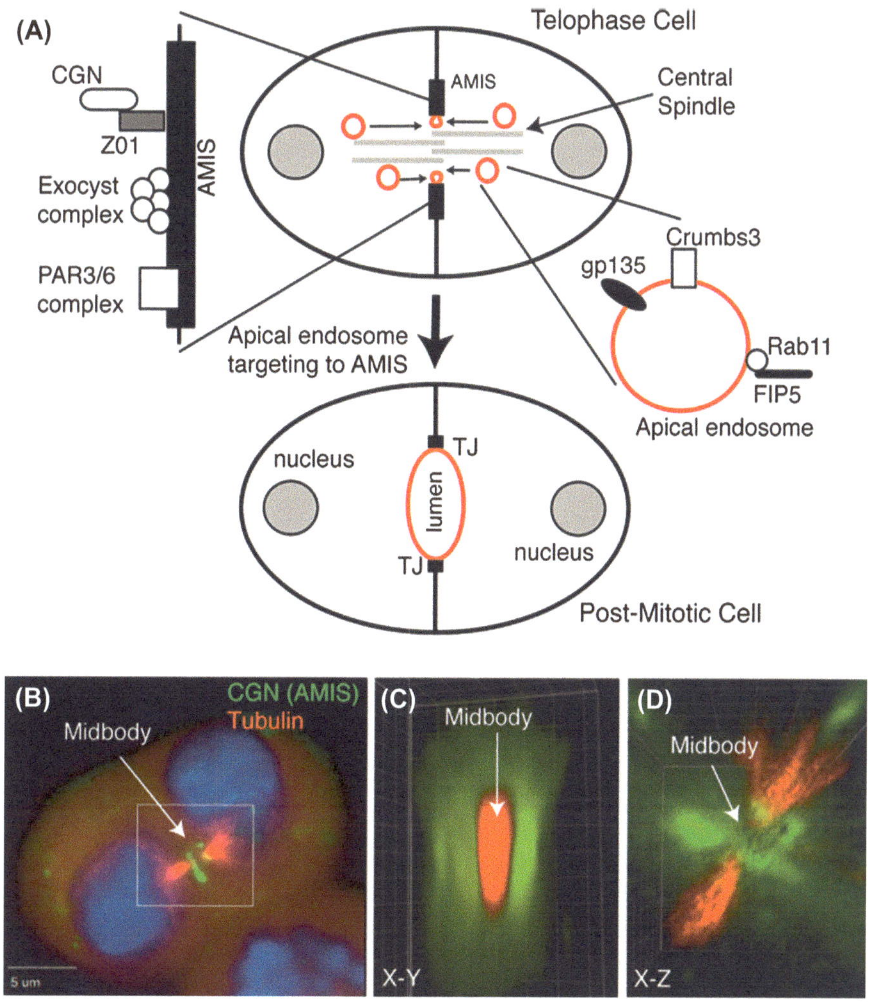

FIGURE 2 Midbody (MB)-dependent recruitment of apical plasma membrane proteins determines the site of nascent apical lumen formation.

(A) Schematic model depicting the role of MB during lumen formation. Red marks apical endosomes and apical plasma membrane. AMIS stands for apical membrane initiation site. (B–D) Madin–Darby canine kidney epithelial cells grown in Matrigel/Collagen matrix were fixed and immunostained with anticingulin (CGN; AMIS marker, antibody generated in the Prekeris lab) and antiacetylated microtubule (central spindle marker) antibodies. Panel B show a single image plane. Panels C and D show 3D rendering based on the entire Z-stack.

Reproduced from Li, D., Mangan, A., Cicchini, L., Margolis, B., & Prekeris, R. (2014). FIP5 phosphorylation during mitosis regulates apical trafficking and lumenogenesis. EMBO Reports, *15(4), 428–437. http://dx.doi.org/10.1002/embr.201338128 with permission from EMBO Reports.*

Rab8/Rabin8. The delivery of these apical cargo proteins to the site of the forming lumen is required for polarized epithelial cyst formation in vitro (Blasky et al., 2015).

Recent work has demonstrated that MB formation during telophase plays a major role in epithelial polarization. It was shown that during fly development MBs associates with TJs, thus providing polarization cues for newly formed daughter cells. Similarly, we have shown that during apical lumen formation in vitro, the MB serves as the first "symmetry breaking" event that establishes the apical lumen site, and that the lumen formation is initiated by AMIS assembly around the MB (Li, Kuehn, & Prekeris, 2014; Li, Mangan, et al., 2014). Thus, it is becoming clear that the MB regulates spatiotemporal properties of AMIS/TJ formation. What remains unknown is how and why the AMIS forms around the MB and how the MB-associated AMIS mediates targeting of apical endosomes during epithelia morphogenesis.

1.1 ANALYZING THE ROLE OF MB DURING EPITHELIA LUMENOGENESIS

1.1.1 Overview

Epithelial cell interactions that create apical lumens occur in three dimensions (3D) but most of our information about lumen formation is derived from the use of two dimensional (2D) models, leaving significant gaps in our understanding of the cellular and molecular mechanisms that drive lumen morphogenesis. Recently, 3D epithelial cultures have emerged as an in vitro technique that allows dissection of the molecular mechanisms governing lumenogenesis. The major strength of these 3D assays is the ability to perform high-resolution imaging during different stages of epithelia polarization while retaining the in vivo-like complexity of lumen formation in a 3D extracellular matrix. Here, we will describe the 3D in vitro lumenogenesis assays using the Madin–Darby canine kidney (MDCK) cell line. Also, similar assays can be performed using other spontaneously polarizing epithelial cell lines, such as Caco-2 (human intestinal epithelial cells), MCF10A (human mammary epithelial cells), or WIF2B (human hepatocyte cell line).

1.1.2 Procedure

1.1.2.1 Embedding MDCK cells in 3D matrix

Carry out steps 1–8 in a tissue culture hood.

1. Plate MDCK cells on 10 cm tissue culture plate and let grow for 24 h at 37°C. The plated cells should not be fully confluent since cells need to be in growth phase. Typically we want to have cells at 50–70% confluency after 24 h incubation. Thus, we usually plate cells at 30% confluency.
2. A couple of hours before plating cells, thaw Matrigel (Corning, #354230) on ice. It is very important to keep Matrigel cold even while thawing, since it rapidly solidifies at room temperature. While Matrigel alone is sufficient to grow polarized MDCK cysts, Matrigel is a very soft extracellular matrix. Thus, in some cases it may be beneficial to supplement it with purified collagen I. We

typically add collagen I to a final 3 mg/mL concentration to increase Matrigel stiffness.

3. Aspirate media and rinse cells with 10 mL phosphate-buffered saline (PBS). Add 2 mL 0.25% Trypsin–EDTA and let sit at 37°C for 10–15 min. MDCK cells are usually difficult to dislodge. Thus, if needed, they can be incubated for 20–25 min.
4. Dislodge cells by gently tapping at the side of 10 cm dish. Harvest cells by adding 8 mL MDCK media (Dulbeco's modified Eagle media supplemented with 10% fetal bovine serum) to plate and transferring cells to 15 mL tube. Sediment cells by centrifugation at 1000× rpm for 3 min.
5. Aspirate media and resuspend cells in 1 mL of MDCK media by pipetting up and down 50–100 times using 1 mL blue pipette tip. Pipetting up and down repeatedly helps to separate cells and is crucial for this assay to embed individual cells in Matrigel. Embedding cell clumps will lead to formation of cell aggregates with multiple lumens. We found that pipetting cells 40 times with a 1-mL pipette usually gives a maximum number of individual cells. However, this number may vary, thus it is advisable to initially pipette cells varying number of times, while monitoring the efficiency of cell separation.
6. Count cells to determine number of cells per milliliter and add 20,000 cells to 25 μL MDCK media in 1.5 mL tube. The optimum number of cells to plate may vary. Plating too many cells may cause clumping of dividing cells and overlap of fluorescence, making it difficult to distinguish single cells and observe them over time. Plating too few cells will make it more difficult to find single cells that are entering metaphase. We recommend testing a range of 10,000–50,000 cells to find the optimum number of cells for each assay.
7. Make a 75% Matrigel/Collagen solution by adding 75 μL Matrigel/Collagen to the 25 μL cell solution from step 6. Gently mix by pipetting up and down. Immediately plate the cell/Matrigel mix by placing a drop onto the center of either an eight-chamber dish (immunofluorescent microscopy) or a 5 cm gridded glass-bottom dish (time-lapse microscopy). This step needs to be done quickly, since the Matrigel-cell mixture solidifies readily at room temperature.
8. Place cells in the incubator and let solidify in the tissue culture incubator at 37°C for 30 min. Add serum-supplemented media.
9. Incubate the cells for 12–18 h. If grown in Matrigel, the majority of MDCK cells will undergo first cell division within 24 h of embedding. Thus, cells analyzed 12–18 h postembedding will either just have completed first cell division or will be at different stages of cytokinesis. Since AMIS is usually recruited to the MB at late telophase, at 12–18 h of incubation postembedding will enhance the chance to find cells at the initial stages of AMIS recruitment of lumenogenesis.

1.1.2.2 Analyzing the lumenogenesis by immunofluorescence microscopy.

1. Gently aspirate media and fix cells with 3% paraformaldehyde for 30 min.

2. Permeabilize cells with 0.5% Triton X-100 for 10 min, followed by 2× 20 min quenching with 100 mM Glycine in PBS.
3. Block nonspecific binding by incubating cells in immunofluorescence buffer (PBS with 0.05% Triton X-100, 0.1% bovine serum albumin, 10% fetal bovine serum, and 10 mM NaN_3) for 4 h at room temperature.
4. Block nonspecific mice IgG by incubating cells in immunofluorescence buffer with 20 mg/mL of goat antimouse $F(ab)_2$ fragments (Jackson ImmunoResearch, #115-006-006) for 40 min. This is a key step if mice primary antibodies will be used for immunofluorescence, since Matrigel is rich in mice IgG. This block then needs to be followed with 3 × 20 min washes with immunofluorescence buffer to remove any nonbound antimouse $F(ab)_2$ fragments.
5. Incubate with primary antibodies diluted in immunofluorescence buffer overnight. To visualize MB and central spindle, we typically use mice antiacetylated tubulin antibodies (Sigma #T7451). To visualize AMIS, any previously published anti-ZO1, antioccludin or anticingulin antibodies can be used. The AMIS also contain other membrane transport and polarity proteins, such as Par3/6 complex, the Exocyst complex, Syntaxin 3, Slp4a (Blasky et al., 2015). However, all these proteins, while somewhat enriched at the AMIS, are also present at the basolateral plasma membrane. As a result, staining using these markers leads to difficulties identifying the AMIS, especially at the early stages of its formation.
6. Incubate 3× 20 min with immunofluorescence buffer to remove nonbound primary antibodies, followed by incubation with appropriate fluorescent tag-conjugated secondary antibodies (in immunofluorescence buffer) for 60 min. Wash again 3× 20 min with immunofluorescence buffer.
7. Aspirate liquid, remove chamber walls, and dry slide for at least 1.5 h before mounting on any suitable mounting medium, such as Vectashield.

1.1.2.3 Analyzing the lumenogenesis by time-lapse microscopy

1. Mount dish on fluorescent microscope and use a gridded chart to label the location of several individual cells, especially those that may be in metaphase (starting division). After every time point, cells can be returned to incubator. The grid coordinates will help to find cells for following time-points. This is a preferred method when long (2–8 h) time-lapses are used, since the entire time-lapse set will take 1–4 days. If short time-lapses (under 1 h) are used, a fluorescent microscope equipped with environmental control chamber needs to be used. In that case, several cells can be picked (if microscope has X–Y motorized stage) and imaged without moving cells back to incubator.
2. Adjust focus to set top and bottom of each cell and take initial 0.2-μm-step z-stack images. To be sure to include the whole cell in the range of imaging from top to bottom, it is best to set the focus a little beyond the point when the cell first begins to blur. It is crucial for the experiment to pick cells that are fully suspended within the matrix and do not touch the bottom of the dish. Any

contact with glass bottom will set a competing polarity axis and usually leads to disorganized and multiluminal cysts.

3. Repeat imaging according to time frame for desired observation. This will result in taking a mini z-stack for every time point. Generally, avoid taking more than 50 time-lapse z-stacks. Taking fewer time-points or fewer images in each mini z-stack will decrease photo-damage to cells. Excessive imaging can be a problem, since mitotic cells are especially sensitive to photo-damage. Shorter exposures will also reduce photobleaching.
4. Taking mini z-stacks at every time point allows the use of postacquisition image analysis to generate three-dimensional images of MDCK cells at each time point during lumen formation. Alternatively, individual images that best represent the lumen formation dynamics can be selected and displayed/analyzed for every time point.

1.1.3 Data interpretation

Immunofluorescence microscopy (IFM) and time-lapse analysis are two great complementary approaches to analyze AMIS formation around MB during lumenogenesis. IFM allows analysis of the presence of various polarity and protein transport regulators during initial AMIS recruitment and formation. However, to truly understand the spatiotemporal dynamics of AMIS formation and lumen initiation, IFM findings need to be confirmed by time-lapse microscopy. Since at 12–18 h postembedding, cells will be at different stages of division and lumenogenesis, it can sometimes be challenging to determine the order and timing of protein recruitment to the MB and AMIS. This issue is complicated due to the fact that cells at late telophase look very similar to cells that already completed first division and have established a nascent lumen. Thus, arbitrarily chosen fixed cell images may not always represent a real order of lumen formation. Consequently, time-lapse analysis is the only approach that can fully define the spatiotemporal dynamics and regulation of nascent lumen formation.

2. ANALYSIS OF MIDBODY UPTAKE BY INTERPHASE CELLS

Now that it has come to light that MBs can either be asymmetrically inherited by one daughter cell or can be uptaken from the extracellular environment, it is important to develop techniques that allow further examination of the potential postmitotic MB functions. Recent work using *C. elegans* suggests that MBs may be engulfed by surrounding cells due to the presence of phosphatidylserine on the outer leaflet of the plasma membrane remnant surrounding the MB. Phosphatidylserines may act similarly to phagocytic "eat me" signal for cells to uptake the MB through an actin-dependent mechanism. Several techniques, including MB purification and MB uptake analysis, recently emerged to allow initial analysis of the machinery and functional consequences of MB accumulation in postmitotic cells.

2.1 MIDBODY PURIFICATION

2.1.1 Overview

This protocol has been modified to get the best yield of MBs using HeLa cells stably expressing MKLP1-GFP (Kuo et al., 2011). Based on recent studies (Ettinger et al., 2011), it is likely that this protocol needs to be adjusted to different certain cell types, as cell lines can release MBs at different frequencies. Additionally, it is possible that MBs derived from different cells can contain different proteins and lipids, thus, the sucrose gradient centrifugation conditions (described later) may need to be adjusted for each cell type.

2.1.2 Procedure

1. Grow ten 15 cm dishes of HeLa cells to subconfluency (~70%) in Dulbecco's modified Eagle media supplemented with 10% fetal bovine serum.
2. Once 70% subconfluency is reached, collect conditioned media (referred to as CM) from all 10 plates and place into 50 mL Corning conical tubes.
3. Centrifuge CM at 300× *g* for 5 min at 4°C. Remove the supernatants and place into fresh 50 mL conical tubes, and perform another 300× *g* centrifugation for 5 min at 4°C. This centrifugation step will remove majority of intact cells or cell debris.
4. Remove the supernatants and place into fresh 50 mL conical tube. Centrifuge the CM at 10,000× *g* for 30 min at 4°C.
5. Discard the supernatants. Combine the pellets and resuspend in 1 mL sterile PBS. Place into a fresh 1.5 mL microcentrifuge tube.
6. Centrifuge the combined pellets at 10,000× *g* for 30 min at 4°C. Discard the supernatant.
7. Resuspend the pellet in nuclease-free water, 6 mM $MgCl_2$, 250 mM sucrose, and 250 mM NaCl to a final volume of 200 uL.
8. Prepare a step-density gradient in a 1.3-mL ultracentrifuge tube consisting of 400 uL of 1 and 2 M sucrose placed on top of 300 uL of a 40% glycerol cushion. For our sucrose gradients, we typically use a buffer that consists of 20 mM HEPES, pH 7.4, 6 mM $MgCl_2$, 1 mM Dithiothreitol, and 100 mM NaCl.
9. Layer the resuspended pellet fraction on top of the step gradient. Ultracentrifuge using a TLS-55 rotor by spinning at 3000× *g* for 20 min at 4°C.
10. Immediately following centrifugation collect the visible opaque band at 2 M sucrose/glycerol Interphase and place into a fresh 1.5 mL microcentrifuge tube.
11. Centrifuge at 10,000× *g* for 30 min at 4°C. Discard the supernatant, and resuspend the pellet in 20–40 uL of sterile PBS. Here, the pellet may not be visible, so it is important to keep track of which side of the tube the pellet may be on. Proceed immediately to MB uptake experiment.

2.2 MIDBODY UPTAKE ANALYSIS

2.2.1 Overview

Here, we describe the protocol and methods for analysis to assess MB uptake by HeLa cells. As stated before, it is likely that MBs isolated from different cell types may have distinct composition and properties. Additionally, various cell lines may have different capacities to uptake MBs. Both of these compounding factors may lead to different results when isolating and "feeding" MBs to various cell types. We will use the term "feeding" to describe the process in which we physically pipette MBs onto the cell population.

2.2.2 Procedure

1. One day prior to MB "feeding," plate HeLa cells on collagen-coated 18 mm coverslips placed in 12 well dishes. These cells should be grown to subconfluency (~70%).
2. Immediately prior to MB "feeding," remove media from wells and add 300 uL of fresh media. From the MB isolation protocol, pipette 10–20 uL of isolated MBs onto HeLa cells.
3. Centrifuge the plate at 1050 rpm for 45 min to assist the MBs in sinking to the cells. Add 200 uL fresh media to each well following centrifugation. Incubate cells for 1–2 h. From our experience, cells need a couple of hours to internalize the MBs.
4. Wash cells with PBS and fix for 30 min in 4% paraformaldehyde.
5. Fix and immunostain cells using a standard immunofluorescence protocol.
6. Collect images using a 63× or 100× oil objectives and a fluorescent microscope. To ensure nonbias in collecting images, random sets of cells were imaged without first examining if those sets of cells had MBs in the viewing field. Since MBs can be present at various subcellular locations, z-mini-stack (1 μm step size) needs to be imaged for every cell analyzed.
7. To assess the efficiency of MB uptake, all individual z-planes should be examined. Since our HeLa cells stably express a known MB marker MKLP1-GFP, the uptaken MBs can be readily identified. Alternatively, MBs can be marked using antibodies that recognize endogenous MB-resident proteins, such as Cep55, Cyk4, or MKLP1. The number of MBs internalized versus the number of cells in the viewing field can be directly compared to achieve an "MB accumulation frequency" at different time-points (Fig. 3). If needed, cells can be costained for other various cellular markers. For example, costaining with anti-CD63 (lysosomal marker, gift from Dr. Andrew Peden) antibody will allow to determine whether the MB is present in cytosol or is targeted to lysosomal degradative pathway (Fig. 3).

2.2.3 Data interpretation

It is becoming clear that MBs are engulfed via phagocytosis-like machinery. As a result, it is possible that some of the MBs are quickly routed to the degradative

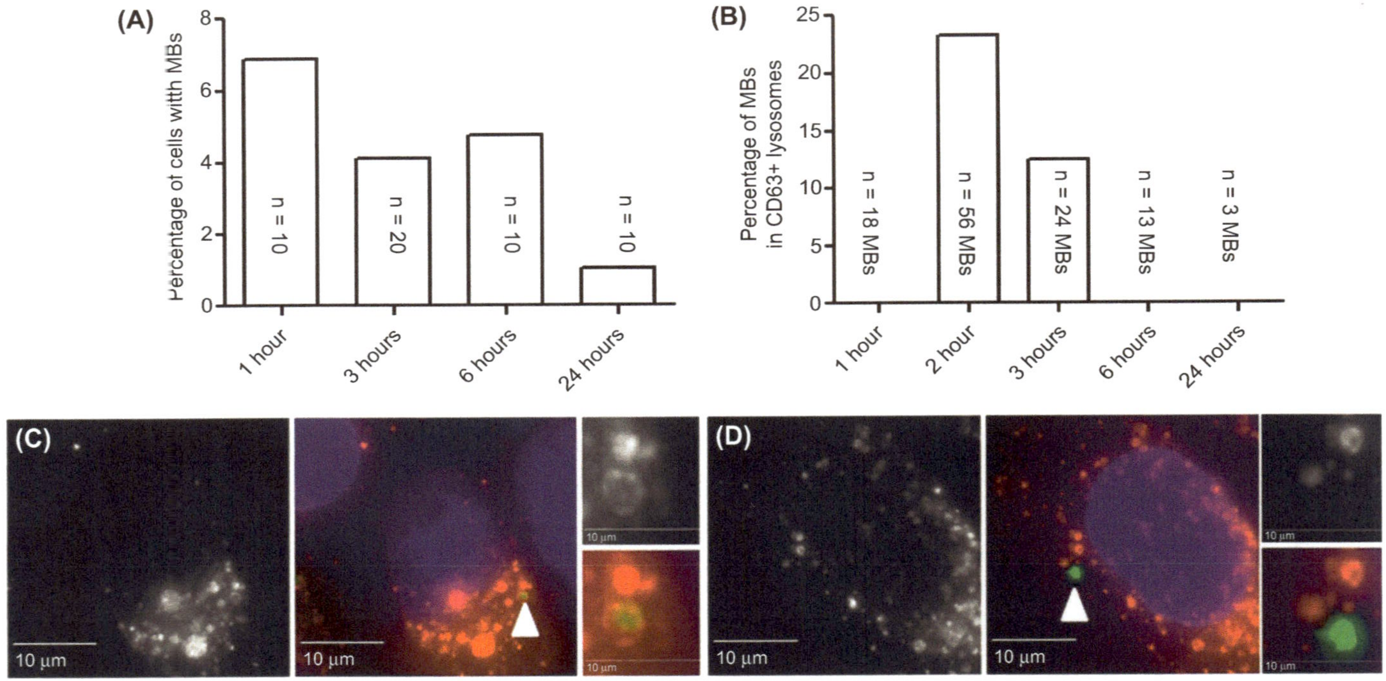

FIGURE 3 Quantification of midbody (MB) uptake.

(A) Time course of MB feeding. The percentage on the y-axis represents the number of MBs per nuclei in a random viewing field of cells. N represents a number of MBs counted. (B) Number of MBs present in CD63-positive lysosomes. Colocalization of MKLP1 and CD63 was scored to identify MBs undergoing autophagy. The percentage on the y-axis represents the number of CD63-positive MBs per the total number of MBs counted for that time point. (C) Representative image of MB present in CD63-positive lysosome. Left panel shows CD63 (red) only. Middle panel shows merged image of CD63 (red) MKLP1 (green) and DNA (blue). *White arrowhead* points to engulfed MB. Right panel shows inset images from middle panel. Right top panel shows CD63 (red) only. Right bottom shows merged image of CD63 (red) and MKLP1 (green).
(D) Representative image of MB within a cell but not within a CD63-positive lysosome. Left panel shows CD63 (red) only. Middle panel shows merged image of CD63 (red) MKLP1 (green) and DNA (blue). *White arrowhead* points to MB. Right panel shows inset images from middle panel. Right top panel shows CD63 (red) only. Right bottom shows merged image of CD63 (red) and MKLP1 (green).

lysosomal pathway and do not participate in signaling. Consistent with this, we do see that a fraction of internalized MBs costain with lysosomal markers, such as CD63, during first 2 h of postfeeding (Fig. 3C). Furthermore, as shown in Fig. 3A, the number of MBs remaining in cells 24 h postfeeding is markedly decreased compared to first hour, indicating that they are likely targeted to the lysosomes for degradation. Interestingly, after 3 h of postfeeding, the majority of intracellular MBs did not colocalize with CD63-positive lysosomes, potentially indicating that these MBs have evaded phagocytic degradation (Fig. 3B), although the machinery governing this process remains completely unknown. These evidence suggest that MBs can be inherited from the extracellular environment and retained in cells, without being degraded, for at least 24 h postinternalization. What remains to be determined is whether these MBs actually function as a signaling platform and whether they affect the properties and proliferation of cells that retain and accumulate postmitotic MBs.

SUMMARY

MBs have recently emerged as putative regulators of several postmitotic cell functions and properties, including cell polarization and proliferation. Furthermore, recent data indicate that MBs can also be uptaken from the extracellular milieu, thus potentially serving as platforms for parallel signal transfer between cells. Further studies, however, will be needed to further understand the regulation and functional significance of MB inheritance, uptake, and accumulation. Here, we have introduced several new techniques that will allow researchers to start investigating postmitotic MB functions.

ACKNOWLEDGMENTS

We are grateful to Abitha Jacob and Dr. Alexander Blasky for critical reading of the manuscript. We thank Dr. Andrew Peden, University of Sheffield, for the CD63 antibody. I apologize to all colleagues whose work could not be cited due to space limitations. Work in Rytis Prekeris' laboratory is supported by NIH (DK064380) and Cancer League of Colorado.

REFERENCES

Blasky, A. J., Mangan, A., & Prekeris, R. (2015). Polarized protein transport and lumen formation during epithelial tissue morphogenesis. *Annual Review of Cell and Developmental Biology, 31*, 575–591. http://dx.doi.org/10.1146/annurev-cellbio-100814-125.

Chai, Y., Tian, D., Yang, Y., Feng, G., Cheng, Z., Li, W., & Ou, G. (2012). Apoptotic regulators promote cytokinetic midbody degradation in *C. elegans*. *Journal of Cell Biology, 199*(7), 1047–1055. http://dx.doi.org/10.1083/jcb.201209050.

Crowell, E. F., Gaffuri, A., Gayraud-Morel, B., Tajbakhsh, S., & Echard, A. (2014). Engulfment of the midbody remnant after cytokinesis in mammalian cells. *Journal of Cell Science, 127*, 3840–3851. http://dx.doi.org/10.1242/jcs.154732.

Dionne, L. K., Wang, X. J., & Prekeris, R. (2015). Midbody: from cellular junk to regulator of cell polarity and cell fate. *Current Opinion in Cell Biology, 35*, 51–58. http://dx.doi.org/10.1016/j.ceb.2015.04.010.

Elia, N., Sougrat, R., Spurlin, T. A., Hurley, J. H., & Lippincott-Schwartz, J. (2011). Dynamics of endosomal sorting complex required for transport (ESCRT) machinery during cytokinesis and its role in abscission. *Proceedings of the National Academy of Sciences of the United States of America, 108*(12), 4846–4851. http://dx.doi.org/10.1073/pnas.1102714108.

Ettinger, A. W., Wilsch-Bräuninger, M., Marzesco, A.-M., Bickle, M., Lohmann, A., Maliga, Z., … Huttner, W. B. (2011). Proliferating versus differentiating stem and cancer cells exhibit distinct midbody-release behaviour. *Nature Communications, 2*, 503. http://dx.doi.org/10.1038/ncomms1511.

Kuo, T.-C., Chen, C.-T., Baron, D., Onder, T., Loewer, S., Almeida, S., … Doxsey, S. (2011). Midbody accumulation through evasion of autophagy contributes to cellular reprogramming and tumorigenicity. *Nature Cell Biology, 127*(13), 1214–1223. http://dx.doi.org/10.1038/ncb2332.

Li, D., Kuehn, E. W., & Prekeris, R. (2014). Kinesin-2 mediates apical endosome transport during epithelial lumen formation. *Cellular Logistics, 4*, e28928. http://dx.doi.org/10.4161/cl.28928.

Li, D., Mangan, A., Cicchini, L., Margolis, B., & Prekeris, R. (2014). FIP5 phosphorylation during mitosis regulates apical trafficking and lumenogenesis. *EMBO Reports, 15*(4), 428–437. http://dx.doi.org/10.1002/embr.201338128.

Pollarolo, G., Schulz, J. G., Munck, S., & Dotti, C. G. (2011). Cytokinesis remnants define first neuronal asymmetry in vivo. *Nature Neuroscience, 14*(12), 1525–1533. http://dx.doi.org/10.1038/nn.2976.

Schiel, J. A., Simon, G. C., Zaharris, C., Weisz, J., Castle, D., Wu, C. C., & Prekeris, R. (2012). FIP3-endosome-dependent formation of the secondary ingression mediates ESCRT-III recruitment during cytokinesis. *Nature Cell Biology, 14*, 1068–1078. http://dx.doi.org/10.1038/ncb2577.

Singh, D., & Pohl, C. (2014). Coupling of rotational cortical flow, asymmetric midbody positioning, and spindle rotation mediates dorsoventral axis formation in *C. elegans*. *Developmental Cell, 28*(3), 253–267. http://dx.doi.org/10.1016/j.devcel.2014.01.002.

Wilcock, A. C., Swedlow, J. R., & Storey, K. G. (2007). Mitotic spindle orientation distinguishes stem cell and terminal modes of neuron production in the early spinal cord. *Development (Cambridge, England), 134*, 1943–1954. http://dx.doi.org/10.1242/dev.002519.

Yang, D., Rismanchi, N., Renvoise, B., Lippincott-Schwartz, J., Blackstone, C., & Hurley, J. (2008). Structural basis for midbody targeting of spastin by the ESCRT-III protein CHMP1B. *Nature Structural and Molecular Biology, 15*(12), 1278–1286. http://dx.doi.org/10.1038/nsmb.1512.

CHAPTER

Micromanipulation of daughter cells for the study of cytokinetic abscission

11

J. Lafaurie-Janvore*,[a], C. Lafaurie[§], M. Piel*,[1]

*Institut Curie, Paris, France

[§]Versailles, France

[1]Corresponding author: E-mail: matthieu.piel@curie.fr

CHAPTER OUTLINE

[a]Current address: Ecole Polytechnique CNRS-EP UMR 7646, Palaiseau, France

Methods in Cell Biology, Volume 137, ISSN 0091-679X, http://dx.doi.org/10.1016/bs.mcb.2016.04.012

Abstract

The last step of cytokinesis, abscission, consists in the severing of the intercellular bridge connecting the two daughter cells. Because daughter cells move randomly on regular cell culture substrates, the use of adhesive micropatterns facilitates the observation of the intercellular bridge and its severing. Here we propose general rules to design micropatterns optimized to study this process. In particular, these micropatterns allow a good stabilization of the daughter cells and a predictable positioning of the intercellular bridge. We suggest a series of micropatterns controlling various cellular parameters such as distance between daughter cells or daughter cells polarization. We give recommendations for videomicroscopy acquisition during cell division and propose automated image analysis methods using kymograph analysis or bridge detection. Finally, we detail methods to artificially cut the intercellular bridge using UV-based laser ablation or using two-photons laser ablation.

INTRODUCTION

Toward the end of cytokinesis, after furrow contraction, the two daughter cells are connected by an intercellular bridge filled with microtubules. A complex series of molecular mechanisms regulate the severing of this bridge, leading to abscission (Agromayor & Martin-Serrano, 2013; Chen, Hehnly, & Doxsey, 2012; Fededa & Gerlich, 2012). In particular, the extracellular environment has been shown to regulate the abscission timing (Lafaurie-Janvore et al., 2013). A major challenge to study cytokinesis dynamics is the difficulty to observe the intercellular bridge due to the heterogeneity of daughter cells behaviors. *In vitro* cultured cells show a high variability in cell shape, cell motility, and cell migration direction. Depending on daughter cells behaviors, the intercellular bridge can adopt various length and shape, might be hidden if the daughter cells are too close, or even broken by another cell migrating in between the two daughter cells. Moreover, abscission occurs at variable times after anaphase (Dambournet et al., 2011; Steigemann et al., 2009) and is a rapid process once started (10–15 min) (Steigemann & Gerlich, 2009). Thus it is challenging to observe abscission at high resolution unless waiting a long time for each observed cell. Methods for controlling the spatial and temporal components of the abscission are thus needed to facilitate the study of this important process.

In this article we present various methods to manipulate the daughter cells. The use of adhesive micropatterns allows a standardization of daughter cells positioning thus facilitating the observation of the intercellular bridge and its severing. Quantitative studies can be performed by controlling specifically some cellular parameters by varying the geometry of the micropatterns. The stereotypical positioning of daughter cells in turn allows the use of automated image analysis to detect sequential cell division steps including abscission. Finally, laser ablation of the intercellular bridge on one side of the midbody is an efficient means to induce abscission on the other side of the midbody at a precise time.

1. ADHESIVE MICROPATTERNS

Adhesive micropatterns can be produced on glass or plastic surfaces using various techniques (Théry, 2010). In this article, we use the UV method previously described in great details (Azioune, Carpi, Tseng, Théry, & Piel, 2010; Azioune, Storch, Bornens, Théry, & Piel, 2009). This simple microfabrication technique allows a robust control of size and shape of the micropatterns with a resolution of about 1 μm. Briefly, glass coverslips are coated with PLL-g-PEG (SuSoS), a cell repellent, before being exposed to deep UVs (Jelight) through a chromium synthetic quartz mask (Toppan) on which the desired patterns are transparent. The surface is subsequently coated with fibronectin (F1141, Sigma Aldrich), which fills in the UV-ablated zones to form the cell-adhesive micropatterns. As a consequence, seeded cells can adhere only on fibronectin-treated regions corresponding to the desired patterns. Interestingly, cells are able to cover nonadherent regions included in a pattern by forming actin stress fibers over the nonadherent zone and by strengthening the focal adhesions attaching the stress fibers to the substrate (Théry, 2010).

1.1 MICROPATTERNS DESIGN

In this first section, we expose general rules to design micropatterns optimized to study cytokinesis.

1.1.1 Separation of the daughter cells

A good micropattern for cytokinesis contains two distinct adhesive regions to separate the two daughter cells after division. We call this design a "double micropattern" (Fig. 1). For single interphase cells, a crossbow-shape pattern induces a good stabilization of the cell with a strong subcellular polarization (Théry et al., 2006): formation of a lamellipodium-like structure over the curved section and a contractile part over the straight section of the pattern. To stabilize each daughter cell on its part, it is thus possible to use a double micropattern containing two crossbows head to tail.

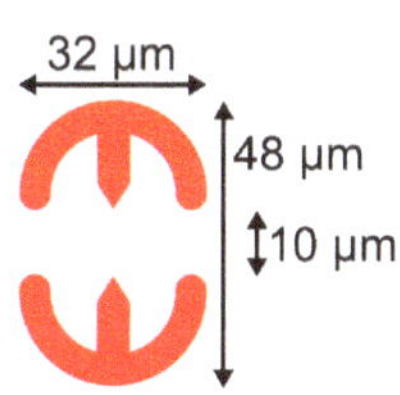

FIGURE 1

Design of a double pattern to study cytokinesis. Dimensions are adapted for HeLa cells.

An important point in the design of micropatterns for cytokinesis is the dimension of the double micropattern. To have the two daughter cells well stabilized on the two parts of the pattern, the mother cell has to be able to spread over the two parts of the pattern before division. If the mother cell stays on one part of the pattern and divides on this part, the daughter cells will stay only on this same part (Fig. 2A). Sometimes, one daughter cell might reach the second part of the double pattern but in an uncontrolled fashion, either before or after the intercellular bridge severing, abolishing the interest for a good observation of cytokinesis and abscission. On the contrary, when the mother cell spreads on both parts of the double pattern, the two daughter cells systematically go back to their own part, probably thanks to retraction fibers attaching the mother cell body to the two parts during mitosis (Théry et al., 2005). In that configuration the two daughter cells are well stabilized, each on their own part of the double pattern with the intercellular bridge clearly observable in between them until abscission occurs (Fig. 2C). On the other hand, if the two parts of the pattern are too close from one another, the daughter cells can easily move closer to each other and even rotate or change their position, preventing observation of the cytokinetic bridge (Fig. 2B). Thus it is crucial to adapt the dimension of the double micropattern to the size of the mother cell prior to mitosis, to allow both the spreading of the mother cell on the entire double pattern (two parts not too far) and to avoid the movement of the daughter cells between the two sides (a large enough gap between the two parts is needed). We never reached 100% of well-stabilized daughters, but it is possible to optimize the design, for a given cell type, to obtain a large number of correct and stable positioning of the daughter cells.

1.1.2 Stabilization of the intercellular bridge

On the double crossbow pattern presented earlier, or on double arcs (brackets), the daughter cells are stabilized and stay on a given side of the pattern, but we observed that the intercellular bridge could be positioned at any place between the daughter cells (Fig. 3 left). On such patterns, the distance between the convex envelopes of the two parts is constant across the width of the pattern. After division, the intercellular bridge often forms at the center of the pattern but can move randomly toward the left or the right of the pattern (Fig. 3 left). To improve the stabilization of the bridge at one reproducible and predictable position, a double pattern with a minimum distance between the convex envelopes of the two parts can be used. In that configuration, the bridge is always stabilized at the position of the minimal distance (Fig. 3 right).

Considering these constraints, here is a summary of the recommendations to design micropatterns allowing a good stabilization of the daughter cells and a predictable positioning of the intercellular bridge:

- Design a double micropattern with two distinct parts.
- The dimensions should allow the mother cell to spread over the entire double pattern: adjust the distance between the two parts (small enough) and the total spreading surface (small enough).

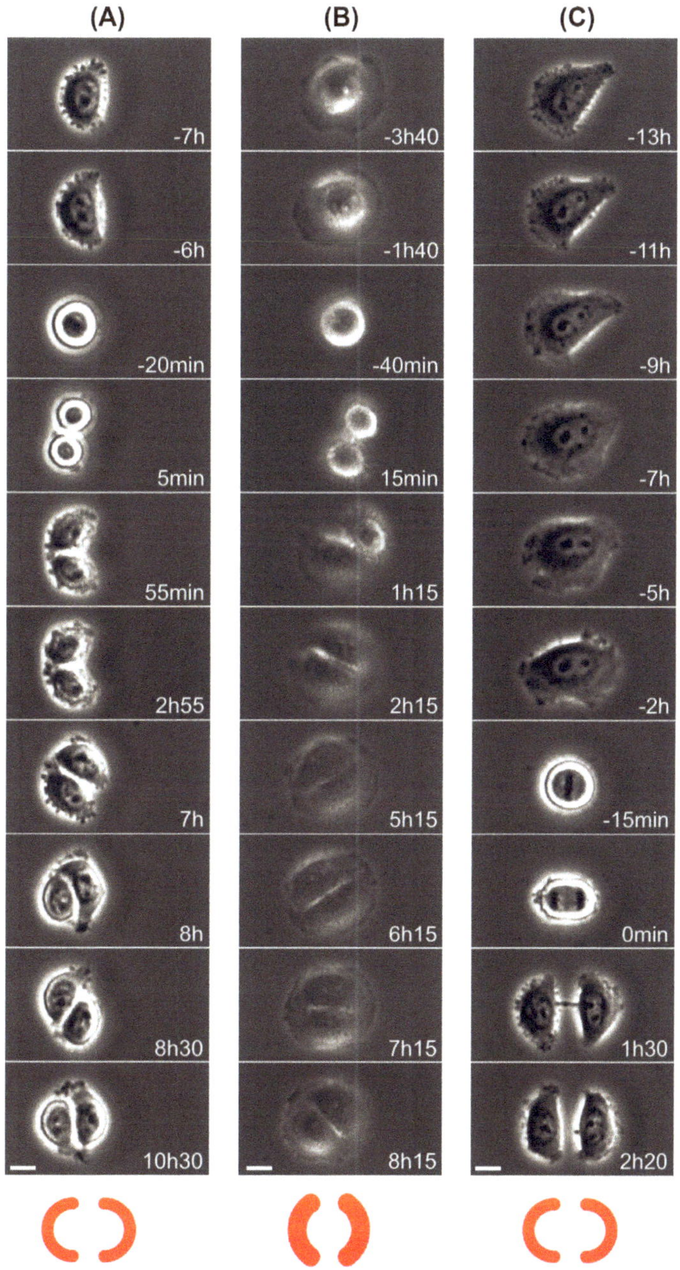

FIGURE 2

Phase contrast videomicroscopy of HeLa cells cultured on three different double patterns represented in red below the microscopy images. (A) The mother cell and the daughter cells stayed only on one part of the double pattern. (B) The daughter cells moved closer to each other and rotated. (C) The mother cell spread on both parts of the double pattern and the daughter cells were stabilized each on their part of the double pattern. Time is in hours and minutes, 0 min is at anaphase, and scale bar is 10 μm.

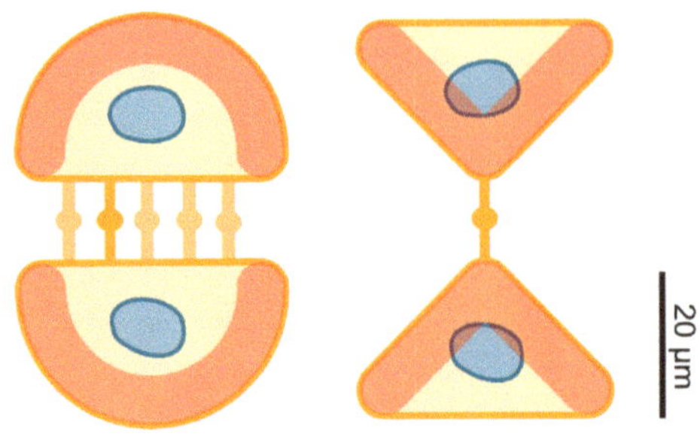

FIGURE 3

Position of the intercellular bridge depending on the design of the double pattern. On the left, the length of the bridge is constant over time but its position varies. On the right, the position of the bridge is perfectly controlled over time.

- The distance between the two parts of the double micropattern should be large enough to avoid the two daughter cells to rotate, return, or change their positions.
- The dimensions of the pattern must be adapted to each cell type depending on cell size and postmitotic behaviors (more or less motile). Dimensions presented in this article are optimized for HeLa cells.
- The distance between the convex envelopes of the two parts of the double pattern should have a unique minimum to induce the positioning of the bridge at this place.

1.1.3 Control of various cellular parameters

Adhesive micropatterns are powerful tools to control various cell parameters and cell processes via the control of adhesion surface area, spreading surface area, or geometry of cell adhesion. In particular, it is possible to reproducibly control intracellular organization of internal cell structures, such as positioning of the nucleus, centrosome, actin stress fiber, Golgi apparatus (Théry et al., 2006), or even cell division axis (Théry et al., 2005). Taking advantage of this knowledge, it is possible to design patterns to combine effects on daughter cell positioning and on internal organization of each daughter cell. We propose a collection of double micropatterns designed to control various cellular parameters during cytokinesis (Fig. 4).

The distance between daughter cells can be modulated by varying the distance between the two parts of the double pattern while maintaining a constant global spreading surface area for the mother cell. Adhesion between daughter cells can be controlled by varying the region between the two parts of the double pattern, while adhesions of the bridge to the substrate can be modulated to impose adhesion only at the extremities of the bridge or only at its center. Because daughter cells adopt different shapes and behavior depending on the adherent region between the two parts of the double pattern, having or not adhesion in this region induces the cells to compress the bridge (with adhesion) or to stretch the bridge (without adhesion). An interesting property of adhesive micropatterns is the possibility to

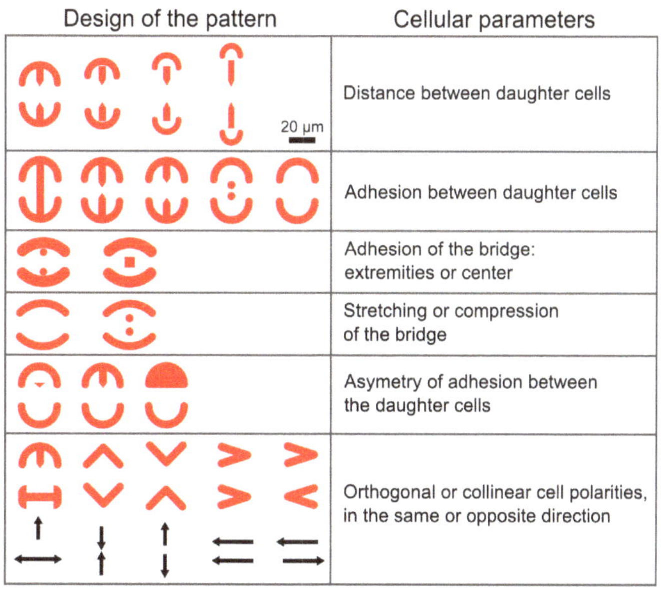

FIGURE 4

Control of various intercellular and intracellular parameters with the double pattern design.

impose asymmetry between the two daughter cells by simply designing a double pattern with two different subparts. It is possible to vary the amount of adhesion of the daughter cells, the contractility of the daughter cells, or even the front–rear polarity, having them collinear or orthogonal, in the same direction or in opposite direction (bottom of Fig. 4).

Thus, this technique opens a large set of possible configuration for the study of abscission, mimicking various potential tissue contexts. Moreover, using adhesive micropatterns provides a normalization of *in vitro* cultured cells—in term of shape, size, and even internal organization and polarization—allowing a more accurate comparison between various treatments.

1.2 CYTOKINESIS MONITORING USING VIDEOMICROSCOPY

General methods to perform videomicroscopy of cell division using fluorescent reporter cell lines have already been described (Schmitz & Gerlich, 2009). Nevertheless, here we highlight important points to achieve successful videomicroscopy of dividing cells, particularly during cytokinesis and abscission.

1.2.1 Preparation of the sample

- Use a stable cell line expressing fluorescent tubulin and DNA: in this article, we used a cell line generated by Daniel Gerlich's group, HeLa Kyoto α-tubulin-EGFP Histone2B-mCherry. It is very convenient to monitor many steps of the cell division process: the DNA fluorescence is used to detect easily prophase, metaphase, and anaphase, while the microtubules fluorescence is used to follow the evolution of microtubules in the intercellular bridge during cytokinesis and to determine easily abscission (Fig. 5A). An important point is that this cell line allows also to verify the quality of DNA segregation during anaphase and to remove from the study any cell presenting a DNA bridge during cytokinesis, which is known to affect abscission.

- Use glass bottom dishes (Fluorodish, FD35-100) or magnetic chamber (Chamlide, CM-B25-1) to hold micropatterned coverslips. It offers optimal fluorescence transmission without autofluorescence or light absorption and is thin enough to use a short working distance objective (for example a 20× 0.75 NA objective has a working distance of 1 mm).
- Ensure a good preparation of the sample. Clean the bottom of the glass with water then ethanol to remove any trace of proteins from the cell culture medium. Fix very well the sample on the microscope's stage to allow good multiposition acquisition. Change the cell culture medium before the beginning of the movie to remove dusts and dead floating cells. Be sure that the sample is horizontal to obtain a homogenous focus all over the field of view (especially important at low magnification).
- Use cell culture medium optimized for fluorescence imaging—that do not contain phenol red. Indeed, working at low magnification (20×), the depth of focus is important and the autofluorescence of the medium affects a lot the signal-to-noise ratio. For example, Dulbecco's Modified Eagle's Medium (DMEM) without phenol red (Gibco, 11054-020) or CO_2-independent medium Leibovitz's L-15 (Gibco, 21083-027) can be used.
- If the microscope is not equipped with a humidified chamber, limit the evaporation of water. Use a CO_2-independent cell culture medium and seal the sample chamber with parafilm or by adding a layer of mineral oil (Sigma Aldrich, M5904) on top of the cell culture medium.

1.2.2 Reducing phototoxicity

Another important point for videomicroscopy is the toxicity due to cell illumination with a strong light source, known as phototoxicity. It is particularly important when performing fluorescent videomicroscopy of dividing cells because cells are even more sensitive to the stress induced by illumination during division and because movies need to be long with a good time resolution. For example, some cells could suffer without dying rapidly but still never divide during the course of the experiment. During metaphase, cells are particularly sensitive and illumination could even block a cell in metaphase, thus never entering cytokinesis. Here are some tips to reduce phototoxicity during acquisition:

- By setting properly the exposition time and the power of the fluorescence source, reduce as much as possible the fluorescent illumination of the cells to have just enough signal (check the gray levels, not the visual quality of the image). The acquired image should be good for quantification even if not particularly

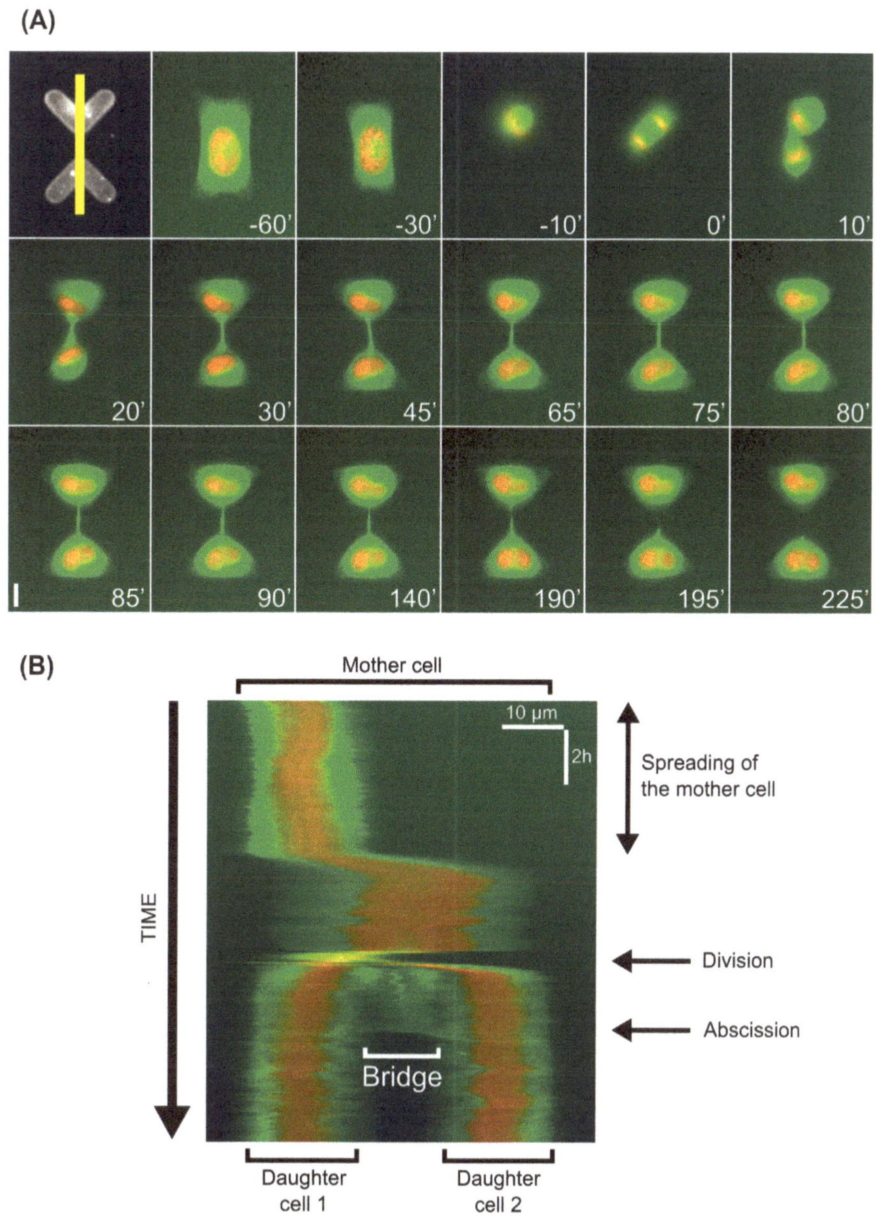

FIGURE 5

Videomicroscopy of a cell dividing on an optimized double pattern. HeLa Kyoto stable cell line expressing histone2B-mCherry (red) and α-tubulin-EGFP (green). (A) Time-lapse visualization, time is in minutes, 0 min is at anaphase. The first image shows the double micropattern (fluorescent fibronectin). The *yellow line* indicates the region used to calculate the kymograph below. Scale bar is 10 µm. (B) Kymograph of the time lapse presented above. Scale bars are 10 µm (horizontal) and 2 h (vertical).

Adapted from Lafaurie-Janvore, J., Maiuri, P., Wang, I., Pinot, M., Manneville, J.-B., Betz, T., ... Piel, M. (2013). ESCRT-III assembly and cytokinetic abscission are induced by tension release in the intercellular bridge. Science, 339*(6127), 1625–1629. http://doi.org/10.1126/science.1233866.*

"beautiful." Because mechanical shutters have a certain delay to respond, it can be better to increase the exposition time (to several 100 ms) and to decrease the intensity of the fluorescence source (with neutral density filters). The total amount of light will be less. What is important is to optimize the time during which cells are exposed to the light compared to the time during which the camera is acquiring. This should be as close as possible to 100%. For faster responding light sources (lasers or LEDs), a short illumination time with a stronger light might be better, but only if the time ratio described above stays close to 100%.

- Because phase contrast objectives are less bright than Differential Interference Contrast (DIC) objectives, use a phase contrast objective only if you really need the phase contrast images.
- If the resolution needed does not require binning 1 acquisition, use binning 2 acquisition to increase the signal to noise ratio and thus decrease illumination time.

1.2.3 Imaging setup

- Use an automated videomicroscope that allows multiposition acquisition of long-term (20–24 h) movies. Use a time lapse of 3–5 min between two consecutive images. Avoid cell synchronization to limit its side effects. For example, thymidine blocks can induce DNA segregation defects affecting abscission.
- Use a robust laser-based autofocus to correct the Z-drift of the stage. The intercellular bridge is only 1–2 μm in diameter, so even a slight drift in Z might compromises the imaging of the fluorescent microtubules of the bridge. Image-based autofocus is slow, not accurate enough for cytokinesis and abscission studies when performed on transmitted light images, and is phototoxic when performed on fluorescence channels. On the contrary, laser-based autofocus uses an infrared laser to detect interfaces between air and glass and between glass and cell culture medium. It is not toxic for the cells and very accurate. It is essential to follow abscission in live cells.
- Use a low-magnification objective (20×) with high numerical aperture (0.75 NA). The low magnification has two advantages: first having a large number of cells per field of view, thus increasing the statistics per image; and second having a higher depth of focus allowing the visualization of the bridge during the entire cytokinesis without the need to acquire too many images on the Z-axis (the bridge is about 10 μm above the substrate during early cytokinesis, but only 1 or 2 μm above the substrate when the daughter cells are completely spread, which is often the case when abscission occurs). The high numerical aperture is essential to obtain a good visualization of the microtubules of the bridge (only 1 μm in diameter) and to collect enough fluorescence signal with a low illumination to preserve the cells from phototoxicity.
- Depending on what is monitored during the acquisition, use a Z-offset to focus on the spread cells (during interphase or late cytokinesis) or on round cells (during metaphase, anaphase, and early cytokinesis) and to image different structures in different color channels.

1.3 AUTOMATED IMAGE ANALYSIS

Another advantage of using adhesive micropatterns to study cytokinesis is the ability to develop automated image analysis software to analyze long-term movies. Many steps of cell division can be automatically detected: metaphase, anaphase, evolution of microtubules in the intercellular bridge, and abscission. A very accurate software has already been proposed to detect most cell division events (Held et al., 2010), but here we focus on the analysis of cells cultured on adhesive micropatterns and on cytokinesis.

1.3.1 Identifying isolated dividing cells on micropatterns

Using a 20× objective to acquire long-term movies, the field of view is generally large enough to contain 8 to 10 double micropatterns. Because these patterns are not all necessarily occupied by a cell at the beginning of the movie and because some cells might not divide during the time course of the movie, the first part of any kind of analysis is to crop substacks containing individual dividing cell. Here are the main steps to be done:

- Using DNA image of the first frame of the movie, detect isolated nuclei. The minimal distance between an isolated nucleus and its nearest neighbor is set in accordance with the gap between two micropatterns. A radius defines a region of interest for each isolated cell. This step allows to keep regions where micropatterns are filled with one cell at the beginning of the movie and to ignore empty micropatterns or micropatterns filled with more than one cell.
- For each region of interest identified in the previous step, select the ones that contain exactly two nuclei in the DNA image of the last frame of the movie. This step allows to ignore cells that did not divide during the movie.
- The regions of interest are finally cropped to create substacks of DNA and microtubules images for further analysis.

We have developed an implementation of this method as an ImageJ plugin which is available at the following address: "http://umr144.curie.fr/en/our-research-groups/systems-cell-biology-cell-polarity-and-cell-division-matthieu-piel/our-links/our."

1.3.2 Complete cell division analysis using kymograph

The stabilization of the intercellular bridge at a reproducible and predictable place on a double micropattern offers the possibility of monitoring various cell division events using kymograph analysis. This type of analysis is very useful to condense the temporal information of a movie into a unique image. Drawing a single line through the center of the double pattern (Fig. 5A) to reconstruct the kymograph allows to detect on a single image the mother cell spreading, the division, the daughter cells spreading, the assembly of the intercellular bridge, and the abscission (time when the bridge is no longer visible between the two daughter cells) (Fig. 5B). It is also possible to quantify various parameters on these kymographs: length of the bridge, distance between daughter cells or between daughter cells' nuclei, evolution over time of microtubules fluorescence in the bridge, and of course, the time between all the different detected cell division steps. By detecting the shape of the double micropatterns (using fluorescent fibronectin), the line used for the kymograph can even be drawn automatically.

1.3.3 Abscission detection using bridge reconstruction

The aim of this analysis is to automatically detect the intercellular bridge and the abscission. To facilitate the detection, the position of the bridge is guessed to be inside a rectangular zone bounded by the nuclei of the two daughter cells. Given an image sequence of a single dividing cell, the main steps of the analysis are the following (Fig. 6):

- Using the DNA stack, detect the first image with two nuclei. This corresponds to anaphase.
- For the following frames, determine the position of the two nuclei.
- Define a given number of segments that are orthogonal to the nucleus–nucleus axis. The higher the number, the better the resolution of detection.
- Using corresponding microtubules images, compute fluorescence intensity profiles along these segments.
- Fit each profile with a Gaussian curve. The position of the bridge is expected to create a peak in the fluorescence signal. If the correlation with the Gaussian curve is good and the variance of the Gaussian curve is small, then the center of the Gaussian curve gives the position of the bridge on the segment. If the correlation with the Gaussian curve is low or if the variance is too high, then no bridge is detected.
- The global shape of the bridge is reconstructed by all the points detected on the orthogonal segments.
- The abscission is defined when the bridge is no longer detected on a given number of successive frames.

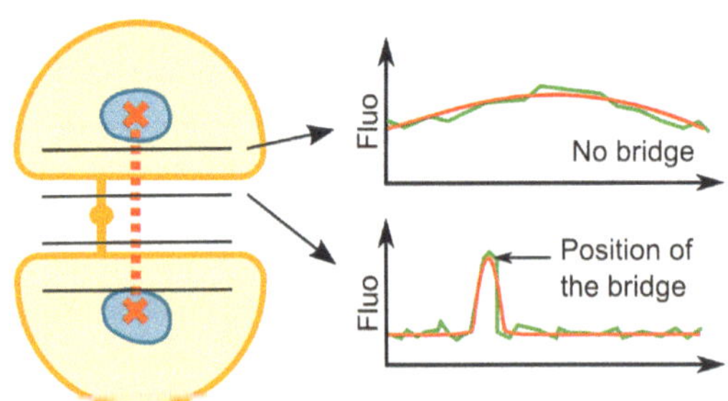

FIGURE 6

Schematic of the method used to automatically detect the position of the intercellular bridge between two daughter cells adhered to a double micropattern. The *green lines* represent the fluorescence intensity profiles along the black lines. Each profile is then fitted with a Gaussian curve (*red lines*).

An implementation of this method has been developed in the form of an ImageJ plugin and a MATLAB script that are available upon request. It is to be noted that this method does not need the detection of the micropattern's shape and can be applied to patterns of any kind of shape. To go further, it could also detect intercellular bridges and abscission of daughter cells that are not cultured on micropatterns but free to move on the substrate. The only prerequired step is mother cells and corresponding daughter cells tracking (cell lineage).

2. LASER ABLATION OF THE INTERCELLULAR BRIDGE

Because in *in vitro* cultured cells the intercellular bridge is often clearly visible between the daughter cells, it is relatively easy to cut it by laser ablation. It is noninvasive, instantaneous, and produces a clean cut compared to micromanipulation that would rather break the bridge. A UV-based ablation system allows the complete severing of the bridge (microtubules and plasma membrane) whereas a two-photons ablation system allows enough resolution to sever only the microtubules while preserving the integrity of the membrane tube. These two methods are described hereafter.

2.1 CUTTING THE INTERCELLULAR BRIDGE

The main reason to artificially cut the bridge is to induce the abscission at a precise moment. We have shown previously (Lafaurie-Janvore et al., 2013) that a laser-induced severing of the bridge on one side of the midbody induces the abscission of the other side of the bridge within 10–15 min, just as normal abscission usually occurs *in vitro*: first severing on one side of the midbody followed shortly after by the severing on the other side, with the same molecular mechanisms involved (Crowell, Gaffuri, Gayraud-Morel, Tajbakhsh, & Echard, 2014; Elia, Sougrat, Spurlin, Hurley, & Lippincott-Schwartz, 2011; Guizetti et al., 2011). Another reason to artificially cut the intercellular bridge is to assess the functional role of this bridge, both for the study of *in vitro* cultured cells (behavior of daughter cells such as polarity, migration, or cell cycle regulation) or of tissues (integrity, morphogenesis).

2.1.1 Equipment

- Inverted confocal spinning disk (Yokagawa CSU-X1 spinning head on a Nikon Eclipse T*i* inverted microscope) equipped with an EM-CCD camera (Evolve, Photometrics) and controlled by Metamorph software (Molecular Devices).
- A pulsed 355-nm ultraviolet laser interfaced with an iLas software (Roper Scientific). The pulse is 400 ps at 10 mW, 1 μJ per pulse, scanning at 25 μs per pixel.
- A 100× oil immersion objective (Nikon S Fluor 100× 0.5–1.3 NA).

2.1.2 Method

- Find a doublet of daughter cells in which the intercellular bridge is well marked by the microtubules fluorescence signal.

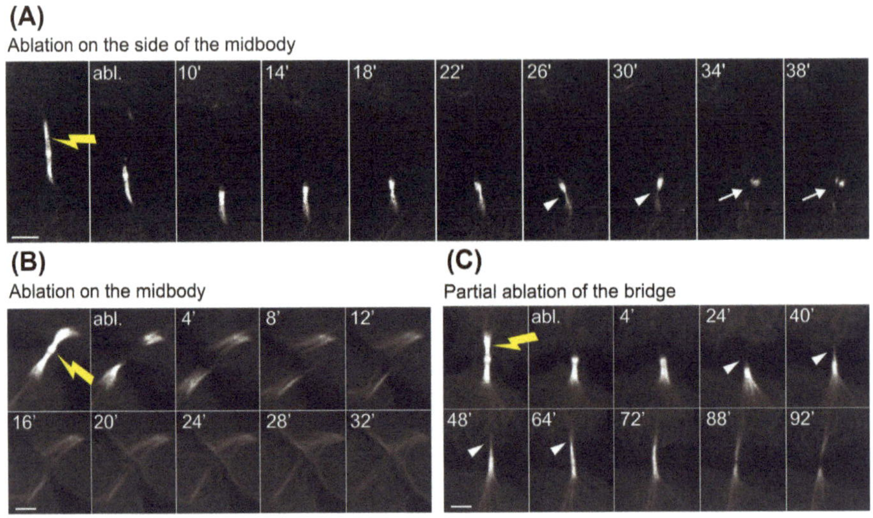

FIGURE 7

Laser-based ablation of the intercellular bridge visualized by α-tubulin-EGFP fluorescence. (A) Ablation performed on the side of the midbody. *Arrowheads* show the pinching on the other side of the midbody induced by the ablation and *arrows* show abscission. (B) Ablation performed on the midbody. (C) Partial ablation of the bridge. *Arrowheads* show the regrowth of the microtubule bundle. Throughout the figure, time is in minutes after ablation and scale bars are 10 μm.

Adapted from Lafaurie-Janvore, J., Maiuri, P., Wang, I., Pinot, M., Manneville, J.-B., Betz, T., … Piel, M. (2013). ESCRT-III assembly and cytokinetic abscission are induced by tension release in the intercellular bridge. Science, 339*(6127), 1625–1629. http://doi.org/10.1126/science.1233866.*

- Adjust the focus on the microtubules fluorescence signal.
- Define a line as the region to be ablated (see next paragraph).
- Perform the ablation. The power and time of exposition of the laser should be optimized to obtain an efficient ablation of the structures (microtubules and membrane) but avoiding side effects (cell damage or cavitation).

The ablation must be performed at a precise region of the intercellular bridge. Indeed, the bridge contains microtubule bundles that are strongly organized. Coming from the two daughter cells, they overlap in the central region, the midbody, where they are highly cross-linked. To maintain the structure of the remaining half bridge after ablation, it is important to cut microtubules of the bridge on the side of the midbody (Fig. 7A). If the ablation is performed on the midbody, it removes the central organization and immediately leads to the complete disassembly of the bridge microtubules on both sides (Fig. 7B). Moreover, if the cutting of the bridge is not complete (partial ablation), the microtubules regrow in the membrane tube and eventually re-form a complete bridge (Fig. 7C).

2.2 CUTTING ONLY THE MICROTUBULES OF THE INTERCELLULAR BRIDGE

To discriminate functions associated with microtubules or with the plasma membrane of the bridge, it might be interesting to cut only the microtubules inside the bridge but not the membrane, keeping an intact membrane tube between the two daughter cells. This can be achieved using a two-photons laser ablation system (Fig. 8).

2.2.1 Equipment

- Inverted laser scanning confocal microscope (Zeiss LSM710NLO) equipped with a two-photons laser Mai-Tai (Spectra Physics). The laser is scalable to 690−1040 nm with pulse width <100 fs.
- The system is controlled by Zen software (Zeiss).
- A 63× oil immersion objective with 1.4 NA.

2.2.2 Method

- Find a doublet of daughter cells in which the intercellular bridge is well marked by the microtubules fluorescence signal.
- Adjust the focus on the microtubules fluorescence signal.
- The region to be ablated must be a circle or a square around the microtubules but included inside the bridge.
- Perform the ablation. The wavelength, the power, and the repetition rate of the laser should be optimized to obtain an efficient ablation of the microtubules without perturbing too much the plasma membrane. If the plasma membrane is slightly damaged during the ablation, it seals rapidly (in few seconds) and re-forms a tube. It is important to verify that the microtubules are well cut and not only bleached. If tension was accumulated in the bridge, a slight movement of the cells right after the microtubules severing should be visible.

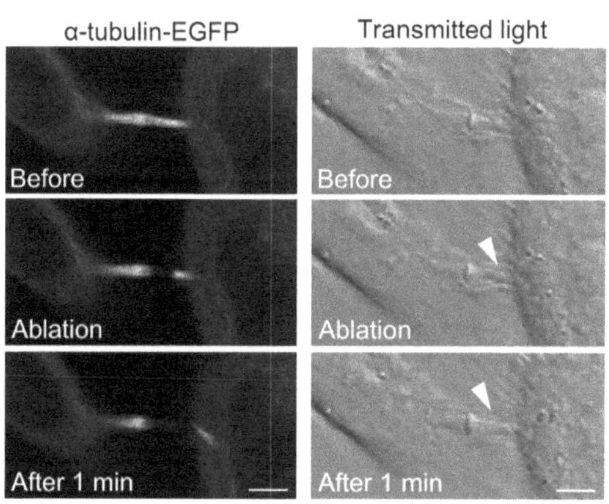

FIGURE 8

Two-photons laser ablation to sever microtubules of the intercellular bridge within an intact membrane tube (*arrowheads*). Scale bars are 10 μm.

ACKNOWLEDGMENTS

We thank the PICT-IBiSA (UMR144 and UMR3215/U934) imaging facilities and the Nikon Imaging Center of the Institut Curie-CNRS Paris for technical support with light microscopy.

REFERENCES

Agromayor, M., & Martin-Serrano, J. (2013). Knowing when to cut and run: mechanisms that control cytokinetic abscission. *Trends in Cell Biology, 23*(9), 433–441. http://dx.doi.org/10.1016/j.tcb.2013.04.006.

Azioune, A., Carpi, N., Tseng, Q., Théry, M., & Piel, M. (2010). Protein micropatterns: a direct printing protocol using deep UVs. *Methods in Cell Biology, 97*, 133–146. http://dx.doi.org/10.1016/S0091-679X(10)97008-8.

Azioune, A., Storch, M., Bornens, M., Théry, M., & Piel, M. (2009). Simple and rapid process for single cell micro-patterning. *Lab on a Chip, 9*(11), 1640–1642. http://dx.doi.org/10.1039/b821581m.

Chen, C.-T., Hehnly, H., & Doxsey, S. J. (2012). Orchestrating vesicle transport, ESCRTs and kinase surveillance during abscission. *Nature Reviews. Molecular Cell Biology, 13*(8), 483–488. http://dx.doi.org/10.1038/nrm3395.

Crowell, E. F., Gaffuri, A., Gayraud-Morel, B., Tajbakhsh, S., & Echard, A. (2014). Engulfment of the midbody remnant after cytokinesis in mammalian cells. *Journal of Cell Science, 127*, 3840–3851. http://dx.doi.org/10.1242/jcs.154732.

Dambournet, D., Machicoane, M., Chesneau, L., Sachse, M., Rocancourt, M., El Marjou, A., … Echard, A. (2011). Rab35 GTPase and OCRL phosphatase remodel lipids and F-actin for successful cytokinesis. *Nature Cell Biology, 13*(8), 981–988. http://dx.doi.org/10.1038/ncb2279.

Elia, N., Sougrat, R., Spurlin, T. A., Hurley, J. H., & Lippincott-Schwartz, J. (2011). Dynamics of endosomal sorting complex required for transport (ESCRT) machinery during cytokinesis and its role in abscission. *Proceedings of the National Academy of Sciences of the United States of America, 108*(12), 4846–4851. http://dx.doi.org/10.1073/pnas.1102714108.

Fededa, J. P., & Gerlich, D. W. (2012). Molecular control of animal cell cytokinesis. *Nature Cell Biology, 14*(5), 440–447. http://dx.doi.org/10.1038/ncb2482.

Guizetti, J., Schermelleh, L., Mäntler, J., Maar, S., Poser, I., Leonhardt, H., … Gerlich, D. W. (2011). Cortical constriction during abscission involves helices of ESCT-III-dependent filaments. *Science, 331*(6024), 1616–1620. http://dx.doi.org/10.1126/science.1201847.

Held, M., Schmitz, M. H. A., Fischer, B., Walter, T., Neumann, B., Olma, M. H., … Gerlich, D. W. (2010). CellCognition : time-resolved phenotype annotation in high-throughput live cell imaging. *Nature Publishing Group, 7*(9), 747–754. http://dx.doi.org/10.1038/nmeth.1486.

Lafaurie-Janvore, J., Maiuri, P., Wang, I., Pinot, M., Manneville, J.-B., Betz, T., … Piel, M. (2013). ESCRT-III assembly and cytokinetic abscission are induced by tension release in the intercellular bridge. *Science, 339*(6127), 1625–1629. http://dx.doi.org/10.1126/science.1233866.

Schmitz, M. H. A., & Gerlich, D. W. (2009). Automated live microscopy to study mitotic gene function in fluorescent reporter cell lines. *Methods in Molecular Biology, 545*, 113–134. http://dx.doi.org/10.1007/978-1-60327-993-2.

Steigemann, P., & Gerlich, D. W. (2009). Cytokinetic abscission: cellular dynamics at the midbody. *Trends in Cell Biology, 19*(11), 606–616. http://dx.doi.org/10.1016/j.tcb.2009.07.008.

Steigemann, P., Wurzenberger, C., Schmitz, M. H. A., Held, M., Guizetti, J., Maar, S., & Gerlich, D. W. (2009). Aurora B-mediated abscission checkpoint protects against tetraploidization. *Cell, 136*(3), 473–484. http://dx.doi.org/10.1016/j.cell.2008.12.020.

Théry, M. (2010). Micropatterning as a tool to decipher cell morphogenesis and functions. *Journal of Cell Science, 123*(Pt 24), 4201–4213. http://dx.doi.org/10.1242/jcs.075150.

Théry, M., Racine, V., Pépin, A., Piel, M., Chen, Y., Sibarita, J., & Bornens, M. (2005). The extracellular matrix guides the orientation of the cell division axis. *Nature Cell Biology, 7*(10), 947–953. http://dx.doi.org/10.1038/ncb1307.

Théry, M., Racine, V., Piel, M., Pépin, A., Dimitrov, A., Chen, Y., … Bornens, M. (2006). Anisotropy of cell adhesive microenvironment governs cell internal organization and orientation of polarity. *Proceedings of the National Academy of Sciences of the United States of America, 103*(52), 19771–19776. http://dx.doi.org/10.1073/pnas.0609267103.

CHAPTER

Measuring abscission spatiotemporal dynamics using quantitative high-resolution microscopy

12

O. Gershony, S. Sherman, S. Adar, I. Segal, D. Nachmias, I. Goliand, N. Elia[1]

Ben Gurion University of the Negev, Beer Sheva, Israel

[1]Corresponding author: E-mail: elianat@post.bgu.ac.il

CHAPTER OUTLINE

Methods in Cell Biology, Volume 137, ISSN 0091-679X, http://dx.doi.org/10.1016/bs.mcb.2016.03.032

Abstract

The spatiotemporal characteristics of ESCRT (Endosomal Sorting Complex Required for Transport)-mediated mammalian cytokinetic abscission have been studied in recent years using quantitative high-resolution light microscopy techniques. Here we describe how to apply spinning disk live cell imaging and structured illumination microscopy (SIM) to define the dynamics and structural organization of abscission and of proteins involved in abscission in a quantitative manner. We further provide a protocol to correlate the structural data, obtained by SIM, to the dynamic information obtained by live cell recordings.

INTRODUCTION

MAMMALIAN CYTOKINETIC ABSCISSION

Cytokinesis, the last step of cell division, can be divided into two steps. The first step involves large-scale constriction of the plasma membrane mediated by the actomyosin contractile ring and leads to partitioning of cellular content into two daughter cells (Green, Paluch, & Oegema, 2012). Completion of this step is manifested by the formation of a narrow membrane bridge that connects the cells, called the intercellular bridge. Microtubules, which originally formed the mitotic spindle, remain in the bridge and overlap at the center, giving rise to a structure called the midbody dark zone (see Fig. 1). The second step, termed abscission, involves severing of the membrane and microtubules of the intercellular bridge, leading to complete separation between the newly formed daughter cells. This process occurs sequentially on both sides of the bridge, leading to the release of a midbody remnant (Elia, Sougrat, Spurlin, Hurley, & Lippincott-Schwartz, 2011; Elia, Ott, & Lippincott-Schwartz, 2013). Because the first cleavage event separates the daughter cells and by that finalizes the division process, it is traditionally considered as completion of abscission. Abscission is mediated by the Endosomal Sorting Complex Required for Transport (ESCRT) membrane fission machinery (Agromayor & Martin-Serrano, 2013; Carlton & Martin-Serrano, 2007; Lee, Elia, Ghirlando, Lippincott-Schwartz, & Hurley, 2008; Morita et al., 2007). Acting in concert with the microtubule-severing protein spastin, this protein complex induces the membrane constriction and fission events that execute abscission and terminate cell division (Elia et al., 2011; Guizetti et al., 2011; Yang et al., 2008).

SPATIOTEMPORAL CHARACTERISTICS

SPATIAL CHARACTERISTICS

The intercellular bridge is a several micrometers long narrow bridge (Fig. 1). Diameter of the bridge is relatively conserved within each cell type and ranges between slightly over 0.6 to ~1 μm in different cell lines (Elia et al., 2011; Guizetti & Gerlich, 2012; Mullins & Biesele, 1977) (Fig. 2). The midbody dark zone is

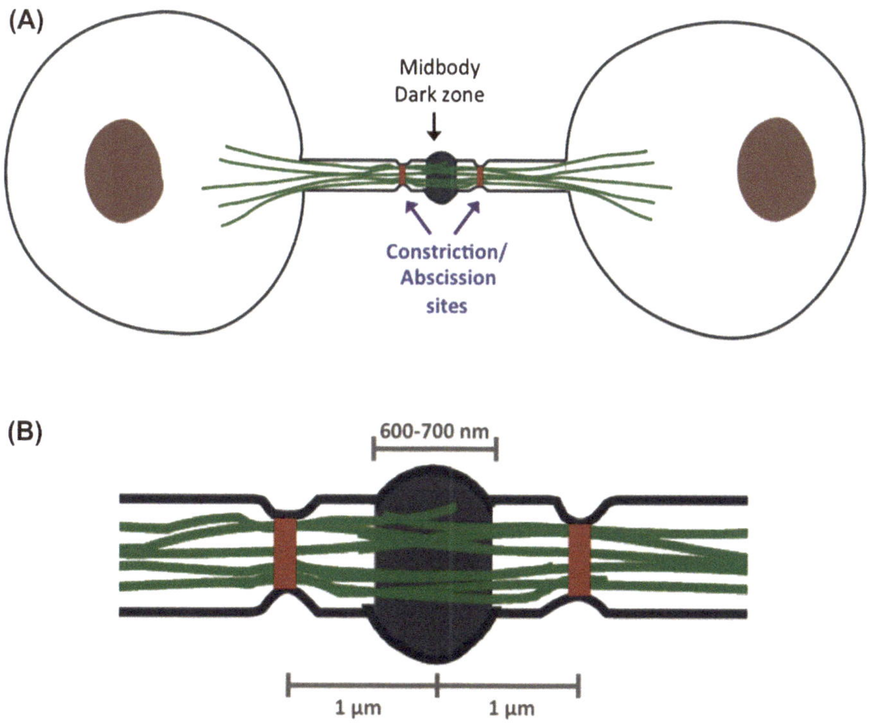

FIGURE 1 The intercellular bridge.

Scheme describing the general morphology of the intercellular bridge that connects two daughter cells at the end of cytokinesis. Plasma membrane, *black*; microtubules, *green*; midbody dark zone, *gray*; constriction/abscission sites, *red*.

Taken from Goliand, I., Nachmias, D., Gershony, O., & Elia, N. (2014). Inhibition of ESCRT-II-CHMP6 interactions impedes cytokinetic abscission and leads to cell death. Molecular Biology of the Cell, 25, *3740–3748.*

positioned at the center of the bridge and accommodates 600–700 nm of the bridge length (Elia et al., 2011; Guizetti & Gerlich, 2012; Mullins & Biesele, 1977) (Fig. 1B). Abscission occurs at about 1 µm away from each side of the center of the bridge in specific regions called constriction/abscission sites. This spatial behavior gives rise to the release of a ~2-µm-long midbody remnant (Elia et al., 2011) (Fig. 3A).

Components of the ESCRT machinery first assemble at the rims of the midbody dark zone forming partially overlapping, large diameter, cortical rings and later accumulate at the abscission sites (Elia et al., 2011). Helical spirals were found at the abscission sites by EM tomography, and it was hypothesized that these spirals are ESCRT based (Guizetti et al., 2011).

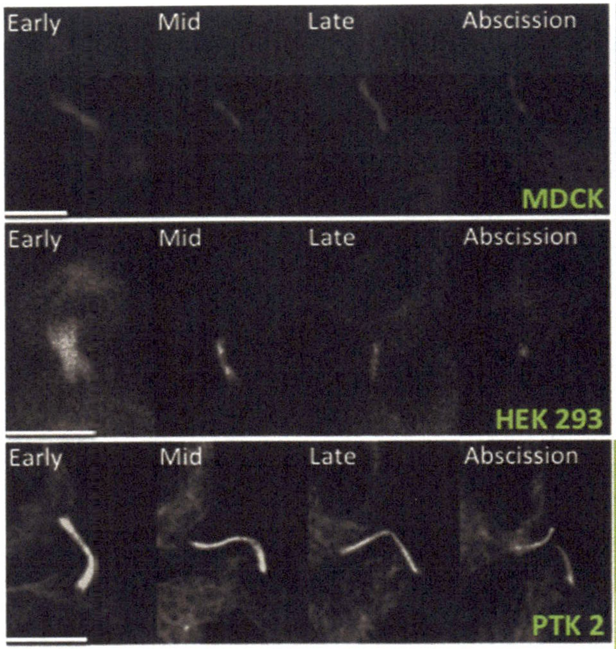

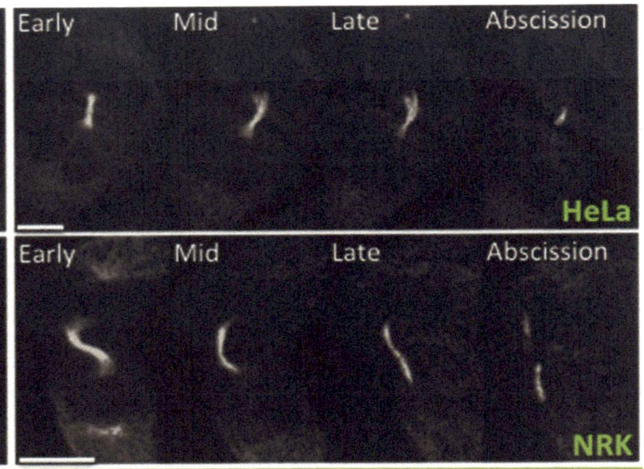

Cell line	Average bridge diameter (μm)	Approximate time anaphase to abscission (hours)
MDCK	0.96	2
HeLa	0.77	2
HEK 293T	0.81	2
NRK	0.63	1.3
PTK 2	0.67	3

FIGURE 2 Characteristics of cytokinetic abscission in different cell lines.

Cells were transfected with fluorescently labeled tubulin and imaged from anaphase to abscission. Shown are maximum intensity projections of selected time points from a 4D movie sequence of the indicated cell line. Note that different cell lines exhibit slightly different bridge morphologies. Table shows characteristic spatiotemporal parameters of abscission in the different cell lines. These parameters should be considered when setting up the imaging experiment. Scale bar, 10 μm.

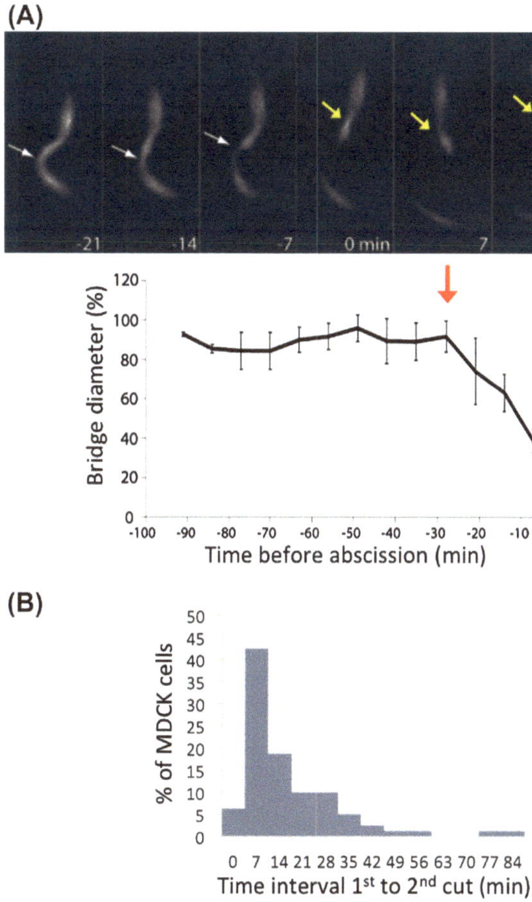

FIGURE 3 Measuring abscission kinetics.

MDCK cells transfected with fluorescently labeled tubulin or stably expressing tubulin-GFP were imaged for 4 h on an SD microscope. Maximum intensity projections of sequential frames from the movie are shown in (A). *Arrows* mark the first (*white arrow*) and second (*yellow arrow*) cuts. Graph in (A) shows the average bridge diameter at different time points before abscission. Image acquisition and data analysis were performed as described in Section 1.1. *Red arrow* indicates the time of abscission onset, characterized by acute reduction in MT diameter. Histogram in (B) shows the distribution of time intervals between the first and second cutting events as measured in MDCK cells ($n = 80$).

Data in A are taken from Elia, N., Sougrat, R., Spurlin, T.A., Hurley, J.H., & Lippincott-Schwartz, J. (2011). Dynamics of endosomal sorting complex required for transport (ESCRT) machinery during cytokinesis and its role in abscission. Proceedings of the National Academy of Sciences of the United States of America, 108, *4846–4851. Copyright 2011 National Academy of Sciences, USA. Scale bar, 2 µm*

TEMPORAL CHARACTERISTICS

While much of mitosis proceeds rapidly (less than 30 min from metaphase to telophase), the intercellular bridge persists for over an hour prior to abscission (Elia et al., 2011, 2013; Steigemann et al., 2009). During this time the cells decondense their chromatin and adhere back to the surface (Florindo et al., 2012; Floyd et al., 2013; Gershony, Pe'er, Noach-Hirsh, Elia, & Tzur, 2014). Measuring the time from anaphase to abscission in different cell lines shows that while in some cases the bridge persists for slightly over 1 h (NRK cells) in other cases the bridge persists for as long as 3 h (PTK2 cells) (Fig. 2). However, the actual act of abscission, characterized by acute constriction and scission of the microtubule bridge, is an acute event that lasts ~20 min (Fig. 3) (Elia et al., 2011; Guizetti et al., 2011). The initiation of microtubule constriction can therefore be defined as abscission onset. Cleavage on one side of the bridge is followed by cleavage on the other side, with less than 15 min separating the two events in the majority of the cases in both HeLa and MDCK cells (Lafaurie-Janvore et al., 2013) (Fig. 3B). ESCRT proteins arrive to the bridge prior to microtubule constriction but redistribute to the abscission sites shortly before abscission onset (Elia et al., 2011) (Fig. 4).

LIVE CELL IMAGING AND STRUCTURED ILLUMINATION MICROSCOPY AS TOOLS TO STUDY ABSCISSION

A direct approach to quantitatively study abscission is by visualizing the process in living cells. This has become possible due to technological advances in light microscopy, particularly superresolution microscopy and high-speed confocal spinning disk microscopy (SD), which allows quantitative characterization of biological processes at close-to-life conditions (Elia et al., 2013; Maddox, Moree, Canman, & Salmon, 2003).

Confocal SD equipped with an EM-CCD camera and temperature incubation offers an ideal tool to study abscission in real time. The sensitivity and long dynamic range of the EM-CCD camera enable to record and quantify a wide range of signal intensities without reaching saturation. We found this parameter to be critical for quantifying the arrival of abscission proteins to the bridge, which in some cases increases by more than 10-fold over the time course of the experiment (Elia et al., 2011). The sensitivity of the EM-CCD camera to dim signals further allows performing live cell imaging experiments on low overexpressing cells, thereby reducing the risk of potential artifacts resulting from overexpression of the fluorescently tagged protein (Elia et al., 2011). The speed, confocality, and temperature control ensure accurate recording of abscission in 4D and at close to physiological conditions.

Structured illumination microscopy (SIM) is one of the most applicable superresolution techniques for cellular structure analysis (Elia et al., 2013). Using SIM, the three-dimensional (3D) structural organization of multiprotein complexes can be resolved in cells that are up to 20 μm thick at 100 nm lateral and 350 nm axial

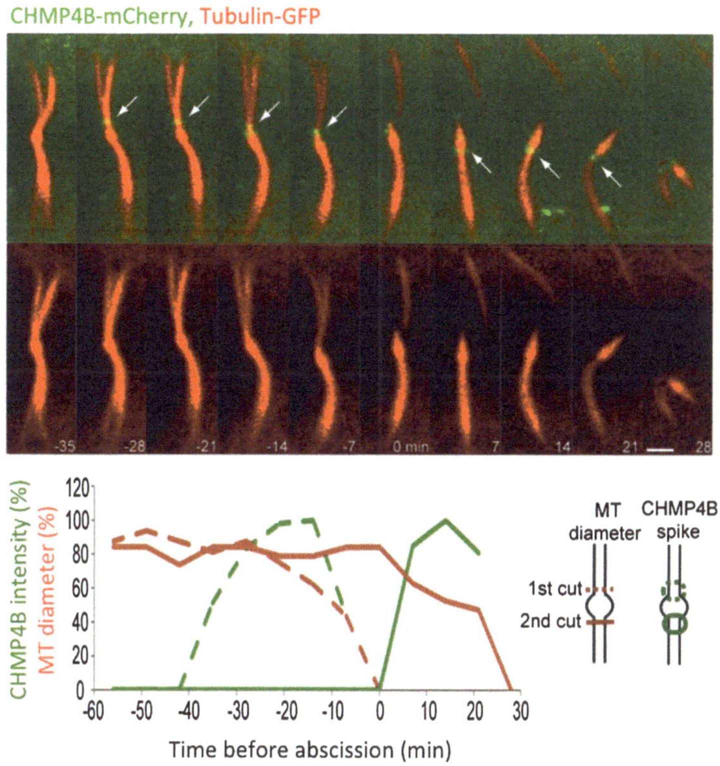

FIGURE 4 Measuring proteins recruitment to the intercellular bridge.

MDCK cells transfected with tubulin-GFP (*red*) and CHMP4B-mCherry (*green*) were imaged during cytokinesis using a spinning disk microscope. Shown are maximum intensity projection images of sequential frames from the movie. Top: CHMP4B + tubulin; bottom: tubulin alone. *Arrows* indicate the arrival of CHMP4B to the abscission sites. Graph shows quantification of CHMP4B intensity relative to the change in MT diameter on both sides of the bridge. Image acquisition and data analysis were performed as described in Section 1.1.

Taken from Elia, N., Sougrat, R., Spurlin, T.A., Hurley, J.H., & Lippincott-Schwartz, J. (2011). Dynamics of endosomal sorting complex required for transport (ESCRT) machinery during cytokinesis and its role in abscission. Proceedings of the National Academy of Sciences of the United States of America, 108, *4846–4851. Copyright 2011 National Academy of Sciences, USA. Scale bar, 2 μm.*

resolution. In SIM, the sample is illuminated through a grid pattern of light, which is rotated three to five times for each focal plane. High-frequency information is obtained in each rotation and is used to reconstruct the final high-resolution 3D image (Gustafsson, 2000; Heintzmann & Cremer, 1999). SIM is ideal for cellular studies as it can be used with any fluorescent dye following traditional protocols for fluorescent labeling of proteins in cells. The twofold increase in resolution in all three dimensions that SIM provides is optimal for the dimensions of the intercellular bridge.

Using SIM, structural organization of multiprotein complexes inside the bridge, which cannot be resolved by conventional light microscopy techniques, can be determined (Elia et al., 2013).

While SIM is capable of providing high-resolution information, its temporal resolution is relatively poor. This is mainly because at least 15 images have to be acquired from each focal plan to generate a high-resolution image. Therefore, SIM experiments are mostly limited to fixed cells. To overcome this limitation and add a temporal dimension to SIM, a live imaging–SIM correlative protocol can be used. In this approach, individual cells expressing a fluorescently tagged protein are imaged live using a confocal SD, fixed at specific time points during abscission and imaged by SIM. By doing so, rare intermediate organizations of abscission proteins can be visualized at high resolution. Integration of the high-resolution structural information, obtained through SIM, with high-speed kinetic measurements of protein dynamics in live cells, obtained through confocal SD microscopy, can be used to map the spatiotemporal organization of different protein complexes during abscission.

1. METHODS

1.1 LIVE CELL IMAGING USING A SPINNING DISK CONFOCAL

1.1.1 Cell preparation

To study abscission, it is important to choose the appropriate cell line for the research question. In our studies, we use MDCK cells—relatively large, flat, and adhesive cells—which enable easy capturing of clear intercellular bridge images. MDCK cells [grown in MEM supplemented with 5% fetal bovine serum (FBS), 2 mM glutamine, 10,000 U/mL penicillin, and 10 mg/mL streptomycin] are plated on chambered, polymer-coated coverslips (we use ibidi 4-well chambered coverslips; Catalog #80426, Martinsried, Germany). The confluency of the cells when plated must be correlated to the time of imaging. We usually aim for ~60% confluency at the time of imaging and therefore usually plate at about 7.5–15% confluency. If required, the cells are transfected 24–48 h prior to imaging. For MDCK, we use Lipofectamine 2000 (Life Technologies, Carlsbad, CA) and carry out the transfection according to manufacturer's guidelines. Synchronization is not required for live cell imaging since only a few cells undergoing cell division are imaged in each imaging session.

Note:

1. The ideal time for posttransfection imaging must be empirically determined. We find that imaging abscission using some constructs, or construct combinations, will be preferable after 24 h while for others 48 h will be essential.

1.1.2 Labeling

When measuring abscission dynamics, several cellular components must be labeled. We find that the most useful label to visualize the intercellular bridge is fluorescently tagged tubulin. This provides a clear and bright signal of the microtubules accommodating the intercellular bridge (Figs. 2 and 3A). For live cell imaging, this can be done by transfection of a tubulin construct that is conjugated to a fluorescent protein such as GFP or mCherry. Transient transfection can be carried out for each experiment, or a stable cell line can be used instead. We often use MDCK cells that are stably expressing tubulin-GFP.

Labeling the plasma membrane can also be useful when studying abscission to directly follow the membrane of the intercellular bridge during cytokinesis. This can be done using membrane-targeted, fluorescently tagged constructs. In our lab, we use a CAAX-GFP construct, which clearly labels the membrane of the intercellular bridge. Notably, we find that locating intercellular bridges based on membrane labeling alone is challenging. We, therefore, recommend cotransfecting cells with tagged tubulin to study the membrane of the intercellular bridge in live cell imaging.

Finally, the protein of interest must be labeled so that it can be tracked and studied during the abscission process. Again, this can be done by transiently transfecting the cells with a construct containing the protein of interest conjugated to a fluorescent protein (Fig. 4). Alternatively, a stable cell line expressing the fluorescently tagged protein of interest can be used.

Abscission dynamics can also be studied in relation to other cellular events. An appropriate marker for the studied process must be chosen and labeled. For example, we used fluorescently tagged lamin to follow nuclear reassembly after mitosis (Fig. 5A). In another case, we used the fluorescently tagged G1 progression marker Cdt1 to study the timing of abscission with respect to the mammalian cell cycle (Gershony et al., 2014) (Fig. 5B and C).

Notes:

1. It is recommended to use stable cell lines as their low expression levels minimize overexpression artifacts. However, these cells might be less receptive for transient transfection.
2. If possible, fluorescent tags that require 405 nm excitation should be avoided since dividing cells often do not survive the duration of imaging.
3. The position of the fluorescent tag must be taken into consideration. CHMP proteins of the ESCRT machinery contain a C-terminal inhibitory loop and therefore must be N-terminally tagged.
4. It is advisable to test several types of fluorescent tags for the protein of interest. For example, we found that expressing mCherry-tagged version of some ESCRT proteins preserves physiological abscission better than the GFP-tagged version of these proteins.

1.1.3 Image acquisition

1.1.3.1 Initial step-up

Approximately 30 min prior to the experiment, the medium in the chamber to be imaged is exchanged with a prewarmed, fresh medium. Once the cells are ready for imaging, the chambered coverslip is mounted on the microscope. We use a fully

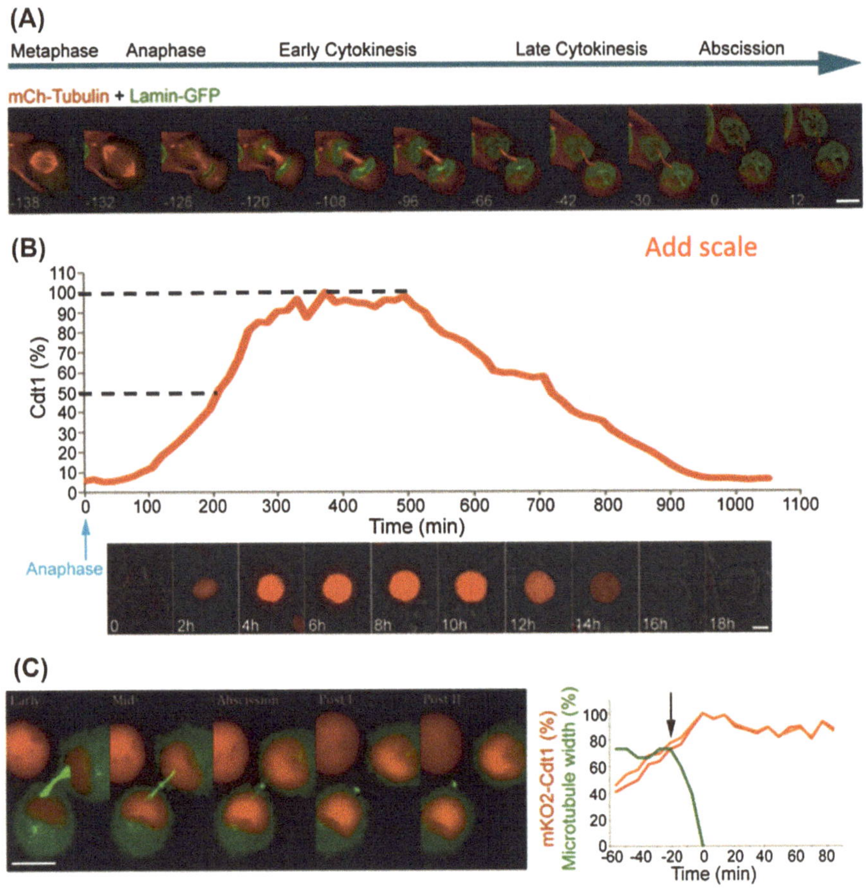

FIGURE 5 Measuring abscission in relation to cell cycle progression.

(A) MDCK cells transfected with tubulin-mCherry (*red*) and Lamin-GFP (*green*) were imaged from early mitosis to abscission using a spinning disk microscope. Shown are maximum intensity projection images of selected frames from the movie. (B) Live cell imaging of BALB/c 3T3 cells stably expressing mKO2-Cdt1 undergoing a complete cell cycle. Plot indicates total Cdt1 levels in the nucleolus as a function of time. Measurements were taken every 15 min. Time 0 was set at anaphase. *Dashed lines* in plot indicate Cdt1 levels that were used for setting the threshold for advanced G1 phase as described in Section 1.1.4. Representative images from the movie sequence are shown below. (C) BALB/c 3T3 cells stably expressing mKO2-Cdt1 (*red*) were transfected with tubulin-GFP (*green*) and imaged during late cytokinesis. Shown are maximum intensity projection images of selected frames from the movie. Graphs on the right show mKO2-Cdt1 levels in the newly formed daughter cells (*red* and *orange lines*) in relation to MT diameter. Image acquisition and data analysis in (A), (B), and (C) were performed as described in Section 1.1.

Data in B and C are taken from Gershony, O., Pe'er, T., Noach-Hirsh, M, Elia, N., & Tzur, A. (2014). Cytokinetic abscission is an acute G1 event. Cell Cycle, *13, 3436–3441. Reprinted by permission of the publisher Taylor & Francis Ltd. Scale bar, 10 μm.*

incubated, temperature-controlled spinning disk confocal microscope with a 63× oil objective (NA, 1.4) and an EM-CCD camera (Evolve; Photometrics).

Dividing cells are located using tubulin fluorescence, which allows the visualization of cells at different stages of mitosis. Cells expressing low levels of the fluorescently tagged proteins are selected for imaging to avoid potential artifacts from overexpression.

1.1.3.2 Acquisition

When setting up a live cell imaging experiment, several parameters must be optimized to allow for normal abscission kinetics. Laser power, exposure time, number of Z-sections, and time intervals must be adjusted to minimize the phototoxicity and photobleaching during the prolonged imaging procedure. For MDCK cells, we find that recording up to 30 Z-sections at 7 min intervals for a total of 3–4 h is optimal for recording the entire abscission process.

Several criteria must be taken into consideration in order for the collected data to be suitable for future quantification:

1. There must be no signal saturation in the entire data set. This may require setting the initial parameters at low intensity values to avoid saturation due to accumulation of proteins at the bridge during abscission progression.
2. The full width of the cells must be recorded in all channels and time points. Notably, although cells round up during mitosis, they adhere back to the surface prior to abscission. Therefore, when imaging mitotic cells, it is advisable to set the Z-parameters so that the entire width of an interphase adherent cell is included. This can be done using a neighboring nondividing cell.

Notes:

1. When recording multiple locations in one experiment, it is preferable to locate cells from the same well for each imaging session. If critical to the experiment, cells from neighboring wells can be imaged, but then they should be located as close as possible to the wells' edge.
2. In some cases, it is recommended to use a 40× objective, which has a larger depth of field. Such an experimental setup will allow capturing thicker cells and will compensate better for Z-drift in long experiments. For example, using a 40× oil objective we were able to track the oscillations of the G1 marker Cdt1 over a course of 40 h.
3. Even adherent cells will move slightly throughout the course of 3–4 h. Therefore, it is advisable to take this minor Z-drift into consideration when setting the Z-stack to ensure that the entire width of the bridge is captured at all time points.
4. Duration of imaging should be adjusted according to the experimental design. PTK2 cells, for example, will require longer imaging time than NRK cells (see table in Fig. 2). The addition of a reagent that is known, or predicted, to inhibit abscission will also require an adjustment in the imaging time to confirm the inhibition phenotype.

5. We find that 5–7 min intervals are optimal for 3D capturing of the entire abscission process in MDCK cells while maintaining physiological conditions. Shorter intervals may be used if subsets of the process are recorded.
6. Cells in early mitotic stages are very sensitive to laser exposure. It is therefore preferable to choose cells that already entered anaphase for live cell recordings of abscission.
7. In some cases, larger Z-steps than the recommended nyquist criteria are used to capture the entire width of the cell while minimizing photobleaching and phototoxicity. In these cases, it is important to avoid undersampling in Z to allow quantification of signal intensities as described in the following section.

1.1.4 Data analysis

Only cells that abide by the above-mentioned parameters, as well as demonstrating typical spatiotemporal abscission kinetics (Fig. 2), are analyzed.

Quantifying abscission onset: This is done by determining the microtubule diameter based on the microtubule fluorescence intensity profile along a line that is positioned perpendicular to the intracellular bridge about 1 μm away from the center of the midbody at the narrowest point along the bridge (from both sides). We routinely use a maximum Z-projection image for this quantification. The time point in which the first fission event occurs is set as time 0 and abscission progression is plotted accordingly (Fig. 3).

Quantifying the protein of interest: Once abscission onset has been determined, analysis on the protein of interest can be performed. Intensity values of the protein at the intercellular bridge are calculated by measuring the total intensity fluorescence of a mask object applied to the sum Z-projection of the movie series. Intensity levels at the intercellular bridge before protein recruitment are set as zero and subtracted from all time points. The highest intensity value is set as 100%, and all intensity values are calculated relatively. These values are then plotted against abscission progression (Fig. 4).

Quantifying abscission in relation to a cellular event: Intensity values of the specific chosen marker are calculated by measuring the total intensity fluorescence of a mask object applied to the sum Z-projection of the movie series. For example, when studying the timing of abscission with respect to cell cycle progression using the G1 marker Cdt1, measurements of the total nuclear mKO2-Cdt1 intensity levels were done on sum projection images of the 3D movie series. Only cells whose mKO2-Cdt1 levels reached a plateau without reaching saturation were analyzed to determine the relative levels of Cdt1 during abscission. Cells with over 50% of Cdt1 intensity at the time of abscission onset were considered at advanced G1 phase (Fig. 5B). These measurements were then correlated with abscission onset (Fig. 5C).

1.2 STRUCTURED ILLUMINATION MICROSCOPY

1.2.1 Sample preparation and labeling

Cells are plated at low density (10%) on number 1.5 high-resolution coverslips (Marienfeld). In the lab, we routinely use MDCK cells for SIM imaging of the intercellular bridge. Whenever possible, labeling endogenous proteins by immunostaining is recommended (see protocol in the next section). This is because secondary antibodies labeled with fluorescent dyes are brighter and more photostable compared to fluorescent proteins and are therefore more suitable for the intensive imaging required in SIM (15–25 images for each focal plane). Labeling endogenous proteins also avoids potential structural artifacts resulting from overexpression. When needed, transfection is carried out as described in Section 1.1.1. To enrich percentage of intercellular bridges, cells are synchronized by aphidicolin (2.5 μg/mL for 16 h) and fixed (4% PFA, 15 min) 10–11 h after aphidicolin wash. Finally, cells are mounted with Fluoromount-G (SouthernBiotech).

1.2.1.1 Immunostaining

Cells are permeabilized with 0.5% Triton-X 100 for 10 min and blocked with 10% FBS for 15 min. Intercellular bridges are routinely labeled by α-tubulin staining (DM1A; Sigma). Protein of interest is labeled with specific primary antibodies. Cells are then subjected to a secondary antibody staining using Alexa-Fluor 405, 488, 594 anti-mouse or anti-rabbit secondary antibodies (Invitrogen).

Notes:

1. SIM imaging demands bright fluorescent signals. Therefore, when using fluorescently tagged proteins, it is recommended to image the cells shortly after fixation (not more than a few days postfixation). Additionally, exposing the slides to light should be avoided during the immunostaining procedure and in storage. If signal intensity is not high enough for performing appropriate SIM acquisition (see the next section), anti-GFP staining can be used (Living color JL-8 clontech). Alternatively, a smaller tag such as flag tag (instead of a fluorescent protein tag) can be used, followed by anti-flag staining.
2. In SIM, the shorter the wavelength, the better the resolution. This should be considered when choosing a fluorescent protein tag or a secondary antibody (for example, Alexa 405 will provide better resolution than Alexa 488).
3. MDCK cells are nicely synchronized with aphidicolin. However, this is not the case for all cells. Synchronization protocol should, therefore, be tested when other cell types are used. Thymidine/double thymidine block can be used as an alternative to aphidicolin.

1.2.2 Image acquisition

For 3D SIM imaging, we use the ELYRA PS.1 microscope (Carl Zeiss MicroImaging). Thin Z-sections (0.15–0.2 μm) of high-resolution images are collected in five rotations and five phases for each channel. For accurate reconstructions and measurements, it is critical to avoid signal saturation and to ensure that the entire width of the bridge is captured in the axial dimension. Images are reconstructed using ZEN software (Carl Zeiss MicroImaging) based on the structured illumination algorithm developed by Heintzmann and Cremer (1999).

Notes:

1. If multichannel acquisition is needed, it is recommended to start with the longest wavelength and end with the shortest wavelength. This will minimize photobleaching throughout acquisition.
2. Exposure time and laser power should be adjusted so that signal intensity is high enough (at least one-third of maximum intensity levels).
3. Using five rotations will result in best resolution. However, if signal intensity does not allow and extensive photobleaching occurs during acquisition, it is recommended to slightly compromise resolution and use three rotations.
4. When imaging proteins inside the intercellular bridge, the signal is often very weak and accumulates in a single condensed region. Such localization pattern can result in artifacts during data reconstruction. It is therefore important to include other parts of the cells where the protein can be found in the captured and reconstructed image.
5. Some of the proteins that localize to the intercellular bridge exhibit a ring/disk morphology. To resolve the entire ring, a Z-stack spanning the entire bridge needs to be taken. We find that using microtubule labeling as a ruler for bridge width is often not enough for fully resolving the rings. It is therefore recommended to use membrane labeling, or the protein itself, when setting the Z-stack parameters. Alternatively, adding three extra Z-steps above and below the microtubule staining is normally sufficient for capturing the entire ring structures.

1.2.3 Data analysis

Once the data are reconstructed, measurements and 3D rendering can be performed.

1.2.3.1 3D rendering

We use Volocity 6 (Perkin-Elmer) for 3D rendering of the intercellular bridge. This allows to rotate the bridge and to observe the protein organization in all dimensions. We find that looking through the bridge (see the third and fourth panels in Fig. 6) is very useful for distinguishing between a ring and a disk organization (top panel, disk; bottom panel, ring).

1.2.3.2 Measurements

All measurements are performed on reconstructed data. We perform most geometrical measurements on a single slice from the central focal plane of the bridge using the intensity profile of a line that is drawn along the bridge (Fig. 7 arrow in image). Several hallmark features of the bridge can be easily detected and measured by analyzing the microtubule intensity profile along such a line (Fig. 7, see MT line intensity profile in plot). The central dark zone appears as a drop in MT signal intensity. In late

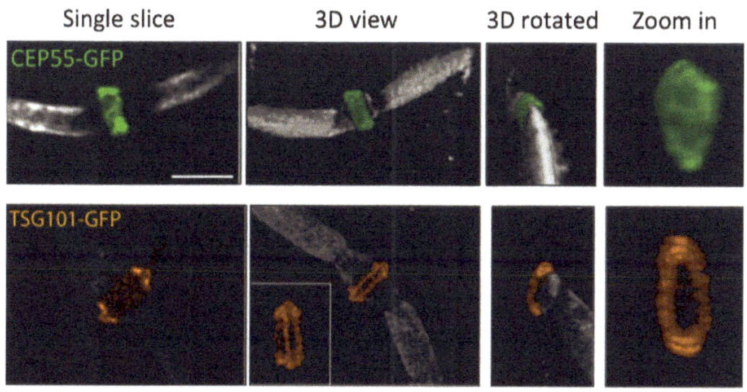

FIGURE 6 Determining the structural organization of intercellular bridge proteins by 3D structured illumination microscopy (SIM).

MDCK cells transfected with either CEP55-GFP (top panel, *green*) or TSG101-GFP (bottom panel, *orange*) were synchronized, stained with anti-tubulin (*white*), and imaged by SIM as described in Section 1.2. Note the filled disk organization of CEP55 versus the hollow ring organization of TSG101 in the 3D rotated images.

Taken from Elia, N., Sougrat, R., Spurlin, T.A., Hurley, J.H., & Lippincott-Schwartz, J. (2011). Dynamics of endosomal sorting complex required for transport (ESCRT) machinery during cytokinesis and its role in abscission. Proceedings of the National Academy of Sciences of the United States of America, 108, *4846–4851. Copyright 2011 National Academy of Sciences, USA. Scale bar, 2 μm.*

intercellular bridges, we additionally observe a loss of MT signal in the constriction sites (Goliand, Nachmias, Gershony, & Elia, 2014). These signatures of MT intensities along the bridge are very useful for determining the localization of different proteins relative to the midbody dark zone and constriction sites (Fig. 7, see CHMP6 line intensity profile relative to MT line intensity profile in plot).

To measure the thickness of a structure, we calculate the distance between the lowest-intensity pixel located at the beginning of the structure and the lowest-intensity pixel located at the end of the structure based on a line intensity profile along the bridge.

Measurements of the net distance between different structural elements are done by calculating the distance between the lowest-intensity pixel located at the end of one structure and the lowest-intensity pixel located at the beginning of the other structure, based on a line intensity profile along the bridge.

Notes:

1. As the bridge is not a straight line, it is often required to use a curved line for plotting the line intensity profile of proteins along the bridge. This step is critical for obtaining accurate measurements.
2. The drop in MT intensity levels along the intercellular bridge described above is characteristic of tubulin antibody staining. These features are not observed when tubulin-GFP fluorescence is used to label the bridge MT.

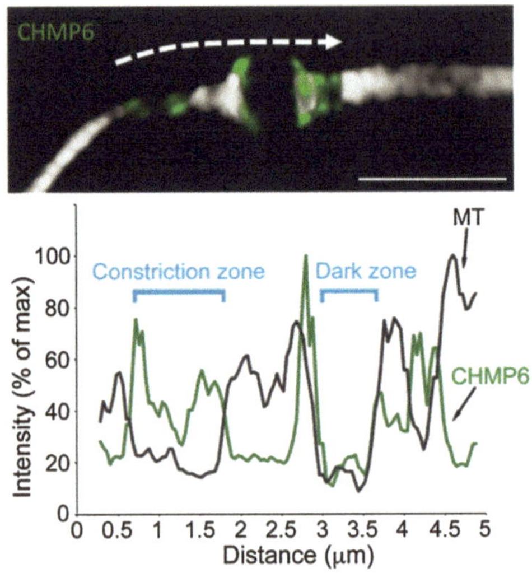

FIGURE 7 Measuring protein localizations along the intercellular bridge in structured illumination microscopy (SIM) images.

MDCK cells transfected with CHMP6-mCherry (*green*) were synchronized, stained with antitubulin (*white*), and imaged by SIM as described in Section 1.2. Shown is a 3D rendered image of reconstructed data. Plot shows tubulin and CHMP6 line intensity profiles along the intercellular bridge (see *dashed arrow* in image for directionality of the line). Note the drop in MT signal intensities in the midbody dark zone and the constriction sites. Image acquisition, data analysis, and measurements were performed as described in Section 1.2.

Taken from Goliand, I., Nachmias, D., Gershony, O., & Elia, N. (2014). Inhibition of ESCRT-II-CHMP6 interactions impedes cytokinetic abscission and leads to cell death. Molecular Biology of the Cell, 25, *3740–3748. Scale bar, 2 μm.*

3. For most measurements, we use the ZEN black software package (Zeiss). These measurements, however, can be easily done with any image analysis tool including ImageJ. Using alternative software often requires exporting the reconstructed data to TIFF files. For conducting appropriate measurements, it is critical to ensure that the pixel size information is accurately represented in the new image file. This information can be usually found in the info page of an image. If not represented accurately, this information can be copied manually from the info page of ZEN to the info page of the alternative image analysis tool.

1.3 CORRELATIVE LIVE–STRUCTURED ILLUMINATION MICROSCOPY

As mentioned previously, using correlative live–SIM imaging can provide high-resolution information on transient intermediate organization of proteins during abscission. Therefore, we recently developed a correlative live–SIM workflow (Fig. 8). The protocol and critical points in the assay are detailed in the following

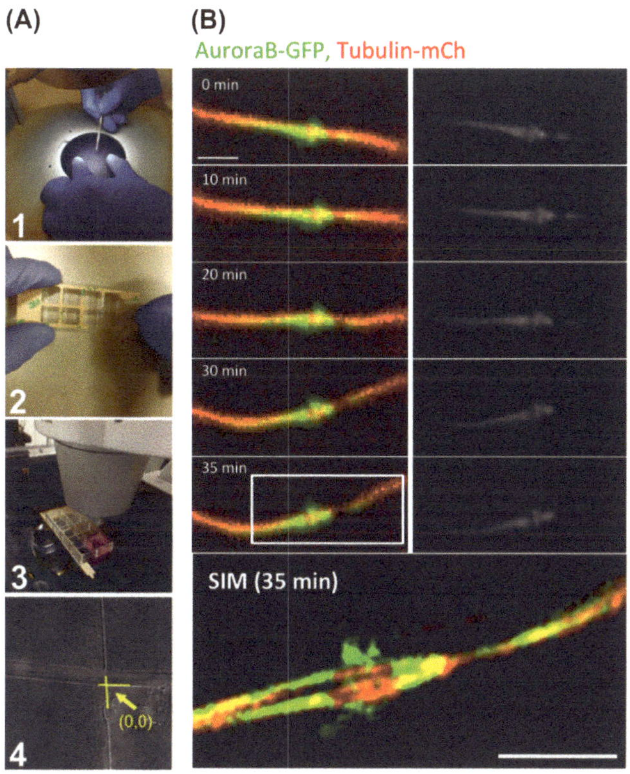

FIGURE 8 Using correlative live–structured illumination microscopy (SIM) to study abscission.

(A) Description of workflow. Numbers indicate sequential steps in the workflow. The entire workflow is described in Section 1.3. (B) MDCK cells were transfected with Aurora B-GFP (*green*) and tubulin-mCherry (*red*) and imaged in 3D by spinning disk for 35 min. 0 min indicates the beginning of the live cell recordings. Maximum intensity projections of sequential time frames from the movie are shown (Left, merged images; right, Aurora B alone). A merged view from a single Z-section of the correlated 3D SIM is shown in the bottom panel. SIM image corresponds to the last time point acquired live (35 min).

Data in A is adapted from Sherman, S., Nachmias, D., & Elia, N. (2015). A simple, straightforward correlative live-cell-imaging-structured-illumination-microscopy approach for studying organelle dynamics. Microscopy Research and Technique, *78, 777–783. Copyright 2015, with permission from John Wiley & Sons. Inc. Scale bar, 2 μm.*

sections. All the guidelines provided for live cell and SIM imaging of the intercellular bridge should be followed when using the correlative assay. Specific modifications are highlighted in the protocol. More information on the approach, including its validity and usability, can be found in Sherman, Nachmias, and Elia (2015).

1.3.1 Chamber preparation

The general workflow is illustrated in Fig. 8A.

1. Using a diamond pen, carve a "plus" mark on number 1.5 18 × 18 mm coverslips.
   a. This is best done under a binocular microscope.
2. Irradiate coverslips for 20 min under UV to ensure sterilization.
3. To form a sealed chamber, glue coverslips to an 8-well sticky slide (ibidi catalog number: 80828) so that the carved mark faces into the chamber.
4. Press gently on the coverslip edges to ensure a tight seal.
5. Plate cells in low density into chamber.
   a. The chamber can hold a maximum volume of 500 μL.

1.3.2 Labeling

Once the cells adhere, they may be labeled. Label cells as described in Section 1.1 before continuing to the following steps.

1.3.3 Live imaging using a spinning disk confocal

6. Using bright-field illumination, search for the carved plus mark. Center on one of the inner intersections and calibrate it as the "zero coordinate."
7. Locate three to four dividing cells (by locating intercellular bridges) and save their coordinates prior to imaging.
   a. Dividing cells were chosen for imaging according to their fluorescence. Dim cells should be excluded from imaging, as the remaining fluorescence will not be sufficient for high-quality SIM imaging.
8. Image dividing cells with minimum laser intensity, exposure time, and number of Z-sections possible to avoid photobleaching. In our experience, cells could be imaged through up to 15 Z-sections for 20 time points or up to 6 Z-sections for 300 time points using a 63× oil objective (NA, 1.4).
9. Fix cells by directly adding 2% PFA (final volume) to imaging chamber, immediately after the last time point. Incubate cells for 15 min and wash twice with PBS.

1.3.4 Structured illumination microscopy imaging

10. Place imaging chamber on a SIM system (we use the Elyra PS.1 (Zeiss)) and relocate the plus mark using bright-field illumination.
11. Calibrate the same inner intersection as the "zero coordinate."
    a. For the Elyra there is no automatic homing. The inner intersection needs to be placed at coordinate (512,512) and then "zeroed."
12. Find the imaged dividing cells by entering the saved coordinates that were acquired on the spinning disk confocal.
13. Image dividing cells with optimal Z-sectioning for SIM images (ie, 126 nm intervals for 561 nm wavelength with a 1.4 NA 63× oil objective; for 488 nm wavelength use 110 nm intervals).

a. When fluorescence intensity is high enough, use five rotations; otherwise, use three rotations.

14. Reconstruct SIM data using ZEN software.

REFERENCES

Agromayor, M., & Martin-Serrano, J. (2013). Knowing when to cut and run: mechanisms that control cytokinetic abscission. *Trends in Cell Biology, 23*, 433–441.

Carlton, J. G., & Martin-Serrano, J. (2007). Parallels between cytokinesis and retroviral budding: a role for the ESCRT machinery. *Science, 316*, 1908–1912.

Elia, N., Sougrat, R., Spurlin, T. A., Hurley, J. H., & Lippincott-Schwartz, J. (2011). Dynamics of endosomal sorting complex required for transport (ESCRT) machinery during cytokinesis and its role in abscission. *Proceedings of the National Academy of Sciences of the United States of America, 108*, 4846–4851.

Elia, N., Ott, C., & Lippincott-Schwartz, J. (2013). Incisive imaging and computation for cellular mysteries: lessons from abscission. *Cell, 155*, 1220–1231.

Florindo, C., Perdigao, J., Fesquet, D., Schiebel, E., Pines, J., & Tavares, A. A. (2012). Human Mob1 proteins are required for cytokinesis by controlling microtubule stability. *Journal of Cell Science, 125*, 3085–3090.

Floyd, S., Whiffin, N., Gavilan, M. P., Kutscheidt, S., De Luca, M., Marcozzi, C., … Lindon, C. (2013). Spatiotemporal organization of Aurora-B by APC/CCdh1 after mitosis coordinates cell spreading through FHOD1. *Journal of Cell Science, 126*, 2845–2856.

Gershony, O., Pe'er, T., Noach-Hirsh, M., Elia, N., & Tzur, A. (2014). Cytokinetic abscission is an acute G1 event. *Cell Cycle, 13*, 3436–3441.

Goliand, I., Nachmias, D., Gershony, O., & Elia, N. (2014). Inhibition of ESCRT-II-CHMP6 interactions impedes cytokinetic abscission and leads to cell death. *Molecular Biology of the Cell, 25*, 3740–3748.

Green, R. A., Paluch, E., & Oegema, K. (2012). Cytokinesis in animal cells. *Annual Review of Cell and Developmental Biology, 28*, 29–58.

Guizetti, J., & Gerlich, D. W. (2012). ESCRT-III polymers in membrane neck constriction. *Trends in Cell Biology, 22*, 133–140.

Guizetti, J., Schermelleh, L., Mantler, J., Maar, S., Poser, I., Leonhardt, H., … Gerlich, D. W. (2011). Cortical constriction during abscission involves helices of ESCRT-III-dependent filaments. *Science, 331*, 1616–1620.

Gustafsson, M. G. (2000). Surpassing the lateral resolution limit by a factor of two using structured illumination microscopy. *Journal of Microscopy, 198*, 82–87.

Heintzmann, R., & Cremer, C. G. (1999). Laterally modulated excitation microscopy: improvement of resolution by using a diffraction grating. *Proceedings of SPIE, 3568*, 185–196.

Lafaurie-Janvore, J., Maiuri, P., Wang, I., Pinot, M., Manneville, J. B., Betz, T., … Piel, M. (2013). ESCRT-III assembly and cytokinetic abscission are induced by tension release in the intercellular bridge. *Science, 339*, 1625–1629.

Lee, H. H., Elia, N., Ghirlando, R., Lippincott-Schwartz, J., & Hurley, J. H. (2008). Midbody targeting of the ESCRT machinery by a noncanonical coiled coil in CEP55. *Science, 322*, 576–580.

Maddox, P. S., Moree, B., Canman, J. C., & Salmon, E. D. (2003). Spinning disk confocal microscope system for rapid high-resolution, multimode, fluorescence speckle microscopy and green fluorescent protein imaging in living cells. *Methods in Enzymology, 360*, 597–617.

Morita, E., Sandrin, V., Chung, H. Y., Morham, S. G., Gygi, S. P., Rodesch, C. K., & Sundquist, W. I. (2007). Human ESCRT and ALIX proteins interact with proteins of the midbody and function in cytokinesis. *The EMBO Journal, 26*, 4215–4227.

Mullins, J. M., & Biesele, J. J. (1977). Terminal phase of cytokinesis in D-98s cells. *The Journal of Cell Biology, 73*, 672–684.

Sherman, S., Nachmias, D., & Elia, N. (2015). A simple, straightforward correlative live-cell-imaging-structured-illumination-microscopy approach for studying organelle dynamics. *Microscopy Research and Technique, 78*, 777–783.

Steigemann, P., Wurzenberger, C., Schmitz, M. H., Held, M., Guizetti, J., Maar, S., & Gerlich, D. W. (2009). Aurora B-mediated abscission checkpoint protects against tetraploidization. *Cell, 136*, 473–484.

Yang, D., Rismanchi, N., Renvoise, B., Lippincott-Schwartz, J., Blackstone, C., & Hurley, J. H. (2008). Structural basis for midbody targeting of spastin by the ESCRT-III protein CHMP1B. *Nature Structural & Molecular Biology, 15*, 1278–1286.

CHAPTER

Analysis of cytokinesis by electron microscopy

13

J. König*, J. Borrego-Pinto[§], D. Streichert*, M. Munzig*, P. Lenart[§], T. Müller-Reichert*[,1]

*Technische Universität Dresden, Dresden, Germany

[§]European Molecular Biology Laboratory (EMBL), Heidelberg, Germany

[1]Corresponding author: E-mail: mueller-reichert@tu-dresden.de

CHAPTER OUTLINE

Methods in Cell Biology, Volume 137, ISSN 0091-679X, http://dx.doi.org/10.1016/bs.mcb.2016.03.029

Abstract

Following up on a chapter on the *Correlative Light and Electron Microscopy of Early Caenorhabditis elegans Embryos in Mitosis* (MCB *79*, 101–119), we present an adaptation of our established protocol for the ultrastructural analysis of either permeabilized or injected embryonic systems. We prepared both drug-treated early *C. elegans* embryos and fluorescently labeled sea urchin embryos of *Lytechinus pictus* for ultrastructural studies on animal cytokinesis. Here we focus on the initial preparation steps of postmitotic embryos for high-pressure freezing and subsequent electron microscopy with an emphasis on electron tomography. The advantages and limitations of our extended protocol will be discussed.

INTRODUCTION

Cytokinesis completes cell division resulting in two separated daughter cells. As shown in Fig. 1, cytokinesis follows chromosome segregation and begins with the assembly of a contractile actomyosin ring (Bringmann & Hyman, 2005; Rappaport, 1985). Constriction of this actomyosin ring then culminates in the formation of an intercellular bridge, consisting of a bundle of antiparallel microtubules associated with the spindle midbody (Glotzer, 2005; Green, Paluch, & Oegema, 2012). This intercellular bridge is the last connection of the two newly formed daughter cells and needs to be resolved in the final steps of cytokinesis (Mierzwa & Gerlich, 2014). As demonstrated by electron tomography, this final step in mammalian cell cytokinesis, called abscission, is caused by a secondary constriction of the plasma membrane, accompanied by a disassembly of the microtubules of the intercellular bridge (Guizetti et al., 2011).

Recent studies on cytokinesis have significantly increased our knowledge of the process in mammalian cells. However, most of the work on cytokinesis and/or abscission has been performed using mammalian tissue culture cells, grown on solid supports (Mierzwa & Gerlich, 2014). In contrast, our current understanding of cytokinesis in either embryonic systems of different phyla or in somatic cells within the tissue context is comparatively very low (Green et al., 2013; Larkin & Danilchik,

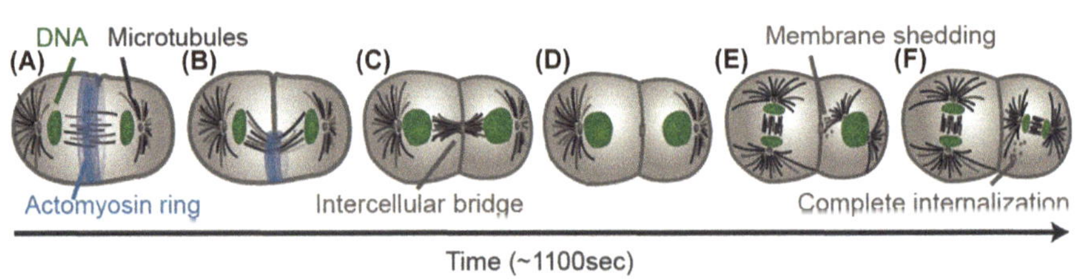

FIGURE 1

Sequence of events during cytokinesis in the early *Caenorhabditis elegans* embryo. (A) Assembly of the actomyosin ring. (B) Furrow ingression. (C) Formation of the intercellular bridge. (D) Microtubule disassembly. (E) Membrane shedding. (F) Internalization of the midbody remnant into the cytoplasm of the P1 cell.

1999). It is possible that the mechanism leading to abscission differs from what was described previously for mammalian tissue culture cells. We therefore aimed to analyze the embryonic ultrastructure of the cytokinesis machinery in nematode as well as in echinoderm embryos in situ.

Several challenges need to be considered for such an analysis of embryonic systems. For light microscopic observations, the intercellular bridge can be difficult to identify between the embryonic cells due to the fact that the bridges are rather small compared to the relatively large size of the zygote. An additional issue to consider is the formation of an eggshell during early embryonic development. The eggshell serves as a protective barrier or shield, thereby significantly reducing the diffusion of reagents into the embryo (Olson, Greenan, Desai, Müller-Reichert, & Oegema, 2012). In addition, the eggshell complicates the addition of drugs to either stop the embryonic development at a specific stage or to manipulate the cytoskeleton of the cytokinetic machinery. A further limitation is often the large size of the embryo itself, reducing the quality of live-cell imaging and/or rendering the staging of embryos very difficult. Limitations in live-cell imaging can be easily circumvented by using fluorescently tagged strains of living organism, such as *Caenorhabditis elegans*. For the vast majority of embryonic systems, however, tagged strains are not available. We therefore applied injection of fluorescently labeled proteins, such as rhodamine-labeled tubulin, into echinoderm embryos to allow unambiguously staging prior to freezing.

To overcome these limitations, we aimed at the application of a correlative light and electron microscopic (CLEM) approach to combine the dynamics information obtained by light microscopy with the high-resolution information of the electron microscope (McDonald, 2009). We decided on adapting an established CLEM procedure, which includes collecting embryos in capillary tubes, staging of developing embryos by light microscopy, fixation by high-pressure freezing, thin-layer embedding in Epon/Araldite, and serial sectioning followed by electron tomography (Müller-Reichert, Srayko, Hyman, O'Toole, & McDonald, 2007; Woog, White, Büchner, Srayko, & Müller-Reichert, 2012). In general, this approach has proved useful for the analysis of wild-type and mutant embryos, as well as the application of RNA-mediated interference (RNAi) for ultrastructural studies in *C. elegans* (Pelletier, O'Toole, Schwager, Hyman, & Müller-Reichert, 2006). Taking advantage of the early *C. elegans* embryo, we describe a method to apply drug treatment in this model organism. In addition, we describe the injection of rhodamine-labeled tubulin into sea urchin embryos to allow a staging of fluorescent samples to investigate abscission also in other organisms. Here, we mainly concentrate on the handling of the samples for high-pressure freezing.

1. METHODS

1.1 HANDLING OF EARLY EMBRYOS

This section describes how we prepare isolated embryos for live-cell imaging and subsequent high-pressure freezing. The use of transparent capillary tubes allows an observation of early embryos under the light microscope, an exchange of medium during imaging, and the easy transfer of samples at a specific developmental stage to the high-pressure freezer (McDonald, Morphew, Verkade, & Müller-Reichert, 2007).

1.1.1 Drug treatment of early Caenorhabditis elegans embryos

For drug treatment experiments, the embryonic eggshell has to be permeable. Therefore, *C. elegans* L4 larvae were cultivated on *perm-1 RNAi* feeding plates for 20–24 h to reduce the eggshell's permeability barrier (Carvalho et al., 2011). We placed two young adult worms in a droplet of Shelton buffer [Shelton Growth Medium (SGM)] (Bowerman & Shelton, 1999) on a high-precision coverslip (0.17 mm) for imaging. SGM is a culture medium to stabilize osmotically unstable embryos. To test the permeability of the eggshell in such drug-treated embryos, we added 2 μg/mL Hoechst and visualized the chromosomes. The worms were then dissected under a stereomicroscope, and single one-cell embryos were collected into capillary tubes as described (Müller-Reichert et al., 2007). The pores of the capillary tubes have a variable molecular weight cut-off from 5 to 20 kDA (Hohenberg, Mannweiler, & Müller, 1994), thus allowing a penetration of the drug into the capillary. We always chose embryos at pronuclear migration to set up the light microscope before the start of live-cell imaging at about onset of cleavage furrow ingression.

To visualize cytokinesis by light microscopy, we used fluorescently labeled *C. elegans* strains, such as TH155 labeled with PH-PLCδ-GFP and tubulin-mCherry to visualize the microtubules and the plasma membrane, respectively. We mounted the coverslip with the embryo submerged in SGM on the stage of an inverted Spinning Disc microscope. Imaging started with the onset of cleavage furrow ingression. In our experiments, this event was defined as time point zero. Because of the movement of the intercellular bridge toward the ventral side of the embryo, a z-stack was imaged with 20 planes and a spacing of 0.6 μm to monitor the development of the permeabilized embryo.

Depending on the experiment, it is often necessary that drug delivery takes place at a specific time point during early embryo development. As an example, we added Latrunculin A to investigate the role of actin during late abscission. For this we transferred early embryos to the SGM buffer containing Latrunculin A at closure of the cleavage furrow. In practice, microscope recording was stopped at the time point of cleavage furrow closure, the capillary with the embryo was then transferred to a new coverslip with SGM containing 20 μM Latrunculin A, and the coverslip was put back on the microscope stage to continue recording. Because of the recording break, we used a second timer to continuously record the early developmental embryonic events. Another possibility would be to use a certain volume of SGM at the beginning of the experiment and then to add SGM containing the drug at high concentration to allow a dilution of the drug to the desired final concentration. At an appropriate embryonic stage as shown in Fig. 2A, the capillary was removed from the microscope stage and its length was reduced to about 1 mm to fit the piece of capillary tubing containing the early embryo into a prefilled

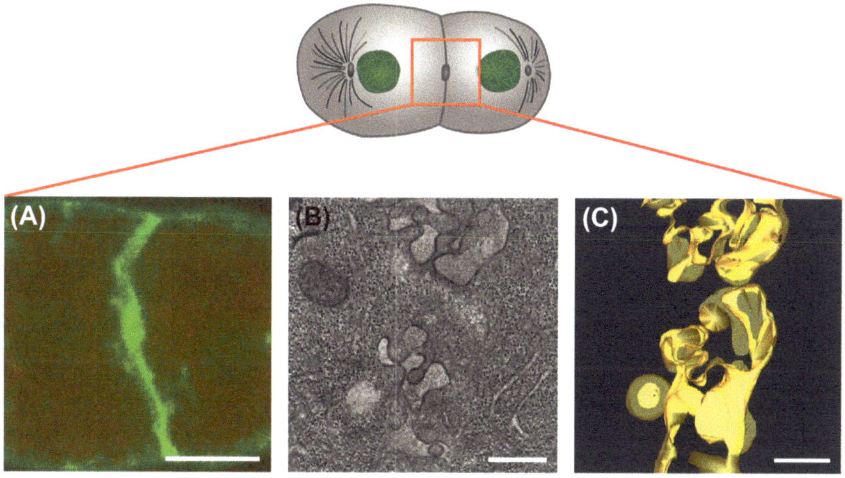

FIGURE 2

Electron tomography of an early Latrunculin A–treated *Caenorhabditis elegans* embryo. (A) Spinning Disc confocal image showing the intercellular bridge of a Latrunculin A–treated embryo. Scale bar, 10 μm. (B) Tomographic slice of a high-pressure frozen, drug-treated embryo. The two daughter cells are still connected. (C) Corresponding three-dimensional model showing the fragmentation of the intercellular bridge due to the disruption of the actin network by Latrunculin A. This fragmentation is visible by multiple membrane ridges at the midbody and leads to a failure in abscission. Scale bars, 500 nm.

membrane carrier. In contrast to the previous publications (Müller-Reichert et al., 2007; Woog et al., 2012), the ends of the capillary were not closed to increase the penetration efficiency of the drug into the lumen of the tubes. The transfer from the light microscope to the high-pressure freezer took about 15–30 s.

1.1.2 Injection of rhodamine-labeled tubulin into sea urchin embryos

As a first step, we collected sea urchin eggs and sperm by injecting 1 mL of KCl (0.5 M) close to the mouth of the animals. The animals released either oocytes (orange) or sperm (white) upon injection. The eggs were collected in filtered sea water (FSW) and then washed several times in FSW. The sperm was collected with a pasteur pipette and kept concentrated in a microcentrifuge tube on ice. The eggs were mounted in microinjection chambers and injected with either rhodamine- or Cy3-labeled tubulin using mercury-containing needles mounted on a CellTram Oil manual injector (Jaffe & Terasaki, 2004) (see also http://mterasaki.us/panda/injection/ for more information). Prior to injection, the protein was loaded into loading glass capillaries together with nonreactive mineral oil. The nonreactive mineral oil prevents protein dissolution into the sea water and allows the identification of injected oocytes. Each time an oocyte was microinjected, the needle was front-loaded from the loading capillary with the combination mineral oil–protein–mineral oil. To obtain zygotes after injection, oocytes were transferred from the microinjection chamber into a 50-mm petri dish with fresh FSW. The collected sperm was diluted 1:5000 and added to the petri dish. Fertilized embryos develop a characteristic fertilization envelope, and only fertilized embryos were collected into capillary tubes and transferred to 8-well dishes. Typically, four embryos were transferred into each 8-well dish and mounted on an inverted laser scanning confocal microscope. We were able to monitor four embryos simultaneously, intervals of 60–90 s. This time frame worked well for documenting cytokinesis because of the time an embryo needs for completion of this stage (approximately 20–30 min).

Optimal imaging of either nematode or echinoderm embryos is performed when the embryo is positioned as close as possible to the coverslip facing the objective. Unfortunately, the capillary tube slightly reduces the quality of imaging because of light distortions. Still, the image quality is good enough to monitor cytological events during early embryonic development (Figs. 2A and 3A). Imaging was stopped at desired stages, and the samples were rapidly frozen. The transfer from the light microscope to the high-pressure freezer took about 15–30 s.

1.2 HIGH-PRESSURE FREEZING AND FREEZE SUBSTITUTION

We applied cryo-immobilization by high-pressure freezing to achieve a fast fixation of the staged embryos. We used a Leica EM PACT2 high-pressure freezer equipped with a rapid transfer system (RTS) as described (McDonald et al., 2007). We moved the freezer next to the light microscope to reduce the transfer time, thus allowing correlative LM/EM experiments. For freezing, the membrane carriers were filled

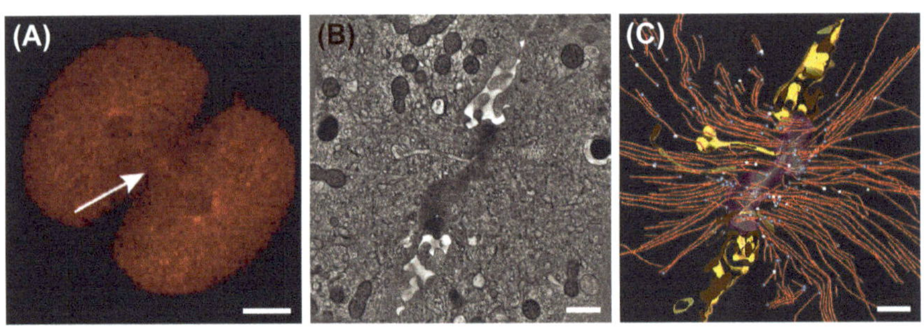

FIGURE 3

Intercellular bridge of an early sea urchin embryo. (A) Fluorescent image of a dividing sea urchin embryo after injection of rhodamine-labeled tubulin. The intercellular bridge is indicated (*white arrow*). Scale bar, 20 μm. (B) Tomographic slice of an intercellular bridge in a high-pressure frozen and freeze-substituted sample. (C) Corresponding three-dimensional model of the intercellular bridge as shown in (B). Color coding: gold, plasma membrane; red, microtubules; *blue spheres*, open microtubule ends; *white spheres*, closed microtubule ends; purple, electron dense region. Scale bars, 500 nm.

with medium containing 20% BSA as a cryoprotectant. Capillary tubes were then placed into the prefilled membrane carriers (100 μm-deep carriers for *C. elegans* embryos, 200 μm-deep membrane carriers for sea urchin embryos). We used M9 buffer for *C. elegans* and FSW buffer for sea urchin embryos. Specimen loading then was carried out essentially as described (Müller-Reichert et al., 2007).

Under liquid nitrogen, the membrane carriers with the samples were transferred to cryo-vials containing anhydrous acetone. The freeze-substitution cocktails contained either 0.5% OsO_4/0.05% uranyl acetate for drug-treated *C. elegans* embryos or 1% OsO_4/0.1% uranyl acetate for sea urchin embryos. Because of the permeabilized eggshell of the embryos, we reduced the osmium concentration for the drug-treated nematode samples. Freeze substitution of both types of samples was carried out using an identical protocol. Samples were placed in the freeze-substitution machine (Leica EMAFS) at −90°C for 8–48 h. The warm up phase was performed with a temperature increase of 5°C/h for 12 h, and the samples were then maintained at −30°C for 5 h. Afterward the temperature was further increased by 5°C/h to 0°C, and the samples were allowed to warm up.

1.3 EMBEDDING, STAINING, AND SECTIONING OF SAMPLES

Following freeze substitution, the samples were washed three times with fresh anhydrous acetone and gradually infiltrated with Epon/Araldite (3:1, 1:1, and 1:3 for at least 1 h, 100% resin for 1 h, then overnight and another hour). Samples were thin-layer embedded as described (Muller-Reichert, Hohenberg, O'Toole, & McDonald, 2003). The resin was then polymerized for 48 h at 60°C. After polymerization it was necessary to prepare the samples for sectioning by remounting the embedded embryos using the following procedure. The embryos embedded in the thin layer of resin were identified using a light microscope and marked using a diamond tip of an objective scribe. The region containing the sample was then cut out under a stereomicroscope using a scalpel and mounted onto a "dummy" block using a two-component fast glue (Muller-Reichert et al., 2003).

For subsequent electron tomography, serial 300-nm-thick sections were collected on FORMVAR-coated, copper slot grids. The sections were poststained using 2% uranyl acetate in 70% methanol, followed by Reynold's lead citrate. Additional 15-nm colloidal gold markers were added to both surfaces of the serial sections to serve as fiducial markers for tilt series alignment. This step needs to be performed before the observation of the sections in the electron microscope to achieve an optimal adherence of the fiducial markers to the surfaces of the serial sections. To attach the fiducial markers, the grids were inserted into an "Hiraoka" spare staining unit and covered with a 15-nm colloidal gold solution for 1 min. To ensure a uniform distribution of the colloidal gold on the surface of the serial sections, the staining device was lifted for some millimeters and dropped back on the table. During the 1-min incubation time, this procedure was repeated about 10 times. Afterward the gold solution was removed, the grids were washed with ddH_2O and finally dried.

1.4 ELECTRON TOMOGRAPHY

To identify those serial sections containing intercellular bridges, the serial sections were screened using a TECNAI12 transmission electron microscope (FEI) operated at 80 kV. Electron tomography was then performed for selected sections to analyze the ultrastructure of intercellular bridges in 3D. We used an intermediate-voltage transmission electron microscope (F30, FEI) operated at 300 kV equipped with a 2k × 2k CCD camera (Gatan US1000) and a eucentric tilting stage. The SerialEM software package was used to automatically acquire images over a range of ±64 degrees with an increment of 1 degree and a pixel size of 1.4 nm. We performed dual-axis electron tomography to reduce the "missing wedge" (Mastronarde, 1997). For this we acquired a tilt series, rotated the grid by 90 degrees and recorded a second tilt series over the same range. The IMOD software package (Boulder Laboratory for 3-D Electron Microscopy of Cells, University of Colorado, Boulder) was used for calculation of the tomograms and modeling of the 3D reconstruction, respectively. We modeled features in the midbody including membranes, microtubules, and vesicles as shown in Fig. 2 for a Latrunculin A–treated *C. elegans* embryo, and in Fig. 3 for an early sea urchin embryo.

2. INSTRUMENTATION AND MATERIALS

2.1 HANDLING OF EARLY EMBRYOS

2.1.1 *Drug treatment of early* Caenorhabditis elegans *embryos*

Instrumentation: Stereomicroscope with light source from the bottom, Spinning Disc Confocal microscope (eg, Zeiss Axio Observer with a Zeiss C-Apochromat 63×/1.2 W objective, Yokogawa CSU-X1 Spinning Disc scan head and a Roper evolve 515 EMCCD camera controlled by ZenBlue software, 20 z-zacks at 0.6 μm, time-lapse intervals of 15–20 s), 488 and 561 laser lines.

Materials: *Caenorhabditis elegans* strain MSN707 (Audhya lab, University of Wisconsin, Madison, USA) expressing ZEN-4::GFP and PH-PLCδ::mCherry to visualize the midbody marker ZEN-4 and the plasma membrane, cellulose capillaries with an inner diameter of 200 μm and a molecular weight cut-off from 5 to 20 kDA (Leica Microsystems), pasteur pipette 2 mL, micropipettor (0.5–10 μL), BD micro lances No. 18 (Becton, Dickinson and Company) high-precision coverslips (0.17 mm ± 0.05 mm thickness, 24 × 60 mm; Roth), No. 23 scalpel tips, pipette tips, nail polish, fine bent tweezers.

Reagents: SGM: 1 mL of 5 mg/mL inulin (Sigma I3754), 50 mg polyvinylpyrrolidone (Sigma P0930), 100 μL BME vitamins (Sigma B6891), 100 μL chemical defined lipid concentrate (Invitrogen 11905-031), 100 μL 100× concentrated Penicillin-Streptomycin (PAA P11-010), 9 mL *Drosophila* Schneiders' Medium (Invitrogen 21720-024), addition of 35% head inactivated FBS (Invitrogen 10270-106).

2.1.2 Injection of fluorescent dyes into sea urchin embryos

Instrumentation: Injection was performed in an inverted transmitted light microscope with movable stage (eg, Nikon Ti-U) equipped with a long working distance 10× objective (eg, Nikon Plan fluor 10× N.A. 0.30 W.D 16.0 mm) and an eyepiece micrometer. Live imaging was performed in a confocal microscope (eg, Leica SP5II equipped with a motorized XY stage, SuperZ Galvo stage, and a 40× HCX PL APO 1.10 NA objective lens), 10–20 Z-stacks were imaged at 2 μm step, time lapse with 60–90 s intervals, 561 laser line.

Materials: U-shaped plastic microinjection chambers (EMBL mechanical workshop, drawings available upon request), double stick tape (Scotch), No.1 22 × 22 mm coverslips, diamond pen, CellTram oil microinjector (Eppendorf), MM-92 motor-drive manipulator (Narishige), microscope adapter (EMBL mechanical workshop), glass capillaries to pull needles (50 μL microcaps, Drummond), Narishige PN-3 glass microelectrode horizontal needle puller, loading capillaries (outer diameter 0.8 mm, inner diameter 0.6 mm, length 100 mm, Drummond), 10 μL Hamilton syringe (Hamilton).

Reagents: KCl solution (0.5 M), high-viscosity silicon grease (VWR), mineral oil (to load the syringe, Sigma), mercury, dimethylpolysiloxane, viscosity: 20 cST (for the loading capillary, Sigma), rhodamine-tubulin (Cytoskeleton, Inc.), and Cy3-tubulin (gift from Nédelec lab, EMBL, Heidelberg).

2.2 HIGH-PRESSURE FREEZING AND FREEZE SUBSTITUTION

Instrumentation: Stereomicroscope with white light source from the bottom and the top, high-pressure freezer (EM PACT2+RTS, Leica Microsystems), freeze-substitution machine (Leica EMAFS, Leica Microsystems).

Materials: Membrane carriers (100 or 200 μm deep, Leica Microsystems).

Reagents: I. SGM + 20% BSA (Roth 8076.2), II. FSW + 20% BSA (Roth 8076.2), acetone (glass distilled; EMS 10015), osmium tetroxide (EMS 19152), uranyl acetate (Plano R1260A), Araldite EMbed 812 (EMS 13940).

2.3 EMBEDDING, STAINING, AND SECTIONING OF SAMPLES

Instrumentation: Oven (60°C) for resin polymerization, objective marker with a diamond tip (Leica), screening microscope (Zeiss Axiolab), stereomicroscope with white light source from the bottom, ultramicrotome (Leica UC6, Leica Microsystems), transmission electron microscope (eg, Zeiss 906) operated at 80 kV.

Materials: Glass slides, tissue, Parafilm, glass slide holders, scalpels, tweezers, "dummy" blocks.

Reagents: Teflon MS-143 solution (Silmid MS143V1USG), Araldite EMbed 812 (EMS 13940), Two-Part Epoxy Adhesive Kit (Miller–Stephenson MS-907), Formvar powder (EMS 15800), methanol, copper slod grids (Plano G2500C), staining matrix Pelco (Plano 22510), uranyl acetate (Plano R1260A), lead citrate (EMS 17810), 15-nm colloidal gold (BBInternational).

2.4 ELECTRON TOMOGRAPHY

Instrumentation: Intermediate-voltage transmission electron microscope operated at 300 kV (FEI TECNAI F30 FEG), high-tilt rotating stage (Gatan model 650), 2k × 2k CCD camera (Gatan US.1000), SerialEM software, 3D reconstruction and modeling software (IMOD; Boulder Lab for the 3D electron microscopy of cells; Boulder, USA).

Materials: Fine-tipped tweezers.

3. DISCUSSION

Here we show that our established approach for CLEM is suitable for drug-treated nematode embryos, as well as for injection of fluorescent probes in sea urchin embryos. As sample loading for high-pressure freezing has been covered in detail previously (Müller-Reichert et al., 2007), we will briefly discuss here the handling of the specimens for staging by light microscopy, the choice of the freeze-substitution cocktail for permeabilized embryos, and the use of the commercially available high-pressure freezers for CLEM experiments.

3.1 ADVANTAGES AND LIMITATIONS IN USING CAPILLARY TUBES

The use of capillary tubes for cryo-immobilization was first introduced by Hohenberg et al. (1994). The main advantage for using capillary tubes is that isolated embryos can be easily handled (Muller-Reichert et al., 2003; Müller-Reichert et al., 2007). In addition, the embryonic development can be monitored by light microscopy, and the permeability of the capillaries allows a diffusion of drugs into the lumen of the tubes and consequently into samples, such as PERM-1-treated early *C. elegans* embryos. Drug treatment experiments in adult nematodes are normally performed using a specific microdevice to trap single embryos (Carvalho et al., 2011). However, for a rapid transfer of single embryos from the light microscope to the high-pressure freezer, such devices are rather unpractical, because the embryos are trapped in small holes to prevent them from floating around during the medium exchange. In addition, the embedding for subsequent electron microscopy and the remounting of specific embryos for serial sectioning would be extremely challenging. The method described in this chapter is more practical to collect early embryos into capillaries by applying established protocols for thin-layer embedding and remounting on dummy blocks (Muller-Reichert et al., 2003; Woog et al., 2012). So far, we did not recognize a difference in the efficiency of drug treatment by using capillary tubes in comparison to the published use of these microdevice chambers (Carvalho et al., 2011). However, the imaging quality is obviously reduced when using a capillary. Another limitation of the tubing is the diameter of the capillary itself. The capillaries have an inner diameter of around 200 μm, thereby restricting the size of the sample that can be frozen. However, this limitation is also given by the maximum depth of good cryo-preservation (Studer, Humbel, & Chiquet, 2008).

3.2 FREEZE SUBSTITUTION OF PERMEABILIZED *Caenorhabditis elegans* EMBRYOS

We routinely use an acetone cocktail containing 1% osmium tetroxide and 0.1% uranyl acetate to provide a secondary fixation and to enhance the contrast of the samples during freeze substitution (Müller-Reichert et al., 2007). For both *C. elegans* and sea urchin embryos, a concentration of 1% OsO_4 and 0.1% uranyl acetate gives good results. Permeabilized *C. elegans* embryos, however, appeared much darker when using this cocktail. In wild-type embryos, the eggshell of the embryo serves as a diffusion barrier. In permeabilized embryos, the reagents of the cocktail may diffuse into the sample much faster, thus leading to an overstaining of the specimen. To avoid an overstaining of the specimens, we reduced the concentration of osmium tetroxide and uranyl acetate to 0.5% and 0.05%, respectively. The samples gave good contrast for electron tomography. However, the osmium concentration could be reduced further. Previous experiments have shown that a concentration of osmium tetroxide as low as 0.01% gives sufficiently contrasted thin sections (Müller-Reichert et al., 2003).

3.3 INJECTION OF RHODAMINE-LABELED TUBULIN INTO SEA URCHIN EMBRYOS

Sea urchin oocytes can be easily injected with fluorescently labeled tubulin following previous protocols established for starfish (Jaffe & Terasaki, 2004) (see also http://mterasaki.us/panda/injection/). With this microtubule labeling, the different embryonic stages can be easily followed by light microscopy, including the abscission stage. However, we noticed that sea urchin oocytes are less robust than starfish oocytes and therefore have to be handled more carefully. First of all, only fresh oocytes should be injected: after a few hours postcollection the rate of fertilization decreases. Moreover, the injection has to be performed with sharper needles, as larger needles will damage the soft cortex of the sea urchin oocyte. It is important that the fluorescently labeled tubulin contains no traces of glycerol, as these oocytes showed to be extremely sensitive to it, and fertilization will be prevented. Fertilization rates were higher when oocytes were fertilized in a petri dish (and not in the microinjection chamber).

3.4 THE USE OF DIFFERENT HIGH-PRESSURE FREEZERS FOR CORRELATIVE LIGHT AND ELECTRON MICROSCOPIC EXPERIMENTS

The EM PACT2+RTS (Leica) is well suited for correlative light/electron microscopic experiments, as the whole machine can be moved in proximity to a light microscope to allow a fast transfer of the samples into the high-pressure freezer. In addition, the time necessary for specimen loading can be significantly reduced compared to other commercially available freezers. The rapid loader automates the closing of the specimen carrier and the insertion of the sample holder into the high-pressure freezer. With the RTS system of the Leica EM PACT2, we could previously reduce the time window for the transfer of the samples to 3–5 s (Pelletier et al., 2006). However, one limitation of the EM PACT2 is in the relatively small size of the sample holders which limits the types of samples that can be frozen. The cavity of the membrane carriers has an inner diameter of only 1 mm and thickness of either 100 or 200 μm. Another aspect to mention here is the availability of this machine on the market, as Leica stopped the production of the EM PACT2 by of the end of 2015. A new high-pressure freezer commercially available will be the EM ICE (Leica, Microsystems, Vienna, Austria). Compared to the EM PACT2, this new machine can handle larger specimens (with a maximum thickness of 200 μm). This machine, however, is larger and heavier and cannot be as easily moved as the EM PACT2. The use of this new machine for CLEM experiments has not been reported yet. An alternative high-pressure freezer for correlative experiments is the Leica HPM100, which is also portable and gives advantages in the freezing of bigger samples.

In conclusion, we have presented two routine methods for the handling of drug-treated, as well as injected, early embryos for high-pressure freezing and electron tomography. Our methods for nematode and echinoderm embryos, however, are also applicable to standard transmission electron microscopy using thin sections for morphological studies on cytokinesis.

ACKNOWLEDGMENTS

The authors would like to thank the members of the EM facility of the EMBL in Heidelberg, especially Rachel Santarella-Mellwig for great support and technical assistance and Dr. Eileen O'Toole for a critical reading of the manuscript. Research in the Müller-Reichert lab was supported by the Deutsche Forschungsgemeinschaft (DFG; grant MU1423/4-1 to TM-R).

REFERENCES

Bowerman, B., & Shelton, C. A. (1999). Cell polarity in the early *Caenorhabditis elegans* embryo. *Current Opinion in Genetics and Development, 9*(4), 390–395. http://dx.doi.org/10.1016/S0959-437X(99)80059-8.

Bringmann, H., & Hyman, A. A. (2005). A cytokinesis furrow is positioned by two consecutive signals. *Nature, 436*(7051), 731–734. http://dx.doi.org/10.1038/nature03823.

Carvalho, A., Olson, S. K., Gutierrez, E., Zhang, K., Noble, L. B., Zanin, E., ... Oegema, K. (2011). Acute drug treatment in the early *C. elegans* embryo. *PLoS One, 6*(9), e24656. http://dx.doi.org/10.1371/journal.pone.0024656.

Glotzer, M. (2005). The molecular requirements for cytokinesis. *Science, 307*(5716), 1735–1739. http://dx.doi.org/10.1126/science.1096896.

Green, R. A., Mayers, J. R., Wang, S., Lewellyn, L., Desai, A., Audhya, A., & Oegema, K. (2013). The midbody ring scaffolds the abscission machinery in the absence of midbody microtubules. *The Journal of Cell Biology, 203*(3), 505–520. http://dx.doi.org/10.1083/jcb.201306036.

Green, R. A., Paluch, E., & Oegema, K. (2012). Cytokinesis in animal cells. *Annual Review of Cell and Developmental Biology, 28*(1), 29–58. http://dx.doi.org/10.1146/annurev-cellbio-101011-155718.

Guizetti, J., Schermelleh, L., Mantler, J., Maar, S., Poser, I., Leonhardt, H., … Gerlich, D. W. (2011). Cortical constriction during abscission involves helices of ESCRT-III-dependent filaments. *Science, 331*(6024), 1616–1620. http://dx.doi.org/10.1126/science.1201847.

Hohenberg, H., Mannweiler, K., & Müller, M. (1994). High-pressure freezing of cell suspensions in cellulose capillary tubes. *Journal of Microscopy, 175*(Pt 1), 34–43. Retrieved from http://eutils.ncbi.nlm.nih.gov/entrez/eutils/elink.fcgi?dbfrom=pubmed&id=7932676&retmode=ref&cmd=prlinks.

Jaffe, L. A., & Terasaki, M. (2004). Quantitative microinjection of oocytes, eggs, and embryos. *Methods in Cell Biology, 74*, 219–242. Retrieved from http://eutils.ncbi.nlm.nih.gov/entrez/eutils/elink.fcgi?dbfrom=pubmed&id=15575609&retmode=ref&cmd=prlinks.

Larkin, K., & Danilchik, M. V. (1999). Microtubules are required for completion of cytokinesis in sea urchin eggs. *Developmental Biology, 214*(1), 215–226. http://dx.doi.org/10.1006/dbio.1999.9395.

Mastronarde, D. N. (1997). Dual-axis tomography: an approach with alignment methods that preserve resolution. *Journal of Structural Biology, 120*, 343–352.

McDonald, K. L. (2009). A review of high-pressure freezing preparation techniques for correlative light and electron microscopy of the same cells and tissues. *Journal of Microscopy, 235*(3), 273–281. http://dx.doi.org/10.1111/j.1365-2818.2009.03218.x.

McDonald, K. L., Morphew, M., Verkade, P., & Müller-Reichert, T. (2007). Recent advances in high-pressure freezing: equipment- and specimen-loading methods. *Methods in Molecular Biology (Clifton, N.J.), 369*, 143–173. http://dx.doi.org/10.1007/978-1-59745-294-6_8 (Chapter 8).

Mierzwa, B., & Gerlich, D. W. (2014). Cytokinetic abscission: molecular mechanisms and temporal control. *Developmental Cell, 31*(5), 525–538. http://dx.doi.org/10.1016/j.devcel.2014.11.006.

Muller-Reichert, T., Hohenberg, H., O'Toole, E. T., & McDonald, K. (2003). Cryoimmobilization and three-dimensional visualization of *C. elegans* ultrastructure. *Journal of Microscopy, 212*(Pt 1), 71–80. Retrieved from http://eutils.ncbi.nlm.nih.gov/entrez/eutils/elink.fcgi?dbfrom=pubmed&id=14516364&retmode=ref&cmd=prlinks.

Müller-Reichert, T., Sassoon, I., O'Toole, E., Romao, M., Ashford, A. J., Hyman, A. A., & Antony, C. (2003). Analysis of the distribution of the kinetochore protein Ndc10p in *Saccharomyces cerevisiae* using 3-D modeling of mitotic spindles. *Chromosoma, 111*(7), 417–428. http://dx.doi.org/10.1007/s00412-002-0220-6.

Müller-Reichert, T., Srayko, M., Hyman, A., O'Toole, E. T., & McDonald, K. (2007). Correlative light and electron microscopy of early *Caenorhabditis elegans* embryos in Mitosis. In *Cellular electron microscopy* (Vol. 79, pp. 101–119). Elsevier. http://dx.doi.org/10.1016/S0091-679X(06)79004-5.

Olson, S. K., Greenan, G., Desai, A., Müller-Reichert, T., & Oegema, K. (2012). Hierarchical assembly of the eggshell and permeability barrier in *C. elegans*. *The Journal of Cell Biology, 198*(4), 731–748. http://dx.doi.org/10.1083/jcb.201206008.

Pelletier, L., O'Toole, E., Schwager, A., Hyman, A. A., & Müller-Reichert, T. (2006). Centriole assembly in *Caenorhabditis elegans*. *Nature, 444*(7119), 619–623. http://dx.doi.org/10.1038/nature05318.

Rappaport, R. (1985). Repeated furrow formation from a single mitotic apparatus in cylindrical sand dollar eggs. *The Journal of Experimental Zoology, 234*(1), 167–171. http://dx.doi.org/10.1002/jez.1402340120.

Studer, D., Humbel, B. M., & Chiquet, M. (2008). Electron microscopy of high pressure frozen samples: bridging the gap between cellular ultrastructure and atomic resolution. *Histochemistry and Cell Biology, 130*(5), 877–889. http://dx.doi.org/10.1007/s00418-008-0500-1.

Woog, I., White, S., Büchner, M., Srayko, M., & Müller-Reichert, T. (2012). Correlative light and electron microscopy of intermediate stages of meiotic spindle assembly in the early *Caenorhabditis elegans* embryo. In *Methods in cell biology – Volume 97* (Vol. 111, pp. 223–234). Elsevier. http://dx.doi.org/10.1016/B978-0-12-416026-2.00012-1 (Chapter 12).

CHAPTER

Studying cytokinesis and midbody remnants using correlative light/scanning EM

14

S. Frémont*,§,[1], A. Echard*,§,[1]

*Institut Pasteur, Paris, France

§Centre National de la Recherche Scientifique, Paris, France

[1]Corresponding authors: E-mail: stephane.fremont@pasteur.fr; arnaud.echard@pasteur.fr

CHAPTER OUTLINE

Abstract

Cytokinesis is an essential step of cell proliferation leading to the physical separation of the dividing cells. Cytokinesis relies on both large scale and local scale cell shape changes, and terminates with the final abscission cut that requires close apposition of the plasma membrane. While furrow ingression is a prominent feature of the early phase of cytokinesis and is easy to visualize in all models, from dividing eggs to culture cells, the later steps of cytokinesis until abscission can be much more difficult to visualize. One key issue is to combine live-cell imaging over several hours and detailed, structural analysis of the cell shape changes in 3D, in particular at the time of cytokinetic abscission. Here, we describe the methodologies that we recently developed for studying cytokinetic abscission in human culture cells using live-cell phase-contrast microscopy, combined

Methods in Cell Biology, Volume 137, ISSN 0091-679X, http://dx.doi.org/10.1016/bs.mcb.2016.03.028

with correlative scanning electron microscopy. This allows us to determine the membrane surface and underlying cytoskeleton of the intercellular bridge with unprecedented precision and to determine the fate of the midbody remnant after abscission.

INTRODUCTION

At the end of mitosis, the physical separation of the two daughter cells concludes cell division. Cytokinesis has been studied since the 19th century, and leads to the appropriate segregation of the cytosolic components, intracellular organelles, and genetic material in each dividing cell (Fededa & Gerlich, 2012). Cytokinesis starts in animal cells with the ingression of a furrow, characterized by a large scale deformation of the plasma membrane at the cell equator which is driven by the contraction of an actomyosin ring (Green, Paluch, & Oegema, 2012). While most effort is focused on this initial step of cytokinesis, the sister cells still remain connected by an intercellular bridge that is eventually cut by an increasingly studied process termed abscission (Mierzwa & Gerlich, 2014). Although viewed as a secondary, almost trivial step, it actually requires major changes in cytoskeleton and membrane dynamics. Indeed, at the time of abscission, all cytoskeletal elements (in particular, microtubules and F-actin) must be cleared at the presumptive abscission site and the lipid composition of the plasma membrane has to be remodeled (Mierzwa & Gerlich, 2014). In the past decade, genome-wide RNA-mediated interference (RNAi) screens have been used to identify the genes involved in late steps of cytokinesis, and highlighted the role of membrane trafficking, as well as lipid and cytoskeleton remodeling (Echard, Hickson, Foley, & O'Farrell, 2004; Eggert et al., 2004; Skop, Liu, Yates, Meyer, & Heald, 2004). Contrary to initial thoughts (Gromley et al., 2005), membrane trafficking is not directly responsible for cytokinetic abscission but participate to the maturation of the intercellular bridge. Vesicles trafficking in and out of the bridge can be visualized with high temporal accuracy using cargoes (Goss & Toomre, 2008; Gromley et al., 2005) or specific vesicular markers of the Rab GTPase family, such as Rab8 (Gromley et al., 2005; Guizetti et al., 2011), Rab11 (Fielding et al., 2005; Schiel et al., 2012; Wilson et al., 2005), and Rab35 (Chesneau et al., 2012; Dambournet et al., 2011; Klinkert, Rocancourt, Houdusse, & Echard, 2016; Kouranti, Sachse, Arouche, Goud, & Echard, 2006). It is now established that Rab35 and Rab11 function in parallel to clear F-actin accumulation within the intercellular bridge, which otherwise would impair cytokinetic abscission. At the molecular level, Rab35 promotes the targeting of the oculo–cerebro–renal syndrome of Lowe (OCRL) phosphatase at the intercellular bridge prior to abscission (Cauvin & Echard, 2014; Cauvin et al., 2016; Dambournet et al., 2011). The OCRL phosphatase is a PtdIns(4,5)P_2 lipid phosphatase that locally hydrolyzes PI(4,5)P2 and therefore prevents F-actin accumulation in late cytokinetic bridges. Similarly, Rab11-endosomes transport the p50RhoGAP that controls F-actin clearance, a step required for abscission (Echard, 2012; Schiel et al., 2012). It is now clear that abscission occurs on the side of the central, electron-dense part of the bridge termed the midbody or Fleming body, at the level of secondary ingressions of the plasma membrane (Elia, Sougrat, Spurlin, Hurley, & Lippincott-Schwartz, 2011; Guizetti et al., 2011; Lafaurie-Janvore et al., 2013; Mullins & Biesele, 1977; Schiel et al., 2012). Importantly, the midbody serves as a platform for the assembly of the abscission machinery (Mierzwa & Gerlich, 2014). A major discovery in the abscission field has been the recognition that the endosomal sorting complex required for transport (ESCRT), initially described for intraluminal vesicle formation in late endosomes and HIV budding, likely constituted the final machinery that locally pinches the plasma membrane and drives abscission (Carlton & Martin-Serrano, 2007; Elia et al., 2011; Guizetti et al., 2011; Morita et al., 2007). Indeed, during the intercellular bridge maturation, ESCRT components are recruited in a hierarchical manner and 17-nm ESCRT-dependent filamentous helices assemble approximately 1 μm from the midbody at the site of abscission (Elia et al., 2011; Guizetti et al., 2011).

The understanding of cytokinetic abscission requires the combination of high temporal resolution live-cell fluorescence or phase-contrast imaging and detailed structural description of the plasma membrane deformation at the site of abscission. The latter can be achieved by serial sections through the bridge (Mullins & Biesele, 1977), 3D electron microscopy (EM)-tomography (Guizetti et al., 2011; Schiel et al., 2011), or scanning EM (SEM) (Crowell, Gaffuri, Gayraud-Morel, Tajbakhsh, & Echard, 2014). One challenge is to correlate changes of the distribution of specific, fluorescently labeled proteins involved in abscission and membrane deformations. So far, it has been challenging to visualize the intermediates of membrane constriction at the time of abscission. Here, we present two complementary methods to study cytokinesis abscission, by combining correlative light microscopy in live cells and SEM which gives structural details of the 3D structure of the cytokinetic bridge, before and after abscission.

1. EXPERIMENTAL PROCEDURES

1.1 ANALYSIS OF THE KINETIC OF ABSCISSION USING PHASE-CONTRAST MICROSCOPY IN LIVE CELLS

Failure in cytokinesis can lead to tetraploid cells, which are genetically instable and can promote tumorigenesis in vivo (Fujiwara et al., 2005). Abscission failure leads to two major and distinct phenotypes: an increase of the number of binucleated, tetraploid cells or a delay in abscission completion. This is due to a decrease or an increase of intercellular bridge stability, respectively. These phenotypes can be revealed by studying single cell division in time-lapse video microscopy. With the live-cell procedure described below, we are able to compare the behavior of dividing cells under different conditions, such as protein depletion or overexpression in different cell lines. Depending on the cell types, the intercellular bridge is more or less visible since it can be buried between the two daughter cells if they do not separate enough. We usually use HeLa cells (ATCC CCL-2) or U2OS cells (ATCC) to study the kinetics of abscission by phase-contrast microscopy.

1.1.1 Materials

Dulbecco's modified eagle medium (DMEM) (Gibco)
Fetal bovine serum (FBS) (PAN biotech GmbH)
35-mm glass bottom dishes (P35G-1.5-20-C, MatTek Corporation)
Inverted video microscope (Nikon Ti-E)

1.1.2 Methods

1.1.2.1 Cell culture

Details of cell culture, transfections, or treatments prior to the video microscopy are not described, as they are standard procedures. However, it is important to get precise and reproducible cell density at the time of video microscopy, since cell density influences the duration of cytokinesis after furrow ingression (Lafaurie-Janvore et al., 2013). Cells have to be transferred in the glass bottom dishes (see later) at least one day before making movies. We also noticed that lipofectant reagents used for plasmid transfection can sometime interfere with the frequency of division and the timing of abscission. It can be necessary to use calcium phosphate-based transfection to circumvent this issue.

1.1.2.2 Live-cell imaging by time-lapse phase-contrast and fluorescence microscopy

1. Grow HeLa cells in DMEM medium supplemented with 5% FBS in glass bottom dishes 2 days before the start of the time-lapse video to reach about 40–70% of confluence.
2. Transfer the cells under the light microscope. We use a fully automatized inverted microscope Nikon Ti-E piloted by Metamorph software. The microscope should be encaged into a temperature box (LIS) to maintain stable temperature at 37°C and 5% CO_2 throughout the acquisition.
   Note 1: The microscope and the surrounding box must be preheated before starting live-cell imaging. In our experience, stable temperature of the microscope, cell chamber, and box is reached after least 4 h.
   Note 2: For long-term imaging, it is essential to use the perfect focus system (PFS, Nikon), which periodically correct for Z-drift.
3. Use either a Plan Fluor 20× NA = 0.45 dry PFS objective (Nikon) or a Plan Apo 40× NA = 0.95 dry PFS objective (Nikon).
4. Define at least 25 positions for each condition in order to capture enough cell divisions during the acquisition. With 25 positions, about 100 cell divisions can be analyzed after 48 h of acquisition. At least 75 cell divisions for each condition are required for statistical analysis.
   Note: We systematically avoid using synchronization methods (nocodazole block or thymidine block release), as we observed that these treatments sensitize for cytokinetic defects. This is experimentally circumvented by

using multiple fields of view over long period of time in nonsynchronized cell populations.

5. Image the cells according to the temporal requirement of the cytokinetic phase of interest. In HeLa cells a 10-min interval frequency is sufficient for studying abscission. Acquire the images with transmitted light illumination for 48–72 h.

   Note 1: When cells are transfected with a plasmid encoding a fluorescently tagged protein, expose the cells to the fluorescent light only every 10 time points to avoid adverse effects of long-term exposure.

   Note 2: The depth of field of 20× and 40× objectives is sufficient to track the behavior of the bridge while only one Z plane is acquired. The bridge is at bout 1 μm above the substrate in HeLa cells. Focus is thus set on interphase cells at this Z plane and maintained throughout the videos; thanks to the PFS.

1.1.2.3 Determination of abscission time and data analysis

To convert the acquired images into movies, we routinely use the ImageJ software. Fig. 1A shows different time points of HeLa cell during division. At time point t0 the cell is in interphase and enters into mitosis at time point t20. Twenty minutes after the mitotic entry (t40), the cell is in metaphase and its chromosomes are aligned in the center of the mitotic spindle. During anaphase, time point t50, the chromosomes separate to opposite poles and furrow ingression starts. At time point t100, the midbody between the two daughter cells is clearly seen (*arrow*). Finally the intercellular bridge is cut at time point t310, and the midbody moves towards one of the daughter cells (*arrow*). This event completes the cell division. Abscission time is measured between the initiation of furrow ingression (t50) and the cut of the intercellular bridge (t310): 260 min in this example. The abscission event can be determined by following the elastic retraction of the midbody (easily followed by its black appearance in phase contrast, or by GFP-CEP55 fluorescence if needed), once the first abscission event occurred on its side. Using this method, we can accurately determine when abscission happens, with better accuracy than looking at the interruption of diffusion between cells or monitoring microtubule cut that takes place before actual abscission.

The results are presented as a cumulative fraction plot. This graphical display represents the cumulative percentage of cells that completed abscission (Y axis) as a function of time (X axis) following the start of furrow ingression. At least 75 dividing cells per condition are analyzed to allow statistical analysis. Fig. 1B shows the distribution of the abscission times in control siRNA- and Rab35 siRNA-treated cells. After Rab35 depletion, abscission is greatly delayed and about 25% of the cells needed more than 15 h to complete abscission. Additionally, an increase by almost twofold of the mean abscission time in Rab35 depleted cells is measured (5.6 and 9.6 h for the control and Rab35 RNAi, respectively, Fig. 1C).

For constructing cumulative plot and statistical data analysis, we use the nonparametric and distribution-free Kolmogorov–Smirnov test (KS test), which determines if two data sets differ significantly. This test has the advantage of making

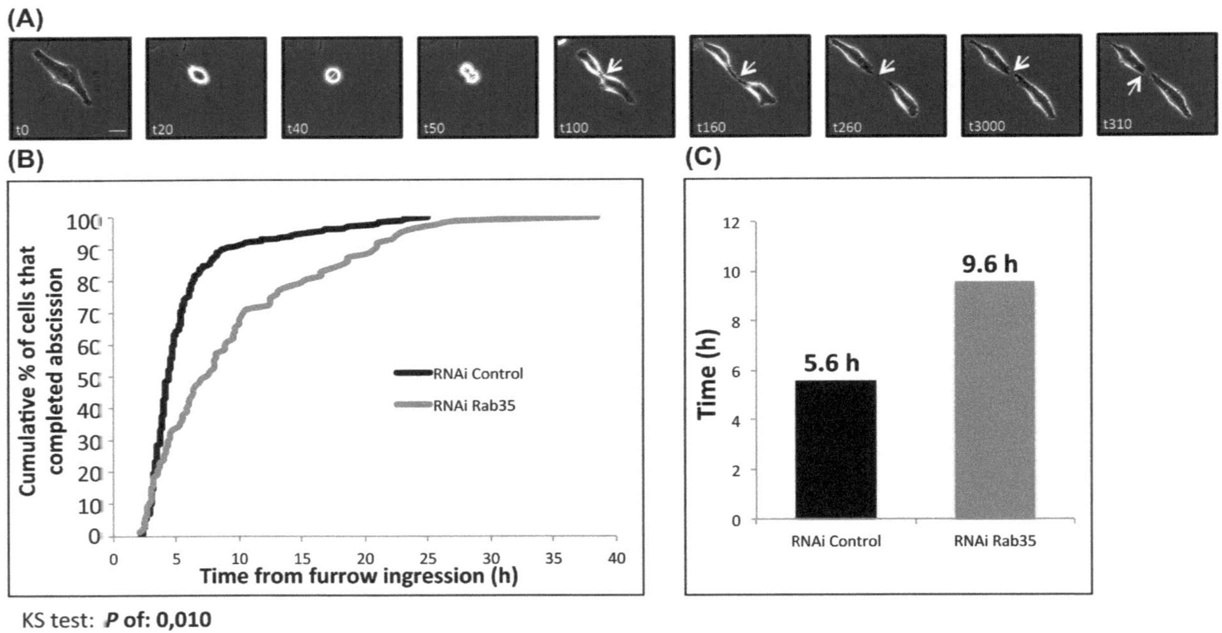

FIGURE 1

Study of abscission kinetics using live-cell imaging. (A) Snapshots of time-lapse phase-contrast microscopy movie of HeLa cells transfected with control RNAi. *Arrow*: midbody within the intercellular bridge. Scale bar: 10 μm. (B) Distribution of the abscission times in cell populations treated with control or Rab35 siRNAs (cumulative % plot). (C) Mean abscission times in cell populations treated with control or Rab35 sRNAs.

no assumption of the distribution data. We routinely use the following Website http://www.physics.csbsju.edu/stats/KS-test.html to perform this analysis. In the example, the abscission time distribution after Rab35 depletion differs from the control by $P = 0.010$.

1.2 SCANNING ELECTRON MICROSCOPY OF THE INTERCELLULAR BRIDGE

1.2.1 Materials

Dulbecco's modified eagle medium (DMEM) (Gibco)
FBS (PAN biotech GmbH)
Phosphatase-buffered saline (PBS) (Gibco)
35-mm glass bottom dishes with grid (P35G-1.5-14-C-GRD, MatTek Corporation)
Taxol (Invitrogene)
Phalloidin (molecular probes)
Glutaraldehyde (Sigma–Aldrich)
PHEM buffer: 60 mM PIPES, pH 6.9; 25 mM HEPES; 2 mM $MgCl_2$; 10 mM EGTA
PEM buffer: 100 mM PIPES, pH 6.9; 1 mM $MgCl_2$; 1 mM EGTA
Osmium tetroxide (Electron Microscopy Science)
Molecular sieve 0.4 nm (Merk)
Critical point dryer (Leica EM CPD300)
Metal coating device (high-resolution ion beam coater, Gatan Model 681)
Inverted video microscope (Nikon Ti-E)
Scanning electron microscope (JEOL6700)

1.2.2 Methods

Correlative light combined with SEM provides an excellent opportunity to investigate both 3D surface changes and cytoskeleton remodeling during maturation of the bridge and at the time of abscission. Associated with live-cell imaging or with the selection of fluorescently labeled fixed cells, the method described later allows to select intercellular bridges of appropriate age using light microscopy, and subsequently to relocalize the same cells in SEM in order to analyze the details of either the intact membrane surface or underlying cytoskeleton.

SEM is carried out either directly on cells that have been fixed to visualize the intact plasma membrane of the bridge, or on cells that have been preextracted by a detergent prior to fixation to reveal the underlying cytoskeleton. The detailed description for preextraction methodology can be found in Svitkina & Borisy (1998). To remove the cell membrane, this group used 1% of the nonionic detergent Triton X-100. We improved the cytoskeletal preservation of the bridge by a quick drug treatment before preextraction. The preextraction steps (2.2.3.3) are skipped when the plasma membrane must be kept intact.

1.2.2.1 Methods for selecting cells of interest

Cells of interest are fixed after time-lapse phase-contrast and/or fluorescence microscopy (see Section 1.1) in order to determine the age of the cytokinetic bridge after furrow ingression. Alternatively, fixed cells can be selected based on specific fluorescence markers present in the bridge without the need of live-cell imaging. Importantly, proteins of interest must be directly tagged with fluorescent tag, since immunochemical labeling requires membrane permeabilization, which is incompatible with the preservation of intact membranes. In the example provided, we used a HeLa cell line that expresses endogenous level of GFP-Actin after TALEN excision and homologous recombination (Cellectis), which allows for the detection of actin in the intercellular bridge. Cells of interest are selected with wide-field microscope or spinning-disk confocal microscope using marked coverslip (see later).

Details of cell culture, transfections, or treatments prior to fixation are not described here, as they are standard procedures. For high-resolution light microscopy, cells are grown in glass bottom coverslips. Internal reference marks are needed on the coverslips to facilitate the relocalization of the selected cells in SEM.

1.2.2.2 Chemical fixation compatible with subsequent scanning electron microscopy steps

Chemical fixation with glutaraldehyde is generally used for cytoskeleton preservation and is the most commonly employed fixative for electron microscopy. Importantly, we do not use methanol or PFA fixations since both treatments interfere with structure preservation of the plasma membrane for SEM.

1.2.2.3 Procedure for fixation and selection of cells of interest

1. Grow HeLa cells in DMEM medium supplemented with 5% FBS in glass bottom dishes 2 days prior to fixation to obtain about 70% confluence.
2. Remove culture medium and rinse cells once with PBS 1× or with a serum-free medium.
3. For visualizing the bridge cytoskeleton, immediately put the extraction solution: 1% Triton X-100, 10 μg/mL Taxol, and/or 10 μM phalloidin in PEM buffer (100 mM PIPES, pH 6.9; 1 mM MgC12; 1 mM EGTA). Incubate cells for 3–5 min at room temperature.

   Note: Taxol and phalloidin are added for preservation of microtubules and actin filaments, respectively. This step has to be skipped in order to preserve intact plasma membrane.
4. Rinse the cells 3× with the PEM buffer.
5. Fix with 2.5% glutaraldehyde in PHEM buffer for at least 30 min at room temperature.

   Note: Fixation can be carried out after time-lapse microscopy (as described in Section 1.1) to get the cell history at the moment of the fixation. The age of the bridge can be determined retrospectively by analyzing the movies.
6. Rinse the cells 3× with the PHEM buffer.

7. Transfer the cells to a light/fluorescent microscope. Here we use an inverted wide-field microscope Nikon Ti-E, as described in Section 1.1. We routinely use a Plan Apo 60× NA = 1.4 oil immersion PFS objective (Nikon) to detect the intercellular bridge and the actin filaments.
8. Spot as many positions as you need for each condition in order to get a panel of different intercellular bridges. We routinely choose 20 different bridges.
9. Postfix with 2% osmium tetroxide (OsO_4) in 1M HEPES buffer at room temperature for a maximum of 1 h.
   Note: Since OsO_4 is toxic and volatile, this step should be performed under fume hood. Gloves as well as protective clothing should also be used.
10. Rinse 3× with distilled water.

1.2.2.4 Sample drying for scanning electron microscopy

The sample has to be dried before the metal coating. The critical point drying method preserves the structure of the cells. This method requires preceding dehydration with an organic solvent. A graded bath of ethanol is routinely used.

1. Dehydrate the sample by transferring into successive baths of ethanol: 25, 50, 75 then 95%, 5 min each, and two times 100%, 10 min each.
   Note: Use molecular sieves to obtain 100% dried ethanol.
2. Detach the coverslip from the dish's bottom with a razor blade and quickly transfer the coverslip into the critical point holder previously put on a petri dish filled with 100% dried ethanol.
3. Place the holder into the critical point dryer's chamber filled with 100% dried ethanol.
4. Follow the instructions for critical point dryer. We routinely use the Leica EM CPD300 critical point.

1.2.2.5 Metal coating for scanning electron microscopy

Metal coating covers the cells with a thin layer of conducting material to supply a contrast to the biological specimen. We cover the sample with 8 nm of gold/palladium by following the instructions on the metal coating device (high-resolution ion beam coater, Gatan Model 681).

1.2.2.6 Scanning electron microscopy image acquisition and data analysis

Select an intercellular bridge using light microscopy and locate this bridge on the coverslip with the internal reference marks. After sample preparation, we routinely use the JEOL6700 microscope for SEM acquisition. The same cells are relocated under the electron microscope using the reference marks.

Fig. 2 shows two daughter cells acquired by transmitted light (A) and fluorescent light (B). The *arrow* indicates the midbody at the centre of the intercellular bridge. Fig. 2C shows the same bridge in SEM after membrane extraction (×15000 magnification). The *arrow* indicates the midbody, and the arrowhead the bundles of microtubules.

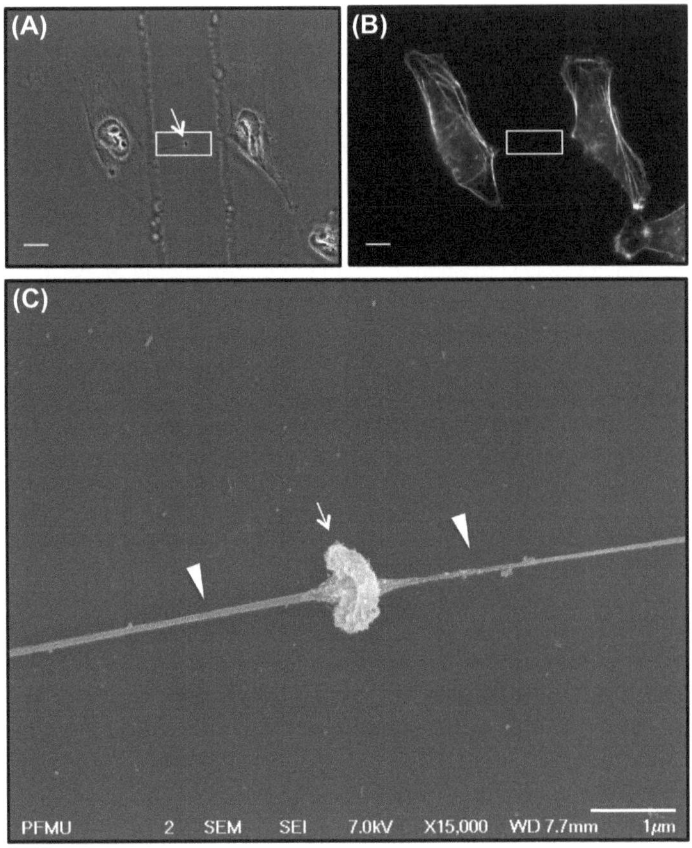

FIGURE 2

Correlative light/scanning electron microscopy pictures of the intercellular bridge. Dividing HeLa cells that express endogenous level of GFP-Actin are visualized in transmitted light (A) and fluorescent light (B) (scale bar, 10 μm). The intercellular bridge of the same cells is visualized in scanning electron microscopy (C) (scale bar, 1 μm). *Arrow*: midbody; *arrowheads*: microtubules of the bridge.

During bridge maturation, the actin cytoskeleton is entirely removed to allow the abscission event. Fig. 2 shows an example of a late/mature intercellular bridge without any actin filaments.

CONCLUSION

We have described two procedures to study cytokinetic abscission in HeLa cells: time-lapse video microscopy and correlative SEM. In principle, these methods can be utilized for other mammalian cell lines with some minor adaptations regarding cell size or cell cycle kinetics. These techniques are useful for characterizing the changes in membrane morphology during late cytokinetic stages, and to study the overall structure of the cytoskeleton prior to abscission. Correlative SEM is simpler and a more accessible technique than EM tomography (Guizetti et al., 2011; Schiel et al., 2011) or serial sections within the intercellular bridge (Dambournet et al., 2011; Mullins & Biesele, 1977), and allows the reconstruction of the 3D surface of the bridge. It revealed fine details of the midbody rim (Crowell et al., 2014), which is unexpectedly a very dynamic structure that sheds material during its maturation (El Amine, Kechad, Jananji, & Hickson, 2013). In addition, correlate light/scanning EM demonstrated that the bridge is cut successively on both sides of the midbody to generate a midbody remnant in a variety of cells (HeLa, U2OS, RPE, primary MEF, or extracted muscle stem cells) (Crowell et al., 2014). Therefore, correlate light/scanning EM is also a powerful technique to study the fate of the midbody after abscission.

ACKNOWLEDGMENTS

We thank the Echard Lab members for helpful discussions and E. Crowell for the initial development of the SEM technique in the lab. We thank the imaging facility Ultrapole, Institut Pasteur. This work has been supported by Institut Pasteur, CNRS, FRM (Equipe FRM DEQ20120323707), INCa (2014-1-PL BIO-04-IP1), iXCore foundation for Research, and ANR (AbsCyStem).

REFERENCES

Carlton, J. G., & Martin-Serrano, J. (2007). Parallels between cytokinesis and retroviral budding: a role for the ESCRT machinery. *Science, 316*(5833), 1908–1912.

Cauvin, C., & Echard, A. (2014). Phosphoinositides: lipids with informative heads and mastermind functions in cell division. *Biochimica et Biophysica Acta*. http://dx.doi.org/10.1016/j.bbalip.2014.10.013.

Cauvin, C., Rosendale, M., Gupta-Rossi, N., Rocancourt, M., Larraufie, P., Salomon, R., … Echard, A. (2016). Rab35 GTPase triggers switch-like recruitment of the Lowe Syndrome Lipid Phosphatase OCRL on Newborn Endosomes. *Current Biology, 26*(1), 120–128. http://dx.doi.org/10.1016/j.cub.2015.11.040.

Chesneau, L., Dambournet, D., Machicoane, M., Kouranti, I., Fukuda, M., Goud, B., & Echard, A. (2012). An ARF6/Rab35 GTPase cascade for endocytic recycling and successful cytokinesis. *Current Biology, 22*(2), 147–153. http://dx.doi.org/10.1016/j.cub.2011.11.058. pii:S0960-9822(11)01378-9.

Crowell, E. F., Gaffuri, A. L., Gayraud-Morel, B., Tajbakhsh, S., & Echard, A. (2014). Midbody remnant engulfment after cytokinesis abscission in mammalian cells. *Journal of Cell Science, 127*(17), 3840–3851. http://dx.doi.org/10.1242/jcs.154732.

Dambournet, D., Machicoane, M., Chesneau, L., Sachse, M., Rocancourt, M., El Marjou, A., … Echard, A. (2011). Rab35 GTPase and OCRL phosphatase remodel lipids and F-actin for successful cytokinesis. *Nature Cell Biology, 13*(8), 981–988. http://dx.doi.org/10.1038/ncb2279. pii:ncb2279.

Echard, A. (2012). Connecting membrane traffic to ESCRT and the final cut. *Nature Cell Biology, 14*(10), 983–985. http://dx.doi.org/10.1038/ncb2598.

Echard, A., Hickson, G. R., Foley, E., & O'Farrell, P. H. (2004). Terminal cytokinesis events uncovered after an RNAi screen. *Current Biology, 14*(18), 1685–1693.

Eggert, U. S., Kiger, A. A., Richter, C., Perlman, Z. E., Perrimon, N., Mitchison, T. J., & Field, C. M. (2004). Parallel chemical genetic and genome-wide RNAi screens identify cytokinesis inhibitors and targets. *PLoS Biology, 2*(12), e379.

El Amine, N., Kechad, A., Jananji, S., & Hickson, G. R. (2013). Opposing actions of septins and Sticky on Anillin promote the transition from contractile to midbody ring. *The Journal of Cell Biology, 203*(3), 487–504. http://dx.doi.org/10.1083/jcb.201305053.

Elia, N., Sougrat, R., Spurlin, T. A., Hurley, J. H., & Lippincott-Schwartz, J. (2011). Dynamics of endosomal sorting complex required for transport (ESCRT) machinery during cytokinesis and its role in abscission. *Proceedings of National Academy Sciences of the United States of America, 108*(12), 4846–4851. http://dx.doi.org/10.1073/pnas.1102714108. pii:1102714108.

Fededa, J. P., & Gerlich, D. W. (2012). Molecular control of animal cell cytokinesis. *Nature Cell Biology, 14*(5), 440–447. http://dx.doi.org/10.1038/ncb2482. pii:ncb2482.

Fielding, A. B., Schonteich, E., Matheson, J., Wilson, G., Yu, X., Hickson, G. R., ... Gould, G. W. (2005). Rab11-FIP3 and FIP4 interact with Arf6 and the exocyst to control membrane traffic in cytokinesis. *EMBO Journal, 24*(19), 3389–3399.

Fujiwara, T., Bandi, M., Nitta, M., Ivanova, E. V., Bronson, R. T., & Pellman, D. (2005). Cytokinesis failure generating tetraploids promotes tumorigenesis in p53-null cells. *Nature, 437*(7061), 1043–1047.

Goss, J. W., & Toomre, D. K. (2008). Both daughter cells traffic and exocytose membrane at the cleavage furrow during mammalian cytokinesis. *The Journal of Cell Biology, 181*(7), 1047–1054.

Green, R. A., Paluch, E., & Oegema, K. (2012). Cytokinesis in animal cells. *Annual Review of Cell and Developmental Biology*. http://dx.doi.org/10.1146/annurev-cellbio-101011-155718.

Gromley, A., Yeaman, C., Rosa, J., Redick, S., Chen, C. T., Mirabelle, S., ... Doxsey, S. J. (2005). Centriolin anchoring of exocyst and SNARE complexes at the midbody is required for secretory-vesicle-mediated abscission. *Cell, 123*(1), 75–87.

Guizetti, J., Schermelleh, L., Mantler, J., Maar, S., Poser, I., Leonhardt, H., ... Gerlich, D. W. (2011). Cortical constriction during abscission involves helices of ESCRT-III-dependent filaments. *Science, 331*, 1616–1620. http://dx.doi.org/10.1126/science.1201847. pii: science.1201847.

Klinkert, K., Rocancourt, M., Houdusse, A., & Echard, A. (2016). Rab35 GTPase couples cell division with initiation of epithelial apico-basal polarity and lumen opening. *Nature Communications, 7*, 11166. http://dx.doi.org/10.1038/ncomms11116.

Kouranti, I., Sachse, M., Arouche, N., Goud, B., & Echard, A. (2006). Rab35 regulates an endocytic recycling pathway essential for the terminal steps of cytokinesis. *Current Biology, 16*(17), 1719–1725.

Lafaurie-Janvore, J., Maiuri, P., Wang, I., Pinot, M., Manneville, J. B., Betz, T., ... Piel, M. (2013). ESCRT-III assembly and cytokinetic abscission are induced by tension release in the intercellular bridge. *Science, 339*(6127), 1625–1629. http://dx.doi.org/10.1126/science.1233866.

Mierzwa, B., & Gerlich, D. W. (2014). Cytokinetic abscission: molecular mechanisms and temporal control. *Developmental Cell, 31*(5), 525–538. http://dx.doi.org/10.1016/j.devcel.2014.11.006.

Morita, E., Sandrin, V., Chung, H. Y., Morham, S. G., Gygi, S. P., Rodesch, C. K., & Sundquist, W. I. (2007). Human ESCRT and ALIX proteins interact with proteins of the midbody and function in cytokinesis. *EMBO Journal, 26*(19), 4215–4227.

Mullins, J. M., & Biesele, J. J. (1977). Terminal phase of cytokinesis in D-98s cells. *The Journal of Cell Biology, 73*(3), 672–684.

Schiel, J. A., Park, K., Morphew, M. K., Reid, E., Hoenger, A., & Prekeris, R. (2011). Endocytic membrane fusion and buckling-induced microtubule severing mediate cell abscission. *Journal of Cell Science, 124*(Pt 9), 1411–1424. http://dx.doi.org/10.1242/jcs.081448. pii:jcs.081448.

Schiel, J. A., Simon, G. C., Zaharris, C., Weisz, J., Castle, D., Wu, C. C., & Prekeris, R. (2012). FIP3-endosome-dependent formation of the secondary ingression mediates ESCRT-III recruitment during cytokinesis. *Nature Cell Biology, 14*, 1068–1078. http://dx.doi.org/10.1038/ncb2577. pii:ncb2577.

Skop, A. R., Liu, H., Yates, J., 3rd, Meyer, B. J., & Heald, R. (2004). Dissection of the mammalian midbody proteome reveals conserved cytokinesis mechanisms. *Science, 305*(5680), 61–66.

Svitkina, T. M., & Borisy, G. G. (1998). Correlative light and electron microscopy of the cytoskeleton of cultured cells. *Methods in Enzymology, 298*, 570–592.

Wilson, G. M., Fielding, A. B., Simon, G. C., Yu, X., Andrews, P. D., Hames, R. S., … Prekeris, R. (2005). The FIP3-Rab11 protein complex regulates recycling endosome targeting to the cleavage furrow during late cytokinesis. *Molecular Biology of the Cell, 16*(2), 849–860.

CHAPTER

Analysis of postcytokinetic roles of cytokinetic components in *Caenorhabditis elegans*

15

Y. Chai[1], D. Tian, W. Li, G. Ou
Tsinghua University, Beijing, China
[1]*Corresponding author: E-mail: yongpingchai@yahoo.com*

CHAPTER OUTLINE

Methods in Cell Biology, Volume 137, ISSN 0091-679X, http://dx.doi.org/10.1016/bs.mcb.2016.03.040

Abstract

Cytokinesis requires the interplay between the cytoskeleton and plasma membrane. Emerging evidence indicates that some cytokinetic components are essential for the postcytokinetic events such as epithelium organization and neural development. We have recently developed live cell imaging and conditional knockout techniques to visualize cytokinetic proteins in *Caenorhabditis elegans* Q neuroblasts and separate their postcytokinetic functions from cytokinetic ones. Here we describe how the fluorescent reporter strains and conditional knockout *C. elegans* are generated and how live cell imaging of Q neuroblast development are performed in our laboratory. Using these protocols, we uncovered a novel role of Anillin in stabilizing the actin network during neuronal migration and neurite outgrowth, and the postcytokinetic fate of midbody, which is released into the extracellular space and degraded by the adjacent macrophage using an apoptotic mimicry. These protocols could also be applicable to study other cellular processes in *C. elegans* or adapted to other model organisms, to provide better insights into their developmental basis.

INTRODUCTION

The success of cytokinesis relies on sophisticated coordination between the actin cytoskeleton, microtubule, and membranes. An actomyosin cytoskeleton-dependent contractile ring generates a contractile force to pull the plasma membrane into the cell interior. The microtubule cytoskeleton from the anaphase spindle delivers position cues to specify the cell equator where contractile ring assembles (Canman et al., 2003). During membrane invagination, the contractile ring must be attached to the membrane and be stabilized within the equator (Green, Paluch, & Oegema, 2012). Various genetic and molecular studies identified the repertoire of essential cytokinesis genes, with ~20 conserved components constituting the core cytokinetic machinery (Eggert, Mitchison, & Field, 2006; Glotzer, 2005).

Despite progresses concerning the spatiotemporal regulation of cytokinetic proteins to complete cytokinesis, several lines of evidence have indicated that some of cytokinetic components play previously unknown roles in other cytoskeleton-based processes during interphase (del Castillo, Lu, Winding, Lakonishok, & Gelfand, 2015; Fotopoulos et al., 2013; Reyes et al., 2014; Tian et al., 2015). Anillin, a multidomain scaffold protein that connects actomyosin, microtubule, septin, RhoA, and plasma membrane, was primarily considered to be a central organizer for the attachment and stabilization of the contractile ring during furrow ingress (Hickson & O'Farrell, 2008a, 2008b; Piekny & Maddox, 2010). Recent studies have demonstrated that Anillin regulates the integrity of cell–cell junctions by organizing the junctional enrichment of RhoA-GTP and actomyosin in the epithelium of *Xenopus* embryos and that Anillin is recruited to the leading edge of the *Caenorhabditis elegans* migrating Q neuroblasts and stabilizes F-actin to ensure coherent cell motility and growth cone extension (Reyes et al., 2014; Tian et al., 2015). The challenge now is to understand the potential roles of cytokinetic components outside of cytokinesis.

The *C. elegans* Q neuroblasts appear to be a valuable model system to study the postcytokinetic function of cytokinesis genes. Initially described by Sulston and Horvitz in their pioneering observation of *C. elegans* postembryonic lineages (Sulston & Horvitz, 1977), each Q neuroblast on the left and right side of L1 larvae undergoes three rounds of asymmetric cell divisions to generate three neurons (the PQR, PVM, and SDQL produced by QL; and the AQR, AVM, and SDQR produced by QR) and two apoptotic cells (Fig. 1). After Q neuroblast cytokinesis, Q cell progenies undertake long-distance migration, apoptosis, and neuritogenesis to differentiate into sensory neurons and interneurons (Fig. 1), all of which are important cellular processes that require the interplay of the cytoskeleton and plasma membrane. We have recently developed fluorescence live cell imaging methodologies to document the behavior of cytokinetic proteins during Q cell development (Chai, Li, et al., 2012). We have also devised somatic TALEN and somatic CRISPR-Cas9 techniques to generate conditional mutations within Q cell lineages, which turns out to be crucial technical advancements to dissect the function of cytokinetic components after cytokinesis (Cheng et al., 2013; Shen et al., 2014).

The combination of imaging and genetic approaches enabled us to discover a new type of asymmetric cell division that requires the polarized distribution of myosin-II during Q cell cytokinesis (Ou, Stuurman, D'Ambrosio, & Vale, 2010). For postcytokinetic studies, we illustrated the fate of midbody after asymmetric cell division and we demonstrated the roles of Anillin in organizing the actin

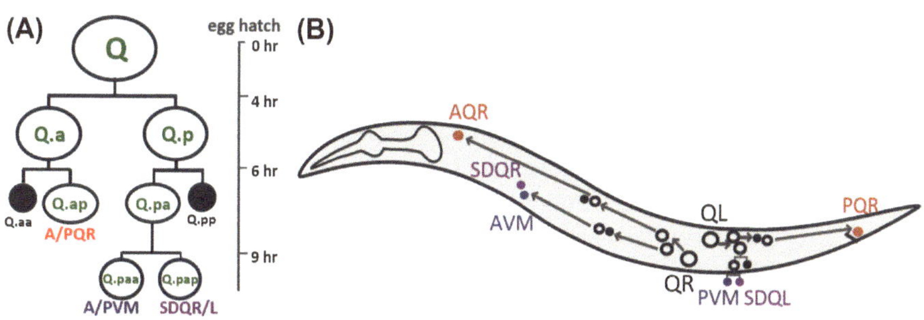

FIGURE 1 ***Caenorhabditis elegans*** **Q neuroblast lineages and Q cell migration.**

(A) A schematic of *C. elegans* Q neuroblast lineages. Q neuroblasts undergo three rounds of asymmetric divisions to generate three different neurons and two apoptotic cells (black). QL (Q neuroblast on the left side of the animal) produces PQR, PVM, and SDQL; and QR on the right side generates AQR, AVM, and SDQR. PQR and AQR are oxygen sensory neurons; PVM and AVM are mechanosensory neurons; and SDQL and SDQR are interneurons. Time on the right shows the time at which the Q cell divisions occur after egg hatching. (B) A diagram of the migration of Q neuroblasts and their descendants in the L1 larval stage. QR and its progenies migrate toward the anterior, and QL and its progenies migrate toward the posterior.

cytoskeleton during neuroblast migration and neurite outgrowth (Chai, Tian, et al., 2012; Tian et al., 2015). Here we will first outline how to construct the transgenic *C. elegans* with multiple fluorescence to label cytokinetic components. We will then outline how to apply the somatic CRISPR-Cas9 method to generate conditional knockout *C. elegans* to study the postcytokinetic roles of cytokinesis genes. Finally, we will describe the imaging procedure for *C. elegans* larvae to visualize Q cell division, migration, and neuritogenesis using spinning disk confocal microscopy.

1. GENERATION OF TRANSGENIC *CAENORHABDITIS ELEGANS* EXPRESSING FLUORESCENT CYTOKINETIC PROTEINS

1.1 PREPARATION OF PLASMID CONSTRUCTS ENCODING FLUORESCENT CYTOKINETIC PROTEINS

1. We construct transgenes by "SOEing" PCR or the In-Fusion cloning protocol. The "SOEing" PCR procedure requires a two-step PCR (Hobert, 2002). In the first step, we amplify the promoter sequence from N2 genomic DNA, the fluorescent protein coding sequence from corresponding plasmid, and the gene of interest from cDNA or N2 genomic DNA by Phusion DNA polymerase premix (New England BioLabs, cat. No. M0530L). These fragment ends contain 20–25 nt overlapping homologous sequences. The second PCR step is performed by using the mixture of the first PCR fragments as template, generating a DNA fragment including the promoter, fluorescent protein, and the gene of interest.
2. The In-Fusion cloning protocol (Clontech, Cat. No.638911) bases on the homologous sequences at the ends of the vector and linearized fragments. For In-Fusion cloning, we have generated a series of vectors with different promoters that can express fluorescent cytokinetic proteins within the specific cells of Q cell lineages or in an inducible manner by the heat shock promoter (Fig. 2). All the vectors have identical 15 nt sequences at the ends of fragments for homologous recombination, facilitating the high-throughput plasmid construction.

1.2 PRODUCTION OF TRANSGENIC *CAENORHABDITIS ELEGANS* EXPRESSING FLUORESCENT CYTOKINETIC PROTEINS

1. Germ line transformation. Microinjection is used for transgene expression in *C. elegans* somatic cells. We followed the standard microinjection protocol for *C. elegans* germ line transformation (Mello & Fire, 1995). The aforementioned transgene plasmids or PCR products, together with an injection marker plasmid, such as P*odr-1::dsRed*, *rol-6(+)* or *unc-76(+)* can be injected into the syncytial

(A)
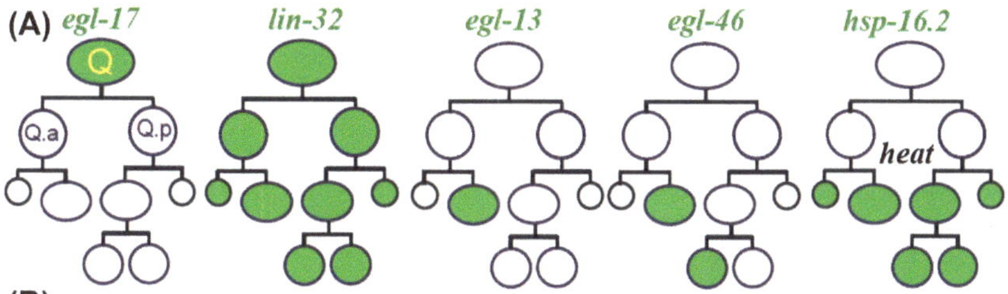

(B)

Promoter	Length (nt)	Forward Primer	Reverse Primer
Pegl-17	4595	cagatggatgtttactgccaactgg	agctcacatttcgggcacctgaaaaatc
Plin-32	3023	acggtcgaaatctcagtctttgc	ggttggtctgactgaaaacgacg
Pegl-13	2148	tcacctgccccgacatta	tcgtctacggctcatgctgg
Pegl-46	3111	ttccagatgttttccttccg	ggccttctgaaatcaaaacgа
Phsp-16.2	419	cagatccagtgagttcgtccaag	gattatagtttgaagatttctaatttc

FIGURE 2 Promoters for the study of postcytokinetic events in Q cells.

(A) Schematics of *Caenorhabditis elegans* Q neuroblast lineage-specific promoters and a heat-shock inducible promoter. Green indicates the expression of the promoter. (B) A table of the promoter length and primers to amply the promoters by PCR reactions.

gonad of N2 or *unc-76* mutants. To avoid the potential artifacts caused by the overexpression of fluorescent cytokinetic components, we inject the DNA plasmid at the lowest possible concentration below which the fluorescence cannot be detected. We usually inject 20–30 young adult animals with DNA constructs at the concentration of 2–20 ng/μL.

2. Transformant screen. F1 hermaphrodites with injection marker are singled and maintained on plates. The plates of F2 progeny with heritable injection marker are considered as independent transgenic lines. The transgenes are inherited as extrachromosomal arrays, which can be further integrated into the *C. elegans* genome using γ-irradiation. For live cell imaging, we choose three transgenic lines whose transmission rates of extrachromosomal arrays are >50%. For further transgene integration, we use the protocol as follows: 100 young adult worms are irradiated by γ rays. The radiation dose is ~50 Gy. Five irradiated worms are picked onto each of 20 Nematode growth medium (NGM) plates after recovery at 20°C for one night. Then worms are incubated at 20°C until starvation and transferred to new NGM plates by cutting one chunk. After starvation three times, chunk the worms to new NGM plates for recovery. Pick 24 young adult healthy worms with transgene from each plate to a 24-well plate (1 worm per well). Screen integrated lines with 100% transgenic progeny after 3–4 days. Outcross integrated lines more than three times to remove background mutations.

2. CREATION OF CONDITIONAL MUTATIONS OF CYTOKINETIC GENES IN *CAENORHABDITIS ELEGANS* USING THE SOMATIC CRISPR-Cas9 TECHNIQUE

2.1 PREPARATION OF CONSTRUCTS TO CREATE CONDITIONAL MUTATIONS OF CYTOKINETIC GENES

1. sgRNA sequence design. We select the sgRNA sequence with the CRISPR design tool (http://crispr.mit.edu). The target sites conform to G(N)N19NGG (N = A, C, G, or T). The initial G optimizes the transcription process, and the NGG (PAM) is essential for Cas9 activity. We usually design two sgRNAs since some sgRNA may not work and two independent sgRNAs may confirm the phenotype.
2. Plasmids construction. We use In-Fusion cloning strategy to generate conditional CRISPR-Cas9 expression vectors by replacing the *eft-3* promoter (pDD162, Addgene: #47549) with the *C. elegans* heat shock promoter P*hsp-16.2* or tissue-specific promoter (Figs. 2 and 3). The expression of sgRNA is driven by P*U6* promoter (Fig. 3).

2.2 GENERATION OF TRANSGENIC *CAENORHABDITIS ELEGANS* CARRYING CONDITIONAL MUTATIONS OF CYTOKINETIC GENES

We microinject the mixture of CRISPR-Cas9 plasmid and an injection marker (eg, *rol-6(+)* or P*odr-1::dsRed*) into the germ line of young adult animals. The concentration of the CRISPR-Cas9 plasmid is 20–50 ng/μL, and the injection marker plasmid is 10–40 ng/μL. We often inject 50 animals to get more than five independent transgenic lines. We noticed that the efficiency of the same construct varies among different transgenic lines, and we usually select at least two lines with the highest efficiency for the molecular and phenotypic analysis.

2.3 ANALYSIS OF THE GENOTYPE AND PHENOTYPES OF CONDITIONAL MUTATIONS

1. Molecular analysis. To detect the molecular lesion and verify the sgRNA efficiency, the transgenic animals with P*hsp-16.2::Cas9* are used to perform T7 endonuclease I (T7EI) assay. After the heat shock at 33°C in a water bath for 60 min, 5–10 transgenic animals at the L4 stage are lysed with PCR buffer at 60°C for 1 h and then at 95°C for 15 min. The DNA fragment containing the sgRNA target sequence is amplified with *Pfu* DNA polymerase (TransGen Biotech, Cat. No. AP221) for 30 cycles. Then the PCR product is purified by the PCR Purification Kit (QIAquick, Cat. No.28106). The purified DNA fragment is melted at 95°C for 15 min and reannealed to form heteroduplexed DNA at room temperature. The reannealed DNA is digested with T7E1 (New England Biolabs, Cat. No. M0302L) at 37°C for 30 min. The digested products are separated

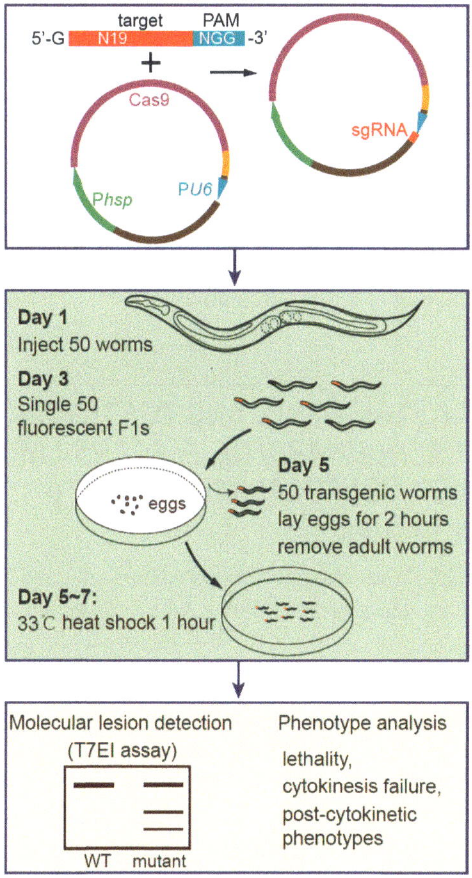

FIGURE 3 Workflow of the somatic CRISPR-Cas9 protocol to generate conditional knockouts after Q cell cytokinesis.

The somatic CRISPR-Cas9 method includes the construction of the Cas9-sgRNA expression plasmids; transgenesis by germ line injection; and molecular and phenotypic analysis of conditional knockout animals. See text for details.

by 2% agarose gel electrophoresis for analysis. The same transgenic animals that were cultured at 15°C are used as the negative control. Indel is quantified by the formula $100\times\ (1\text{-}(b+c)/(a+b+c)^{1/2})$ in which a corresponds to the intact band intensity and b and c present the overall intensities of the digested small bands. We use the ImageJ software (http://rsbweb.nih.gov/ij/) to measure the relative intensities of DNA bands. The PCR fragment is further cloned into a pEASY-T1 cloning vector for Sanger sequencing.

2. Phenotype analysis. We quantify the following phenotypes in conditional knockout animals. (1) Lethality: 100 eggs (4 h after laying) are subjected to the heat-shock treatment at 33°C for 1 h and then maintained at 20°C until the L3 larval stage. The lethality is calculated as 100 minus the number of hatched worms. (2) Cytokinesis failure: To determine the cytokinesis failure, we examine the number of neurons that are produced from Q neuroblasts in conditional knockout mutants. For example, we marked the AQR and PQR (A/PQR) neurons by using the *gcy-32* promoter to express mCherry in these neurons. A/PQR neurons could not be detected in 22% of *ani-1-sgRNA* conditional knockouts, indicating the failure of cytokinesis. Even though the penetrance of the phenotype is consistent across generations once the transgenic lines are established, we quantify the phenotype from three different generations of conditional knockout animals.

3. IMAGING OF CYTOKINESIS AND POSTCYTOKINETIC EVENTS IN *CAENORHABDITIS ELEGANS* LARVAE USING SPINNING DISK CONFOCAL MICROSCOPY

3.1 MICROSCOPY SETUPS

The laser spinning disk confocal microscopy setups that we use to perform the imaging experiments are described as follows, while other considerate setups would be suitable as well.

1. We use Zeiss Observer Z1 (Carl Zeiss MicroImaging, Inc.) or Olympus IX83 (Olympus, Inc.) inverted microscopes. We equip them with Flat-Top XYZ Automated Stage and MS-2000-WK Multi-Axis Stage controller (Applied Scientific Instrumentation, Inc) for multiple-position acquisition.
2. The objectives are 100×, 1.46 N.A. and 10×, 0.25 N.A.
3. To obtain high spatial and temporal resolutions, confocal imaging apparatus includes a spinning disc confocal scanner unit CSU-X1(Yokogawa, Inc.) and an EMCCD digital camera Andor iXon + DU-897D-C00-#BV-500 (Andor Technology, Inc.).
4. We use high-precision spectroscopic and tunable solid-state lasers for fluorescence protein excitations: the 405 nm (100 mW, for TagBFP-tagged protein), 488 nm (100 mW, for GFP-tagged protein), and 568 nm (100 mW, for mCherry-tagged protein) lines of a Sapphire CW CDRH USB Laser System (Coherent, Inc.)

3.2 IMMOBILIZATION OF *CAENORHABDITIS ELEGANS* LARVAE

1. The day before imaging, we collect 100 well-fed young adult fluorescent *C. elegans* on an NGM plate and allow them to lay eggs for 2 h. We remove adult animals to get synchronized progenies.
2. On the experimental day, we make fresh "sandwich" agarose pads (Chai, Li, et al., 2012) with two 12-mm round glass coverslip outside and 3% agarose solid gel in between (Fig. 4).

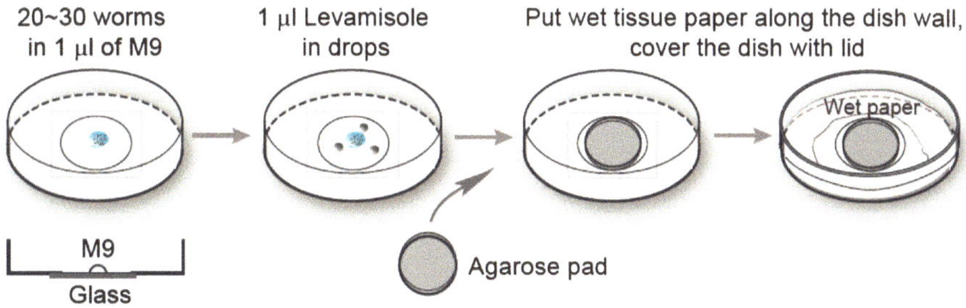

FIGURE 4 Preparation of *Caenorhabditis elegans* larvae for live imaging.

Pipette M9 buffer in the center of the glass bottom of a microwell dish and transfer *C. elegans* larvae into the solution. Pipette 1 mg/mL levamisole in M9 buffer into drops and place the agarose pad on the top. Keep the dish moist with wet paper and a lid.

3. To start an imaging sample setup, we pipet 1 μL of M9 solution (Stiernagle, 2006) in the center of the glass bottom of a microwell dish (MatTek, Cat. No. P35G-1.5-14-C), then transfer 20–30 synchronized L1 larvae into the M9 droplet (Fig. 4).
4. Drop 1 μL of 1 mg/mL levamisole (Sigma Aldrich, Cat. No. 31742) in M9 buffer around the animals containing puddle, avoid any contact between levamisole and animals at this moment.
5. Open one "sandwich" agarose pad, the agarose gel will stay with one of the two coverslip, keep that one as the agarose pad for use. Dry the pad for 10 s in the air, then place it with agarose gel down-facing, onto the center of the microwell plate.
6. The M9 droplet containing animals and levamisole will be pushed to mix by the weight of agarose pad. The animals should be anesthetized within 1 min.
7. Put strings of wet tissue paper inside the sample microwell dish to keep moist and cover the dish with a lid.
8. Under the 100× oil objective, examine the postembryonic developmental stage of fluorescent larva, mark the positions of which at a suitable stage for observation into the position list in the acquisition software. Start the time-lapse acquisition and keep the imaging plane on focus throughout the recording.

3.3 IMAGING CONDITIONS

1. Set the imaging room temperature to be 20°C, unless the worm of interested is a temperature-sensitive mutant.
2. We usually use an exposure time of 200 ms, and set the laser intensity as low as possible to reduce photobleaching and phototoxicity, as long as the imaging quality is satisfactory.
3. The time-lapse interval would be 30 s for cell division studies and 1 min for cell migration studies. The XY movement speed of the automated stage should be set as intermediate for time-lapse imaging.

CONCLUSION

We presented the genetic and live imaging methods to investigate the function and dynamic behaviors of the cytokinetic components after cell division. These assays have provided invaluable support for studying the dynamic behavior and function of Anillin and midbody after cytokinesis (Figs. 5 and 6). First, we uncovered a novel pathway in which Anillin transduces the RhoG signal to the actin cytoskeleton during Q cell migration. Our time-lapse microscopy illustrated that Anillin is redistributed to the leading edge of migrating Q cells and the growth cone during neurite outgrowth (Fig. 5A and B). Using the somatic CRISPR-Cas9 method to bypass the requirement for Anillin in cytokinesis, we showed that Anillin stabilizes F-actin at the leading edge by antagonizing the F-actin severing activity of Cofilin and that Anillin is recruited to the leading edge through a physical association with an active form of RhoG (Fig. 5C and D). Previous studies concerning actin dynamics at the leading edge focus on actin nucleation or F-actin disassembly, and our study indicates that the stabilization of the actin network by the RhoG-Anillin pathway is also important for the coherent cell motility. Second, we showed that, after cytokinesis, midbodies from Q cell lineages were released to the extracellular space and subsequently internalized and degraded by the phagocyte that digests apoptotic Q cell corpses (Fig. 6) (Chai, Tian, et al., 2012; Ou, Gentili, & Gonczy, 2014). We found that midbody degradation was defective in apoptotic cell engulfment mutants (Fig. 6B) and that externalized phosphatidylserine, an engulfment signal for corpse phagocytosis, also mediated the signaling between midbody and phagocyte, demonstrating an "apoptotic mimicry" in midbody degradation (Chai, Tian, et al., 2012). Interestingly, we found that the midbody generated by the first division of *C. elegans* embryo applied a similar mechanism for degradation, implying a common fate for midbody in *C. elegans* (Ou et al., 2014).

Although these genetic and imaging assays are powerful for illustrating the novel function and behavior of cytokinetic components during interphase, there remain areas for improvements. First, it would be useful to construct fluorescent reporter using CRISPR-Cas9-assisted knockin technique. In the germ line transformation experiments, we injected the plasmid DNA at the lowest concentration below which we could not detect the fluorescence, however, these transgenes carry multiple copies of the reporter, raising the question about the distribution of the endogenous proteins. Using CRISPR-Cas9 cleavage to increase the rate of homologous recombination, GFP knockin technique has been recently developed in *C. elegans* and can be applied for the future studies (Dickinson, Ward, Reiner, & Goldstein, 2013). Second, we should further address the potential caveats of somatic CRISPR-Cas9 method: we could alleviate the off-target concern of Cas9 by rescue experiments with the use of synonymous mutations in the sgRNA recognition site (Li, Yi, & Ou, 2015); and we could accurately determine the cells carrying engineered mutations by introducing a GFP knockin of the target gene as an indicator of endogenous gene modification in somatic CRISPR-Cas9 animals. Finally, it may be useful to extend the length of time-lapse imaging in live *C. elegans*.

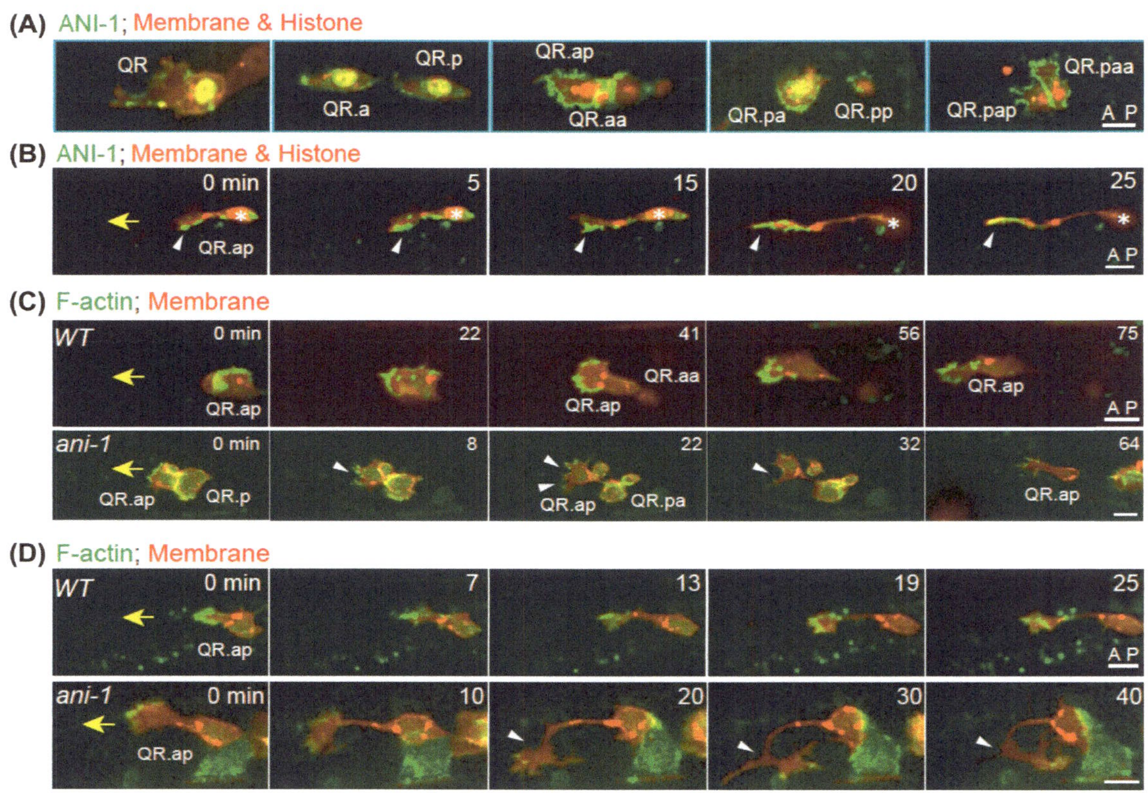

FIGURE 5 The dynamic behavior and function of Anillin during Q cell migration and neuritogenesis.

(A) Fluorescence images of GFP-labeled Anillin (ANI-1; green) and mCherry-tagged membrane and histone (red) in QR cells. (B) Fluorescence time-lapse images of GFP-tagged ANI-1 during the dendritic outgrowth in QR.ap. *Yellow arrow*, the direction of neurite growth; *white arrowheads*, the enrichment of GFP::ANI-1 in the growth cone; asterisks, cell body. (C and D) Images of GFP-tagged F-actin and mCherry-tagged plasma membrane during QR.ap cell migration (C) and neurite growth (D) in wild-type strain (WT) (upper) and *ani-1* conditional mutants (lower). *Yellow arrows* indicate the correct direction of cell migration or neurite growth in WT, and *white arrowheads* show the neurites with branched leading edges or defective guidance in *ani-1* conditional knockouts. Bars, 5 μm.

From Tian, D., Diao, M., Jiang, Y., Sun, L., Zhang, Y., Chen, Z., Ou, G. (2015). Anillin regulates neuronal migration and neurite growth by linking RhoG to the actin cytoskeleton. Current Biology: CB, 25*(9), 1135–1145. doi:10.1016/j.cub.2015.02.072.*

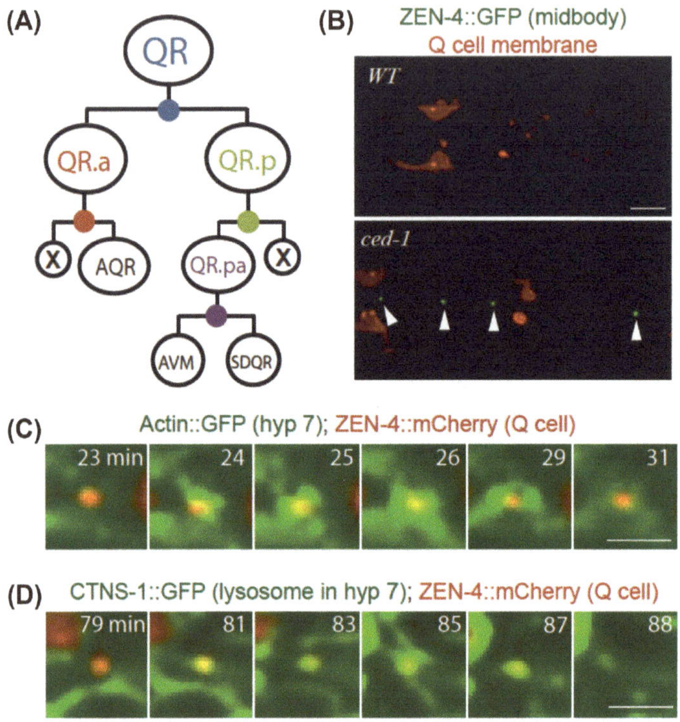

FIGURE 6 The fate of midbody after Q cell cytokinesis.

(A) Four divisions of *Caenorhabditis elegans* Q cells generate four midbodies (*dots in blue, red, green, and purple*. (B) Fluorescence images of Q cell midbody degradation in WT and the *ced-1* mutant which was known to be defective in the engulfment of apoptotic cells. Midbodies (GFP::ZEN-4/MKLP1); plasma membrane (mCherry with a myristoylation signal); and histone (his-24::mCherry). *Arrowheads* show persistent midbody remnants. (C and D) Fluorescent time-lapse images of the recruitment of GFP-tagged actin cytoskeleton (actin::GFP in C) and lysosome (CTNS-1::GFP in D) from the hyp 7 cell to midbody (ZEN-4::mCherry). Time in minutes; movie starts (0 min) at the birth of midbody or the end of cytokinesis of the mother cell. Bar in (B), 5 μm; Bars in (C and D), 2.5 μm.

From Chai, Y., Tian, D., Yang, Y., Feng, G., Cheng, Z., Li, W., & Ou, G. (2012). Apoptotic regulators promote cytokinetic midbody degradation in C. elegans. The Journal of Cell Biology, *199(7), 1047–1055. doi:10.1083/jcb.201209050.*

It usually takes 4 h for Q cells to complete their migration and neurite outgrowth after cytokinesis, however, the maximum length of imaging time is limited to 2–3 h due to the phototoxicity from the frequent laser illumination. During our time-lapse recording, we could extend the length to more than 4 h by increasing the imaging interval from 1 to 3 min. Although these assays were initially developed in *C. elegans*, they could, in principle, be applied to any model organisms to uncover the previously unknown functions of cytokinesis genes in postcytokinetic events, provided that transgenesis and live imaging techniques are available.

ACKNOWLEDGMENTS

This study was supported by the Junior Thousand Talents Program of China to G.O., the National Basic Research Program of China to G.O. and W.L. (973 Program, 2013CB945600, 2012CB966800, and 2012CB945002), the National Natural Science Foundation of China to G.O., W.L., and Y.C. (31201009, 31222035, 31101002, 31171295, and 31190063).

REFERENCES

Canman, J. C., Cameron, L. A., Maddox, P. S., Straight, A., Tirnauer, J. S., Mitchison, T. J., ... Salmon, E. D. (2003). Determining the position of the cell division plane. *Nature, 424*(6952), 1074–1078. http://dx.doi.org/10.1038/nature01860.

del Castillo, U., Lu, W., Winding, M., Lakonishok, M., & Gelfand, V. I. (2015). Pavarotti/MKLP1 regulates microtubule sliding and neurite outgrowth in *Drosophila* neurons. *Current Biology: CB, 25*(2), 200–205. http://dx.doi.org/10.1016/j.cub.2014.11.008.

Chai, Y., Li, W., Feng, G., Yang, Y., Wang, X., & Ou, G. (2012). Live imaging of cellular dynamics during *Caenorhabditis elegans* postembryonic development. *Nature Protocols, 7*(12), 2090–2102. http://dx.doi.org/10.1038/nprot.2012.128.

Chai, Y., Tian, D., Yang, Y., Feng, G., Cheng, Z., Li, W., & Ou, G. (2012). Apoptotic regulators promote cytokinetic midbody degradation in *C. elegans*. *The Journal of Cell Biology, 199*(7), 1047–1055. http://dx.doi.org/10.1083/jcb.201209050.

Cheng, Z., Yi, P., Wang, X., Chai, Y., Feng, G., Yang, Y., ... Ou, G. (2013). Conditional targeted genome editing using somatically expressed TALENs in *C. elegans*. *Nature Biotechnology, 31*(10), 934–937. http://dx.doi.org/10.1038/nbt.2674.

Dickinson, D. J., Ward, J. D., Reiner, D. J., & Goldstein, B. (2013). Engineering the *Caenorhabditis elegans* genome using Cas9-triggered homologous recombination. *Nature Methods, 10*(10), 1028–1034. http://dx.doi.org/10.1038/nmeth.2641.

Eggert, U. S., Mitchison, T. J., & Field, C. M. (2006). Animal cytokinesis: from parts list to mechanisms. *Annual Review of Biochemistry, 75*, 543–566. http://dx.doi.org/10.1146/annurev.biochem.74.082803.133425.

Fotopoulos, N., Wernike, D., Chen, Y., Makil, N., Marte, A., & Piekny, A. (2013). *Caenorhabditis elegans* anillin (ani-1) regulates neuroblast cytokinesis and epidermal morphogenesis during embryonic development. *Developmental Biology, 383*(1), 61–74. http://dx.doi.org/10.1016/j.ydbio.2013.08.024.

Glotzer, M. (2005). The molecular requirements for cytokinesis. *Science, 307*(5716), 1735–1739. http://dx.doi.org/10.1126/science.1096896.

Green, R. A., Paluch, E., & Oegema, K. (2012). Cytokinesis in animal cells. *Annual Review of Cell and Developmental Biology, 28*, 29–58. http://dx.doi.org/10.1146/annurev-cellbio-101011-155718.

Hickson, G. R., & O'Farrell, P. H. (2008a). Anillin: a pivotal organizer of the cytokinetic machinery. *Biochemical Society Transactions, 36*(Pt 3), 439–441. http://dx.doi.org/10.1042/BST0360439.

Hickson, G. R., & O'Farrell, P. H. (2008b). Rho-dependent control of anillin behavior during cytokinesis. *The Journal of Cell Biology, 180*(2), 285–294. http://dx.doi.org/10.1083/jcb.200709005.

Hobert, O. (2002). PCR fusion-based approach to create reporter gene constructs for expression analysis in transgenic *C. elegans*. *Biotechniques, 32*(4), 728–730.

Li, W., Yi, P., & Ou, G. (2015). Somatic CRISPR-Cas9-induced mutations reveal roles of embryonically essential dynein chains in *Caenorhabditis elegans* cilia. *The Journal of Cell Biology, 208*(6), 683–692. http://dx.doi.org/10.1083/jcb.201411041.

Mello, C., & Fire, A. (1995). DNA transformation. *Methods in Cell Biology, 48*, 451–482.

Ou, G., Gentili, C., & Gonczy, P. (2014). Stereotyped distribution of midbody remnants in early *C. elegans* embryos requires cell death genes and is dispensable for development. *Cell Research, 24*(2), 251–253. http://dx.doi.org/10.1038/cr.2013.140.

Ou, G., Stuurman, N., D'Ambrosio, M., & Vale, R. D. (2010). Polarized myosin produces unequal-size daughters during asymmetric cell division. *Science, 330*(6004), 677–680. http://dx.doi.org/10.1126/science.1196112.

Piekny, A. J., & Maddox, A. S. (2010). The myriad roles of Anillin during cytokinesis. *Seminars in Cell & Developmental Biology, 21*(9), 881–891. http://dx.doi.org/10.1016/j.semcdb.2010.08.002.

Reyes, C. C., Jin, M., Breznau, E. B., Espino, R., Delgado-Gonzalo, R., Goryachev, A. B., & Miller, A. L. (2014). Anillin regulates cell-cell junction integrity by organizing junctional accumulation of Rho-GTP and actomyosin. *Current Biology: CB, 24*(11), 1263–1270. http://dx.doi.org/10.1016/j.cub.2014.04.021.

Shen, Z., Zhang, X., Chai, Y., Zhu, Z., Yi, P., Feng, G., … Ou, G. (2014). Conditional knockouts generated by engineered CRISPR-Cas9 endonuclease reveal the roles of coronin in *C. elegans* neural development. *Developmental Cell, 30*(5), 625–636. http://dx.doi.org/10.1016/j.devcel.2014.07.017.

Stiernagle, T. (2006). Maintenance of *C. elegans*. *Wormbook: The Online Review of* C. elegans *Biology*, 1–11. http://dx.doi.org/10.1895/wormbook.1.101.1.

Sulston, J. E., & Horvitz, H. R. (1977). Post-embryonic cell lineages of the nematode, *Caenorhabditis elegans*. *Developmental Biology, 56*(1), 110–156.

Tian, D., Diao, M., Jiang, Y., Sun, L., Zhang, Y., Chen, Z., … Ou, G. (2015). Anillin regulates neuronal migration and neurite growth by linking RhoG to the actin cytoskeleton. *Current Biology: CB, 25*(9), 1135–1145. http://dx.doi.org/10.1016/j.cub.2015.02.072.

CHAPTER

Variations on a theme: imaging cytokinetic and stable rings in situ using *Caenorhabditis elegans*

16

K. Rehain*, R.A. Green§, K.G. Bourdages¶, A.S. Maddox*[,1]

*University of North Carolina at Chapel Hill, Chapel Hill, NC, United States

§University of California, San Diego, CA, United States

¶Université de Montréal, Montréal, QC, Canada

[1]Corresponding author: E-mail: asm@unc.edu

CHAPTER OUTLINE

Abstract

Cytokinesis is an essential event in canonical cell division. In multicellular organisms, cells must divide in the context of neighboring cells in intact tissues. Recent studies have shown that tissue architecture can regulate the dynamics of and molecular requirements for cytokinesis. On the other hand, regulated cytokinesis failure occurs in, and is required for the proper function of, certain cell types and tissues including cardiomyocytes, hepatocytes, and germ lines. One way to build our understanding of cytokinesis in diverse

Methods in Cell Biology, Volume 137, ISSN 0091-679X, http://dx.doi.org/10.1016/bs.mcb.2016.03.039

cell types is to visualize cytokinesis in intact tissues. The nematode *Caenorhabditis elegans* is a powerful system for such inquiries due to the well-characterized, invariant lineage of each of its cells, the ease of genomic modifications including tagging proteins, and many more advantages. The clear cuticle of *C. elegans* allows for live imaging of intact tissues; however, the worm's motility can confound imaging. Here we introduce two *C. elegans* tissues, an epithelial tissue and the germ line, both excellent systems for the study of cytokinesis in the context of an intact animal. Additionally, we present three protocols for overcoming the challenges of live imaging in *C. elegans*.

INTRODUCTION

Cytokinesis is the final stage of cell division, during which nascent daughter nuclei are compartmentalized into separate cells. Early in animal development, completion of cytokinesis is critical for embryo viability. Success of every division is especially crucial for species with determinate development, wherein failure to create a certain cell can result in loss of an entire tissue. However, as development proceeds, many cell types including cardiomyocytes, hepatocytes, and germ cells are created via regulated cytokinesis failure and require high ploidy for proper function (reviewed by Lacroix & Maddox, 2012). Despite the essential nature of cells that are multinucleated, polyploid, and/or syncytial, our understanding of this source of tissue diversity is only rudimentary. This is partially due to the relative inaccessibility of such tissues.

Most of our understanding about the process of cytokinesis has been gleaned from studies of isolated cells and cells in culture; however, in multicellular organisms, cells must divide in the context of surrounding cells within a tissue. Recent studies of cells in intact tissues have shown that certain cellular features, including apical junctions or substrate adhesions, can influence the dynamics of, and molecular requirements for, cytokinesis (Founounou, Loyer, & Le Borgne, 2013; Guillot & Lecuit, 2013; Herszterg, Leibfried, Bosveld, Martin, & Bellaiche, 2013; Morais-de-Sá & Sunkel, 2013; Reyes et al., 2014). For example, the presence of cadherin-based apical junctions in epithelial cells causes polarized furrowing along the apicobasal axis. It is not clear how what we know about cytokinesis from studies of isolated and cultured cells is relevant to the process in these more complex settings. Here we describe three techniques for studying cytokinesis in two intact tissues in the nematode *Caenorhabditis elegans*: an epithelium (Fig. 1A) and the syncytial germ line (Fig. 2A).

MODEL SYSTEM: *CAENORHABDITIS ELEGANS*

Caenorhabditis elegans was first introduced as a model organism in 1965 by Dr. Sydney Brenner (Riddle, Blumenthal, Meyer, & Priess, 1997). Brenner analyzed a number of potential model systems for his research before ultimately choosing

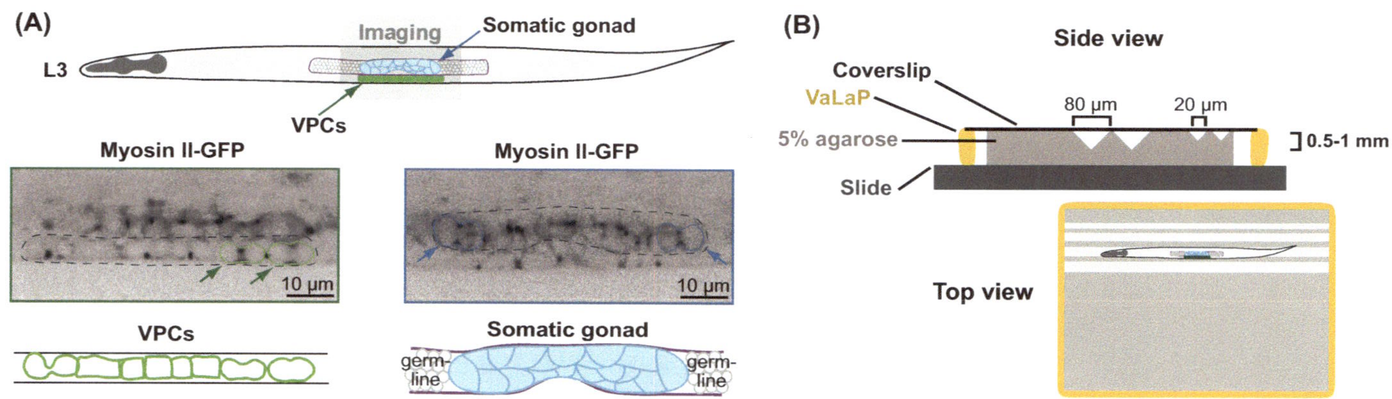

FIGURE 1 Worm mounting and imaging of the vulval precursor cells (VPCs) and cells of the somatic gonad in situ.

(A) Cartoon of a worm at the third larval stage, showing the imaging region encompassing the VPCs [green] and somatic gonad [blue]. Images of worms expressing GFP-tagged nonmuscle myosin II in the VPCs (left) and cells of the somatic gonad (right). *Dotted lines* encircle all VPCs (left) and the somatic gonad (right). Myosin II-GFP is enriched in the contractile ring of dividing VPCs [green arrow] and somatic gonad cells [blue arrow]. Below: schematics corresponding to the images above of the VPCs (left) and cells of the somatic gonad (right). (B) Schematic of the assembled microscopy slide ready for imaging. The agarose pad (light gray), with grooves 20–80 μm wide containing worms is mounted on the slide (dark gray), overlaid with a poly-L-lysine-coated coverslip (black). The coverslip is sealed with VaLaP [yellow]. The bottom panel shows a top view of a worm properly aligned in the grooves of the agarose pad.

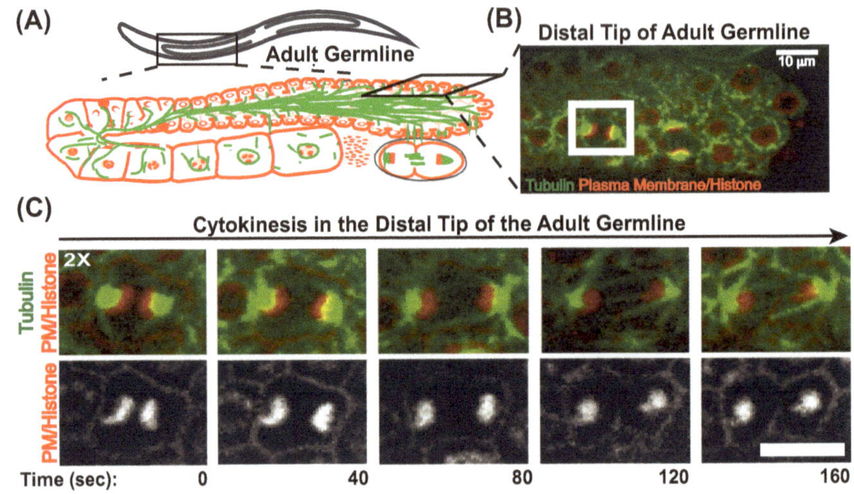

FIGURE 2 Imaging incomplete cytokinesis in the distal tip of the *Caenorhabditis elegans* adult germ line.

(A) Schematic of the *C. elegans* adult germ line. *Boxed area* depicts imaged region of strain expressing mCherry-Histone, mCherry-PH, and GFP-tubulin (OD2127). (B) The surface of the germ line distal tip was imaged with five optical sections separated by 1 μm, at 20 s intervals using 90× magnification (60× objective with 1.5× optivar) on a spinning disk confocal microscope (Andor). A single confocal slice is shown. *White box* marks region shown in C. (C) 2× magnification of the *boxed white region* in B at select time points. Grayscale images (bottom) highlight the changes in cell shape that occur during germ line cytokinesis. Scale bar: 10 μm.

C. elegans, which continues to prove itself an excellent model system today. *Caenorhabditis elegans* are small hermaphroditic roundworms measuring only 1 mm in length when fully grown. They can self-fertilize, develop from embryo to adult in 3 days, and have a short life span of only ~2.5 weeks. Moreover, their cuticle is transparent throughout development, which allows for live imaging of intact organisms (Corsi, 2015). Further, early studies in *C. elegans* determined that cell lineages are invariant (Sulston, Schierenberg, White, & Thomson, 1983). The characterization of these lineages allows researchers to identify specific cells at any point throughout development. *Caenorhabditis elegans* has also been improved as a model system by the generation of research tools including fluorescent protein tagging, gene-product knockdown via RNAi, and endogenous genome editing via a CRISPR/Cas9 (Chalfie, Tu, Euskirchen, Ward, & Prasher, 1994; Dickinson, Pani, Heppert, Higgins, & Goldstein, 2015; Fire et al. 1998; McNally, Audhya, Oegema, & McNally, 2006; Miller et al., 1999). Given the advantages described previously and the ease and low cost of maintaining worms in the lab, the *C. elegans* research community has grown and thrived over the past decades. Moreover, this highly active community has generated a number of resources including an RNAi library covering ~87% of the *C. elegans* genome, a repository of strains, and online resources such as *wormbase.org, wormbook.org, wormatlas.org,* and *Phenobank* (Green et al., 2011; Kamath et al., 2003; Kamath, Martinez-Campos, Zipperlen, Fraser, & Ahringer, 2001; Sönnichsen et al., 2005).

VULVAL PRECURSOR CELLS

The vulval precursor cells (VPCs) comprise an elegant model for the study of cytokinesis in an intact tissue (Bourdages, Lacroix, Dorn, Descovich, & Maddox, 2014). These epithelial cells give rise to part of the egg-laying apparatus of the worm and exhibit many behaviors typical of epithelia. This epithelium is superlatively simple: it is a single layer and begins as a single row (Fig. 1A). The VPCs comprise a subset of the hypodermal cells, which are born during embryonic development and remain nonproliferative until the molt that separates the second and third larval stages. Only three hypodermal cells adopt the VPC fates, based on proximity to the ventral nerve cord and the anchor cell in the somatic gonad, and fate specification signaling among VPCs (reviewed by Schindler & Sherwood, 2013). The Anchor Cell, and thus the VPCs, is centrally located on the worm's anterior—posterior axis (Fig. 1A). The apical surfaces of the VPCs are oriented towards the ventral nerve cord and the basolateral surfaces face a double basement membrane between the VPCs and the somatic gonad. The orientation of VPC divisions is controlled such that the row of three becomes a row of six and then of twelve. During the third round of divisions, certain VPCs divide perpendicular to the worm's long axis, giving the tissue more three-dimensionality. Concurrent with the last round of division, some of the VPCs undergo apical constriction, contributing to the invagination of the tissue (reviewed by Schindler & Sherwood, 2013). Thus, the VPC model allows for study of the three-dimensional challenges associated with cytokinesis in the context of an epithelial tissue.

GERM LINE SYNCYTIUM

In some tissues, cytokinetic rings are arrested to form stable intercellular bridges that link the cytoplasm of adjacent cells, creating cysts or syncytia. Intercellular bridges are an evolutionarily conserved feature of both the male and female germ line and have also been observed in some somatic tissues (reviewed in Haglund, Nezis, & Stenmark, 2011). The *C. elegans* adult germ line is an emerging model to study syncytiogenesis and this important variation of cytokinesis.

At hatching, the *C. elegans* hermaphrodite germ line consists of two cells called the Z2 and Z3. By adulthood, these two cells have built two syncytial gonad arms, each consisting of nearly 1000 cells organized into an interconnected tubular structure (Fig. 2A) (Hirsh, Oppenheim, & Klass, 1976; reviewed in Hubbard & Greenstein, 2000). In this structure, each of the germ cells is connected to a central pool of cytoplasm, called the rachis, via an intercellular bridge. Bridges are proposed to facilitate communication and coordination among cells and are required for fertility in *C. elegans* and most other systems. Structural and regulatory components of the cytokinetic ring are enriched at intercellular bridges (Amini et al., 2014; Maddox, Habermann, Desai, & Oegema, 2005; Zhou, Rolls, & Hanna-Rose, 2013), supporting the long-standing hypothesis that these structures are stabilized cytokinetic rings generated during regulated cytokinesis failure (also referred to as incomplete cytokinesis). While the mechanisms underlying the arrest of cytokinetic rings and formation of these stable intercellular bridges are not known, a large-scale high-content screen has identified genes that regulate the formation and/or maintenance of these germ line compartments (Green et al., 2011), and recent work is beginning to shed light on how key players, such as scaffold proteins ANI-1 and ANI-2, regulate this process (Amini et al., 2014).

In the adult gonad, oocytes progress developmentally from mitosis through the stages of meiotic prophase as they advance spatially from the distal tip to the proximal gonad, adjacent to the spermatheca (Fig. 2A). This linear arrangement makes it possible to examine all stages of oogenesis in a single sample gonad. Throughout germ cell development intercellular bridges are stably open and only close shortly before oocytes become fully mature (reviewed by Hubbard & Greenstein, 2000). Oocyte pseudocells in the distal gonad have relatively stable bridges, and as cells enter the proximal region of the gonad, their intercellular bridges dynamically expand and cytoplasmic flows from more distal cells inflate the oocytes (Wolke, Jezuit, & Priess, 2007). When oocytes are full sized, the bridges close, resulting in cellularization of the now mature oocyte. Proper regulation and maintenance of intercellular bridges is required for *C. elegans* fertility (Green et al., 2011). Premature bridge closure blocks oocyte inflation making oocytes nonfunctional. Despite the essential nature of germ line intercellular bridges in *C. elegans* and throughout phylogeny, little is known about their structure and dynamics. The methods outlined here will facilitate studies of these open questions.

1. METHODS

1.1 IMAGING CYTOKINESIS OF VULVAL PRECURSOR CELLS

A technical challenge confounding all in situ imaging of *C. elegans* is the mobility of the worms. *Caenorhabditis elegans* were named for their "elegant" sinusoidal movement, but this movement confounds the imaging of probes that require long exposure times or processes that occur over extended periods of time. To properly image *C. elegans*, the worms must be immobilized. Techniques to physically inhibit worm movement have been developed and include adhering worms to substrates, the use of microfluidic chambers to clamp down worms, and using agarose and polystyrene beads to block worm movement (Goodman, Hall, Avery, & Lockery, 1998; Hulme, Shevkoplyas, Apfeld, Fontana, & Whitesides, 2007; Kerr et al., 2000; Kim, Sun, Gabel, & Fang-Yen, 2013; Rohde, Zeng, Gonzalez-Rubio, Angel, & Yanik, 2007). Additionally, anesthetics are commercially available that temporarily arrest worm movement. Tetramisole hydrochloride, for example, is an alkaline phosphatase inhibitor that acts as a nicotinic acetylcholine receptor agonist. Treatment of *C. elegans* with tetramisole (or levamisole; Acros Organics #187870100) causes prolonged muscle contraction ultimately leading to paralysis (Lewis, Wu, Levine, & Berg, 1980). We recommend combining several of these techniques to image cytokinesis in intact tissues.

Materials:
22 × 22 mm coverslips
95% ethanol
Poly-L-lysine 0.1%
0.01% tetramisole solution in M9 buffer
M9 buffer (to make up 1 L, add 5.8 g Na_2HPO_4, 3.0 g KH_2PO_4, 0.5 g NaCl, and 1.0 g NH_4Cl to dH_2O and autoclave)
5% agarose in M9
VaLaP (1:1:1 Vaseline, lanolin, and paraffin)
Glass slides
Custom nanofabricated silica plate with 20- to 80-μm-wide pyramidal ridges
Mouth pipette
Eyelash tool (eyelash mounted onto a handle such as a toothpick or pipette tip)

1.1.1 Preparation of poly-L-lysine-coated coverslips

1. Dip 22 × 22 mm coverslips in 95% ethanol.
2. Dry them quickly over a flame.
3. Bathe coverslips in 0.1% poly-L-lysine for 4 h to overnight, on a 2D shaker.
4. Let coverslips dry completely before use.

1.1.2 Worm mounting (Fig. 1B)

1. Pick 10–20 worms at the desired larval stage and place them in a drop of 0.01% tetramisole solution prepared with M9 buffer.
2. Soak worms for a minimum of 10 min to ensure they are completely anesthetized, while ensuring that the worms do not dry out.
3. While the worms soak, melt VaLaP and 5% agarose at 95°C.
4. Place a drop of melted agarose on a microscopy slide positioned between two parallel slides each covered with one piece of lab tape. After allowing the drop to cool slightly, place another clean slide orthogonal to the first two slides, over the drop, compressing it into a flat pad. Alternatively, we use custom nanofabricated silica plates with 20- to 80-μm-wide pyramidal ridges, to create grooves in the agarose.
5. Allow time for the agarose to solidify.
6. Remove the slide or silica plate carefully so as not to break the pad.
7. The agarose pad can be trimmed with a scalpel to ensure it is smaller than the coverslip to be used. When using the silica plate, we ensure that worms will be aligned orthogonally to the microscope stage by creating grooves parallel to the long axis of the slide.

8. Use a mouth pipette to transfer a few anesthetized worms on the agar pad, being careful to transfer only a small amount of the tetramisole solution (not more than 10 μL).
9. Using an eyelash tool, gently arrange the worms, and align them into the grooves if available. During this period, additional tetramisole solution can be added to prevent the worms from drying.
10. Using a mouth pipette or small piece of Whatman paper, remove excess liquid.
11. Place a poly-L-lysine-coated coverslip over the pad. The sticky coating contributes to immobilizing the worms.
12. Carefully inspect the preparation to ensure that the pad is not too dry (a good contact throughout the surfaces of the coverslip and pad indicates proper moisture level), since worm survival is dependent on moisture. However, too much liquid will displace worms upon addition of the coverslip.
13. To seal the coverslip, place a small drop of melted VaLaP on each corner of the coverslip using a 20–200 μL pipette tip.
14. Add more tetramisole solution to fill the entire cavity beneath the coverslip by pipetting the solution at the edge of the coverslip and allowing capillary action to spread it across the sample. Introduce the liquid slowly so as not to disturb the aligned worms.
15. Seal all sides of the coverslip with VaLaP to minimize evaporation.
16. The slide is now ready for imaging. We recommend waiting 10–30 min before beginning imaging, since worms are more likely to remain still during imaging if they have done so during this short delay following slide preparation.
17. VPCs can be identified by transmitted light. They occupy the ventral surface of the worm, midway along the animal's antero–posterior axis, thus centered under the germ line (Fig. 1A). From early L3, almost until the L3/L4 molt, the VPCs form a single row. The anchor cell within the developing somatic gonad may be visible as unusually flared against the paired basement membranes that lie between the somatic gonad and the VPCs; the position of the anchor cell corresponds to the center of the row of VPCs. VPCs in interphase have prominent nuclei and nucleoli. The lateral membranes between VPCs are invisible by transmitted light.

1.2 IMAGING OF THE SYNCYTIAL GERM LINE

Attempts to monitor the germ line during the course of development have been technically challenging. In adults, this tissue is ~50 μm thick, necessitating a high-resolution microscope (confocal, two-photon) for optimal imaging. Mitotic division, and thus cytokinesis of developing oocytes, occurs at the distal tip of the germ line in adult worms, which has a complex three-dimensional structure and can lie at different depths within the body (Fig. 2B and C; Movie 1). Additionally, the *C. elegans* gut, which often covers and obscures the germ line, is autofluorescent in the green channel, obscuring weakly expressed fluorescent probes. Lastly, as elaborated previously, worms are highly mobile. Despite these difficulties, some progress is underway toward longitudinally tracking germ line development, using microfluidic chambers to trap animals (R. Green, E. Gutierrez, unpublished data). Presently, several groups have developed methods for imaging cortical dynamics, incomplete cytokinesis events, and intercellular bridge regulation at a single developmental stage (Amini et al., 2014; Green et al., 2011; Maddox et al., 2005; Wang et al., 2015; Wolke et al., 2007; Zhou et al., 2013). Two techniques for imaging intact *C. elegans* germ lines are outlined here. The first method allows for live imaging of dividing cells in the distal germ line, while the second allows for analysis of stable cytokinetic rings throughout the germ line/germ cell development.

1.2.1 Live imaging in the distal germ line

Materials:
Unseeded NGM plate
Bleach (5%)
M9
2% agarose in M9
Tetramisole hydrochloride (TMHC; Sigma L9756)
Glass slides
Coverslips
Depression slide
Pick or mouth pipette
10 cm petri dish

1.2.1.1 Live imaging protocol

1. Prepare 0.1 mg/mL of tetramisole hydrochloride freshly from powder, in M9 (1X TMHC).
2. Place a depression slide inside a 10 cm petri dish. Pipette a pool of 1X TMHC into a depression slide.
   Note: A 2X mixture will anesthetize faster, but may be toxic to younger worms.
3. Add worms to the pool of TMHC using standard picking techniques, and swish worms off the pick until well separated. Worms will thrash for several minutes. Cover the petri dish to minimize evaporation.
4. Anesthetize for 15–30 min.
   Note: Younger worms require less time (15 min), while adults require longer time (30 min).
5. Periodically check for anesthetization by tapping the depression slide. Worms will not thrash at all when fully anesthetized.
6. During this incubation period, prepare an agarose pad. Place a dime-sized drop of melted 2% agarose on a glass slide, let it cool for 5–10 s and place a second slide on top perpendicular to the slide below, using a single layer of lab tape as a spacer.

7. When worms are anesthetized, remove the top slide to expose the agarose pad and place it on the stage next to the depression slide.
8. Place a 3 uL drop of anesthetic on the agarose pad. Using a pick, transfer the worms one at a time to the agar pad.
   Note: Small worms may be more easily transferred using a mouth pipette, but can get stuck to the walls of the glass capillary; transfer method is a matter of preference.
9. Using an eyelash tool, group the worms together in the center of the pad and cover with a coverslip.
   Note: To improve the chances of finding a worm at the ideal orientation for imaging, mount several worms at a time, since positioning of the worm is random. Select the worms whose germ line distal tip is closest to the coverslip.
10. We recommend imaging a small z-stack (5–10 μm total thickness) with a time interval of 20 s to 1 min to visualize the dynamics of mitotic events at the distal tip.

The mitotic index in the distal tip of younger worms is higher than that in older worms and thus can yield more mitotic events in a given imaging field (approximately 6–9). In addition, since the germ line is not as thick or architecturally complex in young larvae as in adults, it is easier to track divisions in larvae. The ease of identifying worms at L4 stage is another advantage of this strategy. L2–L3 worms offer less complexity still and also have an enhanced number of mitotic cells, but can be harder to stage. This problem can be circumvented by using synchronized worms (described in the next section). Worms of any stage (L2-adult) can be anesthetized via the protocol described here, though the younger worms are more susceptible to toxicity from overanesthetization.

1.2.1.2 Synchronization of young worms

1. Bleach 10–20 adult worms by placing them in a 5 μL droplet of bleach on a small unseeded NGM plate. The bleach will destroy the adult bodies but soak and dilute into the plate by the time the embryos hatch. Alternatively, use bleach to wash adult worms off a small plate, transfer to a 2 mL tube, vortex and pulse spin, remove the supernatant containing bleach and adult carcass parts, and wash the embryo pellet two times with M9.
2. Allow bleached embryos to hatch overnight, and arrest at the first larval stage (L1) by culturing them in the absence of food for 18 h (either on an unseeded plate or in M9 solution).
3. Transfer hatched worms to a plate containing food to allow worms to progress to L2 (24 h) or L3 (36 h).

1.2.2 Static imaging of intercellular bridges in the germ line

Materials:
Worm plates

Glass slides
Poly-L-lysine solution (0.1% w/v aqueous)
18 × 18 mm glass coverslips
M9
25G 5/8 needle
Eyelash tool
Vaseline

1.2.2.1 Preparing worms

1. Identify L4 worms by the ventral white semicircular appearance of the developing vulva.
2. Using standard picking techniques move 15–30 L4s to worm plates seeded with either OP50 bacteria or RNAi bacteria depending on the experiment.
   a. For information on how to conduct RNAi, see Ahringer (2006) lab protocol.
3. Allow worms to develop overnight into young adults.

1.2.2.2 Preparing poly-L-lysine-coated coverslips

1. Apply three 10 μL drops of poly-L-lysine solution (0.1% w/v aqueous) to a glass slide.
2. Place an 18 × 18 mm coverslip over each drop.
3. Heat on heating block set to 125°C until liquid has evaporated and coverslips are dry and appear slightly darker.
4. Let coverslips come to room temperature.

1.2.2.3 Preparing sample (Fig. 3)

1. Remove poly-L-lysine-coated coverslips prepared within one day from slide and place with coated side facing upward on a glass slide.
2. Pipette 10 μL of M9 buffer onto the center of the coverslip.
3. Transfer ~5 young adult worms to the droplet using a worm pick (Fig. 3A).
4. Using a 25G 5/8 needle, cut the worms just posterior to the pharynx, removing the head of the worm (Fig. 3A). This should allow the front gonad to be extruded from the worm (Fig. 3B).
5. Use an eyelash tool to position the gonad against the coverslip and away from the dark gut that was likely also extruded. The gut can also be cut away (Fig. 3C).
   Note: If one or more worm is still moving, remove them so that the gonads are not disturbed during imaging.
6. Apply a drop of Vaseline to each corner of the coverslip (Fig 3D).
7. Flip the coverslip sample side down onto a glass slide. While monitoring the sample through a stereoscope, press the coverslip down gently so that the gonad is only slightly compressed (Fig. 3E).
8. We recommend imaging through the half of the gonad closes to the coverslip, with 0.5 μm z-steps. A maximum intensity projection of the z-stack will allow *en face* visualization of intercellular bridges (Fig. 3F). Transverse sections of many bridges will also be visible in optical sections.

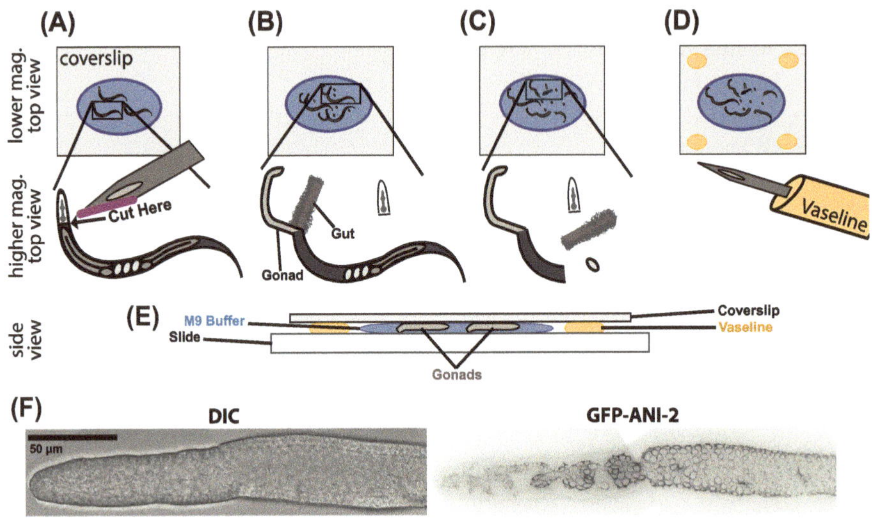

FIGURE 3 Extrusion and mounting of an intact adult gonad.

(A–D) Top: lower magnification top-view schematics of dissections. Bottom: higher magnification views of boxed regions above. (A) The cutting edge [purple] of a 25G x 5/8 needle as well as the optimal cut site for gonad extrusion. (B) Expected results following cutting. Both the gonad and gut have been extruded from the worm body. (C) Sample after the body and autofluorescent gut have been cut away. (D) Vaseline [yellow] applied to the coverslip. (E) Side-view schematic depicting the complete sample with the isolated gonads close to the coverslip and the Vaseline droplets acting as spacers to prevent compression of the gonads. Note: Figures *not* to scale. (F) Stitched DIC and fluorescence images of a gonad extruded from a worm expressing GFP-tagged ANI-2 ^{anillin}. Scale bar: 50 μm.

CONCLUSIONS

To fully understand the regulation and execution of cytokinesis in diverse cell types, we must study it in complex settings. The nematode *C. elegans* has provided a setting for great experimental strides in many areas of research. The three protocols we present for visualizing cytokinesis in the context of intact tissues in *C. elegans* will help continue this trend. Studying cytokinesis in these systems will inform not only how cytokinesis is different among distinct cell types and tissues, but will also provide insight into the general principles of cytokinesis, as well as how this process can be misregulated leading to diseases such as cancer.

ACKNOWLEDGMENTS

KR and ASM are supported by GM102390 from the NIH. KGB is supported by a graduate training fellowship from the Cole Foundation. RAG is supported by Karen Oegema and receives salary and other support from the Ludwig Institute for Cancer Research and NIH grant GM074207.

SUPPLEMENTARY DATA

Supplementary data related to this chapter can be found online at http://dx.doi.org/10.1016/bs.mcb.2016.03.039.

REFERENCES

Ahringer, J. (2006). Reverse genetics. In V. Ambros, & The C. elegans Research Community (Eds.), *Wormbook: the Online Review of* C. elegans *Biology.* http://dx.doi.org/10.1895/wormbook.1.47.1.

Amini, R., Goupil, E., Labella, S., Zetka, M., Maddox, A. S., Labbé, J.-C., & Chartier, N. T. (2014). *C. elegans* Anillin proteins regulate intercellular bridge stability and germline syncytial organization. *Journal of Cell Biology, 206*(1), 129–143. http://dx.doi.org/10.1083/jcb.201310117.

Bourdages, K. G., Lacroix, B., Dorn, J. F., Descovich, C. P., & Maddox, A. S. (2014). Quantitative analysis of cytokinesis in situ during *C. elegans* postembryonic development. *PLoS One, 9*(10), e110689. http://dx.doi.org/10.1371/journal.pone.0110689.

Chalfie, M., Tu, Y., Euskirchen, G., Ward, W. W., & Prasher, D. C. (1994). Green fluorescent protein as a marker for gene expression. *Science (New York, N.Y.), 263*(5148), 802–805. Retrieved from http://www.ncbi.nlm.nih.gov/pubmed/8303295.

Corsi, A. K. (2015). A transparent window into biology: a primer on *Caenorhabditis elegans. Wormbook: the Online Review of* C. elegans *Biology,* 1–31. http://dx.doi.org/10.1895/wormbook.1.177.1.

Dickinson, D. J., Pani, A. M., Heppert, J. K., Higgins, C. D., & Goldstein, B. (2015). Streamlined genome engineering with a self-excising drug selection cassette. *Genetics, 200*(4), 1035–1049. http://dx.doi.org/10.1534/genetics.115.178335.

Fire, A., Xu, S., Montgomery, M. K., Kostas, S. A., Driver, S. E., & Mello, C. C. (1998). Potent and specific genetic interference by double-stranded RNA in *Caenorhabditis elegans. Nature, 391*(6669), 806–811. http://dx.doi.org/10.1038/35888.

Founounou, N., Loyer, N., & Le Borgne, R. (2013). Septins regulate the contractility of the actomyosin ring to enable adherens junction remodeling during cytokinesis of epithelial cells. *Developmental Cell, 24*(3), 242–255. http://dx.doi.org/10.1016/j.devcel.2013.01.008.

Goodman, M. B., Hall, D. H., Avery, L., & Lockery, S. R. (1998). Active currents regulate sensitivity and dynamic range in *C. elegans* neurons. *Neuron, 20*(4), 763–772. http://dx.doi.org/10.1016/S0896-6273(00)81014-4.

Green, R. A., Kao, H.-L., Audhya, A., Arur, S., Mayers, J. R., Fridolfsson, H. N., … Oegema, K. (2011). A high-resolution *C. elegans* essential gene network based on phenotypic profiling of a complex tissue. *Cell, 145*(3), 470–482. http://dx.doi.org/10.1016/j.cell.2011.03.037.

Guillot, C., & Lecuit, T. (2013). Adhesion disengagement uncouples intrinsic and extrinsic forces to drive cytokinesis in epithelial tissues. *Developmental Cell, 24*(3), 227–241. http://dx.doi.org/10.1016/j.devcel.2013.01.010.

Haglund, K., Nezis, I. P., & Stenmark, H. (2011). Structure and functions of stable intercellular bridges formed by incomplete cytokinesis during development. *Communicative and Integrative Biology, 4*(1), 1–9. http://dx.doi.org/10.4161/cib.4.1.13550.

Herszterg, S., Leibfried, A., Bosveld, F., Martin, C., & Bellaiche, Y. (2013). Interplay between the dividing cell and its neighbors regulates adherens junction formation during cytokinesis in epithelial tissue. *Developmental Cell, 24*(3), 256–270. http://dx.doi.org/10.1016/j.devcel.2012.11.019.

Hirsh, D., Oppenheim, D., & Klass, M. (1976). Development of the reproductive system of *Caenorhabditis elegans*. *Developmental Biology, 49*(1), 200–219. http://dx.doi.org/10.1016/0012-1606(76)90267-0.

Hubbard, E. J., & Greenstein, D. (2000). The *Caenorhabditis elegans* gonad: a test tube for cell and developmental biology. *Developmental Dynamics : An Official Publication of the American Association of Anatomists, 218*(1), 2–22.

Hulme, S. E., Shevkoplyas, S. S., Apfeld, J., Fontana, W., & Whitesides, G. M. (2007). A microfabricated array of clamps for immobilizing and imaging *C. elegans*. *Lab on a Chip, 7*(11), 1515–1523. http://dx.doi.org/10.1039/b707861g.

Kamath, R. S., Martinez-Campos, M., Zipperlen, P., Fraser, A. G., & Ahringer, J. (2001). Effectiveness of specific RNA-mediated interference through ingested double-stranded RNA in *Caenorhabditis elegans*. *Genome Biology, 2*(1). http://dx.doi.org/10.1186/gb-2000-2-1-research0002. RESEARCH0002.

Kamath, R. S., Fraser, A. G., Dong, Y., Poulin, G., Durbin, R., Gotta, M., … Ahringer, J. (2003). Systematic functional analysis of the *Caenorhabditis elegans* genome using RNAi. *Nature, 421*(6920), 231–237. http://dx.doi.org/10.1038/nature01278.

Kerr, R., Lev-Ram, V., Baird, G., Vincent, P., Tsien, R. Y., & Schafer, W. R. (2000). Optical imaging of calcium transients in neurons and pharyngeal muscle of *C. elegans*. *Neuron, 26*(3), 583–594. http://dx.doi.org/10.1016/S0896-6273(00)81196-4.

Kim, E., Sun, L., Gabel, C. V., & Fang-Yen, C. (2013). Long-term imaging of *Caenorhabditis elegans* using nanoparticle-mediated immobilization. *PLoS One, 8*(1), e53419. http://dx.doi.org/10.1371/journal.pone.0053419.

Lacroix, B., & Maddox, A. S. (2012). Cytokinesis, ploidy and aneuploidy. *Journal of Pathology, 226*(2), 338–351. http://dx.doi.org/10.1002/path.3013.

Lewis, J. A., Wu, C. H., Levine, J. H., & Berg, H. (1980). Levamisole-resistant mutants of the nematode *Caenorhabditis elegans* appear to lack pharmacological acetylcholine receptors. *Neuroscience, 5*(6), 967–989.

Maddox, A. S., Habermann, B., Desai, A., & Oegema, K. (2005). Distinct roles for two *C. elegans* anillins in the gonad and early embryo. *Development (Cambridge, England), 132*(12), 2837–2848. http://dx.doi.org/10.1242/dev.01828.

McNally, K., Audhya, A., Oegema, K., & McNally, F. J. (2006). Katanin controls mitotic and meiotic spindle length. *The Journal of Cell Biology, 175*(6), 881–891. http://dx.doi.org/10.1083/jcb.200608117.

Miller, D. M., Desai, N. S., Hardin, D. C., Piston, D. W., Patterson, G. H., Fleenor, J., … Fire, A. (1999). Two-color GFP expression system for *C. elegans*. *BioTechniques, 26*(5), 914–8, 920–1.

Morais-de-Sá, E., & Sunkel, C. (2013). Adherens junctions determine the apical position of the midbody during follicular epithelial cell division. *EMBO Reports, 14*(8), 696–703.

Reyes, C. C., Jin, M., Breznau, E. B., Espino, R., Delgado-Gonzalo, R., Goryachev, A. B., & Miller, A. L. (2014). Anillin regulates cell-cell junction integrity by organizing junctional accumulation of Rho-GTP and actomyosin. *Current Biology : CB, 24*(11), 1263–1270.

Riddle, D. L., Blumenthal, T., Meyer, B. J., & Priess, J. R. (1997). *The biological model.* Cold Spring Harbor Laboratory Press. Retrieved from http://www.ncbi.nlm.nih.gov/books/NBK20086/.

Rohde, C. B., Zeng, F., Gonzalez-Rubio, R., Angel, M., & Yanik, M. F. (2007). Microfluidic system for on-chip high-throughput whole-animal sorting and screening at subcellular resolution. *Proceedings of the National Academy of Sciences of the United States of America, 104*(35), 13891–13895. http://dx.doi.org/10.1073/pnas.0706513104.

Schindler, A. J., & Sherwood, D. R. (2013). Morphogenesis of the *Caenorhabditis elegans* vulva. *Wiley Interdisciplinary Reviews. Developmental Biology, 2*(1), 75–95. http://dx.doi.org/10.1002/wdev.87.

Sönnichsen, B., Koski, L. B., Walsh, A., Marschall, P., Neumann, B., Brehm, M., ... Echeverri, C. J. (2005). Full-genome RNAi profiling of early embryogenesis in *Caenorhabditis elegans. Nature, 434*(7032), 462–469. http://dx.doi.org/10.1038/nature03353.

Sulston, J. E., Schierenberg, E., White, J. G., & Thomson, J. N. (1983). The embryonic cell lineage of the nematode *Caenorhabditis elegans. Developmental Biology, 100*(1), 64–119. http://dx.doi.org/10.1016/0012-1606(83)90201-4.

Wang, S., Wu, D., Quintin, S., Green, R. A., Cheerambathur, D. K., Ochoa, S. D., ... Oegema, K. (2015). NOCA-1 functions with γ-tubulin and in parallel to Patronin to assemble non-centrosomal microtubule arrays in *C. elegans. eLife, 4*, e08649. http://dx.doi.org/10.7554/eLife.08649.

Wolke, U., Jezuit, E. A., & Priess, J. R. (2007). Actin-dependent cytoplasmic streaming in *C. elegans* oogenesis. *Development (Cambridge, England), 134*(12), 2227–2236. http://dx.doi.org/10.1242/dev.004952.

Zhou, K., Rolls, M. M., & Hanna-Rose, W. (2013). A postmitotic function and distinct localization mechanism for centralspindlin at a stable intercellular bridge. *Developmental Biology, 376*(1), 13–22. http://dx.doi.org/10.1016/j.ydbio.2013.01.020.

CHAPTER

17

Using fast-acting temperature-sensitive mutants to study cell division in *Caenorhabditis elegans*

T. Davies*[,a], S. Sundaramoorthy*[,a], S.N. Jordan*, M. Shirasu-Hiza*, J. Dumont[§], J.C. Canman*[,1]

**Columbia University, New York, NY, United States*

[§]Institut Jacques Monod, Paris, France

[1]Corresponding author: E-mail: jcc2210@cumc.columbia.edu

CHAPTER OUTLINE

[a]These authors contributed equally and are listed in alphabetical order.

Methods in Cell Biology, Volume 137, ISSN 0091-679X, http://dx.doi.org/10.1016/bs.mcb.2016.05.004

Abstract

Fast-acting temperature-sensitive (ts) mutations are powerful conditional tools for studying transient cellular processes such as cytokinesis. Fast-acting ts cytokinesis-defective mutants are functional at the permissive temperature; yet show a fully penetrant loss-of-function cytokinesis failure phenotype when upshifted to the restrictive temperature. Fast-acting ts mutations thus allow functional tunability and rapid and reversible protein inactivation by simply shifting the temperature at precise times throughout cell division. In this chapter, we describe several techniques and discuss various approaches for harnessing the power of fast-acting ts mutants to study cytokinesis in *Caenorhabditis elegans* using both simple passive heat transfer and more advanced fluidic-based thermal control systems. We also provide detailed protocols for standard dissection, mounting, and imaging of early worm embryos.

INTRODUCTION

Cytokinesis, the physical division of one cell into two (D'Avino, Giansanti, & Petronczki, 2015; Green, Paluch, & Oegema, 2012; Rappaport, 1996), is arguably the most important event in the cell division cycle as without successful cytokinesis, daughter cells do not form and mitosis is bootless. Furthermore, cytokinesis failure is associated with a growing number of human pathologies including neurological disorders, immunological defects, and cancer (Lacroix & Maddox, 2012; Normand & King, 2010; Sagona & Stenmark, 2010).

To ensure each daughter cell inherits a single copy of the genome, cytokinesis is coordinated with nuclear division by both the cell cycle machinery and the mitotic spindle (D'Avino et al., 2015; Green et al., 2012). In animal cells, the mitotic spindle specifies the site of cytokinesis after anaphase onset via signals from the astral and central spindle microtubules, which together specify a zone of active RhoA, a small GTPase, at the division plane (Bement, Benink, & von Dassow, 2005; Jordan & Canman, 2012). Active RhoA then promotes both actin nucleation (via diaphanous formins) and myosin-II motor activation, which form the actomyosin contractile ring. Cytokinesis begins by an initial membrane invagination called the cytokinetic furrow, formed by constriction of the actomyosin contractile ring. Ring constriction proceeds until a dense remnant of the central spindle called the midbody remains at the narrow cytoplasmic bridge between the two daughter cells. The final severing of the cytoplasmic bridge, or abscission, occurs later and is thought to involve the recruitment of microtubule severing proteins, membrane vesicles, and membrane sculpting/transport proteins to the midbody (D'Avino et al., 2015; Green et al., 2012; Mierzwa & Gerlich, 2014).

While traditional forward genetics and RNAi screens have provided a basic understanding of the key genes required for cytokinesis, we still do not understand

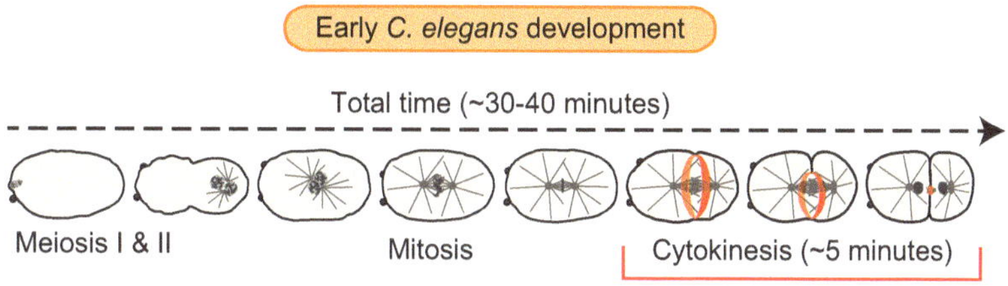

FIGURE 1 Developmental timing of the early *Caenorhabditis elegans* embryo.

The fertilized embryo progresses through meiosis and the first mitotic division in ~30–40 min. As many cytokinesis proteins are required for multiple cellular events, high-resolution temporal control of protein function allows modulation of activity at specific cellular time points.

why the same mutational disruption of a cytokinesis gene does not uniformly affect division of all cell types within a multicellular organism or why some cell types are naturally programmed to fail in cytokinesis during development (Normand & King, 2010). There are two key reasons why traditional gene knockdown is limited in studying the molecular regulation of cytokinesis in vivo: (1) cytokinesis is a dynamic and highly transient process (~5–45 min in most animal cells) and (2) many of the key molecular players in cytokinesis (eg, actin, myosin-II, RhoA) are required at multiple stages of the cell cycle and throughout morphogenesis during organismal development (Fig. 1). To overcome these challenges, researchers have turned to acute protein inactivation to study function specifically during cytokinesis. This has been accomplished with drug treatment, inhibitory antibody injection, dominant-negative protein injection, and fast-acting temperature-sensitive (ts) mutations (Canman et al., 2003; Canman, Salmon, & Fang, 2002; Carvalho et al., 2011; Kurz et al., 2002; Mabuchi & Okuno, 1977; Oegema, Savoian, Mitchison, & Field, 2000; Potapova et al., 2006; Severson, Hamill, Carter, Schumacher, & Bowerman, 2000).

Fast-acting ts mutations are especially powerful because of the ease of functional tunability combined with the power of transient and rapid protein inactivation (~20 s, for example, see Davies et al., 2014). Fast-acting ts mutant proteins are functional enough at the permissive temperature to allow normal development and organismal viability, yet show a fully penetrant loss-of-function cytokinesis failure phenotype when upshifted to the restrictive temperature (Fig. 2). Moreover, ts mutants are "tunable"—that is, use of semirestrictive and semipermissive temperatures can allow varying degrees of functionality. Thus, ts mutants can be used as a sensitized background for synthetic genetic suppression studies by rapid upshift to semirestrictive temperatures at which protein function is weakened but contractile ring constriction is more robust than at fully restrictive temperature (Fig. 2B) (Canman et al., 2008; Jordan et al., 2016; Lewellyn, Carvalho, Desai, Maddox, & Oegema, 2011; Severson et al., 2000). Similarly, synthetic genetic enhancement screening can be done by rapid upshift to semipermissive temperatures at which cytokinesis

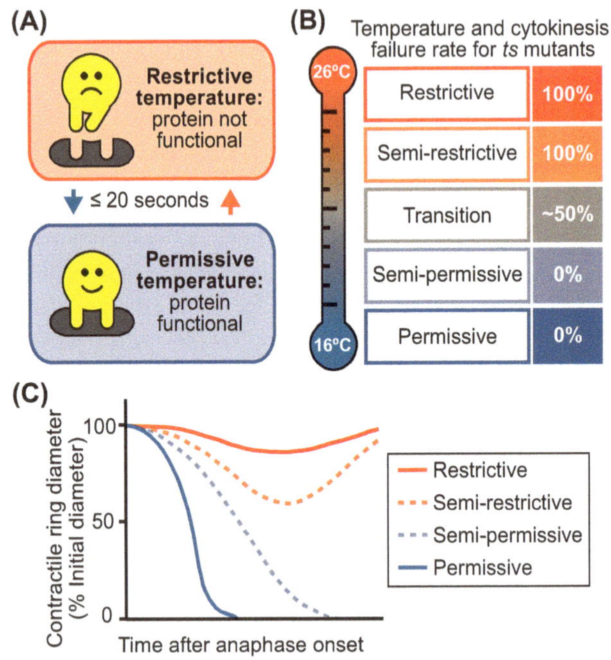

FIGURE 2 Schematic of temperature-sensitive protein activity.

(A) A schematic of a hypothetical temperature-sensitive (ts) mutant protein with function at lower temperatures (permissive), but loss of function at higher temperatures (restrictive). (B) At permissive temperatures, ts mutant proteins are functional and support successful cytokinesis. At restrictive temperatures, ts protein function is reduced and cytokinesis fails. At intermediate semipermissive temperature, cytokinesis is successful, whereas at intermediate semirestrictive temperature, cytokinesis fails. The transition is the pivot temperature between cytokinesis success and failure. (C) Graph showing the tunable kinetics of contractile ring constriction overtime in embryos from a hypothetical ts cytokinesis mutant imaged across various temperatures.

is compromised but still successful without additional perturbations (Fig. 2B) (Canman et al., 2008; Jordan et al., 2016).

It is important to note that not all ts alleles are fast-acting or functional enough at permissive temperature to allow normal development. Each specific ts allele must be tested for rapid modulation of function. Generally, there are two classes of ts mutations: slow-acting and fast-acting (O'Rourke, Carter, et al., 2011). Slow-acting alleles tend to result from mutations in the noncoding sequence, which disrupt gene expression or mRNA processing. Thus at the restrictive temperature, the time required for a loss-of-function phenotype after upshift to restrictive temperature is typically several hours to overnight. This prevents rapid and precise control of protein function. In contrast, fast-acting ts alleles usually result from single amino acid substitutions in the coding sequence that weaken protein structure, such that increased thermal energy (higher temperature) results in a structural change that blocks normal protein folding and function. These ts proteins quickly respond to changes in temperature, allowing rapid control of protein function using the approaches described below.

A collection of *C. elegans* fast-acting ts cytokinesis-defective mutants has been identified in classical forward genetic screens for maternal effect embryonic lethal, sterile, and zygotic lethal mutants (Canman et al., 2008; Encalada et al., 2000; Kemphues, Priess, Morton, & Cheng, 1988; O'Connell, Leys, & White, 1998; O'Rourke, Yochem, et al., 2011; Raich, Moran, Rothman, & Hardin, 1998). *Caenorhabditis elegans* is a powerful genetic system with a rapid generation time (~3–6 days), hermaphroditic self-fertilizing reproduction, optical clarity, and an invariant developmental lineage. Importantly, the core genes required for cytokinesis are conserved from worms to humans. Using fast-acting ts cytokinesis-defective alleles, protein function can be controlled by simply changing the sample temperature either environmentally (with a thermally controlled room or microscope incubation system) or with a thermally precise fluidic control system. In particular, regulating sample temperature with fluidic thermal control allows rapid and reversible modulation of protein function via timed thermal up- and downshifting throughout cell division (Fig. 3).

In this chapter, we describe several techniques for using fast-acting ts alleles to study cytokinesis in the early worm embryo. We describe how to dissect and mount *C. elegans* embryos for live cell imaging and discuss ways to regulate or change sample temperature and thus protein function during imaging. We describe the practical setup required for analysis of any fast-acting ts cytokinesis-defective alleles and several techniques for sample mounting and temperature control, ranging in features and cost. We provide vendors and ordering information for the reagents and equipment we use. Last, we outline a set of criteria to determine if a particular ts allele is suitable to study cell division.

1. MICROSCOPY AND EMBRYO ISOLATION

1.1 MICROSCOPE AND ROOM

To fully harness the power of fast-acting ts mutants, it is ideal to combine temperature control with live cell analysis via fluorescence microscopy. When working with fast-acting ts mutants, it is important to control the temperature of the dissection and imaging room. We use a temperature-controlled room that is allowed to equilibrate at the desired temperature for at least 60 min before imaging (single-zone air conditioner unit with heat pump, eg, Mitsubishi #MSZ-D36NA or

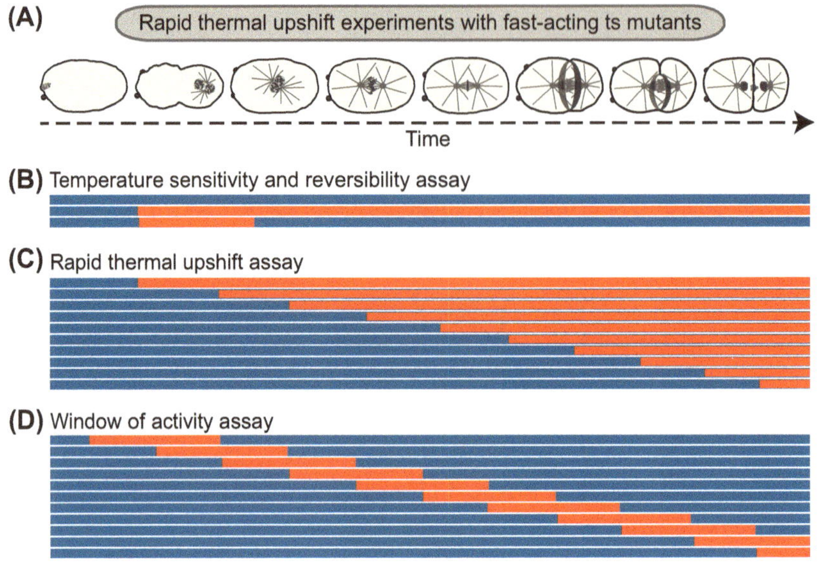

FIGURE 3 Using temperature shifts to study cytokinesis.

(A) Early embryonic development occurs stereotypically from meiosis to cytokinesis. (B–D) Schematic representation of potential experiments combining active temperature control and fast-acting temperature-sensitive (ts) alleles. Upshifts and downshifts between permissive [blue] and restrictive [red] temperatures allow control of protein function at defined stages in the cell cycle. Each line indicates an experiment with an individual embryo at the temperature indicated [red, restrictive; blue, permissive], as it progresses through various cell stages. (B) Upshifting a ts mutant embryo to restrictive temperature immediately before the protein's required function will result in a full loss-of-function (null-like) phenotype if it is fast-acting. Further, a temporary upshift (~3 min) prior to the activity requirement will not cause division failure, if the ts mutant is reversible. (C) Upshifting ts mutant embryos to restrictive temperature at different time points during cell division will indicate when that protein's activity is required during the cell cycle for cytokinesis. Cytokinesis will complete following upshift when the protein function is no longer required. (D) The precise window in which the activity of a specific cytokinesis protein is required can be further determined via temporary upshifts.

Multiaqua #MHWX-24-H-1). We monitor the temperature by averaging the reading of three digital thermometers, directly affixed to the objective using a glue gun and track three or more thermometers spaced throughout the room. To facilitate in vivo analysis of cytokinesis, we use strains that express fluorescent markers, both on their own (controls) and in addition to the desired ts mutation. Typically, we use markers for the chromatin (Histone2B:mCherry (Audhya, Desai, & Oegema, 2007)) and plasma membrane (Phospholipase-C∂^{PH}:eGFP (Audhya et al., 2005)) to monitor the underlying cell cycle and cell shape changes during division, but other fluorescent markers[1] and/or transmitted light can be used as desired.

The precise microscopy setup will vary depending on the application and budget. For simple environmental heat transfer, either an inverted or upright microscope stand may be used. However, the currently available thermally precise fluidic control systems require an inverted microscope stand (see below). We use the following microscope system:

- Nikon Ti inverted microscope with either a 20× 0.75 NA. Plan Apo dry, a 40× 0.95 NA. Plan Apo dry, a 40× 1.3 NA. Plan Fluor oil, or a 60× 1.4 NA. Plan Apochromat oil objective
- Yokagawa CSU-10 spinning disc confocal upgraded with Borealis (Spectral Applied Research)
- Hamamatsu Orca ER camera
- Piezodriven motorized stage (Applied Scientific Instrumentation)
- Perfect Focus auto focus system (Nikon)
- Acousto-optic tunable filter to select the excitation light of two 150 mW lasers for excitation at 488 and 561 nm for eGFP and mCherry, respectively (Spectral Applied Research)
- A filter-wheel for emission wavelength selection (Sutter Instruments)
- Metamorph software (Molecular Devices) to control the microscope components, allowing use of perfect focus, z-sectioning, time-lapse, and multiple stage positions

1.2 WORM CULTURE AND DISSECTION

To obtain healthy single-cell embryos, it is important to ensure a supply of young adults that contain early embryos by chunking the required strains about 2–3 days prior to imaging (older adults will mainly have late-stage embryos). Worms containing ts mutations (and control strains containing the markers alone) should be maintained using standard *C. elegans* husbandry techniques (Brenner, 1974; Riddle, 1997), but grown at permissive temperature to allow growth and propagation of the ts mutant strains. Immediately prior to observation, embryos of the appropriate stage are dissected out of the worm using the following protocol (Fig. 4):

1. Under a dissecting microscope with transmitted illumination in the base (eg, Olympus model SXZ16), pick several young adult hermaphrodites (containing a clear single row of embryos) into a watch glass (Carolina Biological Supply

---

[1]Note: Many transgenically expressed fluorescent markers are transcriptionally silenced at low temperatures in *C. elegans*. Thus, marker-expressing strains should first be tested for low-temperature silencing by raising them at permissive temperature for several generations. For quantitative analysis of fluorescence intensity, control strains (non-ts mutant) should be similarly maintained at permissive temperature (usually 14–16°C) in parallel for comparison to ts mutant strains.

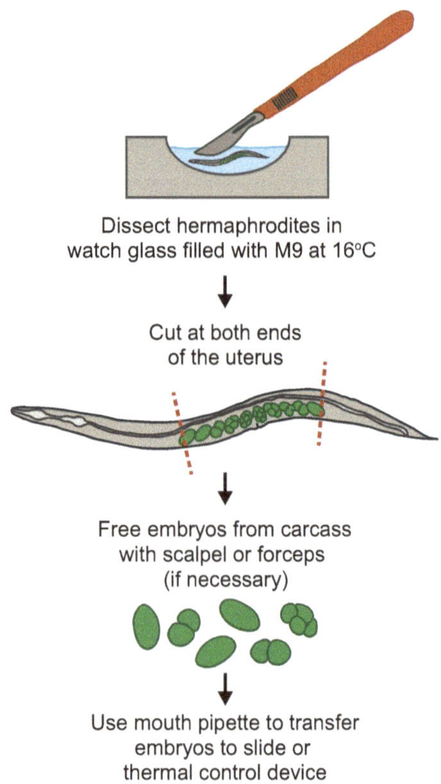

FIGURE 4 Isolating early *Caenorhabditis elegans* embryos for imaging on a light microscope.

Young adults are picked into cold M9 (16°C) and dissected to release embryos by cutting near both ends of the uterus with a scalpel and #15 blade. Embryos of the desired stage can be grouped and transferred to a slide or thermally precise fluidic control system using a mouth pipette.

Company, #742300) containing M9 buffer [85 mM NaCl (Research Products International, #523020), 42 mM Na_2HPO_4 (Research Products International, #523100), 22 mM KH_2PO_4 (Fisher Scientific, #BP362), 1 mM $MgSO_4 \cdot 7H_2O$ (Fisher Scientific, #M63)]. To ensure that the embryos remain at the permissive temperature during dissection, all equipment, M9 buffer, and worm plates should be kept at 16°C prior to use. A small electric wine cooler (Wine Enthusiast #272 02 13) set to 60°F (~16°C) is an inexpensive way to keep worms/watch glass/buffers at the permissive temperature in the microscope room for short periods of time.

a. Alternatively, worms can be directly dissected on the imaging coverslip in a small volume of 16°C M9 (~2–3 μL).

2. To dissect out the embryos, use a #15 scalpel (eg, Bard-Parker #371615) to cut the worm into thirds near both ends of the uterus (red dashed lines, Fig. 4). If multiple embryos are required, dissect up to 10 (or more) worms.
   a. Movement of the worm will eject many embryos from the carcass, usually making further dissection unnecessary, but if embryos remain in the carcass, they can be removed with a scalpel or forceps.
3. Use a mouth pipette to aspirate the collected embryo(s) at the desired developmental stage and transfer to the slide, coverslip, or imaging chamber as described below.
   a. To make the mouth pipette, heat the end of a disposable glass Pasteur pipette (eg, VWR #14673-043) with a flame, and draw the tip out manually, creating a fine ending with ~100 μm diameter. If the diameter of the initial tip is too narrow, the ending can be cut at a wider point. Insert the other (wide) end into a mouth aspirator tube (Sigma Aldrich #A5177).
   b. The shape of the mouth pipette will heavily affect the amount of buffer aspirated with the embryo. Too much buffer can cause the embryos to move, while too little buffer may allow the sample to dry out and/or incorporate air bubbles. With practice, the optimal amount of buffer (and therefore optimal pipette shape) will become clear. Your tongue can be used on the mouth aspirator to minimize suction and control the stop and start of fluid flow.

2. USING FAST-ACTING TEMPERATURE SENSITIVE MUTANT EMBRYOS TO STUDY CYTOKINESIS

Once isolated as described earlier, embryos undergoing cell division can be imaged either: (1) on a standard microscopy slide rapidly shifted to a consistent environmental temperature; or (2) on a fluidic control system that allows precisely timed active temperature shifts. The technique used depends on the type of experiment planned, and each has its own strengths and budget.

2.1 DETERMINING THE PERMISSIVE AND RESTRICTIVE TEMPERATURES FOR A TEMPERATURE-SENSITIVE MUTANT

Throughout this methods paper, we refer to permissive temperature as 16°C and restrictive temperature as 26°C; however, this should be individually determined for each mutant allele. At permissive temperature, the ts mutant strain should show no (or very minor defects) relative to control embryos at the same temperature. This depends on the individual screen and allele, but usually falls within 14–16°C. At restrictive temperature, the ts mutant should show a full loss-of-function phenotype (usually 24–26°C). However, restrictive temperature should not be above 26.5°C, as even control embryos lacking any ts mutations start to show spindle defects at, or above, 27°C—likely due to thermally induced changes in microtubule dynamics.

2.2 ENVIRONMENTAL TEMPERATURE REGULATION

Environmental temperature regulation is relatively inexpensive and can be as simple as moving the slide from an incubator set at permissive temperature to a room or stage incubator set to restrictive temperature. This has the advantage of requiring little specialized equipment but depends on the ability to control the temperature of the room and/or microscope. Moreover, the ability to time the temperature upshift is limited due to the time required to mount the embryos on the microscope, and the sample cannot be imaged during the upshift. If the environmental approach is used, the embryos are mounted on a standard microscope slide (Fig. 5) and kept at 16°C on an ice bucket covered in bubble wrap (Fig. 6) or in an incubator (or electric wine cooler) prior to shifting to the desired temperature by thermal equilibration with the room and microscope set to the desired temperature. Here, we outline two different procedures for mounting

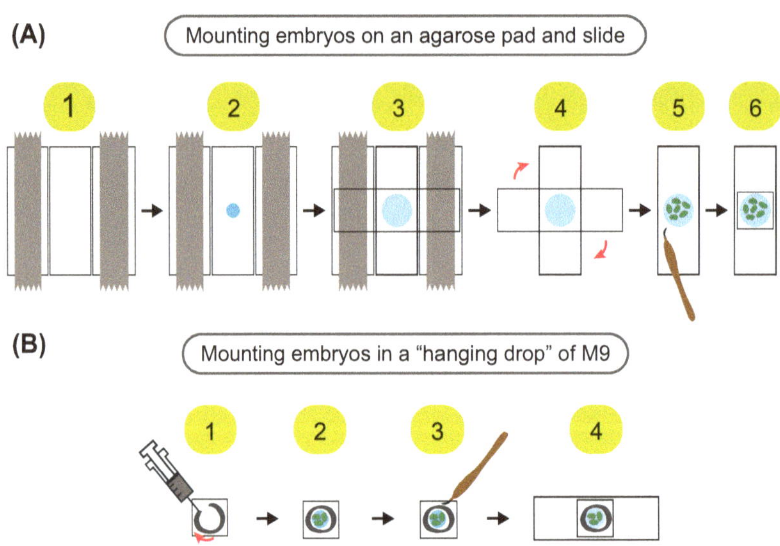

FIGURE 5 Mounting embryos on a slide for environmental temperature upshifts.

(A) To use the agarose pad method, first align three slides side by side and tape down the outer two slides. Next, add a small drop of 2% agarose and cover with a perpendicularly placed slide to make the agarose pad. Once cooled, twist slides relative to each other to expose the pad, add embryos, and top with a coverslip. (B) To mount embryos with the hanging drop method, use a syringe to make a ring of Vaseline on the coverslip. Next, use a mouth pipette to transfer early embryos, cluster embryos with an eyelash tool, and carefully cover with a slide.

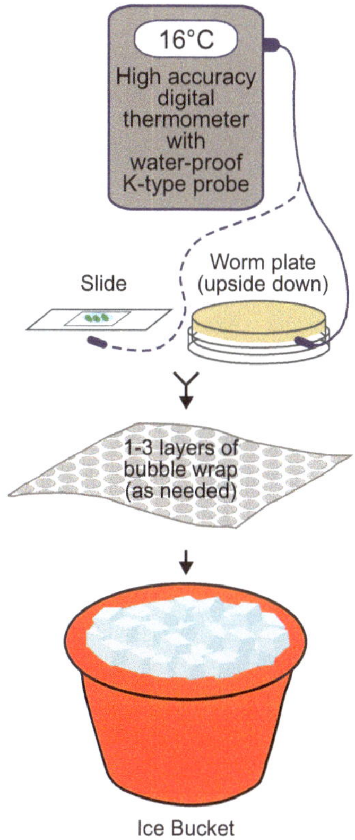

FIGURE 6 An inexpensive method to maintain samples at 16°C.

Worm plates and/or slide-mounted embryos can be maintained at 16°C by placing them on top of an ice bucket covered with bubble wrap, separated from the ice. It is important to monitor the temperature by placing the probe of a digital thermometer as close to the samples as possible on top of the bubble wrap. Depending on the specific bubble wrap, the number of layers may vary, but once you find the correct thickness, this bubble wrap can be saved for future experiments.

embryos on slides for environmental heat transfer, each having its own advantages and disadvantages.

2.2.1 Slide mounting

Traditionally, *C. elegans* embryos have been imaged on standard microscope slides with a 2% agarose pad and a coverslip (Fig. 5A). This has the advantage of being easy to assemble and requiring no special equipment. Moreover, it is possible—with the use of an eyelash tool or ultrafine deer hair—to group embryos together, allowing simultaneous imaging of multiple embryos. Additionally, the process of sandwiching the embryo between the coverslip and the agarose pad results in a slight compression of the embryos, which can improve imaging of fluorescently tagged proteins. However, the geometric compression can rescue mild polarity defects (Tsou, Ku, Hayashi, & Rose, 2003) and may have unexpected consequences on other cell processes. Furthermore, cellular processes that occur perpendicular to the coverslip (such as the division of the ABa and ABp cells in the four cell stage embryo) are difficult to image under compression conditions, as the embryos tend to divide in a more uniform orientation.

An alternate method termed "hanging drop" traps the embryos in a drop of M9 buffer held between the slide and coverslip with a Vaseline petroleum jelly spacer (Fig. 5B). This prevents compression and results in the embryos being more randomly oriented relative to the plane of the slide. This inconsistency in orientation ensures, at least in a subset of the embryos, that imaging occurs at the optimum viewing angle for most cells during embryonic development.

2.2.1.1 Agarose pad method

1. Align three microscope slides (VWR #16004-430) next to each other side by side. Tape down the two outer slides with two pieces of lab tape per slide (VWR #89098), resulting in a double layer of tape acting as a spacer as shown in Fig. 5A.
2. Place a small amount (~25 μL) of molten 2% agarose onto a glass slide using a disposable plastic transfer pipette.
   a. 2% agarose: Dissolve 2 g agarose (Calbiochem #2125) in 100 mL of ddH_20 by heating in microwave. Aliquot into 1.5 mL microcentrifuge tubes for daily use with a heat block near the dissection scope.
3. Immediately place a second slide on top, oriented perpendicularly to the other slides, and allow the agarose gel to set (~20 s).
4. When the agarose has set, carefully separate the two glass slides by sliding them apart; one of the slides should maintain the agarose pad.
5. Transfer the embryos to the pad using a mouth pipette as described earlier. If imaging multiple embryos, carefully cluster embryos together on the coverslip or agarose pad using an eyelash or hair tool.
   a. Eyelash tools are made by attaching a fine eyelash to a toothpick with a small amount of nail polish, or an ultrafine deer hair tool can be purchased (Ted Pella #119).
   b. The appropriate size of each cluster will depend on the magnification used, as increasing the magnification will decrease the number of embryos possible to fit in the field of view. Additional clusters of embryos on the same slide can be used if the imaging parameters allow multiple stage positions. Care must be taken to regulate the volume of buffer used to transfer the embryos as transferring too much buffer will make it difficult to cluster the embryos together.
6. Gently place a coverslip (No. 1.5, 18 × 18 mm, Corning #2850-22) on top of the pad. Lowering the coverslip at an angle, so that one side goes down before the other, will help to prevent air bubbles from being trapped underneath.

a. Invert the slide and locate the embryos using the dissecting microscope. Using a lab marker or sharpie (eg, VWR #52877), carefully draw a circle around the embryos on the slide. It is important not to mark the coverslip side, especially when using an oil objective, as the marker ink could destroy the optics. Hold the slide at an angle on the dissecting scope to prevent squishing the coverslip while marking the position of the embryos on the slide.

2.2.1.2 Hanging drop method

1. Using a syringe and an 18 G needle (eg, BD #309604 and BD #305195), carefully deposit Vaseline petroleum jelly in a ring with approximately 10–15 mm diameter on a coverslip (No. 1.5, 18 × 18 or 22 × 22 mm) (Fig. 5B).
2. Transfer the embryos to the coverslip using a mouth pipette as described earlier.
   a. It is important to cotransfer a sufficient volume of M9 buffer (~5–10 μL) to ensure contact between drop and slide.
   b. Alternatively, worms can be directly dissected on the imaging coverslip in a small volume of M9 and then encircle the drop with a ring of Vaseline.
3. Cluster the embryos using an eyelash tool. With practice, it is possible to rotate the embryos to increase the likelihood that they are in the desired orientation.
4. Carefully place a microscope slide on top of the coverslip, such that the drop of M9 containing the embryo and the Vaseline ring make contact with the slide. In this case, marking the position of the embryos is not required, as embryos are much easier to locate within the confined drop.

2.2.2 Environmental temperature regulation for slide-mounted embryos

To increase the temperature of the slide-mounted embryo, the microscope and room (or microscope incubator) should be preequilibrated to the desired restrictive temperature (usually 26°C). Once the embryos are mounted on the microscope, the temperature of the embryo will quickly match that of the microscope especially when using an oil objective[2]. In many cases, however, the embryo(s) may not yet be at the desired developmental stage for temperature upshift. In this case, it may be useful to let the embryos continue to develop at permissive temperature prior to mounting on the heated microscope. This can be accomplished by maintaining the prepared slide at 16°C in a refrigerating incubator, an inexpensive wine cooler, or an ice bucket topped with a layer of insulating bubble wrap at a thickness to maintain the top of the wrap at permissive temperature (Fig. 6). By periodically making brief observation of the embryo(s) under a relatively high-zoom dissecting scope, the developmental progress can be observed with only momentary increases in

[2]Note: A heating collar attached to an oil immersion objection may be sufficient to maintain sample temperature at restrictive temperature. However, we do not have direct experience with the accuracy of objective collars, and collar installation often requires the manual removal of the outer objective cover provided by the manufacturer. This makes the collar difficult to use with multiple objectives or any objective with a correction collar.

temperature. Once the desired stage is reached, mount the slide on the microscope, which will cause the temperature of the embryo to increase.

An alternative approach is to dissect out the embryo and prepare the slide in a separate room or in a specific dissecting chamber that is cooled to the permissive temperature (eg, Liu, Maduzia, Shirayama, & Mello, 2010; Severson et al., 2000). This allows the embryo to be observed continually on a microscope without disrupting protein function. Once the desired cell cycle stage is reached, the slide is rapidly taken to another room containing the imaging microscope that is at the restrictive temperature. This technique has the advantage that the embryo can be continually monitored at low resolution with no fluctuation in temperature prior to moving to the imaging microscope. However, it requires the use of two thermally isolated and controllable rooms that are located close to each other and does not allow imaging during the upshift.

2.3 THERMALLY PRECISE FLUIDIC CONTROL SYSTEMS

Thermally precise fluidic control systems are more sophisticated and actively control the temperature of the sample throughout the experiment via direct heat transfer to the specimen holder. There are two commercially available devices that we have used: (1) The Therminator (Figs. 7 and 8) and (2) CherryTemp (Fig. 9), both of which allow the sample temperature to be rapidly changed while simultaneously imaging on a light microscope to monitor the effect on cytokinesis.

2.3.1 Therminator (Bioptechs, Inc.)

We worked with Bioptechs, Inc. to develop a device that we call the Therminator. This device enables rapid shifts (~17 s) from permissive to restrictive temperature while simultaneously imaging on an inverted spinning disc confocal microscope (or other light microscope) (Davies et al., 2014). The Therminator imaging unit consists of a sample chamber for embryo mounting on top of the fluid flow chamber separated by a thin layer of glass for rapid heat transfer (Fig. 8). The fluid flow chamber is constantly fed by one of two water baths, each of which can be held at defined temperatures (5–40°C but usually 16°C and 26°C). A mixture of 10% 2-propanol and water is used as the heat conducting fluid to prevent microbial contamination. A switch mechanism determines which water bath feeds the fluid flow chamber and maintains the sample temperature (Fig. 7). The system is constructed to be fully compatible with the imaging light path and with the Perfect Focus (Nikon) autofocusing system. In our hands, high-resolution DIC is not possible with this system, as the device obstructs the required condenser position for Köhler illumination; however, decent transmitted light images can be acquired. The Therminator is designed to accommodate the traditional agarose pad technique, but spacers can be added for noncompressive imaging, if required (Fig. 8A). The imaging unit of the Therminator is composed of two parts: (1) the specimen mounting and fluid flow chamber and (2) the coverslip holder (30 mm diameter, No. 1.5, Bioptechs #30-1313-0319) with a plastic extension and securing latches to hold the entire

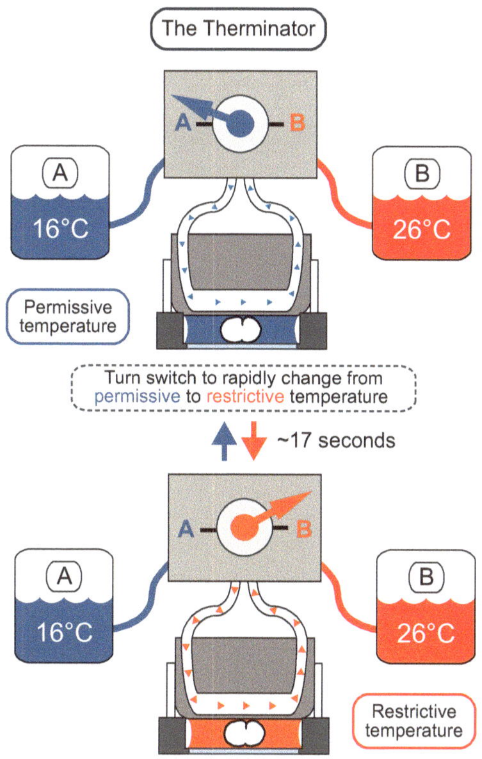

FIGURE 7 Schematic of the Therminator system.

Two water baths, A and B, feed the fluid flow chamber directly above the specimen chamber. By switching between baths A and B (each set to different temperatures), it is possible to rapidly change sample temperature in ~17 s.

imaging unit together (Fig. 8). The fluid flow chamber with mounted specimen is inserted into the coverslip holder so that the specimen is sandwiched between the agarose pad and the glass coverslip. The use of a precisely regulated temperature control system facilitates not just rapid temperature changes but also helps to stably maintain the specimen at a constant temperature with little to no interference from the room temperature.

2.3.1.1 Mounting embryos on the Therminator (Fig. 8B)

1. Prepare the 2% agarose as described earlier (Section 2.2.1.1), and add ~35 μL of molten agarose onto the imaging area of the specimen mounting and fluid flow unit.
   a. Before mounting embryos, ensure that the water bath set to the initial temperature (usually permissive temperature or 16°C) is feeding the flow chamber of the fluid flow unit.

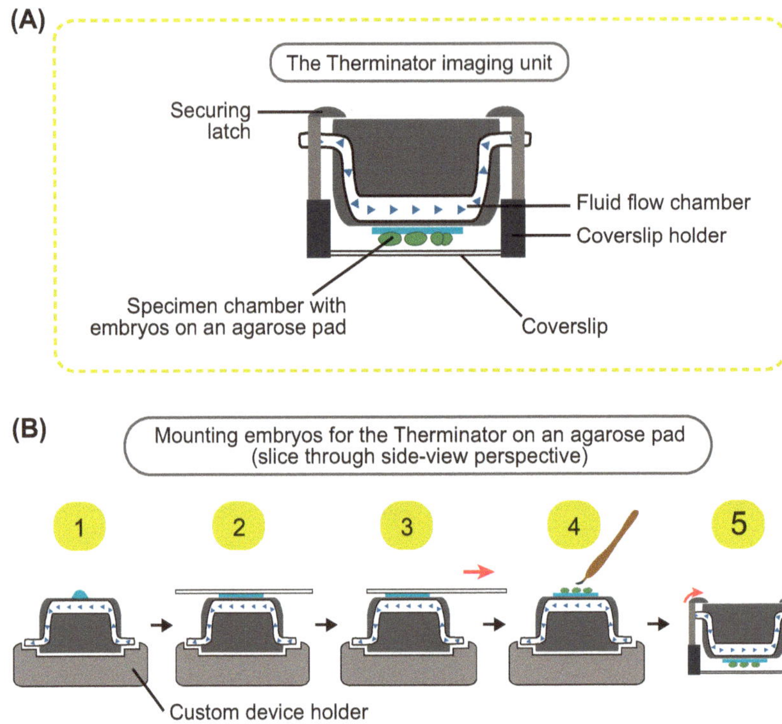

FIGURE 8 Mounting embryos on the Therminator imaging unit.

(A) Schematic diagram of the Therminator imaging unit. Note: Embryos are mounted directly adjacent to the fluid flow chamber allowing for rapid temperature change by convective heat transfer to the glass on which the embryos are mounted. (B) Schematic of protocol for mounting *Caenorhabditis elegans* embryos on an agarose pad for the Therminator. Invert imaging unit onto custom device holder to make the agarose pad. Add molten agarose and use a glass slide to flatten into a pad, allow agarose to solidify, then slide off glass slide carefully so the agarose pad remains on top of the fluid flow chamber. Use a mouth pipette to mount embryos and cluster them with an eyelash tool. Remove the fluid flow chamber from the custom device holder and attach the coverslip holder containing a round coverslip (30 mm). Turn latches to secure the entire imaging unit together.

2. Quickly place a glass slide on top to create a thin agarose pad while the agarose is still molten. Allow the agarose to solidify (~20 s).
3. Carefully slide off the glass slide, leaving the agarose pad on the imaging area of the specimen mounting unit.
4. Transfer the embryos to the agarose pad using a mouth pipette as described earlier. If imaging multiple embryos, carefully cluster embryos together using an eyelash or ultrafine hair tool.

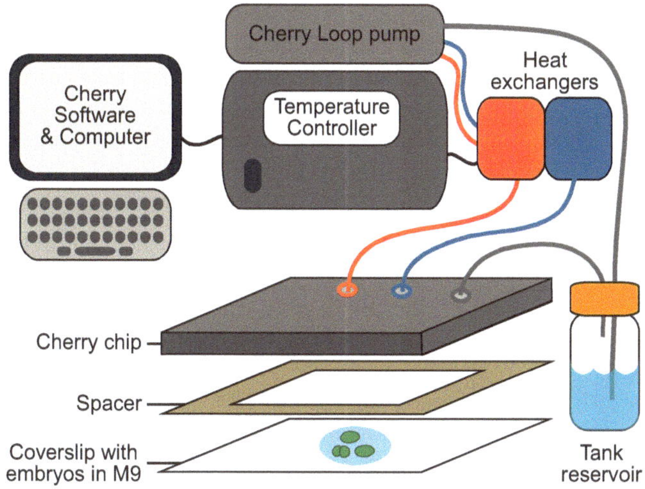

FIGURE 9 Schematic of the CherryTemp system with imaging unit.

Temperature is controlled via the Cherry software, which regulates the heat exchangers and fluid flow into the Cherry chip. Embryos are mounted on a coverslip that fits into the Cherry insert. The Cherry chip is placed on top of the coverslip with a spacer sandwiched in between to avoid compressing the embryos.

5. Prepare the imaging unit sandwich and close the latches to secure the coverslip holder in place. Care must be taken to ensure that a nice seal is established between the coverslip and the agarose pad while confirming that the embryo is not subjected to undue compressive pressure. Mount the assembled device on the microscope stage. It can be secured using stage clips designed for a 35 mm culture dish or manually using small balls of modeling clay (eg, Prang #00740).

Imaging is performed as described earlier to monitor embryonic development. When a change in temperature is experimentally desired, turn the switch to change the source water bath flowing to the fluid flow chamber above the specimen.

2.3.2 CherryTemp (Cherry Biotech; Fig. 9)

CherryTemp is a computer-assisted thermally precise microfluidic control system that maintains sample temperature with constant specimen-level thermal feedback (Fig. 9). The CherryTemp platform is comprised of the CherryTemp dual channel temperature controller box, the CherryLoop fluid pump, the Peltier heat exchangers, the Cherry insert for specimen mounting, and reusable Cherry chips. The platform is controlled through the Cherry software. Using the CherryTemp design, embryos are mounted in M9 buffer sandwiched between a coverslip and the Cherry chip, a microfluidic flow chamber fed by the two Peltier heat exchangers (the agarose pad method is not yet compatible with CherryTemp). The Cherry chip is constructed from transparent polymers that are fully compatible with microscopic imaging methods, such as transmitted light, DIC imaging, phase contrast, or any fluorescent imaging technique. Each heat exchanger can be set and maintained at a specific temperature ranging from 5°C to 45°C by the CherryTemp controller using the Cherry software. The thermalization liquid contained in a glass bottle is continuously pumped to the heat exchangers, then to the Cherry chip, and then recycled back to the bottle in a closed-loop configuration. Before using the CherryTemp platform, all parts of the system must be properly connected, the microscope stage adaptor for the Cherry insert must be installed in place of your regular stage insert, and a Cherry chip should be connected to the heat exchangers. The CherryTemp system has the added safety feature of automatic flow stop if any leak is detected in the system.

2.3.2.1 Mounting embryos on CherryTemp (Fig. 9)

1. Place a coverslip on the Cherry insert (24× 60 mm, No. 1.5). Dissect worms in a watch glass and transfer embryos on the coverslip using a mouth pipette (or dissect directly on the coverslip). Place the provided spacer on the coverslip. The spacer will prevent embryo compression by the Cherry chip and will avoid evaporation of the sample buffer.
   a. Before mounting the embryos, ensure that the temperature has equilibrated in the heat exchangers using the Cherry software and that liquid at the permissive temperature is flowing to the Cherry chip.
2. Place the Cherry insert with the coverslip on the microscope stage adaptor. Install the Cherry chip on top of the spacer so that the drop of medium that contains the embryos is sandwiched between the imaging coverslip and the Cherry chip. Lock the chip in place using the two slide clasps (not shown).
3. Using the Cherry software interface, control the temperature by activating the flow from one heat exchanger to the other (fast shift) or by entering a new set point for the currently flowing heat exchanger (slower shift) as desired throughout the experiment.

3. CHARACTERIZING TEMPERATURE SENSITIVE ALLELES

As described earlier, not all ts alleles behave similarly, as some are fast- and some are slow-acting. Before using a ts allele to investigate when specific protein functions are required during cytokinesis, it is important to check that the allele fulfills specific criteria and is truly a fast-acting ts mutant (Fig. 10). We use the following criteria to determine if a ts mutant is appropriate for rapid upshift experiments:

1. Loss-of-function phenotype
   It is important to determine whether ts mutant embryos behave like their wild-type counterparts at the permissive temperature but show thermal sensitivity and undergo a gradual transition with increasing temperature to mimic full loss of function at the restrictive temperature (eg, Fig. 2). The loss-of-function phenotype may be confirmed by another method (such as RNAi). This is

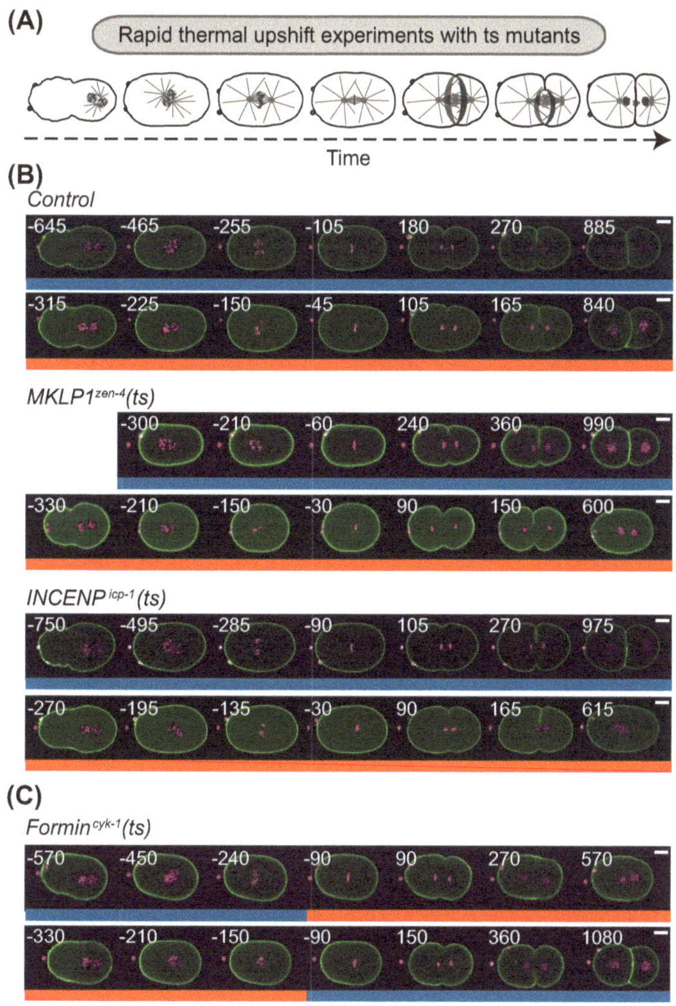

FIGURE 10 Characterizing fast-acting temperature-sensitive (ts) cytokinesis-defective *Caenorhabditis elegans* alleles.

(A) Schematic of early embryonic worm development from pronuclear migration to the end of the first mitotic division. (B) Ideal ts alleles will divide normally at the permissive temperature [16°C, blue], but show the full loss-of-function (null-like) phenotype when rapidly upshifted to 26°C [red], as seen here for *MKLP1^{zen-4}(ts)* and *INCENP^{icp-1}(ts)*. Control embryos divide normally at both permissive and restrictive temperatures. (C) The reversible nature of these ts alleles allows protein inactivation at defined times in the cycle. *Formin^{cyk-1}(ts)* mutant embryos were upshifted to the restrictive temperature either from metaphase until the end of cytokinesis, or from pronuclear meeting until metaphase at which point they were returned to permissive temperature. Inactivation of formin^{CYK-1} prior to metaphase did not disrupt cytokinesis, whereas inactivation after metaphase during cytokinesis blocked successful division. Thus, the *formin^{cyk-1}(ts)* allele is both fast-acting and reversible (See also Davies et al., 2014). Time is in seconds relative to Anaphase Onset (t = 0). Scale bars = 10 μm.

illustrated in Fig. 10B by two different alleles. INCENP$^{ICP\text{-}1}$ is a component of the chromosomal passenger complex (CPC) while MKLP1$^{ZEN\text{-}4}$ is the kinesin component of the centralspindlin complex. At the restrictive temperature, ts alleles of these proteins [respectively, *INCENP*$^{icp\text{-}1}$*(or663ts)* and *MKLP1*$^{zen\text{-}4}$*(or153ts)*] show their well characterized full loss-of-function cytokinesis failure phenotypes and, in the case of *INCENP*$^{icp\text{-}1}$*(or663ts)*, chromosomal segregation defects (Fig. 10B) (Davies et al., 2014; Severson et al., 2000). Importantly, control embryos lacking a ts mutation divide successfully at both permissive and restrictive temperature (Fig. 10B).

2. Fast-acting mutation
   Does upshifting the embryo to the restrictive temperature cause an immediate loss-of-function phenotype? Determining this depends on the gene, phenotype, and cell process being observed. In the case of cytokinesis, we define alleles as fast-acting if temperature upshift in early mitosis causes cytokinesis failure, indicating that function must be lost within minutes (Fig. 3). For example, formin$^{CYK\text{-}1}$ is required for polymerization of contractile ring actin. The *formin*$^{cyk\text{-}1}$*(or596ts)* allele is fast-acting, as upshifting the mutant embryo to the restrictive temperature at metaphase prevents the formation and ingression of the contractile ring which would normally occur just minutes later (Fig. 10C). If cytokinesis does not fail upon rapid upshift or if longer shifts are required to see a loss-of-function phenotype, then the ts mutant would not be deemed fast-acting and/or would be considered a hypomorphic allele.
3. Phenotypic reversibility
   As described previously, fast-acting ts alleles likely lead to temperature-dependant changes in protein structure rather than protein degradation or other long-term effects. Because of this, many ts alleles are reversible and multiple temperature shifts can be used to inactivate and then reactivate function throughout development (Fig. 3). This allows determination of the functional cell cycle window for a protein during cytokinesis with high temporal precision. Function can be blocked during specific cell cycle stages and then reactivated at a later stage in the division cycle. For example, using the *formin*$^{cyk\text{-}1}$*(or596ts)* allele, we found that the actin nucleating protein diaphanous formin$^{CYK\text{-}1}$ is not essential during earlier stages in mitosis prior to anaphase onset (Fig. 10C) (Davies et al., 2014). That is, *formin*$^{cyk\text{-}1}$*(or596ts)* mutant embryos can be maintained at restrictive temperature early in development without disrupting cytokinesis as long as they are returned to permissive temperature prior to anaphase onset. Upon returning these ts mutant embryos to permissive temperature, formin$^{CYK\text{-}1}$ activity is restored and the cell can divide successfully (Fig. 10C) (Davies et al., 2014).

4. CHOOSING A TEMPERATURE CONTROL METHOD

Here we have described current methods to control the temperature of *C. elegans* embryos, allowing acute conditional inactivation of fast-acting ts alleles. Both the

Table 1 Table Comparing the Feature of Different Temperature Control Methods

Upshift Method	Passive Heat Transfer	Therminator	CherryTemp
Temporal resolution of thermal shift	~60 s	~17 s	~7 s
Simultaneous imaging and thermal shift	No	Yes	Yes
Rapid reversibility	No	Yes	Yes
Cost in USD (excluding microscope)	<$1000	~$15000	~$25000
Supplier	–	Bioptechs, Inc. Butler, PA	Cherry Biotech Rennes, France

environmental and the thermally precise fluidic control methods of temperature regulation have distinct advantages and disadvantages, as outlined below. The key advantage of a thermally precise fluidic control system is the ability to simultaneously image the embryo while precisely modulating the protein function with temperature as well as the rapidity and reversibility of temperature changes. This has allowed us to determine the temporal requirements for key cytokinetic proteins with high resolution (Davies et al., 2014). Furthermore, the ability to simultaneously monitor the cell using a light microscope while abrogating protein function allows continual measurement of cellular parameters (such as chromosome position or contractile ring diameter) in a way that would not be possible if switching between a "cool" and "warm" microscope. Additionally, the ability to shift the temperature of the sample multiple times in a single experiment exploits the reversibility of protein inactivation at defined periods of time during the cell cycle, enabling determination of the functional window of activity required for cytokinesis (Fig. 3D) (Davies et al., 2014). However, the financial investment in a device such as the Therminator or CherryTemp system for sample temperature regulation is much greater than that required for environmental temperature regulation (Table 1).

CONCLUSIONS

In this review, we have outlined ways to use fast-acting ts mutants to study cell division in *C. elegans*. The complexity and cost of the approaches outlined earlier vary considerably; however, each allows rapid inactivation of ts mutants for any lab interested in studying cytokinesis (or other complex cellular event). Although environmental temperature change methods have been used successfully (eg, Liu et al., 2010; Severson et al., 2000) and are undoubtedly useful and inexpensive, the ability to reversibly switch temperatures while simultaneously imaging provides a key advantage. For this reason, thermally precise fluidic control systems are likely to be important in the future, both in further study of cytokinesis, but also in examining other dynamic cell processes or developmental events where rapid control of protein function is required. The current limiting factor in using this technology is the availability of fast-acting ts alleles. Previous forward genetic screens for ts alleles have provided a range of useful fast-acting ts mutant strains (Canman et al., 2008; Encalada et al., 2000; Kemphues et al., 1988; O'Connell et al., 1998; O'Rourke, Yochem, et al., 2011; Raich et al., 1998). With the advent of genome engineering techniques such as CRISPR, it now seems possible to design ts alleles of desired proteins. This is complicated by a lack of understanding of the mechanism(s) by which ts mutations give rise to a ts phenotype. However, many ts alleles contain mutations in amino acid residues conserved within protein families and between species, which could guide the design of ts alleles.

The approaches described previously allow dissection of cell processes with high temporal resolution. However, they are limited by lack of spatial specificity, as the whole embryo is subjected to the same temperature changes. Future work will combine ts alleles with techniques that control temperature changes at a subcellular level.

ACKNOWLEDGMENTS

We are grateful to all members of the Canman lab for helpful discussions and feedback on using fast-acting ts mutants to study cytokinesis in *Caenorhabditis elegans*. We thank Sophia Hirsch and Yelena Zhuravlev for critical comments on the chapter. We also thank the *C. elegans* Genetics Center and Bruce Bowerman for worm strains. This work was supported by a postdoctoral fellowship from the Charles H. Revson Foundation (TD); ANR-09-RPDOC-005-01, FRM-AJE201112, and the Emergence Program from Mairie de Paris (JD); NIH-R01-GM105775 and NIH-R01-AG045842 (MSH); and NIH-DP2-OD008773 to (JCC).

REFERENCES

Audhya, A., Desai, A., & Oegema, K. (2007). A role for Rab5 in structuring the endoplasmic reticulum. *Journal of Cell Biology, 178*(1), 43–56. http://dx.doi.org/10.1083/jcb.200701139.

Audhya, A., Hyndman, F., McLeod, I. X., Maddox, A. S., Yates, J. R., 3rd, Desai, A., & Oegema, K. (2005). A complex containing the Sm protein CAR-1 and the RNA helicase CGH-1 is required for embryonic cytokinesis in *Caenorhabditis elegans*. *Journal of Cell Biology, 171*(2), 267–279. http://dx.doi.org/10.1083/jcb.200506124.

Bement, W. M., Benink, H. A., & von Dassow, G. (2005). A microtubule-dependent zone of active RhoA during cleavage plane specification. *Journal of Cell Biology, 170*(1), 91–101. http://dx.doi.org/10.1083/jcb.200501131.

Brenner, S. (1974). The genetics of *Caenorhabditis elegans*. *Genetics, 77*(1), 71–94.

Canman, J. C., Cameron, L. A., Maddox, P. S., Straight, A., Tirnauer, J. S., Mitchison, T. J., … Salmon, E. D. (2003). Determining the position of the cell division plane. *Nature, 424*(6952), 1074–1078. http://dx.doi.org/10.1038/nature01860.

Canman, J. C., Lewellyn, L., Laband, K., Smerdon, S. J., Desai, A., Bowerman, B., & Oegema, K. (2008). Inhibition of Rac by the GAP activity of centralspindlin is essential for cytokinesis. *Science, 322*(5907), 1543–1546. http://dx.doi.org/10.1126/science.1163086.

Canman, J. C., Salmon, E. D., & Fang, G. (2002). Inducing precocious anaphase in cultured mammalian cells. *Cell Motility and the Cytoskeleton, 52*(2), 61–65. http://dx.doi.org/10.1002/cm.10032.

Carvalho, A., Olson, S. K., Gutierrez, E., Zhang, K., Noble, L. B., Zanin, E., … Oegema, K. (2011). Acute drug treatment in the early *C. elegans* embryo. *PLoS One, 6*(9), e24656. http://dx.doi.org/10.1371/journal.pone.0024656.

Davies, T., Jordan, S. N., Chand, V., Sees, J. A., Laband, K., Carvalho, A. X., … Canman, J. C. (2014). High-resolution temporal analysis reveals a functional timeline for the molecular regulation of cytokinesis. *Developmental Cell, 30*(2), 209–223. http://dx.doi.org/10.1016/j.devcel.2014.05.009.

D'Avino, P. P., Giansanti, M. G., & Petronczki, M. (2015). Cytokinesis in animal cells. *Cold Spring Harbor Perspectives in Biology, 7*(4), a015834. http://dx.doi.org/10.1101/cshperspect.a015834.

Encalada, S. E., Martin, P. R., Phillips, J. B., Lyczak, R., Hamill, D. R., Swan, K. A., & Bowerman, B. (2000). DNA replication defects delay cell division and disrupt cell polarity in early *Caenorhabditis elegans* embryos. *Developmental Biology, 228*(2), 225–238. http://dx.doi.org/10.1006/dbio.2000.9965.

Green, R. A., Paluch, E., & Oegema, K. (2012). Cytokinesis in animal cells. *Annual Review of Cell and Developmental Biology, 28*, 29–58. http://dx.doi.org/10.1146/annurev-cellbio-101011-155718.

Jordan, S. N., & Canman, J. C. (2012). Rho GTPases in animal cell cytokinesis: an occupation by the one percent. *Cytoskeleton, 69*(11), 919–930. http://dx.doi.org/10.1002/cm.21071.

Jordan, S. N., Davies, T., Zhuravlev, Y., Dumont, J., Shirasu-Hiza, M., & Canman, J. C. (2016). Cortical PAR polarity proteins promote robust cytokinesis during asymmetric cell division. *Journal of Cell Biology, 212*(1), 39–49. http://dx.doi.org/10.1083/jcb.201510063.

Kemphues, K. J., Priess, J. R., Morton, D. G., & Cheng, N. S. (1988). Identification of genes required for cytoplasmic localization in early *C. elegans* embryos. *Cell, 52*(3), 311–320.

Kurz, T., Pintard, L., Willis, J. H., Hamill, D. R., Gonczy, P., Peter, M., & Bowerman, B. (2002). Cytoskeletal regulation by the Nedd8 ubiquitin-like protein modification pathway. *Science, 295*(5558), 1294–1298. http://dx.doi.org/10.1126/science.1067765.

Lacroix, B., & Maddox, A. S. (2012). Cytokinesis, ploidy and aneuploidy. *Journal of Pathology, 226*(2), 338–351. http://dx.doi.org/10.1002/path.3013.

Lewellyn, L., Carvalho, A., Desai, A., Maddox, A. S., & Oegema, K. (2011). The chromosomal passenger complex and centralspindlin independently contribute to contractile ring assembly. *Journal of Cell Biology, 193*(1), 155–169. http://dx.doi.org/10.1083/jcb.201008138.

Liu, J., Maduzia, L. L., Shirayama, M., & Mello, C. C. (2010). NMY-2 maintains cellular asymmetry and cell boundaries, and promotes a SRC-dependent asymmetric cell division. *Developmental Biology, 339*(2), 366–373. http://dx.doi.org/10.1016/j.ydbio.2009.12.041.

Mabuchi, I., & Okuno, M. (1977). The effect of myosin antibody on the division of starfish blastomeres. *Journal of Cell Biology, 74*(1), 251–263.

Mierzwa, B., & Gerlich, D. W. (2014). Cytokinetic abscission: molecular mechanisms and temporal control. *Developmental Cell, 31*(5), 525–538. http://dx.doi.org/10.1016/j.devcel.2014.11.006.

Normand, G., & King, R. W. (2010). Understanding cytokinesis failure. *Advances in Experimental Medicine and Biology, 676*, 27–55.

O'Connell, K. F., Leys, C. M., & White, J. G. (1998). A genetic screen for temperature-sensitive cell-division mutants of *Caenorhabditis elegans*. *Genetics, 149*(3), 1303–1321.

Oegema, K., Savoian, M. S., Mitchison, T. J., & Field, C. M. (2000). Functional analysis of a human homologue of the *Drosophila* actin binding protein anillin suggests a role in cytokinesis. *Journal of Cell Biology, 150*(3), 539–552.

O'Rourke, S. M., Carter, C., Carter, L., Christensen, S. N., Jones, M. P., Nash, B., … Bowerman, B. (2011). A survey of new temperature-sensitive, embryonic-lethal mutations in *C. elegans*: 24 alleles of thirteen genes. *PLoS One, 6*(3), e16644. http://dx.doi.org/10.1371/journal.pone.0016644.

O'Rourke, S. M., Yochem, J., Connolly, A. A., Price, M. H., Carter, L., Lowry, J. B., … Bowerman, B. (2011). Rapid mapping and identification of mutations in *Caenorhabditis elegans* by restriction site-associated DNA mapping and genomic interval pull-down sequencing. *Genetics, 189*(3), 767–778. http://dx.doi.org/10.1534/genetics.111.134031.

Potapova, T. A., Daum, J. R., Pittman, B. D., Hudson, J. R., Jones, T. N., Satinover, D. L., … Gorbsky, G. J. (2006). The reversibility of mitotic exit in vertebrate cells. *Nature, 440*(7086), 954–958. http://dx.doi.org/10.1038/nature04652.

Raich, W. B., Moran, A. N., Rothman, J. H., & Hardin, J. (1998). Cytokinesis and midzone microtubule organization in *Caenorhabditis elegans* require the kinesin-like protein ZEN-4. *Molecular Biology of the Cell, 9*(8), 2037–2049.

Rappaport, R. (1996). *Cytokinesis in animal cells*. Cambridge, New York, NY, USA: Cambridge University Press.

Riddle, D. L. (1997). C. elegans *II*. Plainview, NY: Cold Spring Harbor Laboratory Press.

Sagona, A. P., & Stenmark, H. (2010). Cytokinesis and cancer. *FEBS Letters, 584*(12), 2652–2661. http://dx.doi.org/10.1016/j.febslet.2010.03.044.

Severson, A. F., Hamill, D. R., Carter, J. C., Schumacher, J., & Bowerman, B. (2000). The aurora-related kinase AIR-2 recruits ZEN-4/CeMKLP1 to the mitotic spindle at metaphase and is required for cytokinesis. *Current Biology, 10*(19), 1162–1171.

Tsou, M. F., Ku, W., Hayashi, A., & Rose, L. S. (2003). PAR-dependent and geometry-dependent mechanisms of spindle positioning. *Journal of Cell Biology, 160*(6), 845–855. http://dx.doi.org/10.1083/jcb.200209079.

CHAPTER

Cytokinesis from nanometers to micrometers and microseconds to minutes

18

P. Kothari, E.S. Schiffhauer, D.N. Robinson[1]

Johns Hopkins University, Baltimore, MD, United States

[1]*Corresponding author: E-mail: dnr@jhmi.edu*

CHAPTER OUTLINE

Methods in Cell Biology, Volume 137, ISSN 0091-679X, http://dx.doi.org/10.1016/bs.mcb.2016.03.038

Abstract

Cytokinesis, a model cell shape change event, is controlled by an integrated system that coordinates the mitotic spindle signals with a mechanoresponsive cytoskeletal network that drives contractility and furrow ingression. Quantitative methods that measure cell mechanics, mechanoresponse (mechanical stress-induced protein accumulation), protein dynamics, and molecular interactions are necessary to provide insight into both the mechanical and biochemical components involved in cytokinesis and cell shape regulation. Micropipette aspiration, fluorescence correlation and cross-correlation spectroscopy, and fluorescence recovery after photobleaching are valuable methods for measuring cell mechanics and protein dynamics in vivo that occur on nanometer to micron length-scales, and microsecond to minute timescales. Collectively, these methods provide the ability to quantify the molecular interactions that control the cell's ability to change shape and undergo cytokinesis.

INTRODUCTION

Cytokinesis is a mechanical and biochemical process, studied since the development of the first microscopes. Because of the dramatic nature of the cell shape change, the condensation of chromosomes, their movement along microtubules, and the pinching of the membrane, the process of cell division has garnered attention from the very earliest phases of cell biology. It was not until the 1940s when Crick recognized that the cytoplasm is thixotropic (viscoelastic) in nature (Crick & Hughes, 1950). This realization facilitated researchers, including Rappaport and Hiramoto, to appreciate that cytokinesis is a force-generating process (Hiramoto, 1963; Rappaport, 1967). The resulting, profound implication is that chemical signaling pathways do not act in isolation to promote furrow contraction; the cytokinetic machinery also requires a mechanosensory and mechanotransducer (Effler, Iglesias, & Robinson, 2007; Srivastava, Iglesias, & Robinson, 2015). In fact, Newton's third law dictates that the force-generating machinery must be both sensitive and responsive to the resulting resistive stresses. Such stresses impact the assembly, contraction, remodeling, and ultimately completion of cytokinesis. Although the field has identified many of the components involved in cytokinesis, much remains to be uncovered about how these components assemble and how their dynamics impact furrow ingression. Describing this network of cytoskeletal and signaling proteins is key to understanding how the cytokinetic machine is built, tuned, and maintained under different environmental stresses and in various genetic backgrounds. Furthermore, cytokinesis is an ideal model process for understanding the general fundamental principles of cell shape change events, including contractility, adhesion, and the effect of Laplace pressures. Thus, studying this single process can help unravel the mechanisms behind cell shape changes in other systems and higher order structures.

Although transillumination microscopy has provided us with a wealth of information regarding dividing cells, this approach fails to capture the spatial and temporal details of this complex mechanical process. The purpose of this review is to describe valuable quantitative techniques that are currently being used to study cytokinesis at the molecular level. While cytokinesis occurs over the timescale of minutes, the process is driven by associations, interactions, and forces that occur on a millisecond-to-second timescale. During cytokinesis, the cell's ability to respond to various external perturbations is dependent upon transient and dynamic interactions that allow for rapid molecular-level responses. Studying these molecular interactions using high spatial and temporal resolution is critical for developing a thorough understanding of the underlying molecular interactions driving cytokinesis. The methods described here are applicable to single-cell systems, and have been used with the social amoeba *Dictyostelium discoideum* (Kee et al., 2012; Srivastava & Robinson, 2015), *Drosophila melanogaster* S2 cells (Kim et al., 2015), as well as mammalian cells (Schiffhauer et al., 2016; Sun et al., 2014).

1. MICROPIPETTE ASPIRATION

1.1 OVERVIEW

During cytokinesis, the division machinery both generates and responds to evolving force vectors in the cell cortex. Therefore, it is essential to observe the cellular response to imposed forces in the physiological range. Methods for applying forces to cells include atomic force microscopy (Matzke, Jacobson, & Radmacher, 2001), optical traps (Ashkin, Dzeidzic, & Yamane, 1987), magnetic beads (Mijailovich, Kojic, Zivkovic, Fabry, & Fredberg, 2002), global cell compression (Kee et al., 2012; Srivastava & Robinson, 2015), and real-time deformability cytometry (RT-DC) (Otto et al., 2015). Although each of these techniques has been useful for addressing specific questions about biochemical and mechanical properties of cells, they often probe smaller or larger force and spatial regimes, with variable accuracy for quantifying the magnitude, direction, and spatial distribution of the force imposed. Micropipette aspiration (MPA) is a powerful system for applying localized forces in the nanoNewton (nN) range to individual cells, closely mimicking the localized forces cells generated and experienced at the cleavage furrow (Gerald, Dai, Ting-Beall, & De Lozanne, 1998; Hiramoto, 1963, 1990; Hochmuth, 2000; Rappaport, 1967; Reichl et al., 2008; Wolpert, 1966; Zhang & Robinson, 2005).

MPA can be used for two major purposes: (1) to measure the intrinsic mechanical properties of cells and (2) to study the accumulation of elements of the cytokinetic machinery in response to applied force, separating force-dependent protein behavior from chemical signals emanating from the mitotic spindle in mitosis.

1.2 CELL MECHANICS

To measure the intrinsic properties of cells, one must consider the cell as a viscoelastic material with both viscous and elastic components. The mechanical properties of a cell in suspension are largely dominated by the cortical cytoskeleton with some contribution from the cytoplasm. Less than 5% of the cell's elasticity is borne by the plasma membrane (Luo, Srivastava, Ren, & Robinson, 2014). If aspirated upon by a fixed pressure into a pipette with radius R_p, a cell will deform into the pipette (length into pipette, L_p, Fig. 1A), until tension in the cortex of the cell, T_c, opposes the force applied. If the cell were perfectly elastic, it would immediately extend to this distance and go no further. By measuring L_p over time on a cell, one can observe that a cell gradually extends into the pipette before stabilizing,

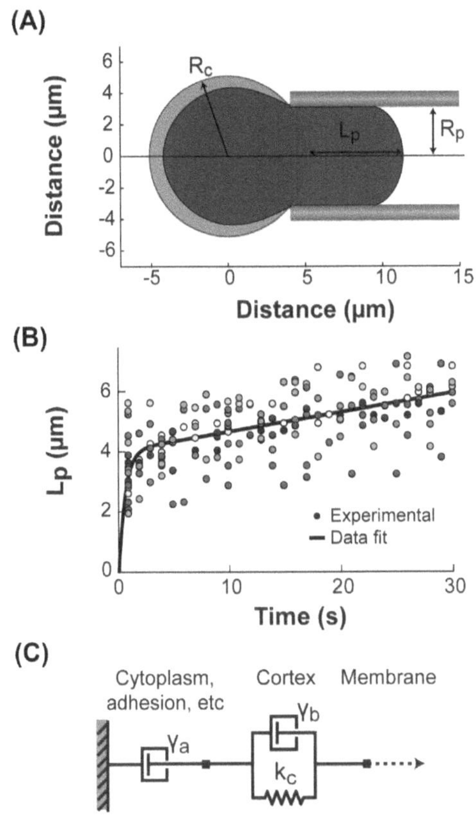

FIGURE 1 Micropipette aspiration to measure cell mechanics.

(A) A schematic of measurements made on a cell aspirated into a pipette where L_p is the distance the cell deforms into the pipette, R_p is the radius of the pipette, and R_c is the radius of the cell. (B) A measurement of a number of *Dictyostelium* cells deforming into the pipette over time. (C) Model of cell mechanics, where γ_b is determined by the initial rate of deformation, γ_a is determined by the slope of the second flow, and k_c defines the elastic component of the cell cortex.

Adapted from Yang, L., Effler, J. C., Kutscher, B. L., Sullivan, S. P., Robinson, D. N., & Iglesias, P. A. (2008). Modeling cellular deformations using the level set formalism. Biomedcentral Systems Biology, 2, *68–84.*

and at pressures greater than critical pressure, P_c, (see later discussion), the cell continues to flow into the pipette slowly (Fig. 1B). The slope of these two phases is determined by the cell's viscous component. We can describe then, the mechanics of a cell using the diagram in Fig. 1C, where the slope of the initial deformation phase is determined by the parallel damper, γ_b, and the slope of the second flow phase, by the damper in series, γ_a (Yang et al., 2008).

A common method for comparing the elasticity of two cells, or of a cell in different phases of the cell cycle, is the measurement of cortical tension (T_c). To get an accurate reading of tension, one must determine the pressure at which the cell is deformed to $L_p = R_p$, or the critical pressure, P_c. It is essential to allow sufficient time for the full relaxation of the parallel damper and to not apply so much pressure as to enter the regime of the damper in series. Then, Eq. (1) can be used to determine T_c:

$$\Delta P_c = 2T_c\left(\frac{1}{R_p} - \frac{1}{R_c}\right) \tag{1}$$

where P_c is the critical pressure, T_c is the cortical tension, R_p is the radius of the pipette, and R_c is the radius of the cell.

1.3 PROTEIN MECHANOACCUMULATION

Similar to using MPA to study cell mechanics, MPA can be used to study the mechanoresponsiveness or accumulation of proteins in response to mechanical stress (force/area). However, the pressure imposed varies by cell type and ranges from 0.5 to 3 nN/μm^2 (Luo, Mohan, Iglesias, & Robinson, 2013; Schiffhauer et al., 2016; Zhang & Robinson, 2005). The accumulation dynamics of proteins to the area of applied stress can be determined by fluorescently tagging a protein of interest and quantifying the ratio of intensity in the pipette to that in other regions of the cell. Two distinct types of deformations are observed in the cortex during MPA. The cortical cytoskeletal network dilates at the tip of the aspirated region, increasing in area. The cortical region along the neck undergoes shear deformation, where the geometry of the network is changing via alterations in the angles of actin filaments to one another. Using a course-grained molecular mechanics model, we have determined that when a cell is aspirated to $L_p = 4R_p$ the increase in area of the tip region of the cell is about twofold, while the network angle change at the neck is about 45 degrees (Luo et al., 2013).

1.4 EXPERIMENTAL SETUP

An epifluorescence microscope, such as the Olympus IX81 with DIC and fluorescence imaging capability, is connected to a motorized water manometer. The manometer is operated through a keypad controller, which allows for the rapid raising or lowering of one of the water tanks to create a height differential (h) between the water tank and the cell. As a result, a precise pressure differential

($\Delta P = \rho g h$, where ρ is the density of water (1 g/cm^3), g is the gravitational coefficient (9.8 m/s^2), and h is the height differential) is applied to the cell through the micropipette. Micropipettes can be purchased (World Precision Instrument, Fire-polished Pre-Pulled Glass Pipettes, TIP5TW1). Alternatively, they may be forged in-house, using thin-wall borosilicate glass tubing (Sutter Instrument, 1 mm outer diameter, 0.75 mm inner diameter, 10 cm length) on a micropipette puller (Micro-Data Instrument, PMP102). We find that micropipettes with a diameter approximately equal to the radius of the cell generally work well for most cell types. The micropipette is spatially controlled on the system by a micromanipulator. For a detailed description of the setup of an MPA system, see Kee & Robinson (2013).

1.4.1 Balancing the system

A micropipette is filled with assay medium and loaded onto the micromanipulator. After locating the micropipette in the field of view at 40× magnification, 1-μm beads are added to an imaging chamber containing the assay medium. With the two water tanks at a level position, the pipette is brought close to the diffusing beads using the micromanipulator. The level of one tank is manually adjusted until no force acts on the beads: they are neither drawn nor pushed away by liquid flow from the micropipette.

1.4.2 Cortical tension (T_c) measurements

Cells in media are placed in an imaging chamber, where the micropipette is then loaded at zero pressure. Pressure is applied to a healthy cell that is entirely in suspension, starting with relatively low pressures (in the 0.1 nN/μm^2 range) and increasing stepwise until the cell is deformed inside the pipette to $L_p = R_p$. The pressure necessary to reach this level of deformation is the critical pressure (P_c), and the cortical tension (T_c) is calculated using Eq. (1). Each cell is measured at least twice to ensure mechanical uniformity for a specific cell. Occasionally the cell membrane will dissociate from the cortex, resulting in rapidly expanding blebs inside the pipette; data for these cells must be discarded.

1.4.3 Quantification for protein mechanoaccumulation

Cells are placed in imaging media such as MES starvation buffer for *Dictyostelium* or Leibowitz L-15 media without phenol red (Invitrogen) for mammalian cells to minimize background fluorescence. Cells are aspirated using a fixed pressure, normally the pressure required by the specific cell type to reach $L_p = 2R_p$. This pressure is reached in the pipette prior to contacting the cell, assuring that the pressure applied is instantaneous and not gradual. The cell is imaged by time-lapse both in DIC and fluorescence channels. The normalized intensity of the protein of interest at the site of stress can be calculated by measuring the mean gray value of the stressed region (I_s), the cortex on the opposite side of the cell from the pipette (I_o), and a background region above or below the cell (I_b), and then using the equation $(I_s - I_b)/(I_o - I_b)$.

2. FLUORESCENCE CORRELATION SPECTROSCOPY

2.1 OVERVIEW

The diffusion and mobility of a protein dictates its lifetime dynamics and the complexes it may be part of. This information is invaluable for developing quantitative models that explain the behavior of the individual protein, and ultimately, the process in which it participates. Macromolecules can passively diffuse or actively move within the cytoplasm of a cell. Passive diffusion is dependent upon the viscosity of the cytoplasm and the size and shape of the molecule, occurring on the order of 5 $\mu m^2/s$. Molecular motors, such as kinesin and myosin, drive-directed transport along microtubules and actin, moving cargo on the order of 1 μm/s. Molecular motors can also "stir" the cytoplasm, actively increasing the apparent diffusion of a protein (Guo et al., 2014). Determining diffusion times of proteins may help decipher mechanisms of a process, and help establish mathematical models that predict biological behavior. Models that describe the mechanosensing behavior of proteins and the cortical viscoelasticity of the cell enhance our understanding of the role of a mechanical meshwork of actin, motors, and cross-linking proteins during furrow ingression (Luo et al., 2013, 2012; Mohan, Luo, Robinson, & Iglesias, 2015; Poirier, Ng, Robinson, & Iglesias, 2012).

Fluorescence correlation spectroscopy (FCS) is a particularly powerful method for measuring diffusion and mobility inside cells. FCS has been used to measure diffusion of molecules in vitro and in vivo, under different stages of the cell cycle and mechanical and pharmacological perturbations. Diffusion and mobility parameters can be extracted from the intensity fluctuations of a fluorescently labeled protein or molecule in a confocal voxel. A correlation function is used to determine the number of fluorescent particles present in the volume over the time course, and single- or multiple-component fits are used to calculate the diffusion time, which is the average time required to diffuse across the confocal volume. The diffusion time extracted is then used to determine diffusion coefficients. This quantitative analysis also provides information on complex assembly and concentration changes. A change in a protein's diffusion time can reflect a change in the cytoskeletal and biochemical interactions in the network. For example, changes in the dynamics of cortexillin I, an actin-bundling protein, and IQGAP2, a scaffolding protein, were studied by FCS in different genetic backgrounds of *Dictyostelium* cells. In an *iqgap2* null background, cortexillin I showed a 30% increase in cytoplasmic diffusion time, suggesting that it engages in different biochemical interactions when IQGAP2 is absent (Srivastava & Robinson, 2015). In addition, FCS was used to demonstrate that global mechanical stresses change protein dynamics, as cortexillin I in compressed cells had twofold slower diffusion rates (Srivastava & Robinson, 2015). The ability to study these dynamics in live cells on the microsecond and nanometer scale is crucial for understanding how changes in the cell cycle and external perturbations affect protein diffusion and mobility, which ultimately affect the cell's biochemistry.

2.2 EXPERIMENTAL SETUP

A laser-scanning confocal microscope, such as the Zeiss 780, can be used with a 40× water objective (typically with a numerical aperture of 1.2) to image a confocal volume of approximately 0.5 fL. An avalanche photodiode detects the fluorescence fluctuations within the volume and the signal is autocorrelated over time (Fig. 2) (Bacia, Kim, & Schwille, 2006; Bulseco & Wolf, 2013). FCS may also be performed with a total internal reflection fluorescence microscope to image membrane proteins, or with a single-plane illumination microscope to image larger three-dimensional samples, such as embryos (Krieger et al., 2015). Having a small number of particles, ideally less than 100, in the volume is essential for accurately identifying small changes. Software, such as Zen, include equations with single- or multiple-component fits to determine correlation values and diffusion times (Carl Zeiss Microimaging, Guide to Basic FCS Experiments). Fluorescence fluctuations are the result of species mobility, as well as various inter- and intramolecular interactions. Fluctuations may also occur due to a change in intrinsic fluorescence, often from a triplet state, in which the fluorophore has transitioned to a dark state. To correct for these fluctuations, a triplet-state component must be added to the model.

Correlating the fluorescence fluctuations over time yields a plot as seen in Fig. 2C. After accounting for the triplet state, the peak correlation amplitude is the autocorrelation value (G_X). The algorithm extracts the particle number ($G_X = 1/N$, where N is the number of particles); a smaller particle number will have a greater autocorrelation value. The diffusion time is extracted from the decay

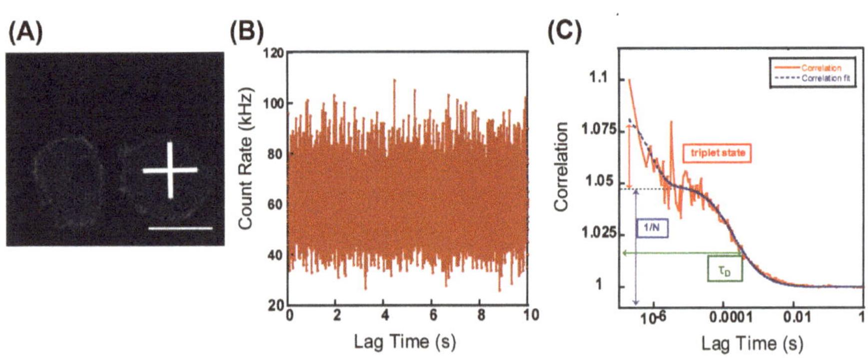

FIGURE 2 Fluorescence Correlation Spectroscopy to assess protein dynamics in live cells: microseconds to milliseconds.

(A) *Dictyostelium* cell expressing GFP-cortexillin I. *Cross* indicates location of the confocal volume. (B) Fluorescence fluctuations recorded over a 10-s time-lapse. (C) Sample autocorrelation plot of fluorescence correlation spectroscopy. *Red line* represents the autocorrelation data, while the *blue dashed line* indicates correlation fit based on a one-component model. A triplet-state component has been added. The peak correlation amplitude is proportional to the number of particles, *N*, while the decay of the slope determines the diffusion time, τ_D. Scale, 10 μm.

of the slope, and a diffusion coefficient may be calculated by Eq. (2), the Stokes–Einstein equation:

$$D = \frac{k_B T}{6\pi\eta r} \tag{2}$$

where k_B is the Boltzmann constant, T is temperature, η is the viscosity, and r is the Stokes radius of the particle.

2.2.1 Calibration of the system

Calibrating the system using a fluorescent molecule with a known diffusion coefficient allows for measurement of the size and shape of the confocal volume created by the laser and pinhole. This calibrating fluorophore should fluoresce at the same wavelength as the fluorescently labeled species of interest to accurately define the confocal volume that will be imaged. An example dye is rhodamine 6G, which diffuses at 426 $\mu m^2/s$ and may be used in a water or buffer solution at 50–200 nM (Bacia et al., 2006; Srivastava & Robinson, 2015). The correct focus is found by scanning from the bottom of the cover glass using a reflected laser light, and moving up 200 μm into the sample. The correction collar of the objective must be adjusted to achieve the highest count rate from the dye. This adjustment corrects for the thickness of the cover glass. Next, the pinhole must be aligned; most software have automatic adjustments in the x and y directions to maximize the count rate from the dye. The pinhole should be set to one airy unit to create the smallest confocal volume possible.

2.2.2 Monitoring fluorescence fluctuations

When imaging the calibration dye or a fluorescently labeled species in solution, images can be acquired continuously for 10 seconds with 10 repetitions. When imaging in live cells, shorter acquisition times with fewer repetitions (we typically use five repetitions of 2 second time-lapses for *Dictyostelium*) should be used to avoid artifacts from active cell migration and photobleaching.

2.2.3 Fitting to models

The fluorescence fluctuations over time are analyzed to yield an autocorrelation value, G_x (Eq. (3)):

$$G_x = \frac{\langle \delta F(t) \cdot \delta F(t+\tau) \rangle}{\langle F(t) \rangle^2} \tag{3}$$

where t is time and $F(t)$ is the fluorescence fluctuation detected. While G_x is a value from 0 to 1, 0 being no correlation and 1 being complete correlation, the software adjusts this range from 1 to 2 to simplify calculations (Bulseco & Wolf, 2013).

To determine an accurate diffusion time, a few parameters must be set. The structural parameter is the ratio of the axial resolution to the radial resolution, and is typically between 5 and 7. This is calculated from the diffusion time of the calibration dye. To account for the triplet-state excitation, a triplet-state component is added to the fit. However, since this state should be rapid, its fit should be restricted to a maximum of 8 μs. In addition, the data can be fit to a single-component or multiple-component 3D diffusion model, depending on the species measured. Because the algorithm may not correctly identify separate groups within the data, we often prefer to focus on simpler models that include single components. The data are then fit to a one-component model to generate a parameter set that may represent a distribution of species.

2.2.4 Data analysis

Once the correlation plots have been fit to the models, the resulting parameters can be analyzed to extract information on the number of species within the volume and their average diffusion time. However, not every trace may fit the model well. The residual trace shows how the data deviate from the fit and is helpful in determining whether the model is an accurate representation of the data. Occasionally the fluorescently labeled species may aggregate, either in solution or in live cells. During live cell FCS, an organelle or large fluorescing species may enter the imaging volume. Such aggregates or disturbances can be detected in the trace as significant deviations or nonuniform fluctuations in the count rate. These traces may be removed from the analysis. Traces that show photobleaching over time should also be discarded. The remaining traces are averaged, the data are fit to the model, and an average diffusion time is extracted. This diffusion time can then be converted into a diffusion coefficient using the dye as calibration for the confocal volume shape and size (Carl Zeiss Microimaging, Guide to Basic FCS Experiments). This analysis is used to measure diffusion times, and also to determine the assembly of complexes, binding stoichiometry, and changes in the network. To measure interactions between two specific proteins or molecules, FCS may be performed with two colors, a technique known as fluorescence cross-correlation spectroscopy (FCCS) (see Section 2.2.4.1).

3. FLUORESCENCE CROSS-CORRELATION SPECTROSCOPY

FCCS involves imaging intensity fluctuations of two different fluorophores within the same focal volume. A significant cross-correlation between the species suggests that the two species might associate based on the probability of the two different fluorophores diffusing in or out of the volume simultaneously. FCCS allows for quantitative interaction measurements between two species in vitro and detection of associative interactions in vivo, yielding information on interaction strengths, dissociation constants, and binding stoichiometries. Traditional methods of studying protein–protein interactions in cells, such as coimmunoprecipitation (co-IP) and Förster resonance energy transfer (FRET), have their own significant limitations. Co-IP requires lysing the cells and a time delay as the extracts are exposed to beads, washed, and eluted; this readily detects very stable associations, but disrupts more transient ones. FRET requires that the donor and acceptor fluorophores reside within a minimal distance in a relatively optimal orientation. FCCS circumvents both types of issues; cells remain intact, and fluorophores do not have to reside within an optimal distance or orientation. Performing FCCS in vivo facilitates analysis of protein–protein interactions in context and allows for the relatively easy process of incorporating acute and genetic perturbations. In the context of cytokinesis, it allows for a comparison between interactions at the cleavage furrow and those that occur globally. In addition, FCCS is sensitive enough to identify transient interactions that may not be detectable by co-IP (Bacia et al., 2006; Bacia & Schwille, 2007).

FCCS requires that two species be labeled with different fluorophores with little overlap in excitation and emission spectra to prevent cross-talk. However, the confocal volumes must have significant overlap to allow for accurate cross-correlation. If the system has one confocal pinhole, then the volume must be adjusted to a single wavelength. However, if two pinholes are used, the two volumes may be set for closer overlap. Although FCCS in vitro can provide reliable measurement of a dissociation constant, K_d, measurement in vivo is limited to relative interaction strengths. This limitation comes from the fact that proteins may be involved in interactions with multiple cellular factors, which complicates the measurement of a true K_d. Further, expression of the fluorescent protein in addition to an unlabeled endogenous population complicates the K_d assessment as well. In addition, the presence of organelles and cytoskeletal networks will also affect cross-correlation. However, the relative interaction strength is still a valuable measurement for characterizing novel, transient interactions that have an impact on biological processes.

The cross-correlation function, G_x, is a measure of how well the fluorescence fluctuations from the two different channels correlate over time, and is calculated using Eq. (4):

$$G_x = \frac{\langle \delta F_A(t) \cdot \delta F_B(t+\tau) \rangle}{\langle F_A(t) \rangle \langle F_B(t) \rangle} \tag{4}$$

where t is time, and $F(t)$ is the fluorescence fluctuation in either the green or red channel. The cross-correlation value ranges from 0 to 1, with 0 being no cross-correlation, and 1 being complete correlation. The scale is again adjusted from 1 to 2, and we find that values generally range between 1.001 and 1.1 in live cells, and between 1.05 and 1.2 for species in solution.

The experimental setup used for FCCS is similar to the one used for FCS, but the additional laser that excites the second fluorophore creates a second confocal volume. When acquiring traces, the two lasers simultaneously excite the sample, and fluorescence fluctuations from both channels are recorded. This results in three plots: two autocorrelation plots (one for each channel), and one cross-correlation plot. The traces with aggregates or photobleaching are discarded, and the correlations are fit to the model. A relative interaction strength (IS^{-1}) may be calculated with Eq. (5):

$$IS^{-1} = \frac{G_x}{V_{eff} G_A G_B} \left(\frac{G_A}{G_x} - 1 \right) \left(\frac{G_B}{G_x} - 1 \right) \tag{5}$$

where V_{eff} is the effective overlap between the confocal volumes (set to 1), G_A and G_B are the individual channel correlation values, and G_x is the cross-correlation value (Bierbaum & Bastiaens, 2013). This equation takes the form of a dissociation constant by comparing the fraction of bound molecules (from G_X) to the free molecules. The values are normalized against the autocorrelations of G_A and G_B.

While FCCS yields information about relative interaction strengths, understanding the dynamic range of the setup is critical. Two fluorescently labeled proteins that are known to bind or two fluorescent proteins linked by a short amino acid sequence may be used as positive controls. As a negative control, two unlinked fluorescent proteins may be used to reveal the dynamic range, which we find is ~10-fold. FCCS is becoming an increasingly valuable technique to directly measure protein–protein interactions in vivo and will continue to allow for the study of transient interactions.

4. FLUORESCENCE RECOVERY AFTER PHOTOBLEACHING

4.1 OVERVIEW

Fluorescence recovery after photobleaching (FRAP) is a commonly used technique to measure turnover dynamics of proteins in cells. A region of a cell expressing the fluorescent protein of interest is photobleached using a laser, and that region is imaged over time to measure the rate of fluorescence recovery by unbleached, fluorescent molecules from outside the region (Jacobson, Derzko, Wu, Hou, & Poste, 1976; Lippincott-Schwartz, Snapp, & Kenworthy, 2001). The resulting recovery curve can be used to calculate the rate of movement, the release time from other binding interactions, and the fraction of the protein that is mobile. For proteins that recover on the order of hundreds of milliseconds to seconds, the recovery time is generally dominated by the unbinding kinetics of the protein from its binding partners. This unbinding may reflect the release time of the bleached protein or, more commonly, release of the unbleached protein from nearby pools. In contrast, a protein with low turnover dynamics may be incredibly stable. Proteins may also fail to recover to the initial fluorescence intensity due to their deep associations in larger complexes and networks. Once the images have been corrected for photobleaching, the percentage recovery represents the population of the protein that is mobile within the cell.

We have shown that during cytokinesis, equatorially enriched proteins, including myosin II, cortexillin I, and IQGAP2, are stabilized under increasing mechanical stress. Cells under compression by agarose overlay demonstrate an increase in the immobile fraction in interphase cells and at the cleavage furrow of dividing *Dictyostelium* cells. The increase in mechanical load stabilizes the cytoskeletal meshwork, therefore increasing the immobile fraction of these proteins. In *myosin II* null cells, cortexillin I and IQGAP2 immobile fractions also increase, although their diffusion times do not. Thus, myosin II promotes distinct cytoskeletal network properties, reflected in these protein mobility differences (Srivastava & Robinson, 2015). Similar increases in the immobile fraction of myosin II have been observed in mammalian cytokinesis as well (Kondo et al., 2011).

With FRAP, one can assess protein dynamics of different regions of the cell, including the cortex and cytoplasm. Imaging these dynamics on the hundreds of milliseconds to seconds timescales with micrometer resolution provides insight into more global changes during cytokinesis, and can reveal alterations in networks and stabilities of complexes.

4.2 EXPERIMENTAL SETUP

A few images of the region of interest are taken to quantify the intensity of the fluorescence before photobleaching. A small region is photobleached with a laser of the excitation wavelength of the fluorescent protein. This region is then imaged over time until recovery has saturated. An unbleached region of the cell is also imaged simultaneously to account for photobleaching due to image acquisition over time. The average intensity of the bleached region, the control unbleached region, and background are quantified (Fig. 3).

After background correction, the intensity is fit to a single exponential form, Eq. (6):

$$\mathrm{NI}(t) = m_1\left(1 - m_2 \cdot \mathrm{e}^{-kt}\right) \tag{6}$$

where m_1 and m_2 are fitting parameters. The immobile fraction may be quantified by Eq. (7):

$$F_{\mathrm{i}} = \frac{1 - m_1}{1 - m_1 + m_2} \tag{7}$$

Quantifying this recovery of intensity can yield insight into the kinetics and complexes of protein dynamics (Srivastava & Robinson, 2015).

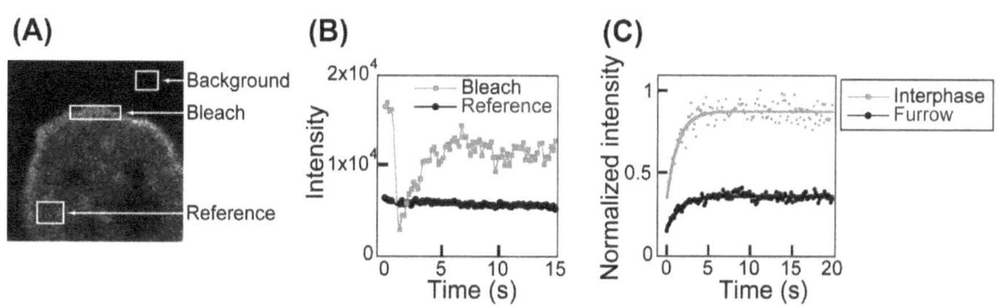

FIGURE 3 Fluorescence Recovery after Photobleaching to determine protein mobility and recovery: milliseconds to seconds.

(A) A sample image of GFP-cortexillin I expressed in *Dictyostelium*, demonstrating the bleach site. (B) Sample fluorescence intensity curves. (C) After background subtraction, the reference intensity remains constant, while the bleached region shows a stereotypical decrease and recovery of fluorescence over time.

Adapted from Srivastava, V. & Robinson, D. N. (2015). Mechanical stress and network structure drive protein dynamics during cytokinesis. Current Biology, 25, *663–670.*

CONCLUSION

MPA, FCS, and FRAP are currently used to measure the biochemical and mechanical aspects of cytokinesis on the time- and length-scales of mechanoresponses, mobility, protein interactions, and network dynamics in live cells. Other techniques, including optical traps and single-molecule imaging, are contributing to our understanding of protein dynamics as well (Ren et al., 2009; Spudich, Rice, Rock, Purcell, & Warrick, 2011). In addition, the introduction of chemically inducible dimerization systems, optogenetics, lattice-light sheet microscopy, and other superresolution microscopy will bring exciting, new quantitative measurements to the field (Chen et al., 2014; Laplant, Huang, Bewersdorf, & Pollard, 2016; Strickland et al., 2012). These techniques can be combined with the tools of genetics, pharmacological agents, and global compression to allow for a rich dissection of individual components involved (Kee et al., 2012; Srivastava & Robinson, 2015). The benefit of these live-cell techniques is the ability to observe and make quantitative measurements of cell mechanics and protein dynamics as the cell undergoes important physiological processes. These mechanics and protein dynamics measurements, especially when combined with other metrics of cell morphological changes, such as furrow ingression kinetics, yield mechanistic insight into pathways and molecular interactions and can even lead to molecular models that predict cellular behaviors (Luo et al., 2012; Mohan et al., 2015; Poirier et al., 2012; Zhang & Robinson, 2005; Zhou et al., 2010). This collection of methods will continue to provide fascinating insight into the mechanical and biochemical mechanisms of cytokinesis.

ACKNOWLEDGMENTS

This work is supported by the National Institutes of Health grants GM66817 and GM109863 to D.N.R. We thank members of the Robinson lab and Vasudha Srivastava for their valuable comments during manuscript preparation.

REFERENCES

Ashkin, A., Dzeidzic, J. M., & Yamane, T. (1987). Optical trapping and manipulation of single cells using infrared laser beams. *Nature, 330*, 769–771.

Bacia, K., Kim, S. A., & Schwille, P. (2006). Fluorescence cross-correlation spectroscopy in living cells. *Nature Methods, 3*, 83–89.

Bacia, K., & Schwille, P. (2007). Practical guidelines for dual-color fluorescence cross-correlation spectroscopy. *Nature Protocols, 2*, 2842–2856.

Bierbaum, M., & Bastiaens, P. H. (2013). Cell cycle-dependent binding modes of the ran exchange factor RCC1 to chromatin. *Biophysical Journal, 104*, 1642–1651.

Bulseco, D. A., & Wolf, D. E. (2013). Chapter 21 – fluorescence correlation spectroscopy: molecular complexing in solution and in live cells. *Methods in Cell Biology, 114*, 489–524.

Carl Zeiss Microimaging, Inc. Guide to basic FCS experiments on Zeiss LSM 780/Confocor 3/BiG systems running ZEN 2010.

Chen, B. C., Legant, W. R., Wang, K., Shao, L., Milkie, D. E., Davidson, M. W., … Betzig, E. (2014). Lattice light-sheet microscopy: imaging molecules to embryos at high spatiotemporal resolution. *Science, 346*, 1257998.

Crick, F. H. C., & Hughes, A. F. W. (1950). The physical properties of cytoplasm. *Experimental Cell Research, 1*, 37–80.

Effler, J. C., Iglesias, P. A., & Robinson, D. N. (2007). A mechanosensory system controls cell shape changes during mitosis. *Cell Cycle, 6*, 30–35.

Gerald, N., Dai, J., Ting-Beall, H. P., & De Lozanne, A. (1998). A role for *Dictyostelium* racE in cortical tension and cleavage furrow progression. *The Journal of Cell Biology, 2*, 483–492.

Guo, M., Ehrlicher, A. J., Jensen, M. H., Renz, M., Moore, J. R., Goldman, R. D., … Weitz, D. A. (2014). Probing the stochastic, motor-driven properties of the cytoplasm using force spectrum microscopy. *Cell, 4*, 822–832.

Hiramoto, Y. (1963). Mechanical properties of sea urchin eggs. II. Changes in mechanical properties from fertilization to cleavage. *Experimental Cell Research*, 76–89.

Hiramoto, Y. (1990). Mechanical properties of the cortex before and during cleavage. *Annals of the New York Academy of Sciences, 582*, 22–30.

Hochmuth, R. M. (2000). Micropipette aspiration of living cells. *Journal of Biomechanics, 1*, 15–22.

Jacobson, K., Derzko, Z., Wu, E.-S., Hou, Y., & Poste, G. (1976). Measurement of the lateral mobility of cell surface components in single, living cells by fluorescence recovery after photobleaching. *Journal of Supramolecular Structure, 5*, 565–576.

Kee, Y.-S., Ren, Y., Dorfman, D., Iijima, M., Firtel, R., Iglesias, P. A., & Robinson, D. N. (2012). A mechanosensory system governs myosin II accumulation in dividing cells. *Molecular Biology of the Cell, 23*, 1510–1523.

Kee, Y.-S., & Robinson, D. N. (2013). Micropipette aspiration for studying cellular mechanosensory responses and mechanics. *Methods in Molecular Biology, 983*, 367–382.

Kim, J. H., Ren, Y., Ng, W. P., Li, S., Son, S., Kee, Y.-S., … Chen, E. H. (2015). Mechanical tension drives cell membrane fusion. *Developmental Cell, 32*, 561–573.

Kondo, T., Hamao, K., Kamijo, K., Kimura, H., Morita, M., Takahashi, M., & Hosoya, H. (2011). Enhancement of myosin II/actin turnover at the contractile ring induces slower furrowing in dividing HeLa cells. *Biochemical Journal, 435*, 569–576.

Krieger, J. W., Singh, A. P., Bag, N., Garbe, C. S., Saunders, T. E., Langowski, J., & Wohland, T. (2015). Imaging fluorescence (cross-) correlation spectroscopy in live cells and organisms. *Nature Protocols, 10*, 1948–1974.

Laplant, C., Huang, F., Bewersdorf, J., & Pollard, T. D. (2016). High-speed super-resolution imaging of live fission yeast cells. J.M. Walker (Series Ed.). In A. Sanchez-Diaz, & P. Perez (Eds.), *Methods in molecular biology: Yeast cytokinesis, methods and protocols* (pp. 45–57). New York: Humana Press.

Lippincott-Schwartz, J., Snapp, E., & Kenworthy, A. (2001). Studying protein dynamics in living cells. *Nature Reviews Molecular Cell Biology, 2*, 445–456.

Luo, T., Mohan, K., Iglesias, P. A., & Robinson, D. N. (2013). Molecular mechanisms of cellular mechanosensing. *Nature Materials, 11*, 1064–1071.

Luo, T., Mohan, K., Srivastava, V., Ren, Y., Iglesias, P. A., & Robinson, D. N. (2012). Understanding the cooperative interaction between myosin II and actin cross-linkers mediated by actin filaments during mechanosensation. *Biophysical Journal, 2*, 238–247.

Luo, T., Srivastava, V., Ren, Y., & Robinson, D. N. (2014). Mimicking the mechanical properties of the cell cortex by the self-assembly of an actin cortex in vesicles. *Applied Physics Letters, 15*, 153701.

Matzke, R., Jacobson, K., & Radmacher, M. (2001). Direct, high-resolution measurement of furrow stiffening during division of adherent cells. *Nature Cell Biology, 3*, 607–610.

Mijailovich, S. M., Kojic, M., Zivkovic, M., Fabry, B., & Fredberg, J. J. (2002). A finite element model of cell deformation during magnetic bead twisting. *Journal of Applied Physiology, 93*, 1429–1436.

Mohan, K., Luo, T., Robinson, D. N., & Iglesias, P. A. (2015). Cell shape regulation through mechanosensory feedback control. *Journal of the Royal Society Interface, 12*, 201501512.

Otto, O., Rosendahl, P., Mietke, A., Golfier, S., Herold, C., Klaue, D., ... Guck, J. (2015). Real-time deformability cytometry: on-the-fly cell mechanical phenotyping. *Nature Methods, 12*, 199–202.

Poirier, C. C., Ng, W. P., Robinson, D. N., & Iglesias, P. A. (2012). Deconvolution of the cellular force-generating subsystems that govern cytokinesis furrow ingression. *PLoS Computational Biology, 8*(4), e1002467.

Rappaport, R. (1967). Cell division: direct measurement of maximum tension exerted by furrow of echinoderm eggs. *Science, 156*, 1241–1243.

Reichl, E. M., Ren, Y.-S., Morphew, M. K., Delannoy, M., Effler, J. C., Girard, K. D., ... Robinson, D. N. (2008). Interactions between myosin and actin crosslinkers control cytokinesis contractility dynamics and mechanics. *Current Biology, 7*, 471–480.

Ren, Y.-S., Effler, J. C., Norstrom, M., Luo, T., Firtel, R. A., Iglesias, P. A., ... Robinson, D. N. (2009). Mechanosensing through cooperative interactions between myosin II and the actin crosslinker cortexillin I. *Current Biology, 19*, 1421–1428.

Schiffhauer, E. S., Luo, T., Mohan, K., Srivastava, V., Qian, X., Griffis, E. R., ... Robinson, D. N. (2016). Mechanoaccumulative elements of the mammalian actin cytoskeleton. *Current Biology* (in press).

Spudich, J. A., Rice, S. E., Rock, R. S., Purcell, T. J., & Warrick, H. M. (2011). The optical trapping dumbbell assay for nonprocessive motors or motors that turn around filaments. *Cold Spring Harbor Protocols, 11*, 1372–1374.

Srivastava, V., Iglesias, P. A., & Robinson, D. N. (2015). Cytokinesis: robust cell shape regulation. *Seminars in Cell & Developmental Biology* (in press).

Srivastava, V., & Robinson, D. N. (2015). Mechanical stress and network structure drive protein dynamics during cytokinesis. *Current Biology, 25*, 663–670.

Strickland, D., Lin, Y., Wagner, E., Hope, C. M., Zayner, J., Antoniou, C., ... Glotzer, M. (2012). TULIPs: tunable, light-controlled interacting protein tags for cell biology. *Nature Methods, 9*, 379–384.

Sun, Q., Luo, T., Ren, Y., Florey, O., Shirasawa, S., Sasazuki, T., ... Overholtzer, M. (2014). Competition between human cells by entosis. *Cell Research, 24*, 1299–1310.

Wolpert, L. (1966). The mechanical properties of the membrane of the sea urchin egg during cleavage. *Experimental Cell Research, 2*, 385–396.

Yang, L., Effler, J. C., Kutscher, B. L., Sullivan, S. P., Robinson, D. N., & Iglesias, P. A. (2008). Modeling cellular deformations using the level set formalism. *Biomedcentral Systems Biology, 2*, 68–84.

Zhang, W., & Robinson, D. N. (2005). Balance of actively generated contractile and resistive forces controls cytokinesis dynamics. *Proceedings of the National Academy of Sciences of the United States of America, 20*, 7186–7191.

Zhou, Q., Kee, Y.-S., Poirier, C. C., Jelinek, C., Osborne, J., Divi, S., ... Robinson, D. N. (2010). 14-3-3 coordinates microtubules, Rac, and myosin II to control cell mechanics and cytokinesis. *Current Biology, 20*, 1881–1889.

CHAPTER

19

An active contour ImageJ plugin to monitor daughter cell size in 3D during cytokinesis

M.B. Smith[1], A. Chaigne, E.K. Paluch[1]

MRC Laboratory for Molecular Cell Biology, University College London, London, United Kingdom

[1]*Corresponding authors: E-mail: dmcbmbs@ucl.ac.uk; e.paluch@ucl.ac.uk*

CHAPTER OUTLINE

Abstract

Controlling relative daughter cell size is key during cytokinesis. Uncontrolled size asymmetries can lead to aneuploidy and division failure. At the same time, precisely regulated size asymmetries are of crucial importance in many divisions during embryonic development. Therefore, being able to monitor daughter cell size is important in cytokinesis studies. However, freely available tools allowing to effectively measure the size of daughter cells in three dimensions during cytokinesis are missing. Here, we describe an open-access plugin for ImageJ or Fiji based on an active contour surface representation of the cells. Our method provides a user-friendly and accurate way to monitor the size of the two daughter cells throughout cytokinesis.

Methods in Cell Biology, Volume 137, ISSN 0091-679X, http://dx.doi.org/10.1016/bs.mcb.2016.05.003

1. BACKGROUND

1.1 ASYMMETRIC CELL DIVISION AND DAUGHTER CELL VOLUME

Cell division is a crucial process for tissue homeostasis, not only because it allows the generation of an increased number of cells, but also because it can produce different cell types. Indeed, embryonic development in many species crucially depends on precisely controlled asymmetric divisions. For example, the first division of the *Caenorhabditis elegans* embryo is strongly asymmetric and not only defines the anteroposterior axis of the embryo but also leads to the generation of daughter cells of different fates, which will give rise either to the ectoderm and the mesoendoderm lineage (for review, see Betschinger & Knoblich, 2004; Gönczy & Rose, 2005). Division can be asymmetric in terms of daughter cell composition only. For instance, in the *Drosophila* male germline, cell division is symmetric in size but segregates the mother and daughter centrosomes differentially (Yamashita, Mahowald, Perlin, & Fuller, 2007). The asymmetry can also be in size, leading to two daughter cells of different volumes (Fig. 1A), for example, in *Drosophila* neuroblasts (Kaltschmidt, Davidson, Brown, & Brand, 2000) or in the extreme case of meiosis where cells typically divide into a large oocyte and a small polar body (for review, see Almonacid, Terret, & Verlhac, 2014; Chaigne, Verlhac, & Terret, 2012).

The relative size of the daughter cells depends on the position of the cleavage furrow during cytokinesis. In most cell types, cleavage furrow position is specified by the position of the spindle that depends on interactions between either the tips or the whole length of astral microtubules decorated with Dynein and Dynactin (Busson, Dujardin, Moreau, Dompierre, & Mey, 1998) and the cell cortex via the LGN/Gαi/NuMA complex (for review, see Kotak & Gönczy, 2013). The cleavage furrow is then positioned midway between the two spindle poles (Bringmann & Hyman, 2005). A second cleavage signal is set by the central spindle midzone. Briefly, the overlapping microtubules of the central spindle recruit the Centralspindlin complex (for review, see Green, Paluch, & Oegema, 2012), which in turn activates RhoA signaling at the cell equator (Bement, Benink, & von Dassow, 2005); this triggers F-actin accumulation and Myosin-II activation, thereby creating the contractile ring. Interestingly, these two signals can compete if the spindle midzone is not midway between the two spindle poles, though the midzone signal appears to ultimately dominate (Bringmann & Hyman, 2005). In some systems, such as *Drosophila* neuroblasts, the spindle can also elongate toward one pole leading to asymmetric midzone positioning (Kaltschmidt et al., 2000). Importantly, the size of the daughter cells can also be influenced by cortex contractility independently of the position of the spindle. Indeed, in case of asymmetry in cortical contractility between the two poles of a dividing cell, one pole can contract at the expense of the other, leading to size asymmetries between daughter cells (Fig. 1B, reviewed in Green et al., 2012). Such a contractility-based mechanism appears to contribute to size asymmetry in *Drosophila* and *C. elegans* neuroblasts (Cabernard, Prehoda, &

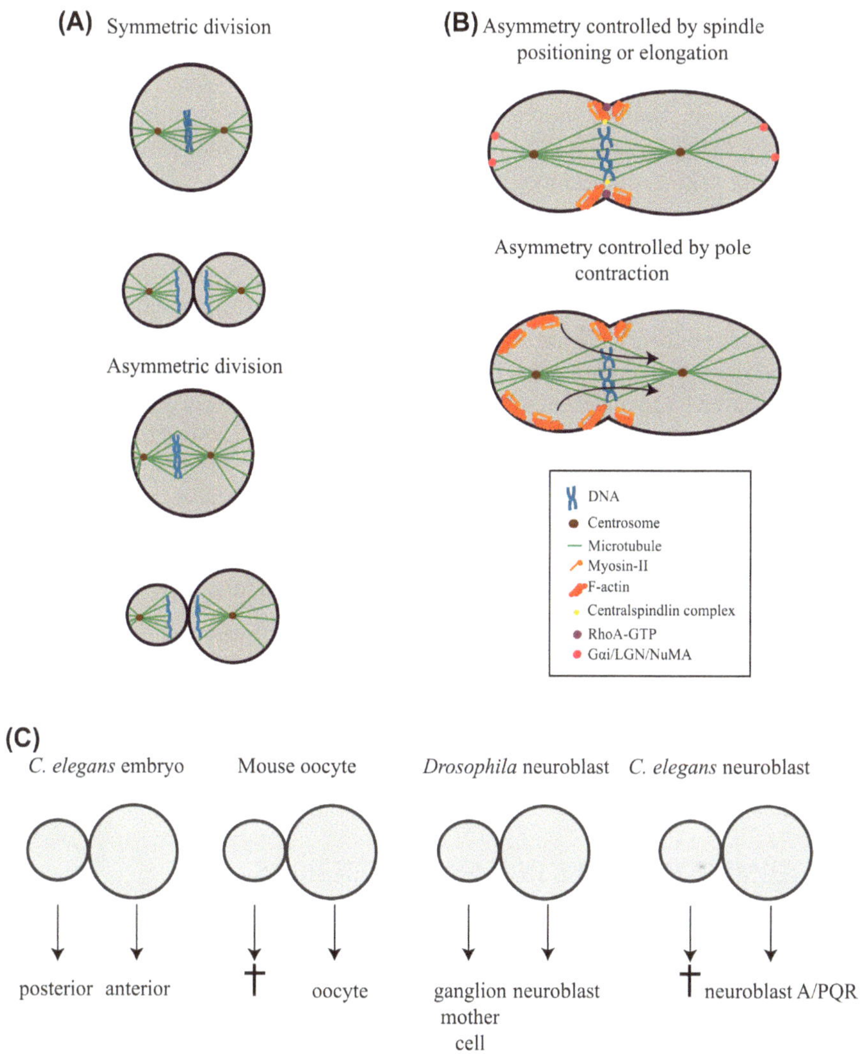

FIGURE 1

Mechanism and roles of divisions asymmetric in size. (A) Schematics of divisions symmetric and asymmetric in size. (B) Possible mechanisms for the generation of size asymmetry during division. Upper panel: via spindle positioning. Lower panel: via pole contractility. (C) Examples of roles of divisions asymmetric in size during embryonic development.

Doe, 2010; Ou, Stuurman, D'Ambrosio, & Vale, 2010) and can lead to aneuploidy and division failure if contractility is not properly controlled (Sedzinski et al., 2011).

Asymmetries in daughter cell size during cell division can be crucial for cell fate. For instance, the smaller cell produced by the asymmetric division of the mouse oocyte or in the *C. elegans* neuroblast lineage will die. It is not always clear if this distinct fate is directly related to size. However, when size asymmetry was inhibited in *C. elegans* neuroblasts, both daughter cells survived and participated in neural formation (Ou et al., 2010). Similarly, *Drosophila* neuroblasts asymmetric division leads to the formation of one neuroblast and one ganglion mother cell, whereas an artificially engineered symmetric division gives rise to two neuroblasts, though this is probably caused by inheritance of apical cues rather than by cell size in itself (Cabernard & Doe, 2009) (Fig. 1C).

To understand the mechanisms and role of size asymmetries during cell division, it is essential to be able to precisely monitor the size of the prospective daughter cells during cytokinesis. As cells are 3D objects, studies of size asymmetry during cell division require 3D measurements of cell volumes, yet cell size analysis is mostly performed in 2D. The main reason for this is the lack of freely available efficient tools to measure daughter cell volumes during cell division. Indeed, the dividing cell is a complex object comprising two roughly ellipsoid connected portions. Most 3D softwares are not effective at properly segmenting and separating such two connected objects and only allow the volume measurements of single or separated objects. Monitoring prospective daughter cell size thus usually requires manually introducing a subjective boundary between the two daughter cells. Furthermore, many of the most effective 3D analysis tools are not open-access (eg, Morphometry analysis in Metamorph, Volocity, Imaris), and the open-access tools often require some coding knowledge to be adaptable for the case of the dividing cell. Here, we describe a new open-access plugin for Fiji or ImageJ that allows monitoring daughter cell volumes in 3D throughout cytokinesis. This plugin provides a user-friendly pipeline to measure volumes of intricate dividing cells for a biologist.

1.2 SEGMENTING CELLS IN THREE DIMENSIONS

Segmenting a cell in three dimensions involves identifying the cell and creating an appropriate representation for measuring the geometric properties of the cell. To measure cell volume in our plugin, we initially considered two representations: a binary volume representation of the voxels occupied by the cell and a parameterized closed surface. The two representations differ primarily in their acquisition and analysis; we outline the general principles of their implementation below.

A binary volume representation involves separating the image into regions, where pixels are labeled as part of the cell or part of the background. The simplest way to accomplish this is to use a threshold. The intensity of each pixel in the image is compared to a chosen threshold value to determine whether the pixel is part of the cell or not. After the threshold is applied, a binary image remains. More specifically each pixel is either 1, meaning part of the foreground or 0, meaning part of the background. The goal is to segment a blob, or continuous region of pixels that represent the cell, so the next step is to use a connected component algorithm, which will separate the different regions into different groups of pixels based on whether the pixels touch each other or not. After this is complete the cell is considered segmented. In an ideal situation, the segmentation is then finished; in practice though, simple thresholding will usually not be sufficient, and more advanced algorithms can be used to improve the results. For example, there are standard binary image operations that can be used to improve thresholding results, such as operations to fill holes in the segmented object or dilate and open operations. There are also more advanced thresholding techniques, such as adaptive thresholding (Pappas, 1992), which can segment regions based on local intensity changes; this is particularly suitable for images where fluorescent intensity varies across the cell, making a unique threshold for the entire image ineffective. Improvements can be made after segmentation by using the blobs to seed a watershed algorithm (Shih, 2010), which grows from the seeded object while keeping the identities of the original segmented blobs. This approach is particularly useful if bright features other than the object of interest are present in the image. Thresholding is an efficient tool for high-throughput analysis because it can be tailored for very specific tasks and used unsupervised. However, for our application, precision and flexibility are more important than high throughput. Thus, instead of pursuing a thresholding protocol specific to dividing cells, we have chosen a surface representation because it is more intuitive to manipulate and rapidly provides measurements.

Using a surface representation relies on the assumption that an object can be enclosed by a continuous closed surface. This is an appropriate assumption for cells undergoing cytokinesis. We used a closed triangulated mesh to represent the cell. The concept of a deformable representation, or active contour, was introduced several decades ago (reviewed in Montagnat, Delingette, & Ayache, 2001). An early study implemented such an approach in 2D using an active contour that deforms to minimize the sum of internal (bending and stretching of the contour) energy and an external "image energy" (Kass, Witkin, & Terzopoulos, 1988). A simple example of an image energy is based on intensity: the brightest regions of the image are regions of lowest energy, and minimizing the image energy will thus move the active contour toward such bright regions. Other types of image energies can be defined, eg, that attract to ridges, edges, and implicit curves. The concept was extended to 3D (Smith et al., 2010; Terzopoulos, Witkin, & Kass, 1988), and the energies have evolved too, going from brightness and gradient-based energies to long-range energies (Xu & Prince, 1998). An example of a long-range energy would be to first use a threshold to segment out features of the image, then treat the distance to a feature as an energy; as a result, the active contour will be attracted to the features identified by thresholding. There are also more complicated energies that can adapt to a known shape (Delingette, 1994) or, when a significant amount of user-segmented data is available, a statistical shape model can be used to improve results (Heimann & Meinzer, 2009).

Active contours work in two stages. The first stage is the initialization, and the second stage is the deformation of the mesh. In our application the mesh deforms iteratively implementing internal energies that mimic a rubber sheet stretching and bending according to external forces (see Appendix and Fig. 2 for detailed description). The mesh continues to deform until the user specifies when to stop. It is possible to impose criteria for stopping the mesh deformation, eg, by setting

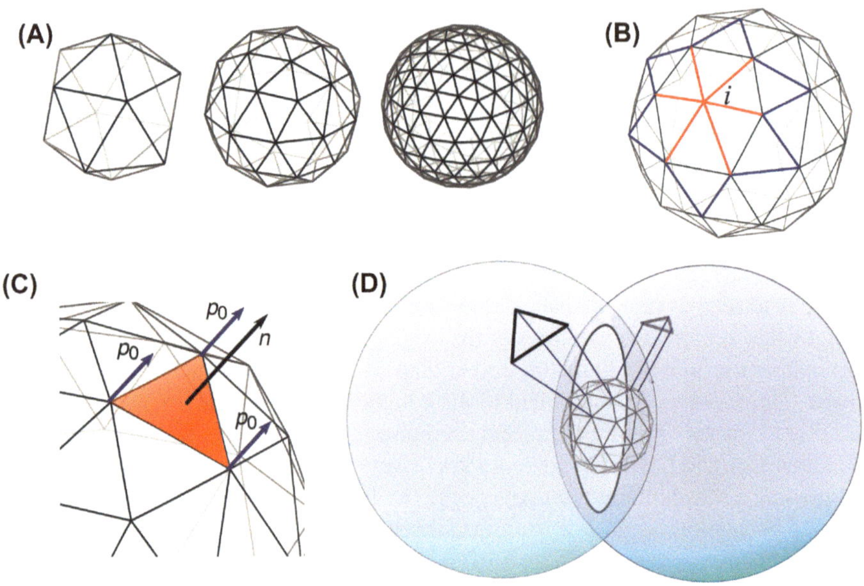

FIGURE 2

Mesh Operations. (A) Mesh on a sphere used for initialization after 0, 1, and 2 divisions. (B) Connections used for calculating the internal energy. Red lines are first order connections, and consecutive red and blue lines are second order connections. See Appendix for details. (C) The pressure force F_p results from the combined effect of the pressure p_0 applied at each node in the direction normal to the local mesh triangle. (D) Example of ray casting a mesh initialized on a sphere at the center of the cell onto the user-defined sphere-based shape outlining the dividing cell.

a minimal amount of change to the mesh between iterations or a minimal variation in energy between iterations. However, our plugin does not include such automatic methods for mesh initialization or deformation cutoff because automation does not increase accuracy but requires the introduction of additional parameters. We have thus omitted these options to keep the program simple and intuitive. The plugin allows a user with 3D microscopy images to explore the merits of active contours quickly and produce accurate results.

2. METHODS

2.1 CELL PREPARATION FOR IMAGING AND ACQUISITION

Our method is based on the detection of the cell contour, thus it is essential to label the cells with a membrane or an actin cortex marker. To get a precise contour for the cells, a good sampling of the cell is required. For monitoring the evolution of daughter cell size, a time-lapse movie should be acquired.

In the examples presented here, HeLa cells stably expressing the chromosome marker Histone 2B (H2B)-mCherry and the actin marker Lifeact-GFP were used. The stable Lifeact-GFP/H2B-mCherry HeLa "Kyoto" cell line was a gift from Murielle Serres (MRC Laboratory for Molecular Cell Biology, University College London, United Kingdom) and was made from H2B-mCherry "Kyoto" HeLa cells (Su, Takaki, & Petronczki, 2011). Z-stacks with one frame every 2 μm were imaged every 3 min using the UPlanSApo 60× NA1.3 silicone objective on an Olympus FV1200 confocal microscope.

2.2 STEPS TO SEGMENT AND MEASURE CELL VOLUME

The workflow for segmenting a cell undergoing cytokinesis to monitor the volumes of the prospective daughter cells using our ImageJ plugin is as follows:

1. Identify a cell that divides (Fig. 3A), crop and rotate the image so that the long axis of the dividing cell is roughly positioned along the x-axis of the image.
2. Initialize the mesh with spherical shapes to roughly fit the shape of the cell (Fig. 3B and C).
3. Deform the mesh to find the surface on the cell based on the image (Fig. 3D).
4. Initialize a plane near the furrow. Autoadjust plane position and orientation, to find the smallest cross section of the mesh (Fig. 3E and F).

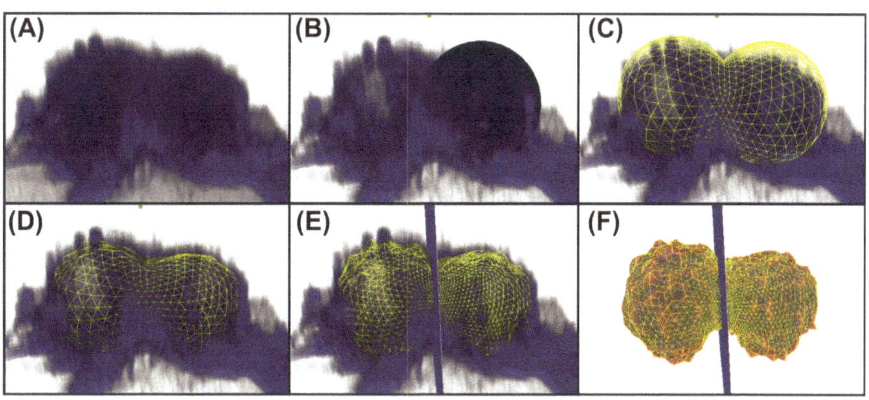

FIGURE 3

Steps for measuring the volume of dividing cells. (A) 3D view of an image stack of a dividing HeLa cell-expressing Lifeact-GFP (blue). (B) Initialization procedure: the user creates spheres around the cell where the mesh will be initialized; only one sphere is represented. (C) Mesh initialized through user interaction based on outline created in B. (D) Deformed mesh to an intensity-based energy. (E) Refined mesh, with a furrow plane added. (F) Segmented cell and adjusted furrow plane.

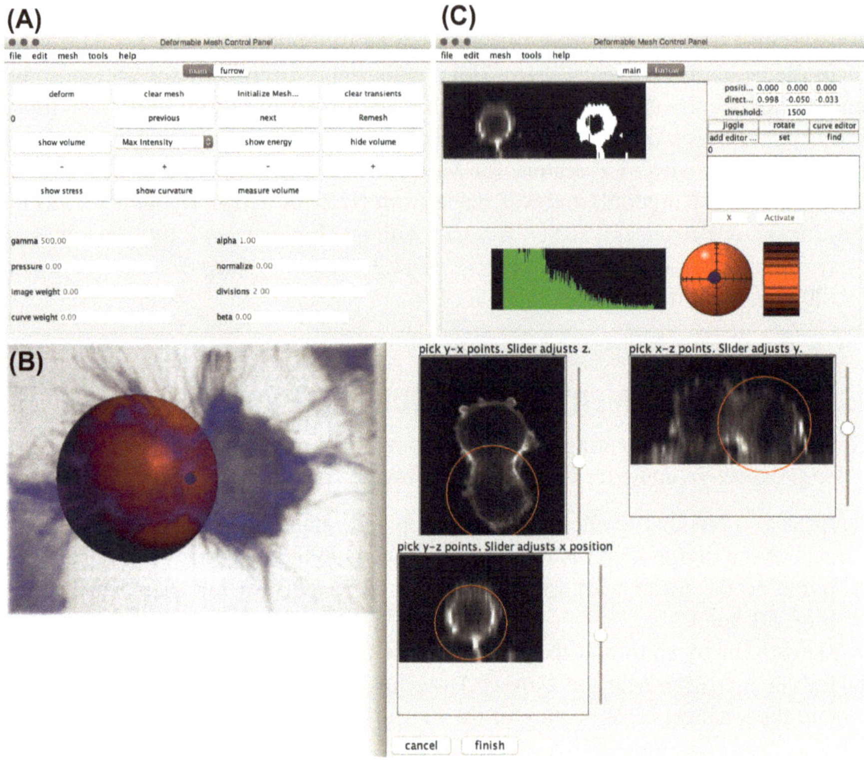

FIGURE 4

User interface of the Plugin. (A) Main control panel with parameters for setting and controlling mesh modifications. (B) Right panel: Initialization dialog window. Left panel: 3D snapshot of the cell being initialized; the initialization sphere is highlighted in red. (C) Furrow control panel for creating and adjusting the location of the furrow plane. Left: cross section of the cell being sliced by the current furrow, binary image (top) and histogram (bottom). By clicking on the histogram, the threshold value for the binary image will change. Right: control panel for automatically finding the furrow plane (top) and controls for adjusting the furrow manually (bottom).

The plugin, source code, and user instructions are freely available at http://www.ucl.ac.uk/lmcb/meshplugin.

The interface of the plugin is presented in Fig. 4A, the mesh initializing dialog on Fig. 4B and the furrow positioning panel on Fig. 4C. In practice, the user opens the image in ImageJ or Fiji, runs the MouseX plugin to crop and rotate the cell accordingly, and then the user runs the Deformable Mesh plugin. The panel presented on Fig. 4A will open and if the user clicks on "Show Volume," the image will be shown in 3D (Fig. 4B). The user should then specify the number of the mesh divisions in the "divisions" box of the control panel (a high amount will be more accurate but a low amount will be faster, see also Fig. 2A). The "Initialize mesh" button will then open the window displayed in Fig. 4B right panel, where the user sees the image of the cell projected in 2D from all sides and is required to draw a circle around each daughter cell to generate the shapes that will initialize the mesh; the "finish" button ends the initialization and the initialization spheres will be shown in the 3D picture (Fig. 4B, left panel). The user should then choose a positive value for the "image weight" (Fig. 4A) that will specify the strength of the attraction exerted on the mesh. This number can be manually tuned to find the best fit. The user should then choose the energy used to attract the mesh: "max intensity," "max gradient," "perpendicular intensity," and "perpendicular gradient" (see next section for details). The "deform" button will then start the mesh deformation. Once the mesh satisfactorily matches the cell surface, the user can stop the deformation, and measurements of the total cell volume can be made. Selecting the "furrow" menu in the control panel (Fig. 4A) will then open the window displayed in Fig. 4C. The "set" button will place a furrow plane, and the user can then move it using "jiggle" (the furrow plane will move along the cell) or "rotate" (the furrow plane will rotate). Alternatively, the user can use the manual control (Fig. 4C, bottom right) to move the plane up and down along the cell (red ladder in Fig. 4C) or to rotate the division plane (red target in Fig. 4C). The user can then go back to the main panel (Fig. 4A) and measure the prospective daughter cells volumes. The mesh is saved as a binary file that stores the node positions and the connections that form the triangles. The furrow can be saved using a plain text file that stores the center of the furrow and the direction normal to the furrow plane. Both of these can be loaded for further use with the plugin. The mesh can also be exported as an STL file, a common file format used in 3D programs, or used to create a binary image. There is an undo button for all commands. The mesh can also be cleared and initialized again, or remeshed if required.

2.3 DETAILED DESCRIPTION OF PARAMETERS AND ANALYSIS SITUATIONS

Some of the challenges our method addresses are the segmentation of cells with extraneous appendages or of cells labeled with markers with large variations in intensity. In this section, we describe some of the approaches to use the plugin effectively.

There are eight numeric parameters in the plugin that can be tuned and adjusted to improve segmentation performance (Fig. 4A). Before creating a new mesh, the "divisions" parameter should be set to a small value. The number of divisions determines how many triangles the mesh will be initialized with (see Appendix). A value of two is a good starting number, it is accurate enough to represent a dividing cell, and creates few enough triangles for the mesh to deform quickly (see Fig. 2A). After a mesh has been created, the image weight needs to be adjusted. The "image weight" determines how strongly the mesh is deformed by the image. If the image weight is too low, the mesh will deform and shrink past desired features of the object being segmented. If the image weight is too high, the mesh will be overconstrained by the image and get stuck in local minima, which will make it appear jagged. Another problem that can occur when adjusting the image weight is that the mesh may deform too much during each step and become unstable. This can be corrected by increasing the parameter "gamma," a damping factor. Small values of gamma enable faster convergence but can result in an unstable mesh. Large values of gamma correspond to small deformations and slower mesh adjustment but stabilize the mesh. After adjusting gamma and image weight until the mesh deforms adequately, the number of divisions can be increased and the mesh "remeshed" to improve the accuracy of the segmentation. The parameters "alpha" and "beta" influence the way the mesh deforms. Alpha is the mesh stiffness. Increasing alpha can cause the mesh to move past unwanted features, while decreasing alpha may prevent the mesh from moving past desired features. Beta is the mesh bending stiffness, and increasing beta can help smooth out a rough or jagged mesh. The last three parameters all add additional forces to the mesh. The "pressure" is a force perpendicular to the surface of the mesh. If the pressure is positive, then the force causes the mesh to swell, similar to a balloon. If the pressure is negative, the mesh will shrink faster. The "normalize" parameter is a force that favors all the mesh triangles to be the same size. It acts parallel to the existing surface and is intended to distribute the triangles in a uniform fashion. Finally, the "curve weight" determines the strength with which the mesh deforms toward a user-defined curve (explained later).

Another way to modulate the behavior of the mesh is to select the image energy being used. The image energies can be selected from a combo-box on the main control panel (Fig. 4A shows "max intensity" selected). There are four available image energies: "max intensity," "max gradient," "perpendicular intensity," and "perpendicular gradient." Fig. 5 illustrates the effects on the contour of using different energies. Fig. 5A is an example of a typical HeLa cell undergoing cytokinesis. Fig. 5B shows the region that a simple threshold followed by a dilate operation would acquire. Fig. 5C shows a contour that was minimized to the "max intensity" energy that attracts the contour to bright ridges. Fig. 5D represents the magnitude of the intensity gradient in the image. The example cell has a bright actin cortex, which causes the gradient intensity to have two maxima. These maxima attract the contour when using the "max gradient" energy, as illustrated in Fig. 5E and F; the specific maximum followed is determined by the initialization. The two other energies, perpendicular intensity and perpendicular gradient, use the "max intensity" and "max gradient" energies but only apply forces perpendicular to the mesh. This helps prevent clustering of mesh nodes.

Many cell types have multiple microvilli, thin actin rich protrusions at their surface, and retraction fibers connecting them to the substrate throughout cytokinesis. When measuring cell volume using simple thresholding, these features will be included because they are as bright as the cell cortex, and as a result, cell volume will be overestimated (Fig. 5C). An active mesh using an intensity-based image energy will deform through these features. This yields a better estimate of the cell volume, especially compared to a 2D segmentation approach, where cell volumes are extrapolated from a 2D image by assuming rotational symmetry and where

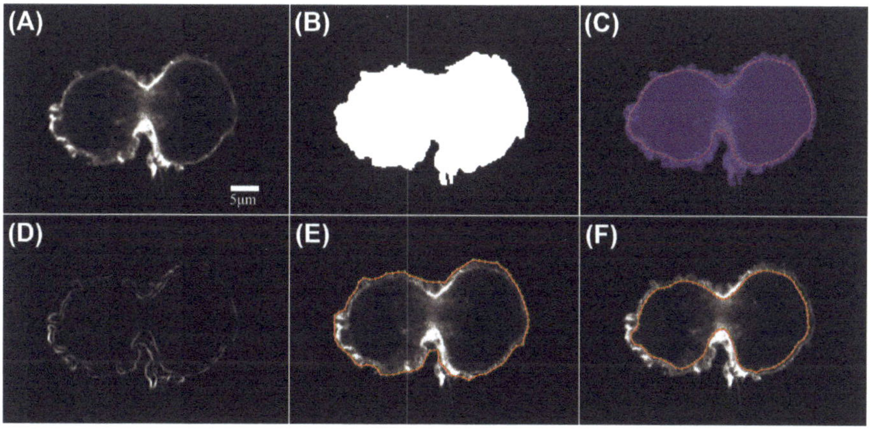

FIGURE 5

Comparison of different segmentation procedures. (A) 2D cross section of a dividing HeLa cell—expressing Lifeact-GFP (white). (B) Segmented image created by thresholding followed by fill holes and open binary image operations. (C) Overlay of binary segmentation (blue) on original cell labeled with the contour found using active contours with an intensity energy (*red line*). (D) Image showing the magnitude of the intensity gradient in the image displayed in A (generated using the "Find edges" plugin in ImageJ). (E) Segmentation of the cell displayed in A using a gradient energy. (F) Alternate curve found using a gradient energy.

microvilli and retraction fibers can lead to a large volume overestimate when extrapolated to 3D.

In images with inhomogeneous intensity distributions, excessively bright regions can pose challenges for segmenting. For example, in cytokinetic cells expressing an actin marker, the contractile ring is often considerably brighter compared to the cortex intensity at the poles; thresholding-based segmentation will lead to the volume of the cell near the ring to be overestimated, while the volumes of the poles will be underestimated. In our plugin, the deforming surface will deform to local maxima of bright ridges, so that the cell boundary will be detected accurately, provided it displays a bright ridge. However, if the variation in intensity is too large, the mesh nodes will cluster to the bright regions. To avoid this, we have included a perpendicular version of the image energy. By selecting a perpendicular energy, the force will only be applied perpendicular to the surface at each node. This reduces node clustering due to bright regions. The normalized parameter can also help reduce node clustering. If the intensity at the poles is so low that the edge no longer forms a ridge, then a gradient energy can be used. The gradient energy is based on the magnitude of the gradient (Fig. 5D). While the gradient energy is a better choice in the absence of bright ridges, when there is a bright ridge, the gradient energy can have two possible local minima and thus should not be used (Fig. 5 E and F).

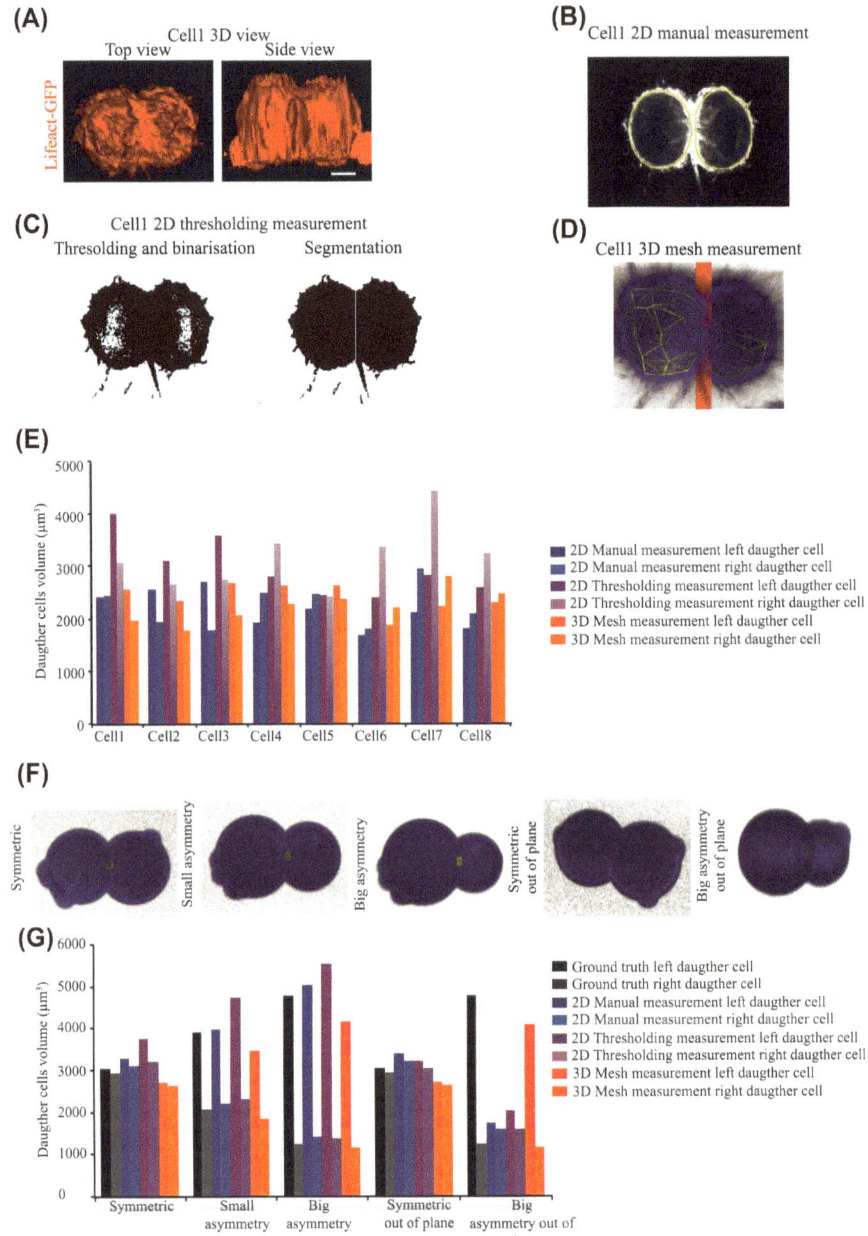

FIGURE 6

Comparison of different methods for the measurement of daughter cell size during cell division. (A) Dividing HeLa cell (Cell1) expressing Lifact-GFP (red). Left panel: top view. Right panel: side view. Scale bar: 5 µm. (B) 2D manual analysis: median plan of Cell1 with the

Finally we have included a way for users to define curves that influence the mesh. By clicking on the "curve editor" button (Fig. 4C), another ImageJ plugin (Smith et al., 2010) will be opened. A 2D curve can be created using the cross section of current furrow plane. Once a curve has been created, clicking the button, "add editor snake" a curve will be added to the list of curves. By selecting the curve from the list, it can be "activated" which means, it will apply a force to the mesh when the mesh is deformed. The scale of the force is determined by the "curve weight" parameter.

At the end of cytokinesis, the two separating daughter cells are only linked via a small connection. At this point the cell(s) will need to be segmented using two separate meshes.

2.4 COMPARISON OF TECHNIQUES

To evaluate the effectiveness of our method, we compared it to other commonly used methods for volume measurements (Fig. 6). We focused on cells toward the end of cytokinesis but before the chromosomes decondense, when the two daughter cells volumes are set, but the cells are still round (Fig. 6A). Our method was compared to two different types of analysis performed in Fiji: 2D manual analysis (Fig. 6B) and 2D thresholding analysis (Fig. 6C). For the 2D manual analysis, the median plan was selected and the cell contour was manually drawn on the two daughter cells (Fig. 6B). For the 2D thresholding analysis, the median plan was manually selected, and a threshold was then manually applied to binarize the image; the *Fill hole* and *Watershed* functions were then used to automatically detect the two daughter cells, and the volume of each cell was calculated from the measured area (Fig. 6C). Cell volumes were then extracted from the 2D analysis by assuming that the two daughter cells displayed spherical shapes. Finally, our plugin was used for 3D mesh analysis, with a division of three and an image weight of 0.01 (Fig. 6D).

regions used to measure the area of the cell drawn in yellow. (C) 2D thresholding analysis in Fiji. Left panel: median plan of Cell1 as a binary mask after thresholding. Right panel: median plan of Cell1 after automatic filling of the holes and watershed. (D) 3D mesh analysis: Cell1 (blue) with the deformed mesh (yellow) and the selected furrow plane (red). (E) Graph showing the volumes of the daughter cells for eight dividing HeLa cells measured using 2D manual analysis, 2D thresholding analysis, and 3D mesh analysis. (F) 3D views of simulated cells of known volumes. From left to right, the two daughter cells are of the same size (symmetric), of slightly different sizes (small asymmetry), of very different sizes (big asymmetry), of the same size but not in the same Z-plane (symmetric out of plane), or of very different sizes and not in the same Z-plane (big asymmetry out of plane). (G) Graph showing the volumes of the daughter cells for the five simulated dividing cells measured using 2D manual analysis, 2D thresholding analysis, and 3D mesh analysis compared to the actual known volumes (ground truth).

We found that the three types of analysis yielded very different results. Overall, 2D thresholding gave very high cell volumes compared to the two other methods (Fig. 6E), presumably because of the presence of multiple microvilli (Fig. 6B) or cell–cell contacts that can perturb automated thresholding. The 2D manual analysis and the 3D mesh analysis gave relatively comparable results (Fig. 6E).

To further compare the different analysis pipelines, we generated simulated cells, which volumes can be used as a ground truth to compare the results of the different methods to. Simulated images of cells were generated by creating an image stack of 60 slices with a height of 250 pixels and a width of 400 pixels. Consistent with our typical microscopy settings, the x and y pixel dimensions were 0.188 μm, and z dimension was 0.47 μm. Cells were simulated using two intersecting spheres to represent the cell body; blebs, often observed at the surface of dividing cells, were then added as smaller spheres intersecting the surface of the cell body. Each pixel was then initialized to either 0, for pixels outside of the cell, or, for pixels inside of the cell, to a value reflecting their position with the respect to the cell cortex, so that the resulting image mimics a cell with a bright cortex and cytoplasmic background. Specifically, initial pixel intensities inside the cell followed a decaying exponential from the surface toward the center of the cell. After the initial representation was created, a Gaussian blur was applied and then white noise was added to the image. The two daughter cells were either of the same size (symmetric) or of different sizes (small or big asymmetry) and were positioned with their long axis either in the plane of observation or out of plane (Fig. 6F). Comparison of the volumes extracted using the different methods with the ground truth values (Fig. 6G, compare colored bars to black and gray bars) showed that for cells dividing in the plane of imaging, the 2D manual and 3D mesh methods gave comparable results, while the 2D thresholding method yielded generally overestimated values. In contrast, both 2D methods proved to be poorly suited to measure daughter cell volumes when the division axis is not parallel to the plane of observation. In summary, our comparison reveals that our assisted plugin yields accurate daughter cell volume values. While it does not perform particularly better than a 2D segmentation procedure for smooth cells dividing in the plane of imaging, it has the advantage of working in 3D, which is particularly important for analyzing cells dividing out of plane.

CONCLUSION

We have described here a plugin to accurately measure the size of the daughter cells during cytokinesis. This plugin is available for Fiji and ImageJ and is based on the precise positioning of a deformable surface (active mesh) that minimizes the energy of an image to best represent the underlying object. It also includes a feature to automatically locate the cleavage furrow. Comparisons of different methods used to measure the size of daughter cells show that this plugin yields very good results even when the cell divides in three dimensions and not strictly parallel to the plane of imaging. The plugin is adaptable to various labeling and imaging situations. While it might not be adequate for high-throughput screening where image resolution might not be high enough, it is appropriate for creating precise datasets that can be used as a metric for further automated analysis.

ACKNOWLEDGMENTS

This work was supported by funding from the Medical Research Council, United Kingdom (core funding to the MRC LMCB) and by the European Research Council (starting grant 311637-MorphoCorDiv to EKP). A.C. acknowledges support from the European Molecular Biology Organisation (ALTF 563-2015) and the Fondation Bettencourt-Schueller.

APPENDIX

We have based our implementation of an active contour mesh on a formulation described previously (Kass et al., 1988; Smith et al., 2010). In brief, we create a connectivity matrix that represents the internal energies, apply the external energies as forces, and then iteratively update the positions of the mesh.

To build the connectivity matrix, we sum the internal energy, elastic and bending energies, at each node. The elastic component of the internal energy for node i is the sum overall of the connections that connect to i:

$$\sum_{n} \alpha_n (r_i - r_j)$$

where each connection, n, connects two nodes, i and j. In the plugin, α_n is set by the parameter alpha.

The curvature is a bit more complicated because it requires pairs connections. For each node, i, we sum overall of the first order connection pairs, $\{m, n\}$, in nonadjacent triangles (red connections in Fig. 2).

$$\sum_{\{n,m\}^1} \beta_n (2r_i - r_j) + \beta_m (2r_i - r_k)$$

where nodes j and k are connected to node i by the n and m connections, respectively.

The second order connections (consecutive red and blue lines in Fig. 4B) are nodes that connect across two triangles that share the same edge. The second triangle does not include the original node. This leads to the following matrix components:

$$\sum_{\{n,m\}^2} \beta_n (r_i - 2r_j) + \beta_m (r_i - 2r_k)$$

In the plugin, first and second order connection stiffness, β, is set by the parameter beta in the plugin. The corresponding equation of motion for each node is thus:

$$\sum_{n} \alpha_n (r_i - r_j) + \sum_{\{n,m\}^1} \beta_n (2r_i - r_j) + \beta_m (2r_i - r_k) + \sum_{\{n,m\}^2} \beta_n (r_i - 2r_j) + \beta_m (r_i - 2r_k) + F(r)_i = 0$$

The internal energy is linear in r and can be used to form a connectivity matrix A such that our problem reduces to satisfying the following equation:

$$Ar + F(r) = 0$$

The solution is found iteratively by assuming F remains constant during an iteration step and including a damping term to prevent too large of step sizes.

$$Ar_t + F(r_{t-1}) = -\gamma(r_t - r_{t-1})$$

where γ is a damping factor called gamma in the plugin. The new positions, r_t, can be found by solving the matrix equation.

$$r_t = (A + \gamma I)^{-1} [\gamma r_{t-1} - F(r_{t-1})]$$

At each step the external energies, F, are calculated as follows. The "pressure" is a force that pushes perpendicular to a triangle's surface. It is an external energy applied in a similar manner to the image energy, and it can be used to speed up convergence. The pressure force on node i is found by summing overall of the triangles that contain the node (Fig. 2C):

$$F_p = \frac{1}{N} \sum_{s} p_0 \widehat{n}_s$$

where p_0 is the "pressure" parameter in the plugin, s indicates the triangle, $\widehat{n}_s$ is the unit vector normal to the triangle, and N is the total number of triangles containing node i. The "normalize" force is calculated by finding the total surface area of the mesh and dividing by the number of triangles to get an average area per triangle, then finding the side length, l_0, of an equilateral triangle with an equivalent area. For each node, the total force is the sum overall of the connections.

$$F_N = \sum_{n} k_N (l - l_0) \widehat{c}$$

where k_N is the parameter "normalize" input by the user, and $\widehat{c}$ is the unit vector along the line connecting the two nodes.

The image energy is the same as previously described (Smith et al., 2010), except for the perpendicular energies. The perpendicular versions of the energies are calculated by only applying the component of the force that is along the normal of the existing surface.

The mesh initialized by user input is obtained using ray casting. First, the user creates spheres that enclose the desired region (Fig. 3B; the closer to the actual shape the faster convergence happens). Once the user has finished an icosahedron is generated at the center of the cell, and the triangles of the icosahedron are subdivided according to the number of divisions selected (Fig. 2A). The points of the resulting mesh are then mapped to the points of the user-defined spheres. The center of the mesh is placed at the center of mass of the user-defined spheres, and the points on the mesh are mapped by ray casting from the center to the user input shape (Fig. 2D).

REFERENCES

Almonacid, M., Terret, M.-É., & Verlhac, M.-H. (2014). Actin-based spindle positioning: new insights from female gametes. *Journal of Cell Science, 127*, 477–483.

Bement, W. M., Benink, H. A., & von Dassow, G. (2005). A microtubule-dependent zone of active RhoA during cleavage plane specification. *Journal of Cell Biology, 170*, 91–101.

Betschinger, J., & Knoblich, J. A. (2004). Dare to be different: asymmetric cell division in *Drosophila, C. elegans* and Vertebrates. *Current Biology, 14*, R674–R685.

Bringmann, H., & Hyman, A. A. (2005). A cytokinesis furrow is positioned by two consecutive signals. *Nature, 436*, 731–734.

Busson, S., Dujardin, D., Moreau, A., Dompierre, J., & Mey, J. R. D. (1998). Dynein and dynactin are localized to astral microtubules and at cortical sites in mitotic epithelial cells. *Current Biology, 8*, 541–544.

Cabernard, C., Prehoda, K. E., & Doe, C. Q. (2010). A spindle-independent cleavage furrow positioning pathway. *Nature, 467*, 91–94.

Cabernard, C., & Doe, C. Q. (2009). Apical/basal spindle orientation is required for neuroblast homeostasis and neuronal differentiation in *Drosophila. Developmental Cell, 17*, 134–141.

Chaigne, A., Verlhac, M.-H., & Terret, M.-E. (2012). Spindle positioning in mammalian oocytes. *Experimental Cell Research, 318*, 1442–1447.

Delingette. (1994). *Simplex meshes: A general representation for 3D shape reconstruction* (pp. 856–859). IEEE Computer Society Press.

Gönczy, P., & Rose, L. S. (2005). Asymmetric cell division and axis formation in the embryo. *WormBook: The Online Review of* Caenorhabditis Elegans *Biology*, 1–20.

Green, R. A., Paluch, E., & Oegema, K. (2012). Cytokinesis in animal cells. *Annual Review of Cell and Developmental Biology, 28*, 29–58.

Heimann, T., & Meinzer, H.-P. (2009). Statistical shape models for 3D medical image segmentation: a review. *Medical Image Analysis, 13*, 543–563.

Kaltschmidt, J. A., Davidson, C. M., Brown, N. H., & Brand, A. H. (2000). Rotation and asymmetry of the mitotic spindle direct asymmetric cell division in the developing central nervous system. *Nature Cell Biology, 2*, 7–12.

Kass, M., Witkin, A., & Terzopoulos, D. (1988). Snakes: active contour models. *International Journal of Computer Vision, 1*, 321–331.

Kotak, S., & Gönczy, P. (2013). Mechanisms of spindle positioning: cortical force generators in the limelight. *Current Opinion in Cell Biology, 25*, 741–748.

Montagnat, J., Delingette, H., & Ayache, N. (2001). A review of deformable surfaces: topology, geometry and deformation. *Image and Vision Computing, 19*, 1023–1040.

Ou, G., Stuurman, N., D'Ambrosio, M., & Vale, R. D. (2010). Polarized myosin produces unequal-size daughters during asymmetric cell division. *Science, 330*, 677–680.

Pappas, T. N. (1992). An adaptive clustering algorithm for image segmentation. *IEEE Transactions on Signal Processing, 40*, 901–914.

Sedzinski, J., Biro, M., Oswald, A., Tinevez, J.-Y., Salbreux, G., & Paluch, E. (2011). Polar actomyosin contractility destabilizes the position of the cytokinetic furrow. *Nature, 476*, 462–466.

Shih, F. Y. (2010). *Image processing and pattern recognition: Fundamentals and techniques.* Piscataway, NJ : Hoboken, N.J: IEEE Press ; Wiley.

Smith, M. B., Li, H., Shen, T., Huang, X., Yusuf, E., & Vavylonis, D. (2010). Segmentation and tracking of cytoskeletal filaments using open active contours. *Cytoskeleton (Hoboken, N.J.), 67*, 693–705.

Su, K.-C., Takaki, T., & Petronczki, M. (2011). Targeting of the RhoGEF Ect2 to the equatorial membrane controls cleavage furrow formation during cytokinesis. *Developmental Cell, 21*, 1104–1115.

Terzopoulos, D., Witkin, A., & Kass, M. (1988). Constraints on deformable models: recovering 3D shape and nonrigid motion. *Artificial Intelligence, 36*, 91–123.

Xu, C., & Prince, J. L. (1998). Snakes, shapes, and gradient vector flow. *IEEE Transactions on Image Processing: A Publication of the IEEE Signal Processing Society, 7*, 359–369.

Yamashita, Y. M., Mahowald, A. P., Perlin, J. R., & Fuller, M. T. (2007). Asymmetric inheritance of mother versus daughter centrosome in stem cell division. *Science, 315*, 518–521.

CHAPTER

Nuclear displacement and fluorescence recovery after photobleaching (FRAP) assays to study division site placement and cytokinesis in fission yeast

20

P. Ullal, P. Bhatia, S.G. Martin[1]

University of Lausanne, Lausanne, Switzerland

[1]*Corresponding author: E-mail: Sophie.Martin@unil.ch*

CHAPTER OUTLINE

Abstract

Cytokinesis is an essential cellular event that completes the cell division cycle. It begins with the assembly of an actomyosin contractile ring that undergoes constriction concomitant with the septum formation to divide the cell in two. Placement of the septum

Methods in Cell Biology, Volume 137, ISSN 0091-679X, http://dx.doi.org/10.1016/bs.mcb.2016.03.031

at the right position is important to ensure fidelity of the division process. In fission yeast, the medially placed nucleus is a major spatial cue to position the site of division. In this chapter, we describe a simple synthetic biology-based approach to displace the nucleus and study the consequence on division site positioning. We also describe how to perform fluorescence recovery after photobleaching to follow the dynamics of cytokinetic proteins at defined time points by live-cell microscopy.

INTRODUCTION

DIVISION SITE POSITIONING AND CYTOKINESIS

The fission yeast *Schizosaccharomyces pombe* serves as an excellent model system to study eukaryotic cell division. These rod-shaped cells show regularity in their growth pattern and divide at a constant cell size medially by assembling an actin–myosin-based contractile ring. The placement of the division site defines whether division is symmetric or asymmetric, and thus plays a key role in specifying daughter cell identities. This process has been extensively studied in fission yeast. In particular, the nucleus plays an important role in defining the position of the future division site, unlike in animal cells, where the mitotic spindle guides the division plane positioning (Almonacid & Paoletti, 2010; Oliferenko, Chew, & Balasubramanian, 2009). In fission yeast, pushing forces from the dynamic interphase MTs place the nucleus at the cell middle (Tran, Marsh, Doye, Inoue, & Chang, 2001). Displacing the interphase/premitotic nucleus to off-center positions by centrifugation (Daga & Chang, 2005) or optical micromanipulation (Tolic-Norrelykke, Sacconi, Stringari, Raabe, & Pavone, 2005) results in cells dividing asymmetrically with the actomyosin ring forming at the displaced position of the nucleus.

The positioning of the nucleus determines the division site placement via an anillin-like protein called Mid1 (Chang, Woollard, & Nurse, 1996; Sohrmann, Fankhauser Brodbeck, and Simanis, 1996), which shuttles between the nucleus and the cell cortex (Paoletti & Chang, 2000). Mid1 forms complexes with several proteins at the cell cortex, resulting in formation of cytokinetic nodes (Lee, Coffman, & Wu, 2012), and recruits components of the actomyosin ring to begin ring assembly (Motegi, Mishra, Balasubramanian, & Mabuchi, 2004). The *mid1* gene is nonessential, but mutant cells show defects in ring positioning (Chang et al., 1996; Sohrmann et al., 1996). The central nuclear position and Mid1 thus act as key positive signals to place division plane at cell middle. In addition, protein complexes at the cell tips of fission yeast provide negative signals that prevent division site placement at cell ends (Celton-Morizur, Racine, Sibarita, & Paoletti, 2006; Huang, Chew, Ge, & Balasubramanian, 2007; Padte, Martin, Howard, & Chang, 2006; Ullal et al., 2015). Pom1, a DYRK-family kinase, is one of the main proteins that act as a negative signal to ensure medial cell division. This kinase contributes to restricting Mid1 localization to the cell middle (Celton-Morizur et al., 2006; Padte et al., 2006; Rincon et al., 2014), and also prevents septation at cell poles independently of Mid1 by negatively regulating the F-BAR protein Cdc15, a key component of the cytokinetic machinery, to prevent cytokinesis at cell ends (Fankhauser et al., 1995; Huang et al., 2007; Ullal et al., 2015). In a *pom1Δ* mutant, Mid1 localization expands over one of the cell poles and therefore shifts the septum positioning toward that pole (Padte et al., 2006). In the absence of Mid1, Pom1-dependent phosphorylation of Cdc15 becomes essential to promote Cdc15 and ring dynamicity and prevent septum formation at cell poles (Ullal et al., 2015).

The assembly of the cytokinetic ring is initiated through the formation of cytokinetic nodes, normally nucleated by Mid1. These nodes, which nucleate and capture actin filaments, coalesce into an actomyosin ring, which constricts until membrane abscission concomitant with formation of primary and secondary septa. The primary septum is eventually digested for cell separation (Oliferenko et al., 2009). Fluorescent tagging of components of the contractile ring has been very useful in understanding the dynamics of the ring structure, and timing of cytokinesis (Ge & Balasubramanian, 2008; Wu, Kuhn, Kovar, & Pollard, 2003). The latter can be monitored by live-cell imaging by marking spindle pole bodies and using their separation as reference time (Wu et al., 2003). One useful technique to study the maturation of the cytokinetic apparatus is fluorescence recovery after photobleaching (FRAP), which allows one to follow the change in dynamics of ring components during cytokinesis progression.

In this chapter, we discuss in detail two protocols to study division plane positioning and cytokinesis in fission yeast. We first describe a novel method to displace the nucleus and study the effects of this displacement on positioning of the division site. This method displaces the nucleus toward a cell pole by inducing, through a synthetic motor chimera, its intracellular transport along actin cables (Lo Presti, Chang, & Martin, 2012). We then describe the use of FRAP to study the dynamics of actomyosin ring components at defined cytokinetic times.

1. METHODS

1.1 SLIDE ASSEMBLY

For all of the experiments described below, unless mentioned otherwise, slide assembly was carried out in the following way before imaging (Fig. 1A).

1. Prepare 2% agarose in EMM medium with appropriate amino acid supplements.
2. Melt one aliquot at 95°C in heat block for 10–15 min.
3. Add 100 μL of agarose onto a microscope slide, wait for a few seconds and place another slide on top to spread the agarose and obtain a pad of around 1 mm thickness.
4. Let the agarose solidify for 1–2 min.
5. During this time concentrate cells by centrifugation (30 s at 1000 g), pipet the supernatant off, leaving about 20 μL of medium in which you resuspend the cells.
6. Remove the upper slide from the agarose pad sandwich.

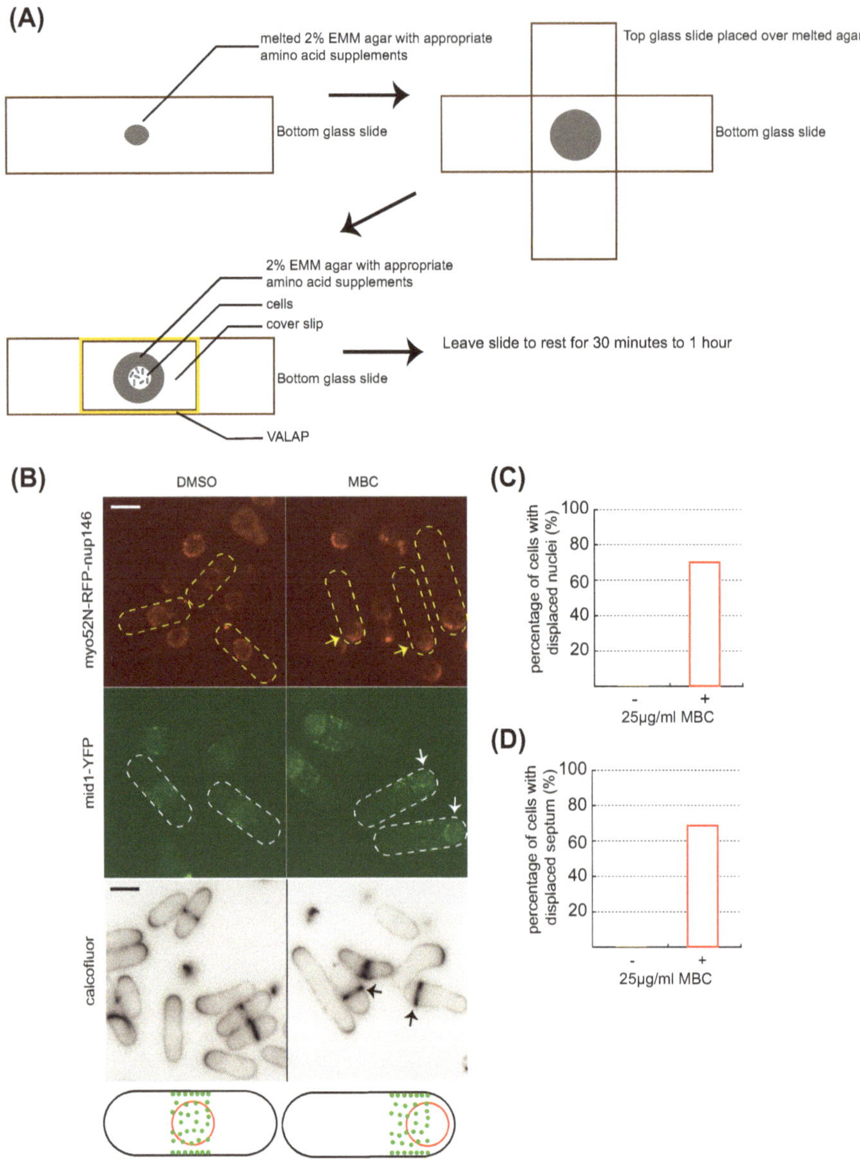

FIGURE 1

Microscope slide assembly and experimental displacement of the nucleus from the cell middle to study division site positioning in fission yeast *Schizosaccharomyces pombe*. (A) A schematic view of slide assembly. (B) Cells expressing the Myo52N-Nup146 chimera showing centered nuclei (left) and displaced nuclei after methyl benzimidazole 2-yl carbamate treatment (right). Top panel shows cells expressing the tagged version of the chimera, Myo52N-RFP-Nup146. *Yellow arrows* show cells with displaced nuclei. Middle

7. Add 1 μL of cell suspension on top of the pad, wait for 30 s for the drop to start drying and cover with a coverslip (24 × 32 mm), spreading the cells evenly across the pad.
8. Seal the pad on all sides with VALAP, a mixture of lanolin, Vaseline (or other petroleum jelly), and paraffin to prevent pad from drying.
9. Let the pad sit for at least 45 min at 25°C before imaging.

Notes:

- During step 3, make sure you do not create air bubbles, which could create unwanted drifts on the pad during imaging.
- Between steps 7 and 8, ensure that the cell suspension is sufficiently dry before sealing to ensure that cells are static on the pad.

Materials:

- Glass slides (76 × 26 × 1 mm, Menzel Gläser, MNJ-250-005L)
- Coverslips (24 × 32 × 0.17 mm, Menzel Gläser, MENZBB024032AC13)
- EMM medium and amino acid supplements (US Biological)
- Agarose (Promega, V3125)
- VALAP (lanolin (Sigma, L7387), Vaseline (Reactolab, 92045-74-4) (or other petroleum jelly), and paraffin (Reactolab, 7005600) in equal weights)

1.2 NUCLEAR DISPLACEMENT ASSAY

The fission yeast nucleus provides spatial information for division site positioning. Several methods, through centrifugation or the use of laser tweezers, have been used to displace the interphase/premitotic nucleus to off-center positions (Daga & Chang, 2005; Tolic-Norrelykke et al., 2005). Here we present a third method to displace the nucleus from the cell center to a cell extremity. This method relies on the transport of the nucleus along actin cables nucleated from the cell poles, through the use of a chimeric protein between the type V myosin motor protein Myo52 and a nuclear pore complex protein Nup146 (Myo52N-RFP-Nup146) (Lo Presti et al., 2012). The *myo52N-RFP-nup146* transgene is integrated in the genome at the *ura4+* locus under control of the *nmt* promoter (see materials below) and is introduced in the strain of interest by genetic crosses. If the red channel is to be used to monitor the localization of a protein of interest, use of an unlabeled transgene, *myo52N-12myc-nup146*, is also possible.

panel shows cells expressing an untagged Myo52N-Nup146 chimera and Mid1-YFP. *White arrows* show off-centered Mid1 localization. The bottom panel shows calcofluor-stained cells to visualize the septum in cells expressing Myo52N-RFP-Nup146. *Black arrows* show cells with off-centered septa. (C) Quantification of displaced nuclei in cells expressing *Myo52N-RFP-Nup146.* (D) Quantification of displaced septum in cells expressing *Myo52N-RFP-Nup146.* Bars are 5 μm.

1. Grow a 3-mL preculture of the *myo52N-RFP-nup146* strain overnight at 25°C in EMM-AL supplemented with thiamine (5 μg/mL) to an OD 600 of 0.6–1.
2. Centrifuge 1.5 mL of culture for 30 s at 1000 g and wash three times with EMM-AL without thiamine.
3. Resuspend in EMM-AL and dilute to appropriate OD 600 in a volume of 10 mL, such that the cells reach mid-log phase after 16 h of culture at 25°C.
4. Grow the strain for 16 h at 25°C in a 100-mL culture flask to induce expression of the Myo52N-RFP-Nup146 chimera. The final OD 600 should be between 0.6 and 0.8. To displace the nucleus, it is necessary to remove microtubules, which provide pushing forces that normally center the nucleus (Tran et al., 2001). Pipet 3 mL of the culture to two distinct culture tubes. To one, add 3 μL DMSO as negative control. To the other, add 3 μL methyl benzimidazole 2-yl carbamate (MBC) 1000× stock. Incubate at 25°C with agitation.
   Note: Microtubule depolymerization is not necessary in mutant strains with defective microtubule cytoskeleton, in which the microtubules are not effective in exerting pushing forces against cell poles. Microtubule depolymerization is also not essential in mutant strains growing in a monopolar manner, in which the actin cables are nucleated from only one cell pole. Such asymmetric organization is sufficient to unbalance the forces, and permit the displacement of the nucleus toward the growing cell end.
5. After 4 h (20 h from the start of *myo52N-RFP-nup146* induction), harvest each culture for slide preparation. For long-term live imaging of the nucleus (labeled by the Myo52N-RFP-Nup146 chimera) and other fluorescent markers, prepare cells on an agarose pad as described in Section 1 above.
6. Acquire microscopy images. We typically use a spinning disk microscope, acquiring optical slices every 0.3 μm, but an epifluorescence microscope is also suitable.
   Note: We used the remainder of the samples to visualize the later stages of cytokinesis using calcofluor white, a fluorophore, which is commonly used to visualize cell wall material and labels the newly formed septum.
7. For calcofluor staining, add 1 μL of a 1 g/L stock of calcofluor to 500 μL of cells before the cells were centrifuged and concentrated (at step 6 above).
8. Concentrate cells by centrifugation (30 s at 1000 g), pipet the supernatant off, leaving about 20 μL of medium in which you resuspend the cells.
9. To prepare slides for calcofluor visualization, pipet 1.5 μL of the resuspended cells onto a slide and cover with a 18 × 18-mm coverslip. This method is also suitable for rapid visualization of fluorescent markers in live cells (within 10 min of slide preparation).

   Notes:

- While preparing slides for calcofluor staining, ensure the cells spread nicely under the coverslip, so that they are immobile. However, be very careful not to press on the coverslip with your fingers as you may crush or distort the cells.

Materials:

- Strain containing *nmt81-myo52N-RFP-nup146-ura4+* transgene. This can be obtained from genetic crosses of your strain of interest with either of (available upon request):
  - h-*ura4-294:nmt81-myo52N-RFP-nup146-ura4+* leu1-32 (YSM2631)
  - h-*ura4-294:nmt81-myo52N-12myc-nup146-ura4+* leu1-32 (YSM2784)
- Incubator/shaker
- Microscope

  Note: We typically use a spinning disk microscope—Leica DMI4000B inverted microscope equipped with HCX PL APO X100/1.46 (NA) oil objective and a PerkinElmer Ultraview Confocal system (including a Yokagawa CSU22 real-time confocal scanning head, and solid-state laser and a cooled 14-bit frame transfer EMCCD C9100-50 camera) to image genetically encoded fluorescence signal. To visualize the septum with calcofluor, any epifluorescence microscope with UV illumination will be appropriate.
- Coverslips (18 × 18 mm Menzel Gläser)
- methyl benzimidazole 2-yl carbamate (MBC) (Sigma, 378674)
- DMSO (Sigma, D4540)
- Calcofluor white (Sigma, 18909)

1.2.1 Data analysis

As proof of concept, we used a wild-type fission yeast strain expressing the Myo52-RFP-Nup146 chimera. Fig. 1B shows wild-type cells expressing the Myo52-RFP-Nup146 chimera. The nucleus is visualized in the red channel. Cells treated with vehicle (DMSO) showed centrally positioned nuclei and septum (calcofluor). By contrast, >60% of cells treated with MBC showed nuclear displacement toward the cell poles (Fig. 1C), with a similar percentage of cells showing septum formation displaced from the cell middle (Fig. 1D).

To visualize the early stages of division site positioning, we also used a strain containing the chimera as well as a tagged version of the Mid1 protein. The *Myo52N-12myc-Nup146 Mid1-YFP* strain showed Mid1 localizing to the cell middle when treated with DMSO, while in cells treated with MBC, Mid1 was clearly seen to localize close to or at the cell poles (Fig. 1B, middle panel).

1.3 MEASURING DYNAMICS OF CYTOKINETIC RING COMPONENTS USING FLUORESCENCE RECOVERY AFTER PHOTOBLEACHING

FRAP is a fluorescence imaging technique used to study the kinetics of protein movement in live cells (Reits & Neefjes, 2001). The protein of interest is tagged with a fluorophore, usually GFP or derivative, and intense illumination is used to photobleach the fluorophore in a defined area of interest. For photobleaching, laser light of appropriate wavelength is typically used, though other setups are possible.

Time-lapse image acquisition pre- and postphotobleaching allows the measurement of fluorescence recovery over time in the bleached zone. The dynamics of the protein under study is then described through two main parameters: the halftime of recovery, which represents the time necessary for the recovery to reach half of its maximal value, and mobile fraction, which represents the fraction of signal that contribute to fluorescence recovery.

FRAP measurements provide a measure of protein diffusion, which can be restricted through interaction with binding partners. Thus, a given protein may exhibit different FRAP dynamics at distinct cell cycle times or distinct cellular locations. Such dynamics changes can be informative to infer changes in the local microenvironment of the protein under study. To study the dynamics of proteins involved in the cytokinesis, it is important to monitor the progress of the division process and ensure results are obtained for defined cytokinesis stages. A good reference time is the separation of the spindle-pole body (SPB), the yeast equivalent of the centrosomes, which marks the entry into mitosis (Wu et al., 2003). The method below provides a detailed description on how to perform and quantify FRAP at defined cytokinetic stages.

1.3.1 Data acquisition

1. Inoculate a 3-mL EMM-ALU culture with a freshly plated strain expressing Sid4-mRFP and your favorite protein tagged with GFP (X-GFP) and let it grow for about 16 h (overnight) at 25°C. The day after, dilute the cells to $OD_{600} = 0.3$ to 0.4 and grow for another 5–6 h to ensure that the cells are in mid-log phase ($OD_{600} = 0.6$–0.8).
2. Harvest 1 mL for slide preparation (see Section 1.1).
3. To choose the cells for photobleaching, start a time-lapse for a field of cells to monitor the SPB and X-GFP in real time. We typically use 4 min time intervals and acquire optical slices every 0.6 μm at each interval. Ensure you minimize signal loss due to imaging by using low exposure times.
   Note: At this point, use the Sid4-mRFP signal as a timer for cells entering mitosis. SPB separation, visualized as the transition from one to two dots of Sid4-mRFP, is T_0. At room temperature (22–23°C), ring formation occurs in the 32 min following SPB separation, while ring constriction take place from approximately 36–70 min
4. At the appropriate time after SPB separation, set the microscope to photobleach the fluorescence signal in the cytokinetic ring in a circular zone of 1 μm diameter.
   Note: the laser intensity required for photobleaching has to be defined in preliminary experiments. The ideal setting will yield a visible, but nonzero bleached zone in the first image immediately after photobleaching.
5. Acquire images before photobleaching and immediately after at regular intervals.
   Note: The optimal time interval between images needs to be defined in preliminary experiments that estimate the time required for maximal fluorescence

recovery in the bleached zone. In the example described later with a recovery halftime of 10–15 s, images were acquired at 1 s interval for 90 s.

1.3.2 Image analysis

Analysis of FRAP data requires measurement of mean fluorescent intensities over time in three regions: (1) the photobleached region, (2) the background outside the cell, and (3) another nonbleached cell. The first of these measurements allows quantification of the loss and recovery of signal upon photobleaching. The two others permit correction of this measure for system noise and imaging-dependent signal loss. We use the open-source image processing software, ImageJ (http://imagej.nih.gov/ij/), to perform this analysis.

1. Export your imaging data to ImageJ.
2. To do so first adjust the brightness and contrast of the image by selecting the "Image" function followed by "Adjust-Brightness/Contrast."
3. Select Analyse > Set measurements...; in the dialog box, select "Mean grey value" and click ok.
4. Select Analyse > Tools > ROI Manager; in the dialog box, click on "Show All."
5. Using the "Oval selections" tool, select the photobleached region and click on "Add" in the ROI Manager. Repeat this for the background outside the cell and the nonbleached cell.
6. To measure the intensities of these zones for all time points, select More > Multi Measure in the ROI Manager. Click "Measure All" slices and "One row per slice" and perform the measurement.
7. Export the fluorescence measurements to a spreadsheet software (Excel or equivalent).
8. Go back to your image and flatten and save it as a TIFF file.
9. In your spreadsheet software, subtract the background fluorescence from the fluorescence intensities of the photobleached region and the nonbleached cell.
10. Correct for loss of signal as a result of imaging by dividing at each time point the adjusted bleached region's intensity by the adjusted intensity of the nonbleached cell.
11. Normalize all values so that the prephotobleaching value equals to 1.
12. Calculate mobile fractions (F_m) and halftime ($t_{1/2}$) values from the best-fit curve equation: $y = a \cdot (1 - e^{-bx})$, where $a = F_m$ and $\text{In}0.5/b = t_{1/2}$ using a curve-fitting program (CurveExpertPro software or equivalent).

Note: The maximal fluorescence recovery may be significantly lower than the prebleach intensity, causing the curve to plateau at a lower intensity than that of the prebleach time point. The derived mobile fraction will be meaningful only if a very small fraction of the total cell intensity has been photobleached (which can be estimated by measuring the total fluorescence intensity of the photobleached cell before and after photobleaching).

Materials:

- Strain of genotype *sid4-mRFP::ura4+ X-GFP*. This can be obtained from genetic crosses of your strain of interest with (available upon request):
  - h-*sid4-mRFP-ura4+* leu1-32 ura4-D18 (YSM1485)
- Incubator/shaker
- Any microscope with a light-focusing device that can be used for photobleaching
- ImageJ, Image processing software (http://imagej.nih.gov/ij/)
- Spreadsheet software (Excel or equivalent)
- Curve-fitting software (CurveExportPro or equivalent)

1.3.3 Data analysis

As proof of concept, we measured the dynamics of Cdc15, a phosphor-regulated protein that is found abundantly in the actomyosin ring and is essential for cytokinesis (Fankhauser et al., 1995; Roberts-Galbraith et al., 2010; Wu et al., 2003). We performed the experiments at early stages of ring formation using FRAP in wild-type and *pom1Δ* strains. Pom1 is a DYRK-family kinase (Bahler & Pringle, 1998), which directly phosphorylates Cdc15 on multiple residues (Ullal et al., 2015). Our aim was to test whether Pom1-dependent phosphorylation of Cdc15 altered Cdc15 dynamics.

Live-cell imaging of the *meGFP-Cdc15 Sid4mRFP* strain showed us the stage of ring formation, which starts immediately after SPB separation. As has been observed before, Cdc15 localization to the actomyosin ring was concomitant with SPB separation (Fig. 2A).

To assess Cdc15 dynamics during ring formation, we chose cells 16 min after SPB separation. Fig. 2B shows a time-lapse of meGFP-Cdc15 in wild-type cells pre- and postphotobleaching. Similar images were acquired for the *pom1Δ* strain (data not shown). Graphical representation of the fluorescence recovery is shown in Fig. 2C. Mobile fraction and half-time calculations showed us that in *pom1Δ*, the mobile fractions were significantly reduced and the halftimes were significantly higher (Fig. 2D and E) compared to the wild-type strain. These data imply that Pom1 promotes the dynamic turnover of Cdc15 within the cytokinetic ring.

CONCLUSION

Here, we have presented techniques to study division site positioning and cytokinesis in fission yeast *S. pombe*. The method we use here to displace the nucleus is an alternative to methods such as centrifugation and micromanipulation, which involve subjecting the cells to physical stress. Our chimera induces a stable and efficient nuclear displacement without subjecting the cells to such stresses. The FRAP method presented here provides a point-by-point protocol, which will be useful to systematically probe the changes in protein microenvironment during cytokinesis.

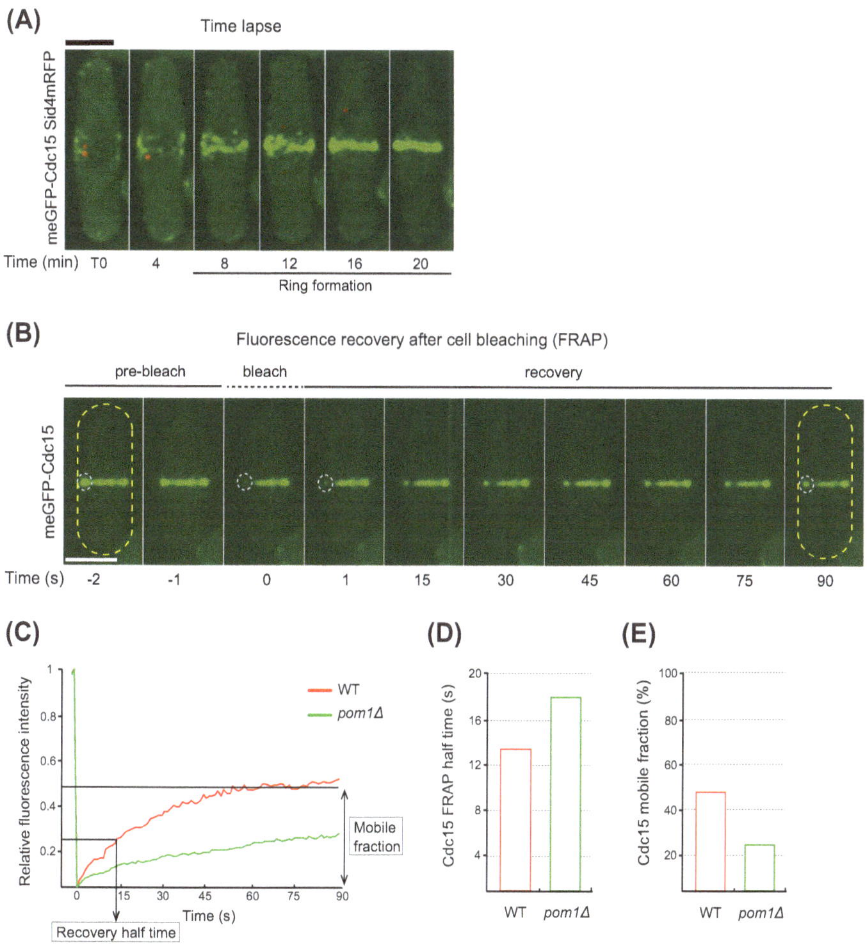

FIGURE 2

Fluorescence recovery after photo bleaching (FRAP) analysis of meGFP-Cdc15 at defined cytokinesis stages. (A) meGFP-Cdc15 (green) was imaged live at 25°C in wild-type cells. Separation of the spindle-pole body, marked by Sid4-mRFP (red), was used as time reference to determine cytokinetic stage. (B) Time-lapse images showing the stages of FRAP analysis including prebleach, bleach, and recovery of meGFP-Cdc15 during ring formation. (C) Fluorescence recovery curves of meGFP-Cdc15 in wild-type and *pom1Δ* strains. The *black lines* schematically indicate the significance of halftime and mobile fraction. (D, E) Recovery halftimes and mobile fractions of meGFP-Cdc15 in wild-type and *pom1Δ* strains as calculated from best-fit curves. Bars are 5 μm.

ACKNOWLEDGMENTS

We thank Laura Merlini for careful reading of the manuscript and Libera Lo Presti for the development of the *myo52N-nup146* transgenes. This work was supported by an ERC Starting Grant (GeometryCellCycle) to SGM.

REFERENCES

Almonacid, M., & Paoletti, A. (2010). Mechanisms controlling division-plane positioning. *Seminars in Cell & Developmental Biology, 21*(9), 874–880. http://dx.doi.org/10.1016/j.semcdb.2010.08.006.

Bahler, J., & Pringle, J. R. (1998). Pom1p, a fission yeast protein kinase that provides positional information for both polarized growth and cytokinesis. *Genes & Development, 12*(9), 1356–1370.

Celton-Morizur, S., Racine, V., Sibarita, J. B., & Paoletti, A. (2006). Pom1 kinase links division plane position to cell polarity by regulating Mid1p cortical distribution. *Journal of Cell Science, 119*(Pt 22), 4710–4718. http://dx.doi.org/10.1242/jcs.03261.

Chang, F., Woollard, A., & Nurse, P. (1996). Isolation and characterization of fission yeast mutants defective in the assembly and placement of the contractile actin ring. *Journal of Cell Science, 109*(Pt 1), 131–142.

Daga, R. R., & Chang, F. (2005). Dynamic positioning of the fission yeast cell division plane. *Proceedings of the National Academy of Sciences of the United States of America, 102*(23), 8228–8232. http://dx.doi.org/10.1073/pnas.0409021102.

Fankhauser, C., Reymond, A., Cerutti, L., Utzig, S., Hofmann, K., & Simanis, V. (1995). The *S. pombe* cdc15 gene is a key element in the reorganization of F-actin at mitosis. *Cell, 82*(3), 435–444.

Ge, W., & Balasubramanian, M. K. (2008). Pxl1p, a paxillin-related protein, stabilizes the actomyosin ring during cytokinesis in fission yeast. *Molecular Biology of the Cell, 19*(4), 1680–1692. http://dx.doi.org/10.1091/mbc.E07-07-0715.

Huang, Y., Chew, T. G., Ge, W., & Balasubramanian, M. K. (2007). Polarity determinants Tea1p, Tea4p, and Pom1p inhibit division-septum assembly at cell ends in fission yeast. *Developmental Cell, 12*(6), 987–996. http://dx.doi.org/10.1016/j.devcel.2007.03.015.

Lee, I. J., Coffman, V. C., & Wu, J. Q. (2012). Contractile-ring assembly in fission yeast cytokinesis: recent advances and new perspectives. *Cytoskeleton (Hoboken, NJ), 69*(10), 751–763. http://dx.doi.org/10.1002/cm.21052.

Lo Presti, L., Chang, F., & Martin, S. G. (2012). Myosin Vs organize actin cables in fission yeast. *Molecular Biology of the Cell, 23*(23), 4579–4591. http://dx.doi.org/10.1091/mbc.E12-07-0499.

Motegi, F., Mishra, M., Balasubramanian, M. K., & Mabuchi, I. (2004). Myosin-II reorganization during mitosis is controlled temporally by its dephosphorylation and spatially by Mid1 in fission yeast. *The Journal of Cell Biology, 165*(5), 685–695. http://dx.doi.org/10.1083/jcb.200402097.

Oliferenko, S., Chew, T. G., & Balasubramanian, M. K. (2009). Positioning cytokinesis. *Genes & Development, 23*(6), 660–674. http://dx.doi.org/10.1101/gad.1772009.

Padte, N. N., Martin, S. G., Howard, M., & Chang, F. (2006). The cell-end factor pom1p inhibits mid1p in specification of the cell division plane in fission yeast. *Current Biology: CB, 16*(24), 2480–2487. http://dx.doi.org/10.1016/j.cub.2006.11.024.

Paoletti, A., & Chang, F. (2000). Analysis of mid1p, a protein required for placement of the cell division site, reveals a link between the nucleus and the cell surface in fission yeast. *Molecular Biology of the Cell, 11*(8), 2757–2773.

Reits, E. A., & Neefjes, J. J. (2001). From fixed to FRAP: measuring protein mobility and activity in living cells. *Nature Cell Biology, 3*(6), E145–E147. http://dx.doi.org/10.1038/35078615.

Rincon, S. A., Bhatia, P., Bicho, C., Guzman-Vendrell, M., Fraisier, V., Borek, W. E., … Paoletti, A. (2014). Pom1 regulates the assembly of Cdr2-Mid1 cortical nodes for robust spatial control of cytokinesis. *The Journal of Cell Biology, 206*(1), 61–77. http://dx.doi.org/10.1083/jcb.201311097.

Roberts-Galbraith, R. H., Ohi, M. D., Ballif, B. A., Chen, J. S., McLeod, I., McDonald, W. H., … Gould, K. L. (2010). Dephosphorylation of F-BAR protein Cdc15 modulates its conformation and stimulates its scaffolding activity at the cell division site. *Molecular Cell, 39*(1), 86–99. http://dx.doi.org/10.1016/j.molcel.2010.06.012.

Sohrmann, M., Fankhauser, C., Brodbeck, C., & Simanis, V. (1996). The dmf1/mid1 gene is essential for correct positioning of the division septum in fission yeast. *Genes & Development, 10*(21), 2707–2719.

Tolic-Norrelykke, I. M., Sacconi, L., Stringari, C., Raabe, I., & Pavone, F. S. (2005). Nuclear and division-plane positioning revealed by optical micromanipulation. *Current Biology: CB, 15*(13), 1212–1216. http://dx.doi.org/10.1016/j.cub.2005.05.052.

Tran, P. T., Marsh, L., Doye, V., Inoue, S., & Chang, F. (2001). A mechanism for nuclear positioning in fission yeast based on microtubule pushing. *The Journal of Cell Biology, 153*(2), 397–411.

Ullal, P., McDonald, N. A., Chen, J. S., Lo Presti, L., Roberts-Galbraith, R. H., Gould, K. L., & Martin, S. G. (2015). The DYRK-family kinase Pom1 phosphorylates the F-BAR protein Cdc15 to prevent division at cell poles. *The Journal of Cell Biology, 211*(3), 653–668. http://dx.doi.org/10.1083/jcb.201504073.

Wu, J. Q., Kuhn, J. R., Kovar, D. R., & Pollard, T. D. (2003). Spatial and temporal pathway for assembly and constriction of the contractile ring in fission yeast cytokinesis. *Developmental Cell, 5*(5), 723–734.

CHAPTER

Single-molecule measurements to study polymerization dynamics of FtsZ-FtsA copolymers

21

N. Baranova, M. Loose[1]

Institute for Science and Technology Austria (IST Austria), Klosterneuburg, Austria

[1]*Corresponding author: E-mail: martin.loose@ist.ac.at*

CHAPTER OUTLINE

Methods in Cell Biology, Volume 137, ISSN 0091-679X, http://dx.doi.org/10.1016/bs.mcb.2016.03.036

Abstract

Bacterial cytokinesis is commonly initiated by the Z-ring, a dynamic cytoskeletal structure that assembles at the site of division. Its primary component is FtsZ, a tubulin-like GTPase, that like its eukaryotic relative forms protein filaments in the presence of GTP. Since the discovery of the Z-ring 25 years ago, various models for the role of FtsZ have been suggested. However, important information about the architecture and dynamics of FtsZ filaments during cytokinesis is still missing.

One reason for this lack of knowledge has been the small size of bacteria, which has made it difficult to resolve the orientation and dynamics of individual FtsZ filaments in the Z-ring. While superresolution microscopy experiments have helped to gain more information about the organization of the Z-ring in the dividing cell, they were not yet able to elucidate a mechanism of how FtsZ filaments reorganize during assembly and disassembly of the Z-ring.

In this chapter, we explain how to use an in vitro reconstitution approach to investigate the self-organization of FtsZ filaments recruited to a biomimetic lipid bilayer by its membrane anchor FtsA. We show how to perform single-molecule experiments to study the behavior of individual FtsZ monomers during the constant reorganization of the FtsZ-FtsA filament network. We describe how to analyze the dynamics of single molecules and explain why this information can help to shed light onto possible mechanism of Z-ring constriction.

We believe that similar experimental approaches will be useful to study the mechanism of membrane-based polymerization of other cytoskeletal systems, not only from prokaryotic but also eukaryotic origin.

INTRODUCTION: THE BACTERIAL CYTOSKELETON

In *Escherichia coli*, the cell cycle can be divided into three distinct, time-separated steps (Aarsman et al., 2005; Den Blaauwen, Buddelmeijer, Aarsman, Hameete, & Nanninga, 1999). First, very early during the cell cycle, the so-called Z-ring assembles at the center of the cell, where it persists until the cell has completely split (Adams & Errington, 2009; Bi & Lutkenhaus, 1991). The Z-ring is a highly dynamic cytoskeletal structure, which is composed of FtsZ, a tubulin superfamily GTPase (de Boer, Crossley, & Rothfield, 1992; Bramhill & Thompson, 1994; Mukherjee & Lutkenhaus, 1998; RayChaudhuri & Park, 1992) and FtsA, a membrane-binding, actin-related protein that recruits FtsZ to the surface of the cytoplasmic membrane (Bork, Sander & Valencia, 1992; Pichoff & Lutkenhaus, 2005). In a second step, the Z-ring recruits about a dozen of other proteins on both sides of the inner cell membrane, where they form the divisome, a multicomponent protein complex. In a final step, once the divisome is completely assembled, the diameter of the division septum continuously decreases until it is fully closed, giving rise to two daughter cells.

As cell division of gram-negative cells depends on the constriction of a three-layered cell envelope, these steps are very difficult tasks, which need to be precisely regulated in space and time. Despite several decades of research, many open questions remain, such as how the FtsZ filaments are organized within the Z-ring, what defines the temporal order of divisome assembly, how FtsZ filaments reorganize during constriction, and how divisome components influence the dynamic properties of the cell division machinery.

Different models for Z-ring constriction dynamics have been suggested (Meier & Goley, 2014; Erickson, Anderson, & Osawa, 2010). According to these models, FtsZ filaments could slide along each other, which would generate a contractile force as the diameter of the Z-ring continuously decreases (Lan, Daniels, Dobrowsky, Wirtz, & Sun, 2009; Szwedziak, Wang, Bharat, Tsim, & Löwe, 2014). Alternatively, filaments would not move but reorganize by constant turnover. In this case, a contractile force could arise either by filament bending (Li et al., 2013; Li, Trimble, Brun, & Jensen, 2007) or due to breakage and resealing of a continuous ring lining the inner cytoplasmic membrane of the bacterial cell (Erickson, 2009). However, the Z-ring could also only provide a spatial signal for the peptidoglycan synthesis machinery. In this case, the diameter of the Z-ring would decrease because it follows the inward growth of the division septum. This would be in agreement with a recent superresolution imaging study, which found that the rate-limiting step during cytokinesis is the synthesis of the cell wall, rather than remodeling of the Z-ring (Coltharp, Buss, Plumer, & Xiao, 2016).

Although recent technological advances have pushed the boundary of optical and electron microscopy toward better resolution (Gahlmann & Moerner, 2014), these techniques could not yet provide sufficient information to fully explain how FtsZ and FtsA are able to perform cell division, showing that superresolution data of fixed cells alone cannot yet provide all the information required to obtain a universal mechanistic understanding of bacterial cell division.

To study the mechanism underlying Z-ring assembly and to obtain more information on the dynamics of FtsZ filaments, we have developed an assay to reconstitute FtsZ polymerization mediated by FtsA on supported lipid bilayers in vitro (Loose & Mitchison, 2014; Nguyen, Field, Groen, Mitchison, & Loose, 2015). Supported lipid bilayers are easy to prepare and robust, and ideal for studying the dynamics of membrane–protein interaction. They have the additional advantage of being amenable to total internal reflection fluorescence (TIRF) microscopy, allowing for high-resolution investigation of transient events down to the single-molecule level. Using this approach, we have found that together with FtsA, FtsZ assembles into highly dynamic cytoskeletal patterns resembling fast moving streams and rapidly rotating rings of FtsZ filament bundles. Furthermore, this experimental setup has allowed us to watch single molecules during the reorganization of the filament network, providing us with information about the underlying filament dynamics.

Here, we describe how to purify the proteins required to reconstitute the initial steps of bacterial cell division, how to select a dye for fluorescence microscopy measurements, and how to use single-molecule imaging to shed light on the polymerization dynamics of FtsZ filaments on a biomimetic membrane.

We believe that this experimental setup not only is a great starting point to understand the self-organization of the bacterial cell division machinery but can also help to understand the dynamics of other polymer systems, such as the actin cortex (Köster et al., 2016; Murrell & Gardel, 2012; Vogel, Petrasek, Heinemann, & Schwille, 2013), intermediate filaments such as septins (Bridges et al., 2014), or the dynamics of membrane-bound microtubules (Nguyen et al., 2014).

1. PROTEIN BIOCHEMISTRY: EXPRESSION, PURIFICATION, AND FLUORESCENT LABELING

1.1 AFFINITY PURIFICATION

To reconstitute cellular functions in vitro, we first need to obtain purified, active proteins. It is generally advised to avoid affinity tags such as hexahistidine (6xHis) in in vitro reconstitution experiments, as they can alter the properties of the purified proteins. Nevertheless, high specificity and ease of use are still important advantages of metal ion affinity chromatography. For our experiments we favor 6xHis-SUMO as an N-terminal fusion partner, which has been found to enhance functional protein production in prokaryotic and eukaryotic expression systems (Panavas, Sanders, & Butt, 2009). Following purification of the fusion protein, a SUMO protease is used to cleave precisely at the C-terminus of the SUMO tag, leaving no additional residues attached to the target wild-type protein.

1.1.1 Purification of 6xHis-SUMO-FtsZ

We expressed 6xHis-SUMO-FtsZ from a pET-derived vector (pTB567) (Uehara, Parzych, Dinh, & Bernhardt, 2010) in C41 cells grown in 2XYT medium supplemented with Ampicillin (50 mg/mL) at 37°C. After growth to an absorbance of 0.8 at 600 nm, protein expression was induced by adding IPTG to a concentration of 1 mM and growing the cells at 37°C for 5 h. Cells were harvested by centrifugation at 6000 *g* for 15 min and resuspended in lysis buffer [50 mM Tris-HCl at pH 8.0, 50 mM KCl, 20 mM imidazole, 10% glycerol, 10 mM β-mercaptoethanol, and protease inhibitors (complete EDTA free, Roche Molecular Biochemicals)] followed by freezing in liquid nitrogen. Cells were lysed in a repeated freeze−thaw cycles of the cell suspension, which was then incubated with lysozyme (10 mg/mL, L-6876, Sigma−Aldrich) and DNase (1 mg/mL, DN25, Sigma−Aldrich) for 10 min before sonification using a tip sonicator. The resulting extract was clarified by centrifugation at 30,000 *g* for 20 min at 4°C to pellet cell debris and membranes. The clarified lysate was then loaded onto a Ni^{2+} agarose resin (HisPur Ni-NTA resin, Thermo Scientific). The column was washed with five column volumes FtsZ lysis buffer (50 mM Tris-HCl at pH 8.0, 50 mM KCl, 30 mM imidazole, 10% glycerol, 10 mM β-mercaptoethanol, and protease inhibitors). The protein was eluted with elution buffer (50 mM Tris-HCl at pH 8.0, 50 mM KCl, 200 mM imidazole, 10% glycerol, and 10 mM β-mercaptoethanol), and peak fractions were pooled.

1.1.2 Purification of 6xHis-SUMO-FtsA

Early attempts to purify wild-type *E. coli* FtsA failed due to its toxicity when overexpressed and its low solubility at high protein concentrations. Using the SUMO-fusion system for overexpression of FtsA, we could circumvent these problems and obtained sufficient amounts of the wild-type protein.

Escherichia coli FtsA was PCR-amplified from MG1655 cells and cloned into a pET-His-SUMO vector. FtsA was then expressed as a His-SUMO fusion in C41 cells. Overnight cultures from single colonies were grown overnight in 2XYT medium supplemented with Ampicillin (50 mg/mL) at 37°C, diluted 1:200 and incubated to an absorbance of 1−1.5 at 600 nm. The protein was then expressed at 18°C overnight after adding IPTG to 1 mM. Collected cells were resuspended in lysis buffer (50 mM Tris-HCl at pH 8, 20 mM imidazole, 500 mM NaCl, 10 mM $MgCl_2$, 10 mM β-mercaptoethanol, and protease inhibitors) and lysed as described in Section 1.1.1. The lysate was loaded onto a Ni^{2+} agarose resin, which was washed with five column volumes FtsA wash buffer (50 mM Tris-HCl at pH 8, 500 mM KCl, 30 mM imidazole, 10 mM $MgCl_2$, 10 mM β-mercaptoethanol, and 1 mM ADP). 6xHis-SUMO-FtsA was eluted in the same buffer supplemented with 200 mM imidazole, and peak protein fractions were pooled.

1.1.3 Purification of 6xHis-Ulp1(403-621)

The *Saccharomyces cerevisiae* SUMO protease Ulp1 is commercially available from various sources (eg, Thermo Scientific, Catalog Number, 12588018) Ulp1 (ubiquitin-like-specific protease 1). Alternatively, the protein can also be purified, for example, using an expression vector available from Addgene (Guerrero, Ciragan, & Iwaï, 2015) (pFGET19_Ulp1, Plasmid #64697).

We purified the C-terminal protease domain Ulp1(403-621) as a 6xHis-tagged protein from pTB145, a gift from Thomas Bernhardt (Harvard Medical School, Boston). The protein was expressed in Rosetta(DE3)/pLysRare cells. Overnight, cultures were grown in 2XYT medium supplemented with Ampicillin (50 mg/mL) and Chloramphenicol (25 mg/mL) at 30°C. The cultures were diluted 1:200 and protein expression was induced by adding IPTG to 1 mM at an absorbance of 0.4 at 600 nm. Cells were collected by centrifugation at 6000 *g* for 15 min and resuspended in Ulp1 lysis buffer [50 mM Tris-HCl (8.0), 300 mM NaCl, 20 mM imidazole, 10% glycerol]. **Important**: **Do not use protease inhibitors for the purification of 6xHis-Ulp1(403-621) as it inactivates Ulp1!** After cell lysis, the sample was loaded onto a Ni^{2+} agarose resin, which was washed with five column volumes Ulp1 lysis buffer [50 mM Tris-HCl (8.0), 300 mM NaCl, 20 mM imidazole, 10% glycerol]. 6xHis-Ulp1(403-621) was eluted in Ulp1 lysis buffer supplemented with 200 mM imidazole, peak protein fractions were pooled and dialyzed into Ulp1 storage buffer [25 mM Tris (pH 8.0), 200 mM NaCl, 50% glycerol, 0.5% NP-40, and 0.5 mM DTT]. The purified protein can be stored in aliquots at −20°C.

1.2 CLEAVAGE OF 6xHis-SUMO TAG

The optimal temperature for cleavage by Ulp1 is 30°C, but the enzyme is active over wide ranges of temperature (2−37°C) and pH values (7.0−9.0). To identify the optimal conditions for cleavage and preservation of target protein activity, it is useful to set up a trial cleavage experiment during which different conditions [protein to enzyme ratio (1:5 to 1:5000), incubation time (1 h to overnight), and temperatures (4 and 30°C)]. Cleavage can then be confirmed by SDS-PAGE. For FtsZ, we used a 6xHis-Ulp1:6xHis-SUMO-FtsZ concentration ratio of 1:100 and performed cleavage and dialysis to remove imidazole simultaneously [overnight at 4°C, while dialyzing into storage buffer (50 mM Tris, pH 8.0, 50 mM KCl, 10% glycerol)]. The protease and released tag were removed from the preparation by passing it over Ni^{2+}-NTA resin and saving the flow trough, which contained the wild-type protein.

In the case of FtsA, the incubation time has to be kept minimal to avoid precipitation of the protein. Therefore, we increased the concentration of Ulp1 to speed up the hydrolysis rate. In the case of 6xHis-SUMO-FtsA, the 6xHis-SUMO tag was removed by adding 6xHis-Ulp1 to a molar ratio of 1:30 to the peak fractions and incubating the mixture on ice for 3.5 h without removing imidazole. However, in some cases imidazole concentrations higher than 150 mM can adversely affect the activity of the protease, and removal of imidazole might be necessary. As we found that native FtsA has the tendency to nonspecifically interact with the Ni^{2+}-NTA resin, we removed the protease and released tag by gel filtration using a Superdex 200 column and FtsA storage buffer (50 mM HEPES at pH 7.5, 500 mM KCl, 10 mM $MgCl_2$, 10% glycerol, and 1 mM ADP). Peak fractions of FtsA were pooled and concentrated to 50 μM (higher concentrations can lead to precipitation of the protein). Finally, we divided the protein into small aliquots and snap froze them in liquid nitrogen before storing them at −80°C.

1.3 ENRICH THE ACTIVE FRACTION OF FtsZ BY POLYMERIZATION/DEPOLYMERIZATION

After removal of the affinity tag, polymerization-competent FtsZ can be enriched by GTP-dependent polymerization (Cho et al., 2011). To facilitate polymerization and bundling of FtsZ, this step is performed at reduced pH and in the presence of $CaCl_2$. The FtsZ preparation was dialyzed against FtsZ polymerization buffer [50 mM piperazine-N,N′-bis(2-ethanesulphonic acid), at pH 6.7, and 10 mM $MgCl_2$]. We then warmed the solution to room temperature for 30 min and induced polymerization by adding GTP and $CaCl_2$ to final concentrations of 10 mM and 5 mM, respectively. After 20 min, we pelleted the polymeric FtsZ by centrifugation at 20,000 *g* for 10 min at room temperature. FtsZ pellets were resuspended in 1/5 volume of FtsZ storage buffer (50 mM Tris, at pH 7.4, 50 mM KCl, 1 mM EDTA, and 10% glycerol) and dialyzed extensively to remove $CaCl_2$ and residual GTP and to ensure complete depolymerization. The protein was then aliquoted, snap-frozen in liquid nitrogen, and stored at −80°C.

1.4 FLUORESCENT LABELING OF PROTEINS

1.4.1 How to choose a fluorescent label

Direct observation of protein dynamics using fluorescence microscopy has led to major advances in our understanding of self-organized behavior of protein systems. To perform biochemical experiments using fluorescence-based techniques, it is important to use a fluorophore that not only has optimal photophysical properties but also does not disrupt the biochemical function of the protein of interest.

Genetically encoded fluorescent proteins are cheap to use and offer the advantage of complete labeling of the protein of interest. However, GFP and its derivatives can dramatically perturb the biochemical properties of the system (Landgraf, Okumus, Chien, Baker, & Paulsson, 2012), as they possess the tendency to dimerize (Zacharias, Violin, Newton, & Tsien, 2002) and are of similar sizes as the protein of interest (26 kDa).

Small organic dyes that can be chemically linked to proteins are often a better choice for in vitro experiments. Due to their high photostability, brightness (product of the molar absorption coefficient at the excitation wavelength and the fluorescence quantum yield), and resistance to blinking (luminescence for limited times, interrupted by dark periods with no emission), they offer superior photochemical properties compared to fluorescent proteins. In addition, due to their small size, organic dyes are less prone to change the biochemical properties of the protein of interest. However, even they can lead to artifactual results: for example, highly negatively charged dyes can cause loss of function of the target protein, and some dyes were found to strongly interact with phospholipid bilayers despite of being hydrophilic (Zanetti-Domingues, Tynan, Rolfe, Clarke, & Martin-Fernandez, 2013). This lipophilicity can lead to a couple of problems: first, it can increase the apparent affinity of a protein to the membrane or in case that the free dye has not been completely removed; it can interact with the membrane directly, leading to false-positive signals in the experiment. For example, Atto 647N is a popular dye due to its outstanding brightness and photostability; however, this dye can perturb the folding of some proteins and shows relatively high levels of nonspecific binding, probably due to its hydrophobicity (Bosch et al., 2014; Hughes, Rawle, & Boxer, 2014; Zanetti-Domingues et al., 2013).

Since it is generally difficult to predict the influence of an organic dye on the protein and its behavior in an in vitro assay, we recommend comparing structurally different dyes with similar spectral properties. For our experiments, we decided to work with AlexaFluor 488 (Life Technologies) and Cy5 (GE Healthcare or Lumiprobe).

1.4.2 How to label proteins

The most frequently used methods for fluorescent labeling of proteins use maleimide or N-hydroxysuccinimide (NHS) derivatives of small-molecule dyes to target surface-exposed reactive groups, cysteine or lysine residues, respectively. As FtsZ does not contain a native cysteine, we modified FtsZ with an amino-terminal cysteine residue by introducing seven additional amino acids (AEGCGEL) to the N terminus of the protein to obtain pTB567-GCG-FtsZ. While many studies use a GSKCK amino-acid labeling, we found that the lysines interfered with the Ulp1 treatment. For our experiments we labeled the FtsZ with AlexaFluor 488-maleimide (Life Technologies) or Cy5-maleimide (GE Healthcare) according to the protocols of the manufacturer and removed the unreacted dye via buffer exchange over a Sephadex G25 column into FtsZ storage buffer (see earlier discussion).

As wild-type *E. coli* FtsA contains nine endogenous cysteines and 21 lysines, we decided to avoid labeling using amino- or thiol-reactive dyes to prevent altering the protein activity due to overlabeling. While genetically encoded SNAP (NEB) and HALO tags (Promega) offer the possibility to specifically label a protein with various organic dyes, they are relatively large (20 kDa). Instead, we decided to apply "sortagging" (Guimaraes et al., 2013; Popp, Antos, Grotenbreg, Spooner, & Ploegh, 2007; Popp, Antos, & Ploegh, 2009; Theile et al., 2013). In this method, an N-terminal truncated version of sortase A from *Staphylococcus aureus* recognizes an LPXTG motif and cleaves the peptide bond between threonine and glycine, thereby releasing the residues C-terminal to the threonine. This peptide is then covalently linked to the N-terminus of a pentaglycine target peptide or protein. Using this trans-peptidase reaction, it is possible to specifically modify either the N- or C-terminus of proteins using a genetically encoded target peptide.

For labeling of FtsA, we introduced a pentaglycine tag (Gly5-FtsA) to the N-terminus of FtsA. As Cy5-labeled LPTEG peptides are usually not commercially available, we modified the N-terminal aminogroup of LPTEG with a Cy5-NHS-Ester (GE Healthcare). For the labeling reaction, we dissolved 2 mg of the lyophilized peptide in 300 μL DMSO and used this mixture to dissolve 1 mg of Cy5-NHS-Ester. Finally, we added 20 μL of DIPEA (diisopropylethylamin) to deprotonate the terminal amine group. To protect the mixture from light, it was covered in aluminum foil and incubated for 12 h at room temperature. Finally, we diluted the mixture to 25% DMSO with H_2O and added an excess amount of glycine to quench any nonreacted dye.

Purified 6xHis-SUMO-Gly5-FtsA (50–150 μM) was then simultaneously incubated with Ulp1 (1–3 μM) and His-SrtA59 (50 μM) in the presence of Cy5-LPTEG (1 mM). Since the activity of sortase requires calcium, we added $CaCl_2$ to 10 mM and incubated the reaction for 4–5 h on ice. The protease, sortase, tag, and nonreacted Cy5 and Cy5-pentapeptide were then removed by a single gel filtration over a Superdex 200 column in FtsA storage buffer to obtain Cy5-GG-FtsA.

2. FLUORESCENCE IMAGING OF PROTEIN DYNAMICS

2.1 GENERAL CONSIDERATIONS REGARDING SINGLE-MOLECULE IMAGING

Fluorescence microscopy assays provide much more detailed information and require less material than bulk experiments using protein gels, fluorimeters, or spec-trophotometers. However, one fundamental problem in single-molecule fluores-cence studies is that fluorescent dyes have a limited lifetime due to irreversible photobleaching, which is mainly caused by molecular oxygen (O_2). Oxygen can directly interact with a fluorophore when it occupies an excited, long-lived triplet states. In this process, it is transformed into highly reactive singlet oxygen, which rapidly reacts with the dye or other organic molecules in solution irreversibly damaging dye and proteins (Aitken, Marshall, & Puglisi, 2008; Benesch & Benesch, 1953; Rasnik, McKinney, & Ha, 2006).

To increase the lifetime of the dye and to prevent oxidative photodamage, oxygen scavenger systems have been developed. These oxygen scavenger systems enzymatically remove oxygen from the buffer, from about 0.5 mM to a few micromolar, which can increase the observation time for some fluorophores by several orders of magnitude (Ha & Tinnefeld, 2012). The most popular oxygen scavenger system is based on the activity of glucose oxidase and catalase in the presence of glucose. D-Glucose is oxidized to gluconic acid by glucose oxidase, depleting oxygen in solution and creating hydrogen peroxide as a side product. In a second reaction step, catalase degrades hydrogen peroxide to H_2O and O_2, restoring half of the oxygen used by glucose oxidase.

Another problem with fluorophores is that they can exhibit high-intensity fluctuations, which is often caused by the population of nonfluorescent triplet state. Oxygen can efficiently quench the triplet state; however, as molecular oxygen is removed from solution to reduce photobleaching, the lifetime of the triplet state increases significantly. To avoid blinking and to depopulate the triplet state, beta-mercaptoethanol or DTT is most commonly added to the buffer, which at the same time provides a reducing environment in the buffer. However, some dyes were found to be more efficiently released from their dark triplet state by alternative triplet-state quenchers. For example, the vitamin E analog Trolox (6-hydroxy-2,5,7,8-tetramethylchroman-2-carboxylic acid) was found to dramatically improve the photochemical properties of Cy5 (Rasnik et al., 2006) and Atto655 (Cordes, Vogelsang, & Tinnefeld, 2009). The presence of ascorbic acid and n-propyl gallate as triplet quencher was found to lead to improved results for AlexaFluor 488 (Aitken et al., 2008).

In our experiments, we used 40 mM D-glucose, 0.040 mg/mL glucose oxidase, 0.016 mg/mL catalase, 20 mM dithiothreitol, and 1 mM Trolox added to the buffer to prevent photobleaching.

2.1.1 Variations of oxygen scavenger systems

As an alternative to glucose oxidase and catalase, a single enzyme system based on protocatechuate-3,4-dioxygenase and protocatechuic acid has been developed, which was found to significantly improve the stability of Cy3, Cy5, and Alexa488 in single-molecule TIRF-M experiments (Aitken et al., 2008).

2.2 IMAGING PROTEIN DYNAMICS

The models for FtsZ filament remodeling during constriction can be divided in two different categories: the Z-ring could either constrict due to filament sliding, which would give rise to a directed movement of single FtsZ subunits. Alternatively, FtsZ filaments could reorganize by polymerization dynamics, in which case single FtsZ subunits would remain static. Using single-molecule fluorescence microscopy, it is possible to observe the dynamics of single FtsZ monomer and to distinguish between these to models.

In TIRF microscopy, the excitation is limited to a thin layer above the reflecting glass substrate, such that only proteins close to the membrane surface will be excited. At the same time the fluorescence background is limited, providing sufficiently high signal-to-noise ratio to detect single molecules. For the experiments described here, we used an Imic TIRF microscope (Till photonics), equipped with 100×/1.49 immersion oil objective (Olympus, Japan) and two Andor 897 (512 × 512) cameras combined with an Andor TuCam beam splitter for dual-color TIRF imaging.

Before starting the experiment, prepare the required protein solutions and have them ready on ice. The proteins can be kept on ice for about 2 h. Ideally and to be able to observe initiation of protein dynamics, the proteins are added next to the microscope, right before imaging.

2.2.1 Sample preparation

- Prepare supported bilayer as described in Nguyen et al. (2015).
- Add oxygen scavenger mix: 1% w/v glucose, glucose oxidase 0.32 mg/mL, catalase 0.055 mg/mL, DTT 1 mM, Trolox 1 mM.
- Add FtsA (to 0.5 μM) and FtsZ (to 2.5 μM) to the buffer. To resolve single molecules, we use 10% of FtsZ labeled with AlexaFluor 488 and 0.001% of FtsZ labeled with Cy5. Mix gently.
- Start the reaction by adding nucleotides (1 mM ATP, 5 mM GTP) and mix gently.
- Image immediately on TIRF microscope.
- To capture the dynamic behavior of proteins, time-lapse movies are typically recorded with a rate of 5 frame/s. To simultaneously observe the reorganization of the FtsZ filament network, the power of the 488 nm laser can be kept low, while we increase the power of the 633 nm laser for single-molecule imaging. Usually, we try to find the minimal laser power that can still resolve single fluorophores to minimize photobleaching.

Using this experimental approach, we found that while the filament network was continuously moving and rearranging, single FtsZ subunits appeared and remained at the same position (Fig.1A–C). This observation rules out a sliding as a mechanism of filament dynamics. Instead, filaments recruited to a flat membrane reorganize solely due to polymerization dynamics, most likely treadmilling.

3. DATA ANALYSIS

In our single-molecule imaging experiments, we found that fluorophores appeared on the surface of the membrane, where they could be observed for several seconds

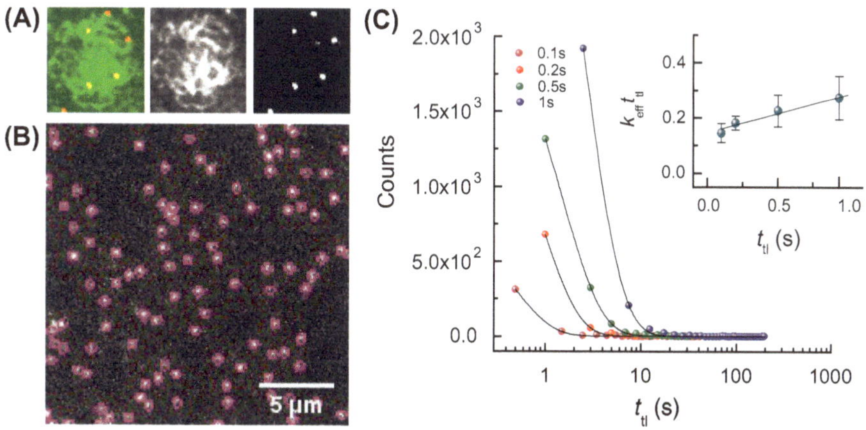

FIGURE 1

Single-molecule tracking of FtsZ. (A) Typical micrograph of FtsZ filaments labeled with Alexa488 (left) single FtsZ monomers labeled with Cy5 (center) and merged into dual-color image (right). (B) Detection of single-particle Cy5-FtsZ in TrackMate based on a Gaussian fit with estimated diameter of 0.7 μm. (C) Quantification of the residence time. Time-lapse movies were acquired at 50 ms exposure time (t_{ex}) with varying acquisition time (t_{tl}) (from 100 ms to 1 s). An effective lifetime (k_{eff}) for single FtsZ subunit was estimated from a monoexponential fit (*black lines*). The insert plot represents a linear dependence between the k_{eff} t_{tl} as function of acquisition time t_{tl}, allowing estimating the intrinsic dissociation rate of FtsZ (k_{off}).

without moving before they disappeared. Since TIRF microscopy only excites fluorophores close to the membrane, the termination of the fluorescent signal corresponds to a detachment of the fluorophore from the membrane into the buffer solution. As FtsZ monomers cannot be recruited to the membrane by FtsA (Loose & Mitchison, 2014), we therefore assume that this observation time corresponds to the time during which FtsZ monomers are incorporated in the FtsZ filament network.

To obtain quantitative information about the lifetime of FtsZ monomers, we used automated particle-tracking software. Several software packages that offer subpixel resolution particle tracking are available (Crocker & Grier, 1996; Jaqaman et al., 2008; Ruhnow, Zwicker, & Diez, 2011). We found the TrackMate plugin in Fiji, the open-source platform for analysis (Schindelin et al., 2012) very powerful and user-friendly. It is based on the particle-tracking algorithm developed by Jaqaman et al. (2008) and allows users to further extend its functionality.

When we analyzed the lifetime distribution of FtsZ subunits, we found that it corresponds to a first-order reaction following an exponential function $f(t) = A\exp(-k_{eff}t)$. For a correct quantification of the residence time, it is important to consider the photobleaching rate; therefore the effective lifetime of the fluorescence is given by

$$k_{\mathrm{eff}} = k_{\mathrm{off}} + k_{\mathrm{pb}}, \tag{1}$$

where k_{off} is the intrinsic detachment rate of the protein from the membrane and k_{pb} is the photobleaching rate. For an estimate of k_{pb}, we used an approach similar to Helenius, Brouhard, Kalaidzidis, Diez, and Howard (2006) and described by Gebhardt et al. (2013). Since the photobleaching rate k_{pb} is proportional to the used light intensity, it depends on the exposure time t_{ex} and the time-lapse time t_{tl}, the time between two frames:

$$k_{\mathrm{pb}} = k_{\mathrm{b}} t_{\mathrm{ex}} / t_{\mathrm{tl}} \tag{2}$$

where k_{b} corresponds to the photobleaching rate constant. From Eqs. (1) and (2), it follows:

$$k_{\mathrm{eff}} t_{\mathrm{tl}} = k_{\mathrm{off}} t_{\mathrm{tl}} + k_{\mathrm{b}} t_{\mathrm{ex}}$$

To obtain the dissociation rate constant k_{off}, we plotted $k_{\mathrm{eff}}\ t_{\mathrm{tl}}$ as a function of the time-lapse time t_{tl} (Fig. 1). In this plot, k_{off} is given by the slope, and the y-intersect corresponds to $k_{\mathrm{b}}\ t_{\mathrm{ex}}$.

After correction of photobleaching, we obtained a residence time of 7.186 ± 1.44 s from five independent experiments.

CONCLUSION

The assay described here allows studying the dynamics and turnover of protein filaments attached to a supported membrane. While we have focused here on the polymerization dynamics of FtsZ and FtsA, we believe that a similar experimental setup will be useful to understand the polymerization and depolymerization of other membrane-bound filaments, such as actin (Köster et al., 2016; Murrell & Gardel, 2012; Vogel et al., 2013) as well as intermediate filaments (Bridges et al., 2014) or microtubules (Nguyen et al., 2014). In combination with physical models describing the filament dynamics, the lifetime distribution of monomers can be used to not only differentiate between different mechanisms of polymerization but also give insight in the spatial architecture of large scale assemblies of biopolymers (Needleman et al., 2010).

ACKNOWLEDGMENTS

We want to thank the Bioimaging Facility at IST Austria for providing excellent service, Jan Brugues (MPI-CBG) Dresden for discussions, Christine Mieck for critical reading of the manuscript, and Thomas Bernhardt (Harvard Medical School, Boston, MA) for plasmids. Natalia Baranova is supported by an EMBO Long-Term Fellowship (EMBO ALTF 1163-2015) and Martin Loose by an ERC Starting Grant (ERCStG-2015-SelfOrganiCell).

REFERENCES

Aarsman, M. E. G., Piette, A., Fraipont, C., Vinkenvleugel, T. M. F., Nguyen-Distèche, M., & den Blaauwen, T. (2005). Maturation of the *Escherichia coli* divisome occurs in two steps. *Molecular Microbiology, 55*(6), 1631–1645. http://dx.doi.org/10.1111/j.1365-2958.2005.04502.x.

Adams, D. W., & Errington, J. (2009). Bacterial cell division: assembly, maintenance and disassembly of the Z ring. *Nature Reviews Microbiology, 7*(9), 642–653. http://dx.doi.org/10.1038/nrmicro2198.

Aitken, C. E., Marshall, R. A., & Puglisi, J. D. (2008). An oxygen scavenging system for improvement of dye stability in single-molecule fluorescence experiments. *Biophysical Journal, 94*(5), 1826–1835. http://dx.doi.org/10.1529/biophysj.107.117689.

Benesch, R. E., & Benesch, R. (1953). Enzymatic removal of oxygen for polarography and related methods. *Science, 118*(3068), 447–448.

Bi, E. F., & Lutkenhaus, J. (1991). FtsZ ring structure associated with division in *Escherichia coli*. *Nature, 354*(6349), 161–164. http://dx.doi.org/10.1038/354161a0.

de Boer, P., Crossley, R., & Rothfield, L. (1992). The essential bacterial cell-division protein FtsZ is a GTPase. *Nature, 359*(6392), 254–256. http://dx.doi.org/10.1038/359254a0.

Bork, P., Sander, C., & Valencia, A. (1992). An ATPase domain common to prokaryotic cell cycle proteins, sugar kinases, actin, and hsp70 heat shock proteins. *Proceedings of the National Academy of Sciences, 89*(16), 7290–7294.

Bosch, P. J., Corrêa, I. R., Sonntag, M. H., Ibach, J., Brunsveld, L., Kanger, J. S., & Subramaniam, V. (2014). Evaluation of fluorophores to label SNAP-tag fused proteins for multicolor single-molecule tracking microscopy in live cells. *Biophysical Journal, 107*(4), 803–814. http://dx.doi.org/10.1016/j.bpj.2014.06.040.

Bramhill, D., & Thompson, C. M. (1994). GTP-dependent polymerization of *Escherichia coli* FtsZ protein to form tubules. *Proceedings of the National Academy of Sciences of the United States of America, 91*(13), 5813–5817.

Bridges, A. A., Zhang, H., Mehta, S. B., Occhipinti, P., Tani, T., & Gladfelter, A. S. (2014). Septin assemblies form by diffusion-driven annealing on membranes. *Proceedings of the National Academy of Sciences of the United States of America, 111*(6), 2146–2151. http://dx.doi.org/10.1073/pnas.1314138111.

Cho, H., McManus, H. R., Dove, S. L., & Bernhardt, T. G. (2011). Nucleoid occlusion factor SlmA is a DNA-activated FtsZ polymerization antagonist. *Proceedings of the National Academy of Sciences*. http://dx.doi.org/10.1073/pnas.1018674108.

Coltharp, C., Buss, J., Plumer, T. M., & Xiao, J. (2016). Defining the rate-limiting processes of bacterial cytokinesis. *Proceedings of the National Academy of Sciences of the United States of America, 113*(8), E1044–E1053. http://dx.doi.org/10.1073/pnas.1514296113.

Cordes, T., Vogelsang, J., & Tinnefeld, P. (2009). On the mechanism of Trolox as antiblinking and antibleaching reagent. *Journal of the American Chemical Society, 131*(14), 5018–5019. http://dx.doi.org/10.1021/ja809117z.

Crocker, J., & Grier, D. (1996). Methods of digital video microscopy for colloidal studies. *Journal of Colloid and Interface Science, 179*(1), 298–310.

Den Blaauwen, T., Buddelmeijer, N., Aarsman, M. E., Hameete, C. M., & Nanninga, N. (1999). Timing of FtsZ assembly in *Escherichia coli*. *Journal of Bacteriology, 181*(17), 5167–5175.

Erickson, H. P. (2009). Modeling the physics of FtsZ assembly and force generation. *Proceedings of the National Academy of Sciences of the United States of America, 106*(23), 9238–9243. http://dx.doi.org/10.1073/pnas.0902258106.

Erickson, H. P., Anderson, D. E., & Osawa, M. (2010). FtsZ in bacterial cytokinesis: cytoskeleton and force generator all in one. *Microbiology and Molecular Biology Reviews, 74*(4), 504–528. http://dx.doi.org/10.1128/MMBR.00021-10.

Gebhardt, J. C. M., Suter, D. M., Roy, R., Zhao, Z. W., Chapman, A. R., Basu, S., … Xie, X. S. (2013). Single-molecule imaging of transcription factor binding to DNA in live mammalian cells. *Nature Methods, 10*(5), 421–426. http://dx.doi.org/10.1038/nmeth.2411.

Gahlmann, A., & Moerner, W. E. (2014). Exploring bacterial cell biology with single-molecule tracking and super-resolution imaging. *Nature Reviews Microbiology, 12*(1), 9–22. http://dx.doi.org/10.1038/nrmicro3154.

Guerrero, F., Ciragan, A., & Iwaï, H. (2015). Tandem SUMO fusion vectors for improving soluble protein expression and purification. *Protein Expression and Purification, 116*, 42–49. http://dx.doi.org/10.1016/j.pep.2015.08.019.

Guimaraes, C. P., Witte, M. D., Theile, C. S., Bozkurt, G., Kundrat, L., Blom, A. E. M., & Ploegh, H. L. (2013). Site-specific C-terminal and internal loop labeling of proteins using sortase-mediated reactions. *Nature Protocols, 8*(9), 1787–1799. http://dx.doi.org/10.1038/nprot.2013.101.

Ha, T., & Tinnefeld, P. (2012). Photophysics of fluorescent probes for single-molecule biophysics and super-resolution imaging. *Annual Review of Physical Chemistry, 63*(1), 595–617. http://dx.doi.org/10.1146/annurev-physchem-032210-103340.

Helenius, J., Brouhard, G., Kalaidzidis, Y., Diez, S., & Howard, J. (2006). The depolymerizing kinesin MCAK uses lattice diffusion to rapidly target microtubule ends. *Nature, 441*(7089), 115–119. http://dx.doi.org/10.1038/nature04736.

Hughes, L. D., Rawle, R. J., & Boxer, S. G. (2014). Choose your label wisely: water-soluble fluorophores often interact with lipid bilayers. *PLoS One, 9*(2), e87649. http://dx.doi.org/10.1371/journal.pone.0087649.

Jaqaman, K., Loerke, D., Mettlen, M., Kuwata, H., Grinstein, S., Schmid, S. L., & Danuser, G. (2008). Robust single-particle tracking in live-cell time-lapse sequences. *Nature Methods, 5*(8), 695–702. http://dx.doi.org/10.1038/nmeth.1237.

Köster, D. V., Husain, K., Iljazi, E., Bhat, A., Bieling, P., Mullins, R. D., … Mayor, S. (2016). Actomyosin dynamics drive local membrane component organization in an in vitro active composite layer. *Proceedings of the National Academy of Sciences of the United States of America*. http://dx.doi.org/10.1073/pnas.1514030113, 201514030.

Lan, G., Daniels, B. R., Dobrowsky, T. M., Wirtz, D., & Sun, S. X. (2009). Condensation of FtsZ filaments can drive bacterial cell division. *Proceedings of the National Academy of Sciences of the United States of America, 106*(1), 121–126. http://dx.doi.org/10.1073/pnas.0807963106.

Landgraf, D., Okumus, B., Chien, P., Baker, T. A., & Paulsson, J. (2012). Segregation of molecules at cell division reveals native protein localization. *Nature Methods*. http://dx.doi.org/10.1038/nmeth.1955.

Li, Y., Hsin, J., Zhao, L., Cheng, Y., Shang, W., Huang, K. C., … Ye, S. (2013). FtsZ protofilaments use a hinge-opening mechanism for constrictive force generation. *Science, 341*(6144), 392–395. http://dx.doi.org/10.1126/science.1239248.

Li, Z., Trimble, M. J., Brun, Y. V., & Jensen, G. J. (2007). The structure of FtsZ filaments in vivo suggests a force-generating role in cell division. *The EMBO Journal, 26*(22), 4694–4708. http://dx.doi.org/10.1038/sj.emboj.7601895.

Loose, M., & Mitchison, T. J. (2014). The bacterial cell division proteins FtsA and FtsZ self-organize into dynamic cytoskeletal patterns. *Nature Cell Biology, 16*(1), 38–46. http://dx.doi.org/10.1038/ncb2885.

Meier, E. L., & Goley, E. D. (2014). Form and function of the bacterial cytokinetic ring. *Current Opinion in Cell Biology, 26*(0), 19–27.

Mukherjee, A., & Lutkenhaus, J. (1998). Dynamic assembly of FtsZ regulated by GTP hydrolysis. *The EMBO Journal, 17*(2), 462–469. http://dx.doi.org/10.1093/emboj/17.2.462.

Murrell, M. P., & Gardel, M. L. (2012). F-actin buckling coordinates contractility and severing in a biomimetic actomyosin cortex. *Proceedings of the National Academy of Sciences of the United States of America, 109*(51), 20820–20825. http://dx.doi.org/10.1073/pnas.1214753109.

Needleman, D. J., Groen, A., Ohi, R., Maresca, T., Mirny, L., & Mitchison, T. (2010). Fast microtubule dynamics in meiotic spindles measured by single molecule imaging: evidence that the spindle environment does not stabilize microtubules. *Molecular Biology of the Cell, 21*(2), 323–333. http://dx.doi.org/10.1091/mbc.E09-09-0816.

Nguyen, P. A., Field, C. M., Groen, A. C., Mitchison, T. J., & Loose, M. (2015). Using supported bilayers to study the spatiotemporal organization of membrane-bound proteins. In *Building a cell from its component parts* (Vol. 128, pp. 223–241). Elsevier. http://dx.doi.org/10.1016/bs.mcb.2015.01.007.

Nguyen, P. A., Groen, A. C., Loose, M., Ishihara, K., Wühr, M., Field, C. M., & Mitchison, T. J. (2014). Spatial organization of cytokinesis signaling reconstituted in a cell-free system. *Science, 346*(6206), 244–247. http://dx.doi.org/10.1126/science.1256773.

Panavas, T., Sanders, C., & Butt, T. R. (2009). SUMO fusion technology for enhanced protein production in prokaryotic and eukaryotic expression systems. *Methods in Molecular Biology, 497*, 303–317. http://dx.doi.org/10.1007/978-1-59745-566-4_20.

Pichoff, S., & Lutkenhaus, J. (2005). Tethering the Z ring to the membrane through a conserved membrane targeting sequence in FtsA. *Molecular Microbiology, 55*(6), 1722–1734. http://dx.doi.org/10.1111/j.1365-2958.2005.04522.x.

Popp, M. W., Antos, J. M., Grotenbreg, G. M., Spooner, E., & Ploegh, H. L. (2007). Sortagging: a versatile method for protein labeling. *Nature Chemical Biology, 3*(11), 707–708. http://dx.doi.org/10.1038/nchembio.2007.31.

Popp, M. W.-L., Antos, J. M., & Ploegh, H. L. (2009). Site-specific protein labeling via sortase-mediated transpeptidation. *Current Protocols in Protein Science/Editorial Board, John E. Coligan… [et al.]*. http://dx.doi.org/10.1002/0471140864.ps1503s56.

Rasnik, I., McKinney, S. A., & Ha, T. (2006). Nonblinking and long-lasting single-molecule fluorescence imaging. *Nature Methods, 3*(11), 891–893. http://dx.doi.org/10.1038/nmeth934.

RayChaudhuri, D., & Park, J. T. (1992). *Escherichia coli* cell-division gene FtsZ encodes a novel GTP-binding protein. *Nature, 359*(6392), 251–254. http://dx.doi.org/10.1038/359251a0.

Ruhnow, F., Zwicker, D., & Diez, S. (2011). Tracking single particles and elongated filaments with nanometer precision. *Biophysical Journal, 100*(11), 2820–2828. http://dx.doi.org/10.1016/j.bpj.2011.04.023.

Schindelin, J., Arganda-Carreras, I., Frise, E., Kaynig, V., Longair, M., Pietzsch, T., … Cardona, A. (2012). Fiji: an open-source platform for biological-image analysis. *Nature Methods, 9*(7), 676–682. http://dx.doi.org/10.1038/nmeth.2019.

Szwedziak, P., Wang, Q., Bharat, T. A. M., Tsim, M., & Löwe, J. (2014). Architecture of the ring formed by the tubulin homologue FtsZ in bacterial cell division. *eLife, 3*. http://dx.doi.org/10.7554/eLife.04601.

Theile, C. S., Witte, M. D., Blom, A. E. M., Kundrat, L., Ploegh, H. L., & Guimaraes, C. P. (2013). Site-specific N-terminal labeling of proteins using sortase-mediated reactions. *Nature Protocols, 8*(9), 1800–1807. http://dx.doi.org/10.1038/nprot.2013.102.

Uehara, T., Parzych, K. R., Dinh, T., & Bernhardt, T. G. (2010). Daughter cell separation is controlled by cytokinetic ring-activated cell wall hydrolysis. *The EMBO Journal, 29*(8), 1412–1422. http://dx.doi.org/10.1038/emboj.2010.36.

Vogel, S. K., Petrasek, Z., Heinemann, F., & Schwille, P. (2013). Myosin motors fragment and compact membrane-bound actin filaments. *eLife, 2*, e00116. http://dx.doi.org/10.7554/eLife.00116.

Zacharias, D. A., Violin, J. D., Newton, A. C., & Tsien, R. Y. (2002). Partitioning of lipid-modified monomeric GFPs into membrane microdomains of live cells. *Science, 296*(5569), 913–916. http://dx.doi.org/10.1126/science.1068539.

Zanetti-Domingues, L. C., Tynan, C. J., Rolfe, D. J., Clarke, D. T., & Martin-Fernandez, M. (2013). Hydrophobic fluorescent probes introduce artifacts into single molecule tracking experiments due to non-specific binding. *PLoS One, 8*(9), e74200. http://dx.doi.org/10.1371/journal.pone.0074200.

CHAPTER

22

Preparation of centralspindlin as an active heterotetramer of kinesin and GTPase activating protein subunits for in vitro structural and functional assays

M. Mishima

University of Warwick, Coventry, United Kingdom

E-mail: m.mishima@warwick.ac.uk

CHAPTER OUTLINE

Methods in Cell Biology, Volume 137, ISSN 0091-679X, http://dx.doi.org/10.1016/bs.mcb.2016.04.005

Abstract

Centralspindlin is a crucial regulator of animal cytokinesis, consisting of MKLP1 kinesin-6 and CYK4 Rho-family GTPase activating protein (RhoGAP). As a microtubule-bundling protein, it plays a crucial role in the formation of the central spindle. Through distinct accumulation to the antiparallel microtubule overlaps at the central spindle and the midbody, it recruits various downstream factors to the site of cell division as well as anchors the plasma membrane to maintain the narrow intercellular channels between the daughter cells until their final separation (abscission). A unique and functionally important feature of centralspindlin as a kinesin-containing protein complex is that the nonmotor component, CYK4, is not a passive cargo of the MKLP1 motor, but an integrated component of a microtubule-organizing machinery. Thus, for in vitro structural and functional assays, it is pivotal to prepare active stoichiometric complexes of the two components. Discussed here are two complimentary approaches, (1) reconstitution of the complex in bacterial extracts (*in extract* reconstitution) and (2) purification of a native complex from a mammalian cell line using a localization and affinity purification (LAP) tag.

INTRODUCTION

Centralspindlin is a crucial regulator of animal cytokinesis (D'Avino, Giansanti, & Petronczki, 2015; Green, Paluch, & Oegema, 2012). Two molecules of a kinesin-6 (KIF23/MKLP1/CHO1 in human, Pavarotti in fly, and ZEN-4 in *Caenorhabditis elegans*) MKLP1, collectively, and two molecules of a Rho-family GTPase activating protein (RhoGAP) (RACGAP1/MgcRacGAP in human, Tumbleweed/RacGAP50C in fly and CYK-4 in *C. elegans*), CYK4, collectively, form a stable heterotetramer (Davies et al., 2015; Mishima, Kaitna, & Glotzer, 2002; Pavicic-Kaltenbrunner, Mishima, & Glotzer, 2007). Both subunits of centralspindlin are essential for in vitro microtubule bundling and the formation of the central spindle and the midbody in dividing cells. Centralspindlin accumulates to the central antiparallel microtubule overlap of the central spindle and the midbody and serves as a key hub of the protein–protein interaction network for cytokinesis signaling (Douglas & Mishima, 2010; Lee, Davies, & Mishima, 2012). Its direct interaction partners include ECT2 Rho GDP/GTP-exchange factor, which is the major activator of Rho GTPase the master regulator of the contractile ring formation (Somers & Saint, 2003), CEP55, which recruits the ESCRT III complex to the midbody for abscission (Lee, Elia, Ghirlando, Lippincott-Schwartz, & Hurley, 2008; Zhao, Seki, & Fang, 2006), and PRC1, another microtubule-bundling and signaling hub in the network (Lee, Esmaeili, Zealley, & Mishima, 2015). Centralspindlin also contributes to maintenance of the narrow intercellular channel by anchoring the plasma membrane through the C1 domain in CYK4 until the abscission happens (Lekomtsev et al., 2012).

For the tetrameric complex to form, CYK4 forms a homodimer through the coiled coil near the N-terminus of the molecule and binds to a globular mass in the atypically long neck domain of the dimeric MKLP1 (Davies et al., 2015). This reconfigures the two motor domains in the MKLP1 dimer to be suitable for antiparallel microtubule bundling. The minimum fragments of *C. elegans* centralspindlin subunits competent for the heterotetramer formation are ZEN-4 435–550 and CYK-4 1–120, respectively. If connected to a peptide capable of coiled-coil dimerization derived from GCN4, a yeast transcription factor, the N-terminal 40 amino acids of CYK-4 is sufficient for binding to ZEN-4 (Davies et al., 2015; Pavicic-Kaltenbrunner et al., 2007). Since complex formation is essential for the functions of centralspindlin, preparation of active centralspindlin as a heterotetrameric complex is important for its biochemical and biophysical characterization. In this article, two complementary approaches, ie, purification of *C. elegans* complexes reconstituted in recombinant bacterial lysates and purification of the human complex from HeLa cells stably expressing localization and affinity purification (LAP)-tagged MKLP1, will be described.

An important feature of centralspindlin to note for successful purification of active centralspindlin is that it is an intrinsically insoluble protein (Hutterer, Glotzer, & Mishima, 2009). Both human and *C. elegans* complex can be purified as soluble heterodimers in the presence of 250 mM KCl or NaCl. However, when the salt concentration is lowered to the physiological level (~150 mM) or lower, they start to form higher-order clusters in variable sizes depending on the salt and protein concentrations. The clustering of centralspindlin is essential for productive interaction of centralspindlin with microtubules to organize the central spindle and the midbody. In dividing cells, the clustering is negatively regulated by 14-3-3 proteins that bind to the conserved RRSRS motif in the C-terminal domain of MKLP1 when the second serine in this motif is monophosphorylated (RRSRpS form, pS refers to a phosphorylated serine residue) (Douglas, Davies, Joseph, & Mishima, 2010). Centralspindlin is released from this negative regulation when the first serine in the motif is phosphorylated by Aurora B kinase (RRpSRS or RRpSRpS forms).

In the case of the *C. elegans* complex, clustering of centralspindlin depends on the C-terminal region of the predicted coiled coil of ZEN-4 motor and the constructs lacking this region (ZEN-4 586–601) are incapable of clustering and behave as soluble proteins even under the physiological or lower salt conditions (Hutterer et al., 2009). This type of soluble preparations can be used for some in vitro applications, such as the surface microtubule gliding assay, in which the motor proteins are immobilized on the coverglass surface and, thus, do not have to be competent for clustering (Davies et al., 2015; White, Raghuraman, Perozo, & Glotzer, 2013). However, to study the activities for which the clustering of centralspindlin plays a key role such as microtubule bundling, a clustering-competent construct needs to be used (Davies et al., 2015). For preparation of such a sample, the salt concentration in the buffers should be kept high throughout the steps of purification.

1. PURIFICATION OF A RECOMBINANT *CAENORHABDITIS ELEGANS* CENTRALSPINDLIN COMPLEX RECONSTITUTED IN BACTERIAL EXTRACT

A popular method to prepare a recombinant protein complex in *E. coli* is to express its components from a bi-/poly-cistronic vector (Selleck & Tan, 2008; Sessa et al., 2005). Coexpression of the components of a complex within the same bacterial cell can facilitate the proper folding, solubility, and activity of the protein complex. However, to examine the combinatorial effects of various mutations in multiple components by expressing them from a bi-/poly-cistronic vector, a new construct has to be prepared for each combination. We have discovered that the complex between ZEN-4 and CYK-4 can be reconstituted by combining the two bacterial lysates containing separately expressed ZEN-4 and CYK-4, respectively (Davies et al., 2015). This approach greatly facilitates preparation of the complexes of ZEN-4 and CYK-4 with various combinations of deletions/mutations. Discussed here is a case of ZEN-4 1−601 fragment tagged with chitin-binding domain (CBD-Z601) and CYK-4 1-40-GCN4 chimera tagged with glutathione S-transferase (GST) (C40G). Sequential affinity column chromatography using chitin beads and glutathione Sepharose selectively purified the complex of ZEN-4 and CYK-4, which was confirmed by co-elution in size-exclusion chromatography.

1.1 MATERIALS

- DNA constructs and bacterial strains for expression
  - A plasmid that produces a fusion protein between GST and a CYK-4 fragment capable of binding to ZEN-4 containing a cleavage site for PreScission protease (LEVLFQ|GP) within the linker. Here we use TD376 plasmid that expresses GST-(linker with a PreScission cleavage site)-(CYK-4 1−40)-(GCN4 254−281) (GST-C40G, 299 aa), which releases a 73 aa CYK-4-GCN4 chimeric peptide (C40G) after cleavage.
  - A plasmid that produces a fusion protein between chitin-binding domain (CBD) and a ZEN-4 fragment capable of binding to ZEN-4 with a cleavage site for tobacco etch virus (TEV) protease (ENLYFQ|G) within the linker. Here we use MM227 plasmid that is derived of a pET28a vector (Novagen) and expresses CBD-(linker with a TEV cleavage site)-(ZEN-4 1−601)-(5 aa linker)-(15 aa biotin-acceptor) (CBD-Z601, 686 aa), which releases a ZEN-4 fragment (Z601, 625 aa with the C-terminal tag) after cleavage. Although the C-terminal biotin-acceptor peptide is biotinylated by bacterial BirA enzyme (Berliner et al., 1994; Mishima, Pavicic, Grüneberg, Nigg, & Glotzer, 2004), it is not essential for the purification procedure described below.

    These plasmids are separately transformed into *E. coli* BL21(DE3) strain with appropriate antibiotics. For the expression of CBD-Z601 under the T7 promoter, the DE3 cassette, which encodes T7 polymerase, is essential.
- Isopropyl-β-D-thiogalactoside (IPTG, Merck Biosciences, 420322) 1 M stock in water, store at −20°C.

- Poly-Prep Chromatography Columns (Bio-Rad, 7311550). A disposal chromatography column for a bed volume 0.5–2 mL with a 10 mL reservoir.
- Chitin beads (New England Biolabs, S6651), store at 4°C.
- AcTEV protease (Invitrogen, 12575) 10 U/μL, store at −20°C.
- Glutathione-Sepharose 4B (GE Healthcare, 17-0756), store at 4°C.
- PreScission protease (GE Healthcare, 27-0843) stock 2 U/μL, store at −20°C.
- Phenylmethylsulfonyl fluoride (PMSF) (Sigma, P7626) 0.2 M stock in ethanol, store at −20°C.
- Lysozyme (Sigma, L6876) 50 mg/mL in 50% (v/v) glycerol/TE, store at −20°C.
- Bacterial lysis buffer, 250 mM NaCl, 10 mM 4-(2-hydroxyethyl)-1-piperazineethanesulfonic acid (HEPES) pH7.7, 1 mM $MgCl_2$, 1 mM ethylene glycol tetraacetic acid (EGTA), 0.1% (w/v) Triton X-100, chill on ice and supplement with 1 mM dithiothreitol (DTT), 0.1 mM adenosine triphosphate (ATP), 10 μg/mL leupeptin, 10 μg/mL pepstatin before use
- Gel filtration buffer, 250 mM NaCl, 10 mM HEPES pH7.7, 1 mM $MgCl_2$, 1 mM EGTA, 0.2 mM DTT, 25 μM ATP

1.2 PROCEDURE

1.2.1 Expression of the recombinant proteins and reconstitution of the complex in the lysate

1. Separately culture the bacterial strains from a single colony at 37°C overnight in 50 mL Luria-Bertani (LB) medium with antibiotics (50 μg/mL ampicillin for pGST-CYK-4 construct and 50 μg/mL kanamycin for pCBD-TEV-ZEN-4 construct, respectively).
2. Dilute the 50 mL precultures with 500 mL of LB medium with an antibiotics and culture at 25°C until OD_{600} reaches ~0.7.
3. Add IPTG to final 0.2 mM and culture at 25°C for additional 6 h.
4. Harvest the bacteria by centrifugation at 3000 rpm for 30 min at 4°C using a rotor such as Sorvall HBB-6 and re-suspend the bacterial pellet with 50 mL of cold PBS, transfer the suspension to a 50 mL Falcon tube and collect bacteria by centrifugation.
5. Discard the supernatant as cleanly as possible, freeze the pellet with liquid nitrogen, and store at −80°C until use.
6. Thaw and resuspend the bacterial pellet on ice by pipetting with 30 mL of chilled bacterial lysis buffer.
7. Add 150 μL of 0.2 M PMSF (final 1 mM) and 0.3 mL of 50 mg/mL lysozyme (final 0.5 mg/mL) and keep on ice for 1 h until the suspension becomes less turbid and highly viscous.
8. Disrupt DNA and bacterial debris by sonication until the lysate becomes fluid.
9. Centrifuge 14,000 rpm for 30 min at 4°C in a Sorvall SS-34 (or an equivalent such as Beckman JA-20) rotor and collect the supernatants (bacterial cell extracts).
10. Mix the two extracts and incubate on ice for 3 h.

1.2.2 Affinity purification by CBD-tag on ZEN-4

1. During the on-ice incubation of the bacterial lysate, equilibrate 1 mL chitin beads on a Poly-Prep column by three times of 10 mL wash with the bacterial lysis buffer.
2. Transfer the equilibrated chitin beads to the mixed extract by a disposal plastic dropper and rotate the suspension ~30 rpm at 4°C for 5 h.
3. Sediment the beads by a brief centrifugation (500 rpm, 3 min) and remove the supernatant by decantation.
4. Transfer the beads back to the chromatography column with 5 mL lysis buffer and wash the column with 10 mL of the lysis buffer.
5. Wash the beads 10 times with 1 mL lysis buffer supplemented with 10 mM ATP by gravity flow. This removes bacterial 60-kDa protein (chaperonin?), which often associates with ZEN-4 constructs containing the motor domain.
6. Wash the beads with 10 mL of the lysis buffer containing 0.1 mM ATP.
7. Add 1 mL of the lysis buffer supplemented with AcTEV protease (final 250 U/mL) and let it go into the beads by gravity flow.
8. Close the outlet with a cap and cover the surface of the bead bed with 100 μL of the same solution and incubate at 4°C for 36–72 h.
9. Elute the TEV-cleaved product with the lysis buffer in 1 mL fractions.
10. Check the purification steps by SDS-PAGE (10 μL).

1.2.3 Affinity purification by GST-tag on CYK-4

1. Equilibrate 0.5 mL of glutathione-Sepharose beads with the lysis buffer on a Poly-Prep column.
2. Load the peak eluates from the chitin beads to the glutathione-Sepharose column by 0.5 mL by gravity flow.
3. Wash the column five times with 0.5 mL of the lysis buffer.
4. Load 0.5 mL of the lysis buffer supplemented with PreScission protease (final 80 U/mL) and let it go by gravity flow.
5. Close the outlet and cover the surface of the bead with 100 μL of the same solution.
6. Keep at 4°C for ~36 h.
7. Elute the PreScission-cleaved product by 0.5 mL fractions of the elution buffer.
8. Check the purity by SDS-PAGE (10 μL).

1.2.4 Gel filtration

1. Equilibrate a Superdex 200 10/300 (24 mL bed volume) column with the gel filtration buffer on an Äkta system (GE Healthcare).
2. Load the peak fraction (0.5 mL) and collect the eluates in 0.5 mL fractions at flow rate 0.25 mL/min.
3. Check the purity by SDS-PAGE (10 μL).

1.3 RESULTS AND DISCUSSIONS

An example of the result of this procedure using CBD-tagged ZEN-4 1−601 fragment and GST-tagged chimera between CYK-4 1−40 and GCN4 254−281 is shown in Fig. 1. The eluates from the chitin beads after cleavage of CBD-Z601 by TEV-protease contained the ~70 and ~35 kDa proteins, which correspond to Z601 and GST-C40G, respectively, as major components (Fig. 1A). They also contained ~15-kDa band with unknown identity (asterisk) and a fainter 27 kDa band, which is probably the TEV protease (TEV). The pooled peak fractions (E1 ~ E5) contained CBD-Z601 in a complex with GST-C40G, which was trapped by the glutathione-Sepharose 4B beads (Fig. 1B), and free CBD-Z601, which was not and thus separated into the flow-through fraction (FT) along with the ~15 kDa contaminant. The proteins bound to the glutathione beads and eluted by cleavage of GST-C40G by PreScission protease (Fig. 1B E1 ~ E5) contained ~70 and ~6 kDa bands, which correspond to Z601 and C40G, respectively. The PreScission protease in the cleavage buffer (Fig. 1B Pre) was retained on the column and did not appear in the elution fractions (E1 ~ E5) as it is tagged with GST. Z601 and C40G in the peak faction were in a stable complex as confirmed by coelution from a size-exclusion chromatography column (Superdex 200 10/300 GL) (Fig. 1C).

The K_D between ZEN-4 and CYK-4 dimers is 10 nM (Pavicic-Kaltenbrunner et al., 2007). Thus, for the successful purification of a stoichiometric complex, it is important to keep the concentrations of each subunit at least one order higher than the K_D all through the steps of the purification. For example, the concentration of ZEN-4 as a dimer (~150 kDa) should be maintained higher than ~15 μg/mL. To avoid the dissociation of the complex during the washing steps, it is also important to

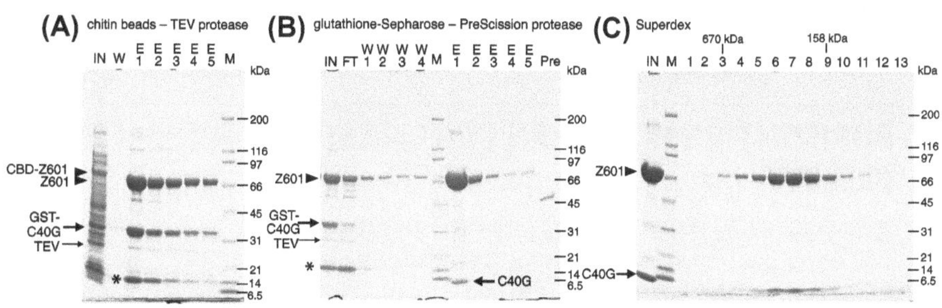

FIGURE 1

Purification steps of the complex of ZEN-4 601 (Z601) and CYK-4 1−40::GCN4 (C40G) reconstituted in bacterial extract. Chitin-binding domain–tagged Z601 (CBD-Z601) and glutathione S-transferase–tagged C40G (GST-C40G) were separately expressed. The complex was reconstituted by combining the two extracts and sequentially purified by (A) affinity chromatography with chitin beads and elution by TEV protease, (B) affinity chromatography with glutathione-Sepharose 4B and elution by PreScission protease, and (C) size exclusion chromatography using Superdex 200. *IN*, input; *W*, wash; *E*, eluate; *M*, protein standard markers; *FT*, flow through; *Pre*, PreScission protease; *Asterisk*, contaminant with unknown identity.

wash the columns as quickly as possible with minimum volumes of buffer. For the time-consuming cleavage process, high protein concentration should be maintained by minimizing the bed volume. In the case of Fig. 1, the concentration of the complex on the column was 1–10 μM.

The essentially same sequential affinity chromatography can also be used to purify the complex from bacterial cells simultaneously expressing both ZEN-4 and CYK-4 from a bicistronic vector (Davies et al., 2015). Although the bicistronic approach ensures the equimolar expression at the level of mRNA, it does not always guarantee the equal translation and posttranslational folding and stability of the individual components. In extract reconstitution described here allows us to fine-tune the relative amount of the component. When there is a large difference in the expression levels of the two components, the relative volumes of the extracts can be adjusted so that the molar concentrations in the combined extract are roughly equal.

The complex can also be reconstituted by combining separately purified ZEN-4 and CYK-4 (Davies et al., 2015; Pavicic-Kaltenbrunner et al., 2007). However, while ZEN-4 in bacteria behaves well and can easily be purified if a high ionic strength (250 mM) is maintained, CYK-4 samples purified from bacteria tend to weakly oligomerize when they contain the N-terminal ZEN-4 binding region (my unpublished observation). In crude extracts, bacterial chaperones probably maintain free CYK-4, both soluble and stable, and facilitate reconstitution of the complex between ZEN-4 and CYK-4. The complex reconstituted in extract behaves very well as a soluble heterotetramer in the presence of high salts. For preparation of a stoichiometric centralspindlin complex to be used in vitro structural and functional analyses such as in vitro motility and microtubule bundling assays, observation by an atomic force microscopy, FRET analysis (Davies et al., 2015), in extract reconstitution described here is currently the first choice.

2. PURIFICATION OF HUMAN CENTRALSPINDLIN FROM HeLa CELLS

For applications that require a more "native" preparation, the centralspindlin complex can be purified from HeLa cells stably transformed with a large human genomic DNA fragment in which *KIF23* gene is tagged with a LAP tag by recombineering in bacteria. For example, to demonstrate that centralspindlin can form clusters independently of other proteins and becomes insoluble under the physiological salt concentration, it was necessary to work with a "native" preparation of centralspindlin purified from animal cells (Hutterer et al., 2009). A similar preparation of native centralspindlin from mitotic HeLa cells was used to examine the effect of 14-3-3-binding on the clustering of centralspindlin and its microtubule-bundling activity (Douglas et al., 2010). In this case, the cells were treated with an inhibitor of Aurora B kinase so that phosphorylation of the first serine in the conserved RRSRS motif found in the C-terminal domain of MKLP1 is inhibited and the second serine is monophosphorylated (RRSRpS form) and thus can be bound by 14-3-3 proteins.

Although it was not explicitly mentioned in the original report (Maliga et al., 2013; Poser et al., 2008), the N-terminal LAP (NLAP)-tag used for bacterial artificial chromosome (BAC) recombineering contains a FLAG-tag in addition to EGFP and S-tag in the linker (EGFP—PreScission site—S-tag—TEV site—FLAG—target protein). For elution of an S-tagged protein from S-protein beads, cleavage by TEV protease or a harsh condition such as 0.2 M glycine pH2 or 3 M guanidine thiocyanate is necessary. Thus, to prepare active EGFP-tagged centralspindlin in the first affinity chromatography, we use the FLAG-tag−anti-FLAG antibody affinity, which allows quick but much milder elution without cleavage between EGFP and KIF23. The GFP-tagged complex is further purified by the second affinity chromatography using anti-GFP antibody beads and cleavage by PreScission protease, which results in purer but non−GFP-tagged centralspindlin.

2.1 MATERIALS

- NLAP-KIF23 HeLa cell line, A HeLa Kyoto cell line stably transformed with a human BAC (CTD-2081C9) tagged with a LAP cassette at the N-terminus of KIF23 (MKLP1/CHO1) was prepared as described (Poser et al., 2008).
- Dulbecco's modified Eagle medium (DMEM) with pyruvate (Invitrogen 21969-035) supplemented with 4 mM L-glutamine and 10% (v/v) fetal calf serum, penicillin, streptomycin, and 0.2 mg/mL G418 (DMEM + G418)
- Triple-layer flasks (Nunc, TKT-130-010E, 500 cm^2)
- Nocodazole (Sigma, M14040) 40 μg/mL in dimethyl sulfoxide (DMSO), store at −20°C
- Microcystin LR (Sigma, M2912) 1 mM stock in DMSO
- ZM447439 (Tocris Biosciences, 2458) 2 mM stock in DMSO
- Anti-FLAG M2 affinity gel (Sigma, A2220)
- 3xFLAG peptide (Sigma, F3290) 10 mg/mL in TBS
- Anti-GFP agarose beads (MBL, D153-8) or GFP-Trap_A (ChromoTek, gta-20)
- PreScission protease (GE Healthcare, 27-0843-01) stock 2 U/μL
- HeLa lysis buffer, 200 mM NaCl, 10 mM NaF, 40 mM β-glycerophosphate, 20 mM HEPES, 2 mM $MgCl_2$, 1 mM EGTA, 0.5% (w/v) Triton X-100, pH7.5, freshly prepare, chill on ice, and supplement with 10 μg/mL leupeptin, 10 μg/mL pepstatin, 1 μM Microcystin LR and 1 mM PMSF before use. To prepare Aurora B site-dephosphorylated centralspindlin add 1 μM ZM447439
- Chromatography buffer, 250 mM NaCl, 10 mM HEPES, 1 mM EGTA, 1 mM $MgCl_2$, pH 7.6, chill on ice and supplement with 10 μg/mL leupeptin, 10 μg/mL pepstatin, 1 μM Microcystin LR and 10 μM ATP

2.2 PROCEDURE

2.2.1 Preparation of the KIF23 BAC HeLa cells

Approximately 10^9 mitotic cells are collected by repeating 8 rounds of culture in 9 triple-layer flasks. For one set,

1. NLAP-KIF23 HeLa cells are grown with 70 mL DMEM + G418 medium in 9 triple-layer flasks (500 cm^2/flask).
2. When they reach ~70% confluence, add nocodazole to final 40 ng/mL and culture the cells for 18 h.
3. If necessary, treat the cells with 2 μM ZM447439 for 30 min.
4. Shake off the mitotic cells from the flasks and collect them by centrifugation in a 500 mL centrifuge tube (Corning 431123) using a Sorval HBB-6 rotor.
5. Remove the medium, resuspend the cells with ~50 mL chilled PBS and transfer the suspension to a 50 mL Falcon tube.
6. Collect the cells by centrifugation, remove the supernatant.
7. Freeze the cell pellet by liquid nitrogen and store at −80°C.

2.2.2 Extraction and anti-FLAG affinity purification

1. Thaw the cell pellets in the 8 Falcon tubes (total ~15 mL cell) with ice chilled 64 mL lysis buffer (8 mL per tube) supplemented with protease and phosphatase inhibitors.
2. Lyse the cells by gentle pipetting with a 10 mL disposable pipet using a pipet-aid.
3. Centrifuge at 14,000 rpm for 30 min at 4°C using a JA20 rotor (Beckman) and collect the ~70 mL supernatant in two 50 mL Falcon tubes.
4. Add 0.5 mL of anti-FLAG beads equilibrated with the lysis buffer to the extract (0.25 mL beads per tube) and rotate the suspensions at 4°C for 5 h.
5. Collect the beads with a gentle centrifugation and transfer them to a Poly-Prep column.
6. Wash the column with 10 and 5 mL of the lysis buffer.
7. Wash the column with 5 mL of the chromatography buffer.
8. Elute centralspindlin with 0.2 mg/mL 3xFLAG peptide in the chromatography buffer in 0.5 mL fractions.
9. Check the steps of purification by SDS-PAGE (10 μL).

2.2.3 Anti-GFP affinity purification

1. Add Triton X-100 to final 0.1% to the peak fraction and incubate it with 50 μL of anti-GFP beads on ice for 12 h.
2. Wash the beads with 2 × 400 μL the chromatography buffer.
3. Resuspend the beads with 60 μL 3 U/mL PreScission protease in the chromatography buffer supplemented with 1 mM DTT and leave on ice for 36 h.
4. Collect the supernatant after brief centrifugation with a desktop centrifuge.
5. Check the fractions by SDS-PAGE (3 μL each).

2.3 RESULTS AND DISCUSSIONS

An example of preparation of HeLa centralspindlin is shown in Fig. 2. The eluates from the anti-FLAG column contained a doublet around 150 kDa (arrowheads) and a ~85-kDa band (arrow) as major bands (Fig. 2A). These polypeptides were retained

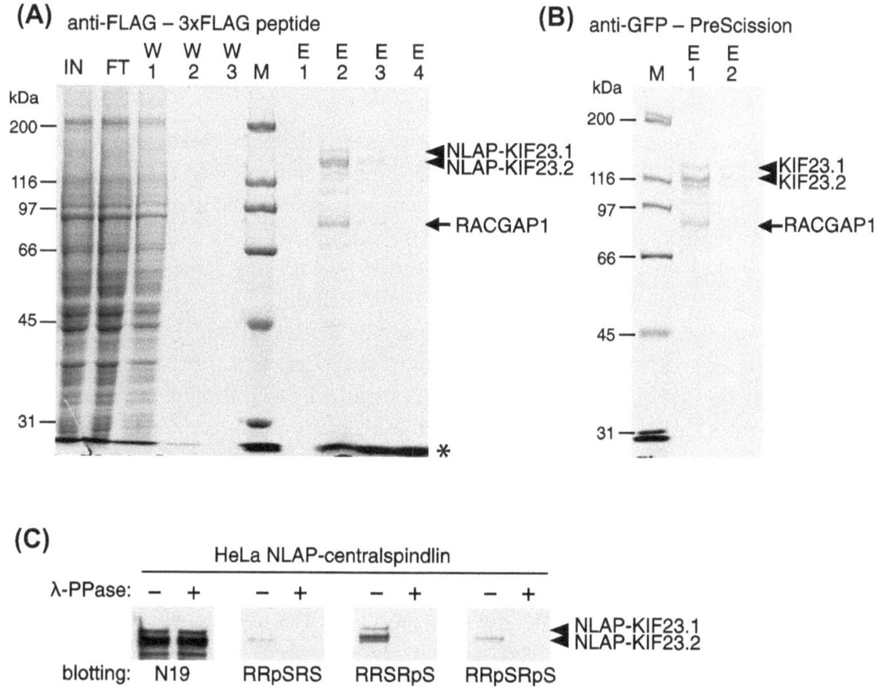

FIGURE 2

Purification steps of human centralspindlin from an NLAP-KIF23 HeLa cell line and evaluation of the phosphorylation state of the RRSRS Aurora B phosphorylation/14-3-3 binding motif in the C-terminal domain of KIF23. Centralspindlin containing NLAP-KIF23 from the mitotic NLAP-KIF23 cells treated with an Aurora kinase inhibitor was sequentially affinity-purified by (A) anti-FLAG antibody beads and elution with 3xFLAG peptide and (B) anti-GFP antibody beads and elution by PreScission protease. (C) The eluate from the anti-FLAG affinity chromatography was treated with λ-phosphatase (λ-PPase) and blotted with a peptide antibody against the KIF23 N-terminal sequence (N19) and the phospho-specific antibodies against the RRSRS motif (RRpSRS, RRSRpS and RRpSRpS) (Douglas et al., 2010). *IN*, input; *W*, wash; *E*, eluate; *M*, protein standard markers; *FT*, flow through; *Asterisk*, 3xFLAG peptide.

on the following anti-GFP column and eluted after cleavage by PreScission protease (Fig. 2B). While the mobility of the ~85 kDa band (arrow) did not change, that of the doublet (arrow heads) was increased after the cleavage (120 kDa thicker band and 135 kDa thinner band). This indicates that the ~150-kDa doublet is NLAP-KIF23, the 120–135-kDa doublet is KIF23 and the 85-kDa band is RACGAP1. This was confirmed by western blotting with anti-GFP and anti-RACGAP1 antibodies (Fig. 2C and data not shown). It is known that KIF23 exists in multiple splicing isoforms and that, in HeLa cells, the isoform 2 (also known as MKLP1) lacking the exon 18 is much more abundant than the isoform 1 containing all the exons (Kuriyama, Gustus, Terada, Uetake, & Matuliene, 2002). This is consistent with the band pattern of the NLAP-KIF23 and cleaved KIF23 doublets, giving a further support for the identity of these bands and demonstrating the physiological regulation of the gene expression from a BAC.

With a closer look, however, we realize that the molecular weights based on the mobilities in SDS-PAGE of both the KIF23 doublet and the RACGAP1 band are larger from the sequence-based predictions. The sequence of the modified BAC predicts that the molecular weights of the NLAP-KIF23 isoforms 1 (1275 aa) and 2 (1171 aa) are 145 and 135 kDa, respectively. However, the values based on the actual mobilities of the KIF23 isoforms were 158 and 143 kDa (Fig. 2A). These are bigger than the predictions by the sequence of the NLAP-tagged BAC, ie, 145 and 135 kDa for the NLAP-tagged isoforms 1 (1275 aa) and 2 (1171 aa), respectively. Since the PreScission cleavage leaves a part of the linker of the NLAP tag at the N-terminus of the KIF23 isoforms, the cleavage products are 61 aa longer than the endogenous counterparts and thus estimated to be 116 and 104 kDa (1021 and 917 aa). However, the mobility-based molecular weights of the cleavage products (135 and 120 kDa) were also bigger than these predictions (Fig. 2B). These discrepancies in the molecular weight estimations (~15 kDa bigger values by the actual mobilities than the sequence-based predictions) are not specific to the BAC-derived KIF23 isoforms, but common with the endogenous ones. While the molecular weights of the endogenous isoform 1 (CHO1) and isoform 2 (MKLP1) are predicted to be 110 kDa (960 aa) and 98 kDa (852 aa), respectively, the mobility-based estimations of the endogenous KIF23 isoforms found in the immunoprecipitates by anti-KIF23 or by anti-RACGAP1 antibody were 127 and 112 kDa (Mishima et al., 2002).

Although phosphorylation often causes a mobility shift in SDS-PAGE and the multiple sites on KIF23 are known to be phosphorylated, phosphorylation of KIF23 does not seem to be the major reason for the discrepancy between the molecular weight and the actual mobility since a phosphatase treatment, which abolishes the phosphorylation at the Aurora B/14-3-3 regulatory motif, does not change the mobilities of the KIF23 isoforms (Fig. 2C). This is consistent with the observations that their mobilities are largely constant through the cell cycle. Their extremely high isoelectric points (>8.5) are likely to be a reason for the molecular weight−mobility discrepancy although the possibilities of unidentified posttranslational modifications cannot be excluded.

The mobility of RACGAP1 in the centralspindlin preparation according to this procedure, which appears at ~85 kDa in SDS-PAGE, was also smaller than expected from the predicted molecular weight (632 aa, 71.0 kDa). In contrast to KIF23, this is likely due to the mitotic phosphorylation. RACGAP1 shows large shift (reduced mobility) during prometaphase and metaphase, which returns to the interphase level after anaphase onset (Wolfe, Takaki, Petronczki, & Glotzer, 2009; Zhao & Fang, 2005). The upshift of the RACGAP1 band in the nocodazole-arrested lysate is canceled by purvalanol A, an inhibitor of CDK1, indicating that the mitotic upshift of RACGAP1 is due to the phosphorylation by CDK1

(Wolfe et al., 2009). Consistent with this, RACGAP1 immunoprecipitated from a microtubule-binding fraction prepared from HeLa cells without any phosphatase inhibitor appears at the mobility expected by the molecular weight (73.2 kDa) (Hutterer et al., 2009).

The HeLa centralspindlin prepared from mitotic but Aurora B–inhibited HeLa cells as described assembles into higher-order clusters and bundle microtubules in a 14-3-3-sensitive manner (Douglas et al., 2010). In other words, in the absence of 14-3-3, Aurora B phosphorylation on the C-terminal domain of KIF23 is dispensable for microtubule bundling by centralspindlin. This is in good agreement of the in vivo knockdown and rescue experiments with the mutations in the Aurora B phosphorylation/14-3-3 binding RRSRS motif (Douglas et al., 2010; Joseph, Hutterer, Poser, & Mishima, 2012). This centralspindlin preparation was also used for a structural study by high-speed atomic force microscopy and demonstrated the conserved molecular architecture of centralspindlin (Davies et al., 2015).

A large limitation of the current protocol is the yield of the final sample, which is ~50 μg from ~15 mL cell pellet (~10^9 cells). This is not very bad if we consider that it is roughly 1/3 of 0.01% of the total protein in 15 mL cell pellet assuming the total protein concentration in the cell is 100 mg/mL. It is roughly equivalent to 100% yield of a protein that exists at 10 nM (~K_D between CYK-4 and ZEN-4). However, for further purification, we need to scale up the starting material by, for example, culturing the HeLa cells in suspension using a spinner flask.

CONCLUSIONS

In this chapter, two complementary approaches for preparation of active centralspindlin heterotetramers have been discussed. Although they focus on centralspindlin, both the methods are in principle applicable to other protein complexes as well. Although these have been useful for various in vitro studies of centralspindlin functions and regulations, there remain major limitations as the recombinant complex produced in bacteria lacks eukaryotic posttranslational modifications and that the yield and the purity/homogeneity of the preparations from HeLa cells are not comparable to the bacterial preparations. Novel methodologies to overcome these limitations are awaited.

ACKNOWLEDGMENTS

I thank Tim Davies, Max Douglas, Andrea Hutterer, Kian-Yong Lee, Nimesh Joseph, Julia Mason, Noriyuki Kodera, Toshio Ando, Gabriele S. Kaminski Schierle, Eric Rees, Miklos Erdelyi, and Clemens F. Kaminski for their contributions to the works mentioned in this article as applications of the described centralspindlin preparations, and Tim Davies for his helpful comments on the manuscript. This work was supported by Cancer Research UK Programme Grant (C19769/A11985).

REFERENCES

Berliner, E., Mahtani, H. K., Karki, S., Chu, L. F., Cronan, J. E., Jr., & Gelles, J. (1994). Microtubule movement by a biotinated kinesin bound to streptavidin-coated surface. *The Journal of Biological Chemistry, 269*, 8610–8615.

Davies, T., Kodera, N., Kaminski Schierle, G. S., Rees, E., Erdelyi, M., Kaminski, C. F., … Mishima, M. (2015). CYK4 promotes antiparallel microtubule bundling by optimizing MKLP1 neck conformation. *PLoS Biology, 13*, e1002121.

D'Avino, P. P., Giansanti, M. G., & Petronczki, M. (2015). Cytokinesis in animal cells. *Cold Spring Harbor Perspectives in Biology, 7*, a015834.

Douglas, M. E., Davies, T., Joseph, N., & Mishima, M. (2010). Aurora B and 14-3-3 coordinately regulate clustering of centralspindlin during cytokinesis. *Current Biology: CB, 20*, 927–933.

Douglas, M. E., & Mishima, M. (2010). Still entangled: assembly of the central spindle by multiple microtubule modulators. *Seminars in Cell & Developmental Biology, 21*, 899–908.

Green, R. A., Paluch, E., & Oegema, K. (2012). Cytokinesis in animal cells. *Annual Review of Cell and Developmental Biology, 28*, 29–58.

Hutterer, A., Glotzer, M., & Mishima, M. (2009). Clustering of centralspindlin is essential for its accumulation to the central spindle and the midbody. *Current Biology: CB, 19*, 2043–2049.

Joseph, N., Hutterer, A., Poser, I., & Mishima, M. (2012). ARF6 GTPase protects the post-mitotic midbody from 14-3-3-mediated disintegration. *EMBO Journal, 31*, 2604–2614.

Kuriyama, R., Gustus, C., Terada, Y., Uetake, Y., & Matuliene, J. (2002). CHO1, a mammalian kinesin-like protein, interacts with F-actin and is involved in the terminal phase of cytokinesis. *The Journal of Cell Biology, 156*, 783–790.

Lee, H. H., Elia, N., Ghirlando, R., Lippincott-Schwartz, J., & Hurley, J. H. (2008). Midbody targeting of the ESCRT machinery by a noncanonical coiled coil in CEP55. *Science, 322*, 576–580.

Lee, K. Y., Davies, T., & Mishima, M. (2012). Cytokinesis microtubule organisers at a glance. *Journal of Cell Science, 125*, 3495–3500.

Lee, K. Y., Esmaeili, B., Zealley, B., & Mishima, M. (2015). Mechanical resilience of the central spindle. *Nature Communications, 6*, 7290.

Lekomtsev, S., Su, K. C., Pye, V. E., Blight, K., Sundaramoorthy, S., Takaki, T., … Petronczki, M. (2012). Centralspindlin links the mitotic spindle to the plasma membrane during cytokinesis. *Nature, 492*, 276–279.

Maliga, Z., Junqueira, M., Toyoda, Y., Ettinger, A., Mora-Bermudez, F., Klemm, R. W., … Hyman, A. A. (2013). A genomic toolkit to investigate kinesin and myosin motor function in cells. *Nature Cell Biology, 15*, 325–334.

Mishima, M., Kaitna, S., & Glotzer, M. (2002). Central spindle assembly and cytokinesis require a kinesin-like protein/RhoGAP complex with microtubule bundling activity. *Developmental Cell, 2*, 41–54.

Mishima, M., Pavicic, V., Grüneberg, U., Nigg, E. A., & Glotzer, M. (2004). Cell cycle regulation of central spindle assembly. *Nature, 430*, 908–913.

Pavicic-Kaltenbrunner, V., Mishima, M., & Glotzer, M. (2007). Cooperative assembly of CYK-4/MgcRacGAP and ZEN-4/MKLP1 to form the centralspindlin complex. *Molecular Biology of the Cell, 18*, 4992–5003.

Poser, I., Sarov, M., Hutchins, J. R., Hériché, J. K., Toyoda, Y., Pozniakovsky, A., … Hyman, A. A. (2008). BAC TransgeneOmics: a high-throughput method for exploration of protein function in mammals. *Nature Methods, 5*, 409–415.

Selleck, W., & Tan, S. (2008). Recombinant protein complex expression in *E. coli*. *Current Protocols in Protein Science* (Chapter 5), Unit 5 21.

Sessa, F., Mapelli, M., Ciferri, C., Tarricone, C., Areces, L. B., Schneider, T. R., … Musacchio, A. (2005). Mechanism of Aurora B activation by INCENP and inhibition by hesperadin. *Molecular Cell, 18*, 379–391.

Somers, W. G., & Saint, R. (2003). A RhoGEF and Rho family GTPase-activating protein complex links the contractile ring to cortical microtubules at the onset of cytokinesis. *Developmental Cell, 4*, 29–39.

White, E. A., Raghuraman, H., Perozo, E., & Glotzer, M. (2013). Binding of the CYK-4 subunit of the centralspindlin complex induces a large scale conformational change in the kinesin subunit. *The Journal of Biological Chemistry, 288*, 19785–19795.

Wolfe, B. A., Takaki, T., Petronczki, M., & Glotzer, M. (2009). Polo-like kinase 1 directs assembly of the HsCyk-4 RhoGAP/Ect2 RhoGEF complex to initiate cleavage furrow formation. *PLoS Biology, 7*(5), e1000110.

Zhao, W. M., & Fang, G. (2005). MgcRacGAP controls the assembly of the contractile ring and the initiation of cytokinesis. *Proceedings of the National Academy of Sciences of the United States of America, 102*, 13158–13163.

Zhao, W. M., Seki, A., & Fang, G. (2006). Cep55, a microtubule-bundling protein, associates with centralspindlin to control the midbody integrity and cell abscission during cytokinesis. *Molecular Biology of the Cell, 17*, 3881–3896.

CHAPTER

In vitro reactivation of the cytokinetic contractile ring of fission yeast cells

23

I. Mabuchi*[,1], J. Kashiwazaki*, M. Mishra[§]

*Gakushuin University, Tokyo, Japan

[§]Tata Institute of Fundamental Research, Mumbai, India

[1]Corrresponding author: E-mail: issei.mabuchi@gakushuin.ac.jp

CHAPTER OUTLINE

Abstract

Cytokinesis is a process by which a mother cell is divided into two daughter cells after chromosome segregation. In both animal and fungal cells, cytokinesis is carried out by the constriction of the contractile ring made up of actin, myosin-II, and other conserved proteins. Detailed genetic and cell biological analysis of cytokinesis has led to the identification of various genes involved in the process of cytokinesis including the cytological description of the process. However, detailed biochemical analysis of the process is lacking. Critical questions that aim to understand aspects, such as the organization of actin and myosin in the contractile ring, the architecture of the ring, and the molecular process of ring contraction, remain unanswered. We have developed a method to address these aspects of cytokinesis. Using the fission yeast *Schizosaccharomyces pombe*, we present a method whereby cell-ghosts containing functional contractile rings

Methods in Cell Biology, Volume 137, ISSN 0091-679X, http://dx.doi.org/10.1016/bs.mcb.2016.03.037

can be isolated and used to perform various biochemical analysis as well as detailed electron microscopy studies.

INTRODUCTION

Cell division is an essential event for proliferation, development, and maintenance of multicellular organisms. It is composed of mitosis and cytokinesis. In mitosis chromatids are segregated on a mitotic spindle leading to the formation of daughter nuclei. On the other hand, cytokinesis is a distinct process in which a force to divide the cell into two is generated at the division cortex of the cell. The machinery which exerts the force for cleavage at the division site was found by electron microscopy (EM) in the late 1960s to 1970s to be a bundle of actin filaments forming a ring structure in the division cortex of various animal cells (reviewed by Mabuchi, 1986), and the ring was called **the contractile ring**.

It has been established that the contractile ring contracts via interaction of actin filaments and myosin-II ATPase activity (Mabuchi & Okuno, 1977; Straight et al., 2003). However, there are many unsolved questions about cytokinesis: How are actin and myosin recruited to the division site? How are these proteins organized into the contractile ring structure? What is the detailed architecture of the contractile ring? What is the molecular process of ring contraction? How are actin and myosin disassembled during and after contraction?

To solve these important questions, it is desirable to have an in vitro system to study the mechanism of cytokinesis. Pioneering work was done using "glycerinated fibroblasts" (Hoffman-Berling, 1954). The telophase fibroblasts were permeabilized by glycerol at a low temperature. Upon addition of ATP the cleavage furrow advanced in these cells. This method was improved later using a detergent solution (Cande, 1980), but never used or developed further probably because of the difficulty to obtain a high proportion of dividing cells and to visualize the contractile ring during its contraction.

In 1988 we isolated the cleavage furrow from dividing newt eggs by a surgical technique (Mabuchi, Tsukita, Tsukita, & Sawai, 1988) and confirmed that the contractile ring was composed of actin filaments of opposite directionalities. These filaments were attached to the plasma membrane at their plus ends and were cross-linked with each other by cross-linking proteins. We also demonstrated that the isolated furrow contracted upon addition of ATP, which confirmed that the isolated furrow retained the original structure. These were novel findings but the mechanism of contraction was still unclear. Furthermore, this method did not become popular because it was difficult to collect many furrows in a short period, which was required for biochemical analyses or EM.

Next we isolated the cleavage furrow possessing the contractile ring, on a large scale from sea urchin eggs (Fujimoto & Mabuchi, 1997; Yonemura, Mabuchi, & Tsukita, 1991). Using this system we showed that elongation factors were concentrated in the isolated furrow preparation, and EF1α was necessary in progression of the cleavage (Fujimoto & Mabuchi, 2010). However, these furrows did not contract upon addition of ATP suggesting that the structure was damaged during isolation. In spite of this unsatisfactory result, the isolation medium was successfully applied in the following experiments.

More than 10 years after the discovery of the contractile ring in animal cells, a similar ring, which was called the F-actin ring, was found at the division site of the fission yeast *Schizosaccharomyces pombe* (Kanbe, Kobayashi, & Tanaka, 1989; Marks & Hyams, 1985). Since *S. pombe* is genetically tractable and many useful cell division mutants have been isolated in the past, we chose this organism for studying the mechanism of cytokinesis. We have recently developed an in vitro system by which we can study contraction of the contractile ring using *S. pombe* cells (Mishra et al., 2013). We used a cell cycle mutant *cdc25* to synchronize the cells in G2 phase of the cell cycle. These cells were engineered to express myosin-II-GFP to visualize the contractile ring. The cell walls of synchronized cells were removed by enzymic digestion to obtain spheroplasts. When these spheroplasts formed contractile rings, they were permeabilized by a detergent solution. The permeabilized cells thus obtained had some plasma membrane remnants but did not have visible cytoplasms, and they were called cell-ghosts. The myosin ring was retained in the cell-ghosts as seen by fluorescence microscopy (Fig. 1). Proteins which were reported to be essential for cytokinesis were largely localized in the contractile ring.

Upon addition of ATP, the contractile ring in the cell-ghost contracted to completion. The attachment of the ring to the perforated cell membrane seemed to be weak because the ring contracted leaving the membrane behind. The speed of contraction was more than 20 times faster than that in live cells, and it was similar to contraction rates observed in animal cells. Rings contracted slowly in the presence of other nucleotides, but did not show any contraction in the presence of nonhydrolyzable form of ATP, AMP-PNP. The C-shaped immature ring was also present in the ghost preparation but this structure did not contract at all upon ATP addition. The myosin-II ATPase inhibitor Blebbistatin interfered with contraction, and the ring in the ghost obtained from *myo2·myp2/myo3* double myosin-II mutant cells did not show contraction, indicating that myosin-II ATPase activity is required for ring contraction. Rings obtained from *cdc8* (tropomyosin) mutant cells also did not contract at a restrictive temperature but disassembled, suggesting that tropomyosin is required in maintaining the integrity of the ring. Addition of actin filament–cross-linking proteins IQGAP or fimbrin interfered with contraction in a dose-dependent manner, suggesting that excess cross-linking of actin filaments interferes with filament sliding during ring contraction.

1. METHODS

1.1 PREPARATION OF CELL-GHOSTS

1.1.1 Genetic techniques and chemicals

Schizosaccharomyces pombe strain used in this study was MBY 5763, *cdc25*-22 *rlc1::3xGFP-kan*^{r} *leu1*-32 h^{-} (Mishra et al., 2013). *cdc25* gene encodes a

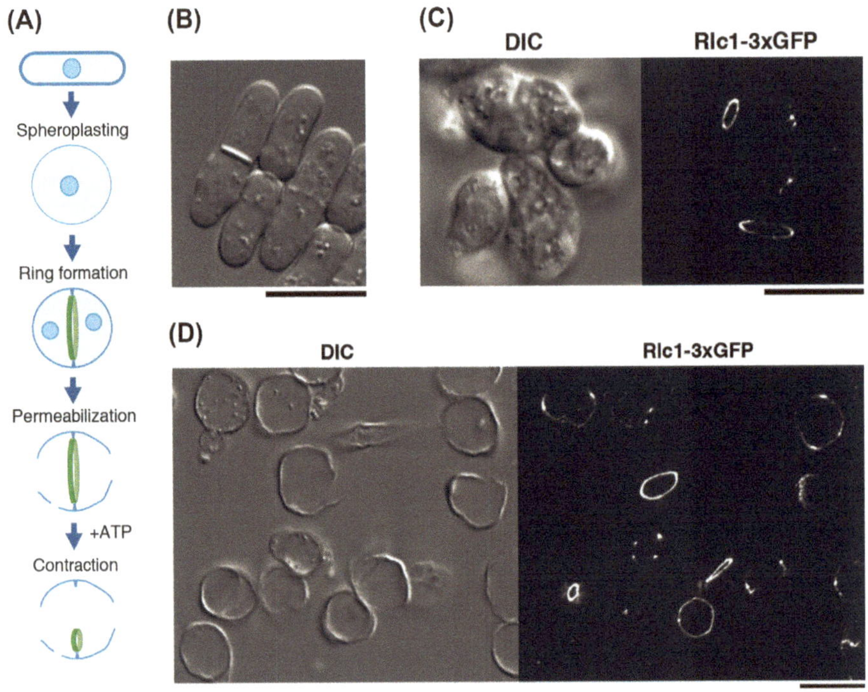

FIGURE 1

Fission yeast cells and cell-ghosts with contractile rings. (A) Schematic illustration of procedure to obtain cell-ghosts. (B) Live fission yeast *Schizosaccharomyces pombe* cells. (C) Live spheroplasts expressing Rlc1-3xGFP. Contractile rings are clearly seen under a fluorescence microscope. (D) Cell-ghosts prepared from spheroplasts expressing Rlc1-3xGFP. Scale bars, 10 μm.

phosphatase which activates cyclin-dependent kinases. Standard procedures for *S. pombe* genetics are described elsewhere (Moreno, Klar, & Nurse, 1991). It is recommended to use the temperature-sensitive cell division mutant *cdc25-22* cells to obtain a synchronous cell culture to enhance the proportion of spheroplasts and cell-ghosts with contractile rings.

Strains expressing epitope-tagged proteins are constructed using a PCR-based approach (Bähler et al., 1998). It is necessary to transform the strain with GFP-tagged contractile ring–specific protein to visualize the contractile ring under a fluorescence microscope before preparation of the cell-ghosts, and during reactivation of the contractile ring. The most convenient protein to be tagged with GFP is myosin regulatory light chain Rlc1 as these strains do not show any defects in cytokinesis.

Actin or myosin in *S. pombe* cells can be perturbed either in live cells or in isolated ghosts. Myosin ATPase inhibitor Blebbistatin (B0560, Sigma) can be used for the in vitro system (the ghosts) at a concentration around 0.1 mM. F-actin stabilizer Jasplakinolide (ALX-350-275, Alexis Corporation) can also be used for both live cells and the in vitro system at a concentration around 0.1 mM. The other F-actin stabilizer phalloidin (P2141, Sigma) and phallacidin (P8157, Sigma) can only be used in vitro experiments at a concentration around 0.1 mM because these cyclic peptides cannot enter the cell membrane. Actin polymerization inhibitor Cytochalasin A (C6637, Sigma) and microtubule polymerization inhibitors methyl benzimidazole-2-yl-carbamate (381586, Sigma) and thiabendazole (T-8904, Sigma) can be used both for live cells and in vitro system at concentrations of 25 and 50 μg/mL, respectively.

1.1.2 Spheroplasting

Spheroplasts are obtained as described previously (Mishra et al., 2012). Cells are first grown at the permissive temperature of 25°C to early log phase in minimal medium (MM) containing 0.5% glucose (Fig. 1). These cells are then shifted to the restrictive temperature of 36°C for 4 h before spheroplasting. Cells (30 mL culture of OD 0.4–0.6) are harvested at 3000 rpm for 3 min, using either a fixed angle or a swing out rotor and washed once with 20 mL 1X E-buffer (50 mM sodium citrate, 0.1 M sodium phosphate pH 6.0) once and resuspended in 1X E-buffer with 1.2 M sorbitol. The cell suspension is adjusted to 5×10^7 cells/mL and incubated with 6 mg/mL lysing enzyme (L-1412; Sigma) at 36°C for 20–50 min. Cells are monitored for spheroplasting using a phase contrast microscope. Cells are then washed with E-buffer containing 1.2 M sorbitol once 90% of cells were spheroplasted, followed by a wash with E-buffer containing 0.6 M sorbitol. Cells were then suspended in MM with 0.8 M sorbitol and grown with gentle shaking (80 rpm) at 25°C (Fig. 1).

1.1.3 Permeabilization of spheroplasts and preparation of cell-ghosts

Contractile ring formation in these spheroplasts is monitored by fluorescence microscopy. For a typical experiment, at the point of spheroplasting, Rlc1-3xGFP signal is detected in a single bright dotlike structure in 100% of the spheroplasts and actomyosin arcs or contractile rings are not seen. After 5 h of incubation at 25°C actomyosin arcs can be detected (in 3–7% of cells) while 25–35% of spheroplasts have fully formed contractile rings (Fig. 1).

Spheroplasts are washed twice with wash buffer (0.8 M sorbitol, 2 mM EGTA, 5 mM $MgCl_2$, 20 mM PIPES-NaOH, pH 7.0) and incubated for 5 min in isolation buffer (0.16 M sucrose, 50 mM EGTA, 5 mM $MgCl_2$, 50 mM potassium acetate, 50 mM PIPES-NaOH, pH 7.0, 0.5% NP-40, 10 μg/mL leupeptin, 10 μg/mL aprotinin, 10 μg/mL pepstatin, 0.5 mM PMSF, 1 mM DTT) on ice. Spheroplasts are homogenized with a Teflon/glass homogenizer. Cell-ghosts thus obtained are washed twice with reactivation buffer (0.16 M sucrose, 5 mM $MgCl_2$, 50 mM potassium acetate, 20 mM MOPS-NaOH, pH 7.0, 10 μg/mL leupeptin, 10 μg/mL aprotinin, 10 μg/mL pepstatin, 0.5 mM PMSF, 1 mM DTT). About 30% of the resultant cell-ghosts retain intact contractile rings (Fig. 1) while another 5–10% have actomyosin arcs or discontinuous or fragmented contractile rings. Majority of intact fully formed contractile rings are competent for in vitro contraction upon ATP addition.

1.2 IN VITRO CONTRACTION ASSAY

1.2.1 Chamber assembly, sample application, and reactivation of contractile ring

The reactivation of contractile ring in the cell-ghost is assessed with the handmade perfusion chamber shown in Fig. 2. The base of the chamber is constructed with a large coverslip (Matsunami Glass Ind., Ltd., Tokyo, Japan; 24 × 36 mm, No. 1) coated with 0.01% poly-L-lysine (Sigma, P1524) and double-sided tape (Nichiban, Tokyo, Japan; NW-5S, 60-μm wall thickness). The coverslip should be used within 2–3 h after adhering poly-L-lysine. Suspension of cell-ghosts are adhered to the base of chamber and then covered by a small coverslip (Matsunami, 18 × 18 mm, No. 1). Volume in the chamber is around 8 μL. Cell-ghosts are washed with reactivation buffer by perfusion to wash away cell-ghosts, which did not attach to the glass surface. Reactivation is performed by perfusion with reactivation buffer containing appropriate concentration of ATP. A majority of intact contractile rings in wild-type cell-ghosts undergo ATP-mediated contraction. If 0.5 mM ATP is perfused, contraction is usually accomplished in 30–60 s at 25°C.

1.2.2 Microscopy

Myosin-II is visualized by Rlc1-3xGFP as described previously. F-actin can be stained with BODIPY-FL-phallacidin, Alexa 568-phalloidin (Molecular Probes-Invitrogen, Eugene, OR) or synthesized tetramethylrhodamine-labeled Lifeact peptide (Operon Biotechnologies, Tokyo, Japan). Conventional fluorescence microscopy, three-dimensional reconstruction, and time-lapse observations are performed using a DeltaVision system (Applied Precision, Issaquah, WA) attached to an Olympus IX-70 widefield inverted fluorescence microscope equipped with Olympus UplanSApo ×100 oil immersion objective lens (NA 1.4, Olympus, Tokyo, Japan), and a Photometrics CoolSNAP HQ camera (Roper Scientific, San Diego, CA). Temperature is

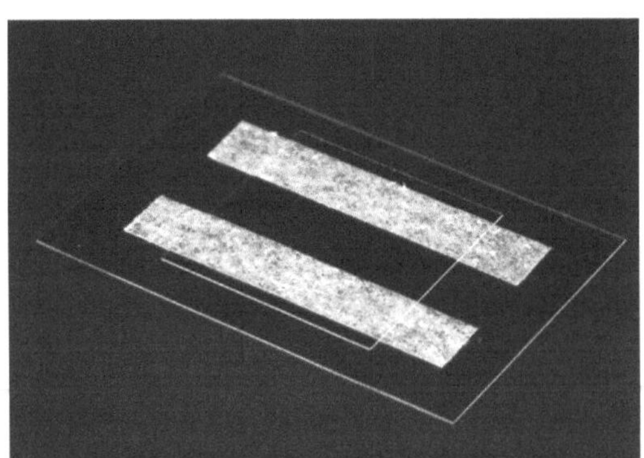

FIGURE 2

Assembly of flow chamber for reactivation analysis. Two strips of double-stick tape are attached on a coverslip (24 × 36 mm) to which poly-L-lysine had been applied. A smaller coverslip (18 × 18 mm) is mounted on the double-stick tapes.

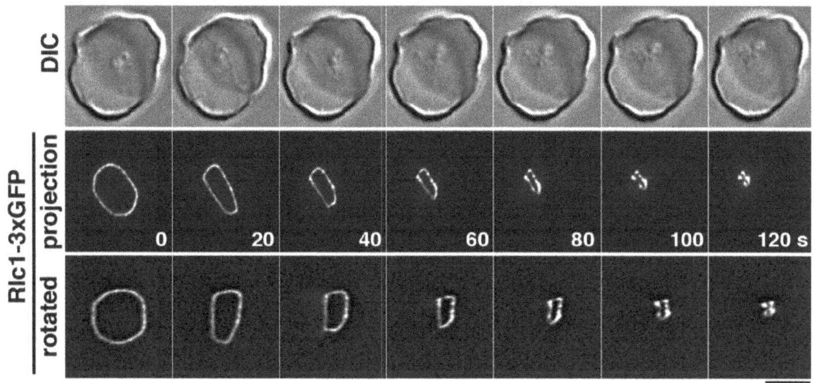

FIGURE 3

Reactivation of a contractile ring with an ATP-containing reactivation medium. (*Pictures are taken by DeltaVision system.*) Numbers indicate time in seconds after addition of 0.5 mM ATP. The lowest images are obtained from the middle images by rotation so that the ring plane is perpendicular to our view axis. Scale bar, 5 μm.

controlled by both, the heated stage and a lens heater (for example, MATS-55RAF20, MATS-LH; TOKAI HIT, Shizuoka, Japan). More than 14 optical sections (0.5 μm spaces) can be acquired every 10 s. Images are captured and processed by iterative constrained deconvolution using SoftWoRx (Applied Precision), and analyzed by SoftWoRx and ImageJ (W.S. Rasband, National Institutes of Health, Bethesda, MD) incorporated with the McMaster Biophotonics Facility "ImageJ for Microscopy" collection of plugins (T. Collins, McMaster Biophotonics Facility, Hamilton, ON, Canada).

For measurement of contraction rate, stacks of optical sections are reconstructed in a horizontal direction using the three-dimensional rotation tool of SoftWoRx and the perimeter of each ring is measured (Fig. 3). Each contraction rate is determined by using values of the most linear phase of contraction. For measurement of fluorescence intensity sum projections of six deconvolved sections were used. Integrated intensity is calculated from mean value and area, after subtracting the background mean value of the same area. Graphical images are established using Prism-5 (Graphpad Software, La Jolla, CA, USA).

CONCLUSION

We presented a method to isolate a cell-ghost from a fission yeast cell and reactivate the contractile ring retained in the cell-ghost. This method enables in vitro manipulation of the contraction of the contractile ring so that the conditions which are required for the contraction of the ring can be studied. This will also enable ultrastructural studies of the contractile ring because the potency of contraction is proof that the contractile ring possesses its normal ultrastructure, and loss of cytoplasm would benefit EM observation of the ring.

ACKNOWLEDGMENTS

We are grateful to Dr. Sarada Bulchand for her critical reading of the manuscript.

REFERENCES

Bähler, J., Wu, J. Q., Longtine, M. S., Shah, N. G., McKenzie, A., 3rd, Steever, A. B., … Pringle, J. R. (1998). Heterologous modules for efficient and versatile PCR-based gene targeting in *Schizosaccharomyces pombe*. *Yeast, 14*, 943–951.

Cande, W. Z. (1980). A permeabilized cell model for studying cytokinesis using mammalian tissue culture cell. *Journal of Cell Biology, 87*, 326–335.

Fujimoto, H., & Mabuchi, I. (1997). Isolation of cleavage furrows from regular sea urchin eggs and identification of furrow-specific proteins. *Journal of Biochemistry, 122*, 518–524.

Fujimoto, H., & Mabuchi, I. (2010). Elongation factors are involved in cytokinesis of sea urchin eggs. *Genes to Cells, 15*, 123–135.

Hoffman-Berling, H. (1954). Die Glycerin-wasserextrahierte Telophasenzelle als Modell der Zytokinese. *Biochimica et Biophysica Acta, 15*, 332–339.

Kanbe, T., Kobayashi, I., & Tanaka, K. (1989). Dynamics of cytoplasmic organelles in the cell cycle of the fission yeast *Schizosaccharomyces pombe*: three dimensional reconstruction from serial sections. *Journal of Cell Science, 94*, 647–656.

Mabuchi, I. (1986). Biochemical aspects of cytokinesis. *International Review of Cytology, 101*, 175–213.

Mabuchi, I., & Okuno, M. (1977). The effect of myosin antibody on the division of starfish blastomeres. *Journal of Cell Biology, 74*, 251–263.

Mabuchi, I., Tsukita, S., Tsukita, S., & Sawai, T. (1988). Cleavage furrow isolated from newt eggs: contraction, organization of actin filaments, and protein components of the furrow. *Proceedings of the National Academy of Science of the United States of America, 85*, 5966–5970.

Marks, J., & Hyams, J. S. (1985). Localization of F-actin through the cell division cycle of *Schizosaccharomyces pombe*. *European Journal of Cell Biology, 39*, 27–32.

Mishra, M., Huang, Y., Srivastava, P., Srinivasan, R., Sevugan, M., Shlomovitz, R., … Balasubramanian, M. K. (2012). Cylindrical cellular geometry ensures fidelity of division site placement in fission yeast. *Journal of Cell Science, 125*, 3850–3857.

Mishra, M., Kashiwazaki, J., Takagi, T., Srinivasan, R., Huang, Y., Balasubramanian, M. K., & Mabuchi, I. (2013). *In vitro* contraction of cytokinetic ring depends on myosin II but not on actin dynamics. *Nature Cell Biology, 15*, 853–859. http://dx.doi.org/10.1038/ncb2781.

Moreno, S., Klar, A., & Nurse, P. (1991). Molecular genetic analysis of fission yeast *Schizosaccharomyces pombe*. *Methods in Enzymology, 194*, 795–823.

Straight, A. F., Cheung, A., Limouze, J., Chen, I., Westwood, N. J., & Mitchison, T. J. (2003). Dissecting temporal and spatial control of cytokinesis with a myosin II Inhibitor. *Science, 299*, 1743–1747.

Yonemura, S., Mabuchi, I., & Tsukita, S. (1991). Mass isolation of cleavage furrows from dividing sea urchin eggs. *Journal of Cell Science, 100*, 73–84.

CHAPTER

Xenopus extract approaches to studying microtubule organization and signaling in cytokinesis

24

C.M. Field*,§,[1], **J.F. Pelletier***,§, **T.J. Mitchison***,§

**Harvard Medical School, Boston, MA, United States*

§*Marine Biological Laboratory, Woods Hole, MA, United States*

[1]*Corresponding author: E-mail: Christine_Field@hms.harvard.edu*

CHAPTER OUTLINE

Methods in Cell Biology, Volume 137, ISSN 0091-679X, http://dx.doi.org/10.1016/bs.mcb.2016.04.014

Abstract

We report optimized methods for preparing actin-intact *Xenopus* egg extract. This extract is minimally perturbed, undiluted egg cytoplasm where the cell cycle can be experimentally controlled. It contains abundant organelles and glycogen and supports active metabolism and cytoskeletal dynamics that closely mimic egg physiology. The concentration of the most abundant ~11,000 proteins is known from mass spectrometry. Actin-intact egg extract can be used for analysis of actin dynamics and interaction of actin with other cytoplasmic systems, as well as microtubule organization. It can be spread as thin layers and naturally depletes oxygen though mitochondrial metabolism, which makes it ideal for fluorescence imaging. When combined with artificial lipid bilayers, it allows reconstitution and analysis of the spatially controlled signaling that positions the cleavage furrow during early cytokinesis. Actin-intact extract is generally useful for probing the biochemistry and biophysics of the large *Xenopus* egg. Protocols are provided for preparation of actin-intact egg extract, control of the cell cycle, fluorescent probes for cytoskeleton and cytoskeleton-dependent signaling, preparation of glass surfaces for imaging experiments, and immunodepletion to probe the role of specific proteins and protein complexes. We also describe methods for adding supported lipid bilayers to mimic the plasma membrane and for confining in microfluidic droplets to explore size scaling issues.

List of Abbreviations

AurkA Aurora kinase A
AurkB Aurora kinase B
CPC Chromosomal passenger complex
ER Endoplasmic reticulum
TIRF Total internal reflection fluorescence microscopy

INTRODUCTION

Cell-free systems provide an experimental bridge between living cells and pure protein systems and allow chemical and physical manipulations that are not possible in living cells. *Xenopus* egg extracts are unique among cell-free systems in providing essentially undiluted cytoplasm with intact metabolism, organelles, and experimental control of cell cycle state. Quantitative proteomics studies in these extracts have established the identity and approximate concentration of the most abundant ~11,000 proteins in *Xenopus* egg cytoplasm (Wühr et al., 2014), which has been extremely useful in designing extract experiments. When air is excluded, eg, when extract is sandwiched between two glass surfaces, mitochondria in the extract remove oxygen down to very low levels within minutes, but ATP production continues by anaerobic metabolism (Niethammer, Kueh, & Mitchison, 2008). This greatly reduces photobleaching and phototoxicity of synthetic fluorophores such as Alexa dyes, making the system ideal for live fluorescence imaging. *Xenopus* eggs are naturally adapted for development in low oxygen environments, and this may be an important reason that the extract prepared from them can recapitulate normal metabolism under conditions of limited gas exchange.

Actin-intact extract is a variant of classic CSF extract, which was developed by Murray for dissecting cell cycle regulation (Murray, 1991). CSF extract recapitulates the cell cycle state of the unfertilized egg, which is arrested in metaphase of meiosis II. CSF extract is normally prepared with cytochalasin D to block actin polymerization and to prevent dramatic gelation-contraction that occurs when actin-intact CSF extract is warmed to room temperature (Field et al., 2011). To prepare actin-intact CSF extract, we leave out the cytochalasin and modify the conditions in the crushing spin to facilitate separation of cytoplasm from yolk and fat. CSF extract, whether actin-intact or cytochalasin treated, has high Cdk1 activity, reflecting the M-phase cell cycle state of the unfertilized egg. It can be easily converted to interphase (I-phase) by addition of 0.4 mM calcium to mimic fertilization (Murray, 1991). The calcium concentration rapidly returns to baseline by pumping into the endoplasmic reticulum (ER). The resulting I-phase extract is stable on ice for at least 2 h. Reentry into M-phase can be permanently blocked by addition of protein synthesis inhibitor to block Cyclin B synthesis. Importantly, I-phase actin-intact extract does not undergo bulk gelation-contraction, so microtubule and actin dynamics can be imaged without this complication.

Until recently, we lacked the ability to reconstitute two important aspects of egg biology in extract: actin dynamics and plasma membrane biology. Recent progress has filled these gaps. Here, we report methods for preparing extract that leaves actin dynamics intact, but still allows clean separation of cytoplasm from yolk and fat in the important crushing and stratifying spin. We also report addition of supported lipid bilayers (SLBs) to model the plasma membrane. Using these tools we recently reconstituted the spatiotemporal regulation of early cytokinesis in egg extracts and probed the role of the chromosomal passenger complex (CPC) and several cytokinesis-associated kinesins in the signaling events contributing to this regulation (Nguyen et al., 2014).

1. ACTIN-INTACT EXTRACT PREPARATION

1.1 REQUIREMENTS

Hormone Stocks

- Pregnant mare serum gonadotropin (PMSG)
  - Calbiochem 367222
  - 1000 units/mL, dissolved in water
  - Store at −20°C, do not freeze/thaw
- Human chorionic gonadotropin (HCG)
  - Prospect HOR-250
  - 1000 units/mL, dissolved in 10 mM $NaPO_4$ pH 7.2
  - Store at 4°C

Buffers and Reagents

- 25× MMR
  - 50 mM Na-HEPES, pH to 7.8 with NaOH
  - 1 mM EDTA
  - 1 M NaCl
  - 20 mM KCl
  - 10 mM $MgCl_2$
  - 20 mM $CaCl_2$
  - Autoclave, store at room temperature (RT)
- 20× XB Salts
  - 2 M KCl
  - 20 mM $MgCl_2$
  - 2 mM $CaCl_2$
  - Sterile filter, store at RT
- 2 M Sucrose
  - Sterile filter, store at 4°C
- 1 M K-HEPES
  - pH to 7.7 with KOH
  - Sterile filter, store at 4°C
- 0.5 M K-EGTA
  - pH to 7.7 with KOH
  - Sterile filter, store at RT
- 1 M $MgCl_2$
  - Sterile filter, store at RT
- Protease inhibitors (PIs, aka LPC)
  - 10 mg/mL leupeptin (Sigma L2884)
  - 10 mg/mL pepstatin A (Sigma P5318)
  - 10 mg/mL chymostatin (Sigma C7268)
  - Dissolve in DMSO; store in 100 μL aliquots at −20°C
- Cysteine, free base (Sigma C7352)
- 100× energy mix, to supplement extract before freezing
  - 900 mM creatine phosphate (Sigma 10621614001)
  - 100 mM ATP (Sigma A3377)
  - 100 mM $MgCl_2$
  - Store in 100 μL aliquots at −20°C

Equipment

- Glass crystallizing dishes, for washing eggs (VWR 89000-294)
  - 150 mm diameter × 75 mm height or equivalent
- 13 × 51 mm Ultra-Clear tubes (Beckman 344057)
- Clinical centrifuge, to pack eggs
  - Low speed, with a swinging bucket rotor
  - Use at 16−18°C
- Vacuum
- Centrifuge and swinging bucket rotor that can go 15,000 × g

- Use at 4°C
- Wide bore Pasteur pipettes
  - Option 1: Polypropylene transfer pipettes. Cut off tips
  - Option 2: Glass Pasteur pipettes. Break tips and fire polish
    Pipettes cut to two different diameters are needed, gardening size: 2 mm inner diameter and transfer size: 3.5 mm inner diameter. Note: A Xenopus egg is 1.2 mm in diameter.

1.2 EXTRACT PREPARATION

This protocol is for two to three frogs; the number someone familiar with the preparation can manage in parallel. Some steps (eg, dejellying, Section 1.2.4) are time sensitive and require constant attention. Therefore, to increase the scale, one can (1) recruit others to help with time-sensitive steps or (2) pool eggs from multiple frogs. The latter is not ideal, because different batches of eggs have different properties.

1.2.1 Several days before

At least 2 days before, prime frogs with PMSG hormone to induce maturation of oocytes. Inject 100 U (100 μL) per frog into the dorsal lymph sac using a 27-gauge needle. Store primed frogs in dechlorinated water with 2 g/L rock salt (aka frog water) in a cool room (ambient temperature of 16–18°C). We wait for at least 2 days after priming before inducing them to lay eggs. The priming lasts for up to 2 weeks. Do not feed frogs for 2–3 days before injecting them to induce ovulation (Section 1.2.2). Frogs fed recently can defecate and foul the laid eggs.

1.2.2 Day before

16–18 h before extract preparation, two to three primed frogs are induced to ovulate by injection with 500 U (0.5 mL) of HCG. After injection, frogs are placed in separate containers with 1 L of 1× MMR for egg laying. Containers are placed in a dark and quiet 16°C incubator overnight. Eggs sitting in MMR longer than 18 h produce inferior extract. Egg quality must take precedence over egg quantity. A good frog usually produces eggs that yield about 0.8 mL of extract. It is best not to use any batches of eggs containing >10% of lysed eggs, activated[1] eggs, or "puffballs" (swollen eggs that are often white and puffy). Strings of eggs that do not have fully separated jelly coats are also inferior. In general, we keep each frog separate and prepare separate extracts from the eggs of each frog.

1.2.3 Hours before

Buffers. All buffers should be at 16–18°C. Prepare just before use unless indicated otherwise; rinse all glassware with double distilled water prior to use.

[1]Activated eggs have lost their metaphase arrest. When this occurs the black pigment at the animal pole will contract. If significant numbers of activated eggs are included in the preparation, the M-phase arrest of the extract will be lost.

- 1.5 L 1× MMR (can make in advance and store at 16°C)
- 600 mL dejellying solution
  - 30 mL 20× XB salts
  - 2.61 mL 10 N KOH, gives pH 7.8
  - 12 g cysteine powder, gives 2% (w/v) cysteine
  - Add cysteine within an hour before and make sure fully dissolved
- 1.5 L CSF-XB
  - 75 mL 20× XB salts, to 100 mM KCl, 1 mM $MgCl_2$, 0.1 mM $CaCl_2$
  - 37.5 mL 2 M sucrose, to 50 mM
  - 15 mL 1 M K-HEPES, to 10 mM
  - 165 μL 10 N KOH, to maintain pH 7.7
  - 15 mL 0.5 M K-EGTA,[2] to 5 mM
  - 1.5 mL 1 M $MgCl_2$, to (additional) 1 mM
- 100 mL CSF-XB + protease inhibitors (PIs)
  - 100 mL CSF-XB (from the 1.5 L above)
  - 100 μL LPC (1 aliquot) to 10 μg/mL, mix immediately

Materials

- Six (two per frog) Ultra-Clear tubes (13 × 51 mm)
  - Add 1 mL CSF-XB plus PIs to each tube.
- Three (one per frog) glass crystallizing dishes (aka egg dishes)
  - Rinse well with ddH_2O then with MMR
- Wide bore Pasteur pipettes, both sizes (Section 1.1)

1.2.4 Extract preparation

Perform all steps at 16−18°C unless noted otherwise. Higher temperatures are a common cause of bad extracts. Eggs are fragile, so handle them gently and always keep them under buffer, especially after jelly coat removal. The quality of dejellied eggs deteriorates over time, so work efficiently. The procedure from washing the eggs in MMR to crushing the eggs should take less than 1 h.

1. Garden eggs, if necessary. Use a wide bore Pasteur pipette to delicately remove sloughed skin, lysed eggs, activated eggs, and puffballs (Section 1.2.2). Gardening may continue throughout but should not delay the preparation.
2. Transfer eggs in MMR to an egg dish by pouring gently.
3. Wash eggs three times in ~150 mL fresh MMR until all debris is removed. Pour buffer gently and not directly onto the eggs. Find a compromise between delicate handling and thorough mixing.

[2] Unfertilized *Xenopus* eggs are arrested at metaphase of meiosis II by a biochemical activity termed cytostatic factor (CSF), partly composed of the APC/C inhibitor, XErp1/Emi2. Upon fertilization, sperm entry induces a calcium rise, which results in degradation of XErp1/Emi2 and release of M-phase arrest (Rauh, Schmidt, Bormann, Nigg, & Mayer, 2005). CSF-XB buffer contains the calcium chelator EGTA, and thus metaphase arrest is maintained during extract preparation.

4. Dejelly eggs.
   4.1. Remove as much MMR as possible. Decant larger volumes of buffer, or use a transfer pipette to aspirate smaller volumes of buffer. When the volume of buffer is small, decanting the buffer may stretch the eggs and damage them.
   4.2. Add ~100 mL dejellying solution. Intermittently, swirl eggs gently in the dejellying solution. After 3 min, pour off most of the dejelly solution and add 100 mL more dejelly solution. As jelly dissociates from the egg, the jelly appears as transparent shells or schlieren streaks in the buffer.
   4.3. After eggs are dejellied (5–7 min total in dejelly solution), they will pack more tightly and orient with their animal poles upward. The volume of settled eggs will decrease about fivefold.
5. Remove dejelly solution.
   5.1. Remove as much dejellying solution as possible (see the previous discussion).
   5.2. Rinse eggs thoroughly, three to four times in ~150 mL fresh CSF-XB. To minimize residual dejelly solution, remove as much buffer as possible each time. Dejellied eggs are quite fragile. Tilt the egg dish to keep the eggs under buffer. Eggs exposed to air will lyse. Pour solutions down the side of the wash vessel or aspirate with transfer pipette, and gently swirl the eggs ensuring maximal buffer exchange between washes.
   5.3. Rinse eggs two times in ~30 mL total CSF-XB plus PIs. Leave eggs in a small volume of CSF-XB plus PIs after the second wash.
6. Collect eggs.
   6.1. Draw eggs up into the wide bore Pasteur pipette, then slowly deposit them into an Ultra-Clear tube with 1 mL CSF-XB plus PIs. Insert pipette below the meniscus of the buffer to break surface tension, and gently release eggs into the solution. Minimize transfer of buffer along with eggs.
   6.2. Fill the tube to very near the top. The first spin is a packing spin (see the next list), so the egg height in the tube will decrease.
   6.3. Aspirate excess buffer from the top of the eggs (see Fig. 1A).
7. Packing spin, to displace buffer and minimize dilution of the extract.
   7.1. Spin tubes in a clinical centrifuge at 700 × g for 30 s then 1400 × g for 15 s. Adjust the speed or duration so eggs remain intact, but eggs are no longer spherical (Fig. 1B) and they do not move if the tube is tilted.
   7.2. Using a vacuum and pipette tip aspirate all remaining buffer.
   7.3. Keep packed eggs on ice before crushing spin, but proceed to this step immediately.
8. Crushing spin, to separate components.
   8.1. Transfer tubes to a swinging bucket rotor at 4°C.
   8.2. Crush eggs in a centrifuge at 15,000 × g for 15 min at 4°C.
   8.3. Remove tubes and place on ice.

FIGURE 1 Preparation of actin-intact extract.

(A) *Xenopus* eggs deposited in a centrifuge tube. (B) After the packing spin, eggs appear tightly packed, with the animal pole facing upward. They are distended but remain intact. Excess buffer is aspirated from the top. (C) After the crushing spin, egg components stratify. The top, yellow layer contains lipid droplets. The second, black layer contains pigment. The third, straw-colored layer is the cytoplasmic extract (*green bracket*). The bottom, dark layer contains yolk, pigment granules, and egg cortex. Lower panels are 3× magnified. Note the distended eggs in panel B.

   8.4. The top, yellow layer contains lipid droplets. The second, black layer contains pigments. The third, straw colored layer is the cytoplasmic extract. The bottom, dark layers contain yolk, pigments, and nuclei. Fig. 1C shows a well-separated extract.

9. Harvest extract.

   9.1. Wipe ice water from the surface of the tube.

   9.2. Puncture tube near the bottom of the cytoplasmic layer with an 18-gauge needle on a 1 cc syringe. Gently pull the extract with the needle bevel upward. Avoid the upper black layer and darker material spread through the cytoplasmic layer. This material decreases extract quality.[3]

9.3. Gently deposit extract in an Eppendorf tube on ice. Immediately add 1/1000 volume of protease inhibitors, gently flicking the tube to mix.

10. Store M-phase extract on ice for 4–8 h without loss of activity.

10.1. Monitor the gelation-contraction activity of the extract[3] (Section 1.3.2).

10.2. For long-term storage, see Section 1.4.4.

1.3 CELL CYCLE CONTROL AND MONITORING

Microtubules, actin, and keratin polymers, as well as endomembrane systems and the cortex, are highly regulated by the cell cycle. When working with *Xenopus* egg extract, it is important to control and monitor its cell cycle state.

1.3.1 Converting M-phase to I-phase

Sperm entry upon fertilization induces a transient rise in cytosolic calcium, triggering degradation of CSF and release of M-phase arrest. We exploit this natural response to calcium to prepare I-phase extracts. $CaCl_2$ from a 50× or 100× stock is added to M-phase extract to a final concentration of 0.4 mM. The ER rapidly uptakes calcium, decreasing the cytosolic concentration, so immediate and thorough mixing is crucial to release CSF arrest. We use 1.5 mL Eppendorf tubes with 25–100 μL extract per tube. Larger volumes may cause problems with mixing and gas exchange. After calcium is added, gently pipette the entire solution up and down three to five times to ensure mixing, or flick the tube about eight times. Do not vortex the extract. Incubate at 20°C for 10 min, flicking occasionally. Then place on ice for at least 2 min. To prevent M-phase reentry, protein synthesis inhibitors (cycloheximide 100 μg/mL or puromycin 50 μM) can be added to block synthesis of cyclin B and cell cycle progression (Field, Nguyen, Ishihara, Groen, & Mitchison, 2014).

Two alternative methods for making I-phase extract have previously been described and are reviewed in Field et al. (2014). Both involve releasing the M-phase arrest just before or during the egg crushing spin. Even on ice, interphase extract changes over time due to cell cycle progression. Therefore, we prefer to maintain M-phase arrest during extract preparation and storage by use of EGTA in CSF-XB buffer.

1.3.2 Monitoring cell cycle state

The classic test for cell cycle state in egg extract was direct measurement of Cdk1 kinase activity using ^{32}P-ATP (Gautier et al., 1990). This may still be the best

[3]Inferior extract does not perform well in the gelation contraction assay (Section 1.3.2) and loses cell cycle arrest earlier. Inferior extract can be partially "rescued" by filtering it through a 0.9–1.2 μm syringe filter (at a loss of ~50% of the volume). Filtered extract maintains its cell cycle state.

method, but radioactive assays are no longer routine in most laboratories. We use several methods to monitor cell cycle state:

1 Gelation-contraction
Actin-intact M-phase extract undergoes bulk gelation-contraction within a few minutes of warming to RT, while I-phase extract does not. This assay is fast and easy and also tests the integrity of the actin cytoskeleton. We apply it to every extract preparation before starting experiments. After several hours storage on ice, we often repeat to check if the extract is still good. Fill a small Petri dish with mineral oil. Pipette several 1-μL droplets of extract under the mineral oil and monitor with a dissecting microscope or just by eye. In a good actin-intact M-phase extract, light scattering particulates will contract toward the center of the droplet within 2–3 min and form a tight ball in the center by ~10 min. I-phase extract, made by adding calcium to CSF extract and incubating at RT for 2–3 min, still forms an actin gel, but it does not contract, or contracts very slowly (Field et al., 2011). We often use this assay as a spot check for conversion to interphase.

2 Microtubule aster morphology
Add fluorescent tubulin probe (see Section 2.1.2) to a 20–30 μL aliquot of extract, then bring the extract to 7.5% DMSO to induce microtubule polymerization. Incubate at 20°C for 20 min and image. In M-phase, microtubules assemble into asters under the influence of dynein and NUMA (Verde, Berrez, Antony, & Karsenti, 1991). In I-phase they form long, interconnected bundles.

3 Chromatin morphology
Add demembranated *Xenopus* sperm nuclei (see Section 3.2) prelabeled with Hoechst at 1 μg/mL (Sigma B2261) to extract at a final concentration of ~100 nuclei/μL. Remove a few microliters of extract over time and image. In M-phase, chromatin is condensed and slowly resolves into mitotic chromosomes. In I-phase, it gradually decondenses, acquires a nuclear envelope, and forms spherical nuclei (Murray, 1991).

4 Phosphoprotein analysis via Western blot
The mitotic phosphoprotein monoclonal antibody (MPM2; Millipore 05-368) recognizes many phosphorylated Cdk1 substrates. Cdc27 protein, recognized by antibody AF 3.1 (Upstate 05–906), undergoes a large mobility shift by SDS page due to loss of M-phase phosphorylation (Field et al., 2011; Rankin & Kirschner, 1997).

1.4 GENERAL CONSIDERATIONS

1.4.1 Temperature

Prepare extract in a cool room at 16–18°C. If the eggs or extract are warm, actin may polymerize and sediment out upon centrifugation. Move packed eggs on ice to the centrifuge used for the crushing spin, and store harvested extract on ice. It is important to realize that, even on ice, biochemical reactions continue; for example, kinases still phosphorylate substrates, albeit more slowly than at 20°C. For imaging experiments, a temperature of 18–20°C is optimal, though a wider range is tolerated.

1.4.2 Dilution

It is important to keep the extract concentrated. The total volume of all aqueous additions should be <10%. Dilution by a larger factor can release the cell cycle arrest and damage cytoskeleton assembly reactions. The buffer used for dilution is important. High salt concentrations can damage the extract. Water, or dilute buffer, is preferred. The total volume of DMSO added must be <0.5% if microtubule dynamics are important. Higher concentrations of DMSO induce microtubule polymerization.

1.4.3 Addition of small molecules/drugs

We find that the EC_{50} for drug action is often 10- to 1000-fold higher in extract than in tissue culture cells, although mechanism and specificity are maintained. We believe there are two reasons: (1) tissue culture cells are surrounded by a large reservoir of drug in the medium and can accumulate drugs to a higher internal concentration and (2) hydrophobic drugs partition into lipid reservoirs and/or acidic compartments present in extract. For these reasons, EC_{50} for drug action in extract must be determined empirically, and we aim to use them at ~5× the EC_{50} to maximize specificity.

If drugs are added from a DMSO stock, it is important to keep the total percent DMSO at <0.5% (see Section 1.4.2). Drug stocks are often prepared as concentrated stocks in DMSO, eg, 1000×. These should be diluted in small volumes of egg extract to give 10–50× final stocks. Dilution in aqueous buffer can cause hydrophobic drugs to precipitate, in which case the final concentration is unreliable.

1.4.4 Storage

It is optimal to work with freshly made extract. This can be stored on ice for up to ~8 h, with best results obtained in the first ~4 h. M-phase extract can be flash-frozen in liquid nitrogen in PCR tubes and stored at −80°C for up to 1 month. If the extract is to be frozen, add sucrose (50 mM final concentration) and energy mix (Section 1.1) prior to freezing. Sucrose protects the organelles, and energy mix may compensate for mitochondrial damage. Upon thawing, frozen extract behaves similar to fresh extract in many assays, eg, gelation-contraction, but is never quite as good as fresh for reconstituting complex microtubule arrays. When working with I-phase extract, it is better to convert M-phase extract to I-phase before freezing (Section 1.3.1).

2. FLUORESCENT PROBES

2.1 FUSION PROTEINS AND FLUORESCENTLY TAGGED PROTEINS

Proteins can be expressed as fusions with fluorescent proteins or expressed unlabeled and tagged with reactive dyes. All probes have the potential to perturb and should be added at the minimum concentrations needed for imaging. Reactions with and without a given probe should be carefully compared to ensure minimal perturbation. We dilute probes as 10× to 100× stocks in actin-intact extract, and often freeze in liquid nitrogen in small aliquots. Adding 10% (v/v) of frozen extract to fresh has no discernable negative effect. We focus on a few of our key protein probes in later sections.

2.1.1 F-Actin

We have mainly used Utrophin-GFP (Burkel, von Dassow, & Bement, 2007; Field et al., 2011) and Lifeact-GFP (Field et al., 2011; Nguyen et al., 2014; Riedl et al., 2008). Directly labeled G-actin gives very low contrast images, presumably due to the large monomer pool (Rosenblatt, Agnew, Abe, Bamburg, & Mitchison, 1997). Labeled phalloidin gives high contrast images and is convenient, but it perturbs actin dynamics. Care is required when using Utrophin-GFP. It can be perturbing, as it tends to induce bundles in a concentration-dependent manner. We prefer Lifeact-GFP, expressed in bacteria as a GFP fusion with a 6xHis tag for purification. Use at a final concentration of 100–200 nM.

2.1.2 Microtubules

We standardly use labeled bovine brain tubulin as a probe, adding it at approximately 200–250 nM final concentration. We prepare it as described (Hyman et al., 1991) using the NHS esters of Alexa 488, 568, and 647 (Life Technologies). See Fig. 2A. High-quality labeled tubulin can also be purchased from Cytoskeleton Inc. Labeled *Xenopus* egg tubulin probes (Groen & Mitchison, 2016) improve signal/noise for imaging, presumably because 20°C is too cold for optimal polymerization of a mammalian tubulin probe. These probes are superior when the quality of the microtubule image is important, or for quantification of polymer levels.

2.1.3 EB1

We use human EB1-GFP in egg extract to quantify the direction, rate, and persistence of microtubule polymerization (Fig. 2A). Human EB1-GFP tagged with 6xHis construct was obtained from Kevin Slep (UNC Chapel Hill), expressed in *Escherichia coli*, and is typically added at 40–60 nM. Endogenous EB1 is ~1.6 μM (Wühr et al., 2014). *Xenopus* EB1-GFP provided inferior images, perhaps due to less efficient partitioning onto microtubules. We used plusTipTracker Matlab software module (Applegate et al., 2011) to perform automated detection and frame-to-frame linking of EB1 to obtain growth tracks (Fig. 2B) (Ishihara, Nguyen, Groen, Field, & Mitchison, 2014; Nguyen et al., 2014).

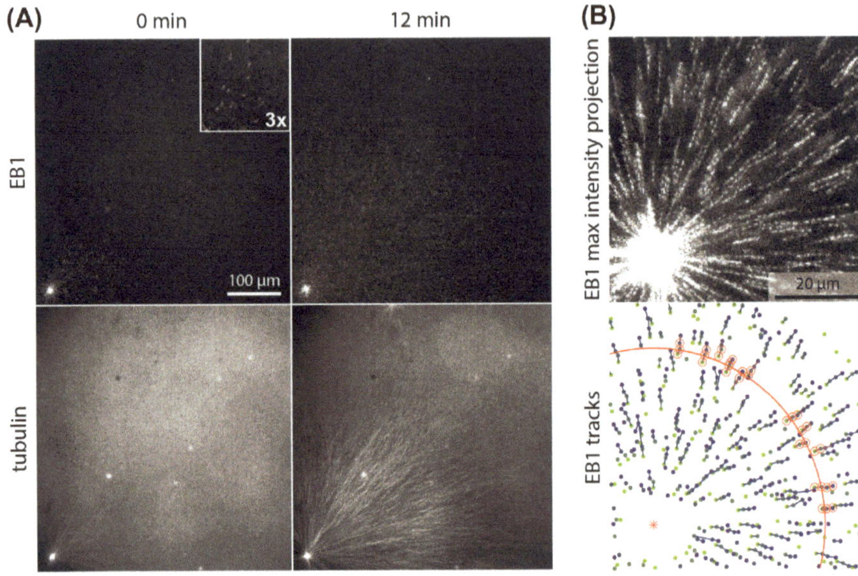

FIGURE 2 Quantitative analysis of interphase aster growth using EB1 tracking.

(A) Two frames from a 20× widefield movie of an aster growing from an anti-AurkA bead in I-phase extract. The extract contains 0.04 mg/mL CC1, a dynein inhibitor which blocks microtubule sliding. The coverslips are passivated with PEG. Note that individual EB1 comets can be clearly visualized and counted (see 3× insert). (B) Top panel: maximum intensity projection of a short movie to illustrate EB1 comet velocity. (B) Bottom panel: automated EB1 tracking (plusTipTracker, Applegate et al., 2011) was used to count the number of microtubules crossing a given radius (*red circle*). This analysis showed that most microtubules are nucleated away from the bead once the aster gets large (Ishihara et al., 2014).

2.1.4 RhoA-GTP reporter

RhoA-GTP at the inner surface of the plasma membrane is a central positive regulator of actin and myosin II during cytokinesis (Piekny, Werner, & Glotzer, 2005). To report on its localization during cytokinesis signaling, we adapted a reporter based on the RhoA-binding domain of Rhotekin that was developed for imaging cleavage furrow induction in *Xenopus* embryos (Bement, Benink, & von Dassow, 2005). The rGBD of mouse Rhotekin fused to mCherry (mCherry-rGBD) tagged with 6xHis was expressed in *E. coli* and added to extract at a final concentration of 140 nM. This probe accumulates on SLBs at the interface between I-phase microtubule asters (Nguyen et al., 2014). RhoA-GTP localization was dependent on activity of the AurkB subunit of the CPC, but was not dependent on the presence of Kif23, the kinesin subunit of Centralspindlin. This unexpected finding suggests the existence of a signaling pathway from the CPC to RhoA that does not depend on Centralspindlin (Nguyen et al., 2014).

2.2 ANTIBODIES

We often use directly labeled antibody probes. They must be used with caution since antibody binding can activate or inactivate protein targets. With this caveat they provide a quick readout of endogenous protein localization. We typically affinity purify IgG from rabbit serum by binding to antigen and eluting with low pH. Many of our antibodies are generated by immunizing with ~20 amino acid peptides from the C-terminus of the protein (Field, Oegema, Zheng, Mitchison, & Walczak, 1998). We currently use two microscale methods for labeling IgG.

2.2.1 IgG labeling with anti-IgG Fab fragments

Rabbit IgG is mixed with Fab fragments of goat anti-rabbit IgG labeled with a fluorophore, such as DyLight 647 (Jackson Immunoresearch 111-607-003). Jackson sells Fab fragments coupled to a wide range of other fluorophores. We typically mix equal mass of IgG and Fab, corresponding to a molar excess of Fab. After 30 min at RT, the mixture is diluted into extract for imaging. Typically, we aim for a final concentration of labeled IgG of ~1 μM. The IgG–Fab mixture can be stored for at least a month at 4°C.

2.2.2 On-resin IgG labeling

This method is very useful for direct labeling of small amounts of antibody and has given reliable results for many antibodies and several fluorophores. The labeling works well using Protein G UltraLink resin (Thermo Scientific 53125) for rabbit and mouse IgG and Affi-Prep Protein A resin (Biorad 156-0006) for rabbit IgG.

1. To hold the resin, a 200 μL polypropylene pipette tip is sealed at the pointed end by brief passage through the flame of a Bunsen burner. To allow liquid egress while retaining resin, a 1–2 mm slit is made parallel to the long axis of the sealed tip by cutting with a thin razor blade. All procedures are performed at room temperature (20–25°C).
2. 40 μL of 50% resin slurry is loaded into the tip and washed three times with 100 μL of 0.15 M NaCl, 10 mM K-HEPES (pH 7.7) to give a packed resin bed of ~20 μL. Use a pipetman to force buffer through the packed resin bed, here and in all subsequent steps.
3. Purified antibody against a protein of interest, typically an affinity-purified rabbit IgG or a commercial monoclonal antibody, is passed through the resin bed three to five times sequentially to bind. For 20 μL of packed resin we typically label 20–100 μg IgG per tip. Alternatively, IgG can be incubated with resin in a tube for ~1 h with gentle rotation, then packed and washed.
4. The resin is washed three times with 100 μL of 200 mM K-HEPES pH 7.7 to remove unbound contaminants.
5. The resin-bound IgG is then labeled with an amine-reactive fluorescent dye. We always use Alexa Fluor NHS esters (Life Technologies), but similar dyes and reactive esters should work. Dye is dissolved in DMSO at ~50 mM and stored as frozen aliquots at −20°C. 0.5 μL of dye stock is diluted into 25 μL of

200 mM K-HEPES pH 7.7 and immediately loaded onto the resin bed. Make sure the resin is saturated with dye (ie, a little comes out the bottom) and leave a small amount of dye solution on top of the resin bed so that it does not dry out. NHS esters hydrolyze with a half-life of a few minutes in aqueous solution above pH 7, so it is important to load the diluted dye onto the resin rapidly after dilution. We use the same amount of NHS ester independent of the amount of labeled IgG on the resin. It is a large molar excess over the IgG, but most reacts with the resin or gets hydrolyzed. The tip is then incubated at room temperature for 10–60 min.

6. A second aliquot of dye is diluted, added to the resin bed, and incubated in the same manner.
7. After the second labeling reaction, the bed is washed five times with 0.15 M NaCl, 10 mM K-HEPES pH 7.7 to remove unbound dye. The wash buffer should be free of color by the third wash.
8. Labeled IgG is then eluted with five times 20 μL aliquots of 200 mM acetic acid. Each aliquot of eluate is immediately mixed with 5 μL of 1 M Tris–Cl pH 9 to neutralize and cooled to 4°C. The colored fractions are pooled. We store antibody in the resulting Tris–acetate buffer for months at 4°C. If necessary, antibody can be exchanged into other buffers by two to three rounds of dilution and reconcentration using a 10 kDa centrifugal concentration device (Sigma Z677108), or by desalting using a small G25 column. For longer-term storage of labeled IgGs, we add glycerol to 50% (v/v) and store at −20°C.

3. MICROTUBULE ORGANIZING CENTERS AND MICROTUBULE NUCLEATION

To nucleate asters we have used four different microtubule organizing center (MTOC) types: anti-AurkA IgG on beads, centrosomes attached to sperm nuclei, isolated centrosomes, and *Tetrahymena* pellicles. During aster growth in I-phase extract, microtubules from adjacent asters do not interpenetrate due to mutual inhibition of microtubule growth. Instead, they form aster–aster interaction zones (AAIZs) where Aurora B kinase (AurkB) localizes. Aster–aster interactions are important for cytokinesis signaling (Nguyen et al., 2014). Due to AAIZ formation, higher MTOC densities result in smaller asters. In practice, we dilute MTOCs in extract, so the final concentration is 1–20 per 20× imaging field. To some extent, adding more nucleating sites mimics decreasing blastomere size during embryo development (Field, Groen, Nguyen, & Mitchison, 2015).

3.1 ANTI-AurkA IgG-COATED BEADS

These were developed by Tsai and Zheng (2005). They are very reliable and easy to store, but require IgG to AurkA. Asters nucleated from AurkA beads are symmetric.

We have used them to explore how a single MTOC can nucleate a microtubule array that fills a millimeter-sized egg (Ishihara et al., 2014) and to reconstitute cytokinesis signaling (Nguyen et al., 2014). See Fig. 2A.

Requirements

- 2.8 μm magnetic Dynabeads coated with Protein A (Thermo 10002D)
- Rack to hold tubes with magnet to pull beads
  - For 1.5 mL Eppendorf tubes (Bangs Laboratories LS001)
- Wash buffer
  - 50 mM KCl
  - 10 mM K-HEPES pH 7.7
  - 1 mM $MgCl_2$
  - 1 mM EGTA
- Storage buffer
  - Wash buffer plus
  - 0.05% sodium azide
  - 100 μg/mL bovine serum albumin (BSA)
- Rotating wheel for incubation

Preparation of MTOCs

1. Rabbits were immunized with full length *Xenopus* AurkA fused to GFP and expressed in bacteria.
2. Antisera were first passed over a column of GFP, to remove antibodies against the fusion protein partner.
3. The antisera were then passed over an AurkB column, to remove Aurora B antibodies before the final purification. This ensures the purified antibody is specific for Aurora A.
4. Dynabeads Protein A are washed three times with Wash Buffer and then coated with saturating amounts of anti-AurkA IgG (30 min, room temperature).
5. Beads with antibody attached are then washed twice with Storage Buffer and stored at 4°C. Beads are stable for months. The BSA prevents bead aggregation and nonspecific adsorption; however, it should not be added prior to antibody binding to the beads.

3.2 DEMEMBRANATED *XENOPUS* SPERM WITH CENTROSOMES ATTACHED

Centrosomes attached to sperm nuclei are the natural MTOC that is introduced to eggs during fertilization. However, asymmetric signals from the chromatin make their nucleation activity complex. Chromatin can nucleate microtubules (Heald et al., 1996), and it also sends an asymmetric signal via AurkB activity. As a result, sperm centrosomes nucleate asters that are polarized in their structure and activity early in the cell cycle (Field et al., 2015). See Fig. 3A. Later sperm form nuclei and centrosomes split and migrate apart (Fig. 3B and C). The following protocol is adapted from Murray (1991).

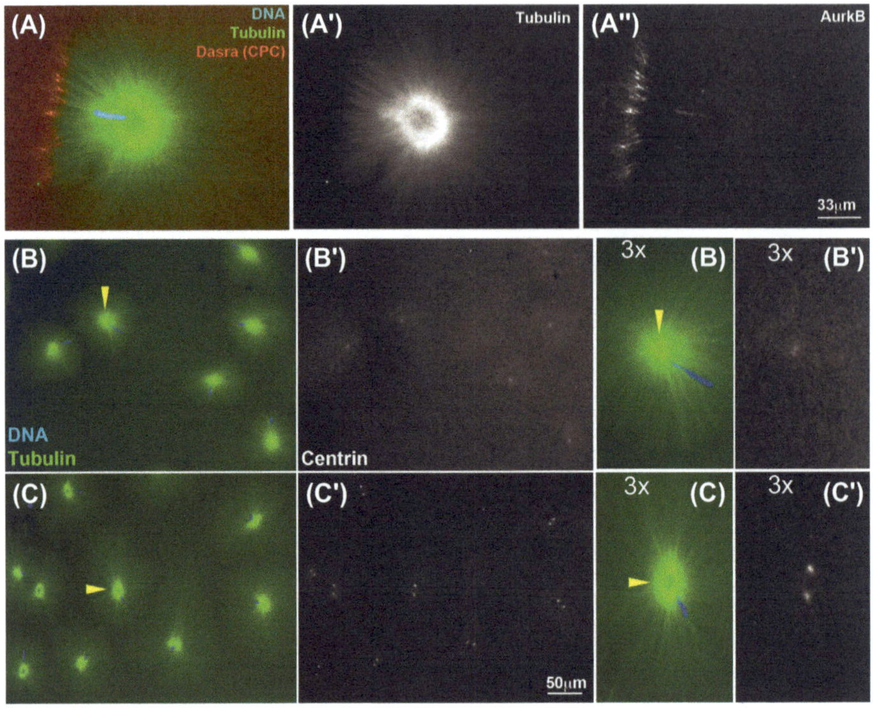

FIGURE 3 Microtubule nucleation by sperm plus centrosomes.

Permeabilized sperm nuclei with attached centrosomes were added to actin-intact interphase egg extract. 20× widefield imaging. (A) Example of asymmetric aster formation. Note that (A') the microtubules on the side of the aster close to the sperm chromatin are shorter and more bundled and (A") recruit chromosomal passenger complex (CPC) to their plus ends (Dasra-GFP). For CPC imaging it is essential to use PEG-passivated glass surfaces (Field et al., 2015). (B and C) Centrosome duplication and separation. (B and B') and (C and C') are the same field in different channels taken 18 min apart. *Yellow arrows* indicate a centrosome shown at 3× magnification. Note centrosome duplication and separation visualized using fluorescent tubulin and anti-Centrin IgG probes. The extent of centrosome separation was sensitive to the coverslip coating, with more separation on plain glass coverslips, which bind more dynein. This example uses partially passivated glass.

Buffers and reagents

All buffers should be prepared and placed on ice before starting the preparation.

- MMR (Section 1.1)
- 2× NPB
    - 500 mM sucrose
    - 30 mM K-HEPES pH 7.4 with KOH
    - 2 mM EDTA pH 8.0
    - 1 mM spermidine trihydrochloride (Sigma S2501)

  - 0.4 mM spermine (Sigma S1141)
  - 2 mM DTT (Sigma D9779)
- Protease inhibitors (LPC) 1/1000 of 10 mg/mL stock (see Section 1.1)
- Lysolecithin (L-α-lysophosphatidylcholine Type 1) (Sigma L4129)
  - 10 mg/mL, made fresh the morning of the preparation
- 10% BSA
- 10 mg/mL Hoechst 33342 stock (Sigma B-2261)
  - Stored in water in a light-tight vessel at −20°C
- Sperm dilution buffer (SDB)
  - 100 mM KCl
  - 10 mM K-HEPES pH 7.7
  - 1 mM $MgCl_2$
  - 150 mM sucrose
  - Can prepare as a 5× stock and store at −20°C
- Sperm storage buffer
  - 1× NPB
  - 30% glycerol
  - 30 μL 10% BSA for 1 mL
- Hoechst stock solution, 10 mg/mL in water

Equipment

- Small Petri dish
- Razor blade or scalpel
- Cheesecloth
- Funnel
- Centrifuge and Beckman JS 13.1 swinging bucket rotor or equivalent
- 14 mL round bottom polypropylene tubes (Falcon 352059)
- Hemocytometer (or another method to quantitate sperm yield)

3.2.1 Isolation of sperm

1. Remove testes from one to two male *Xenopus laevis*.
2. Place in small Petri dish and remove as much fat and blood vessels as possible without causing damage.
3. Rinse three times in 1× MMR and two times in 1× NPB.
4. Transfer testes to a dry Petri dish and macerate thoroughly with a razor blade until they look like a slurry.
5. Add 2 mL NPB + LPC to the macerate. Gently pipette up and down through a fire-polished truncated Pasteur pipette with an opening of ~3 mm in diameter.
6. Pass macerate through about four thicknesses of cheesecloth on a funnel, collecting into a 14 mL round bottom Falcon tube. Rinse dish with 8 mL of NPB + LPC. Pass this rinse through the cheesecloth, collecting it in the tube.
7. Centrifuge at 3000 rpm for 10 min at 4°C in a Beckman JS-13.1 or equivalent swinging bucket rotor.
8. Discard supernatant. Resuspend the pellet in 8 mL NPB + LPC.

9. Centrifuge again, as in step 7.
10. Resuspend pellet in 1 mL NPB + LPC, using a clipped (larger opening) P1000 pipette tip to minimize damage to the sperm due to shear forces.

3.2.2 Demembranation and concentration of sperm

1. Reserve 1 μL sperm in NPB on ice to use as a control later when monitoring the progress of demembranation. Warm the rest of the sample to room temperature, and add 50 μL of 10 mg/mL lysolecithin. Mix by gentle flicking and incubate for 5 min at room temperature.
2. Monitor demembranation.
   2.1. To the control sample from Step 1 and an equivalent sample after lysolecithin treatment, add 9 μL of NPB and 1/100 Hoechst stock. Image 5 μL of each. After demembranation, the Hoechst signal should be much brighter and sperm should look more stringy. It is typical for 1–5% of nuclei to be partially membranated, with part or all of the nuclei remaining globular, circular, or rounded.
3. Add 10 mL cold NPB + 3% BSA + LPC to the tube of demembranated sperm. Mix gently by pipetting up and down and spin at 3000 rpm for 10 min at 4°C.
4. Decant supernatant. The lysolecithin-treated pellet should look slightly more translucent than it did prior to treatment.
5. Resuspend in 5 mL NPB + 0.3% BSA without LPC. Mix gently by pipetting, then centrifuge at 3000 rpm for 10 min at 4°C.
6. Decant supernatant and resuspend pellet in 500 μL Storage Buffer. This is now your demembranated sperm stock. Keep on ice, and determine yield (Section 3.2.3). Aliquot in 20 μL volumes, flash-freeze in liquid nitrogen, and store at −80°C.

3.2.3 Yield determination

1. Add 1 μL of the demembranated sperm stock and 1 μL of 1/100 dilution of Hoechst stock to 98 μL of SDB. Mix the sperm stock very well using a clipped (larger opening) pipette tip just before removing the 1 μL sample.
2. Mix the diluted SDB/Hoechst/sperm very well and allow a small amount to flow into the chamber of a hemocytometer by capillary action.
3. Count nuclei in a square of the hemocytometer under a compound microscope (we usually visualize with Hoechst and bright field combined). From two males, you should obtain counts of at least 200–250 × 10^4 cells/mL for a 1:100 dilution of the 0.5 mL suspension.

3.3 FREE CENTROSOMES

Centrosomes can be purified from tissue culture cells (Reber, 2011). For us it has been easier to prepare them from *Xenopus* sperm, using M-phase extract to naturally separate the centrosome from the nuclear envelope. Free centrosomes nucleate symmetrical asters, unlike sperm, where the chromatin causes strong asymmetry in microtubule organization (Field et al., 2015).

Requirements

- *Xenopus* sperm (Section 3.2)
- M-phase extract (Section 1)
- Cytochalasin D, 10 mg/mL in DMSO (Sigma C8273)
- 1 M $CaCl_2$
- Ficoll 400 (Sigma F9378)
- 1× XB salts (Section 1.1)
- 1× CSF-XB Buffer (Section 1.1)
- Fluorescent tubulin (Section 2.1.2)
- 14 mL round bottom polypropylene tubes (Falcon 352059)
- Microcentrifuge
- Centrifuge and swinging bucket rotor that can run at 25,000 × g

1. Add cytochalasin D (0.2 μL of 10 mg/mL stock) to 100 μL of M-phase extract, or prepare M-phase extract in the presence of cytochalasin. Add sperm nuclei (up to 5000 per μL) and incubate for 10–20 min at 20°C. This incubation naturally dissociates centrosomes from chromatin. Addition of cytochalasin prevents gelation-contraction of the extract.
2. Add calcium to the extract to a final concentration of 0.4 mM. This cycles the extract into interphase (Section 1.3.1). Incubate for 1 h and 20 min at RT. During this time the DNA decondenses and forms nuclei (Section 1.3.2). Additionally, the centrioles replicate and centrosomes assemble.
3. The remaining steps of the protocol should be performed at 4°C, using buffers chilled to 4°C.
4. Dilute the extract 5× in cold 1× XB and centrifuge in a microfuge at 3000 × g for 5 min.
5. Keep the supernatant. The pellet is very soft and should contain most of the nuclei.
6. Dilute the supernatant 5× in cold CSF-XB.
7. Prepare a 3 mL 20% Ficoll cushion in CSF-XB in a 14 mL Falcon tube.
8. To concentrate centrosomes, pellet the supernatant onto the Ficoll cushion. The cushion is important, since pelleting them in a tube leads to irreversible aggregation (Mitchison & Kirschner, 1986). Carefully layer the diluted centrosomes atop the Ficoll cushion, and spin in a swinging bucket rotor at 25,000 × g for 15 min at 4°C.
9. Discard most of the diluted extract far above the cushion, then carefully collect fractions nearer the cushion. Most of the centrosomes will be in the Ficoll cushion, close to the interface with the diluted extract.
10. Assay by adding 1–2 μL of centrosome preparation to 20 μL of M-phase extract and fluorescent tubulin (Section 2.1.2). One should see one aster form per centrosome. Centrosomes can be stored at −80°C for prolonged periods.

11. Modifications
    11.1. To increase the purity of centrosomes, dilute the most concentrated fraction(s) again and centrifuge through an additional cushion. Alternatively, centrosomes can be purified on a sucrose gradient, then stored in high sucrose (Mitchison & Kirschner, 1986).
    11.2. To scale up, start with a larger volume of extract.

3.4 *TETRAHYMENA* PELLICLES

Tetrahymena pellicles are detergent-extracted cortices of deciliated *Tetrahymena* cells. They contain a dense array of basal bodies that nucleate a dense array of microtubules. They provide a convenient local source of polymerized microtubules that is not centrosomal in origin (Lombillo, Stewart, & McIntosh, 1995). We used them to compare microtubule stability inside and outside a mitotic spindle (Needleman et al., 2010) and as a source of microtubules nucleated independently of centrosomes or chromatin (Ishihara et al., 2014). See Coue, Lombillo, and McIntosh (1991) for pellicle isolation.

3.5 SPONTANEOUS NUCLEATION

Microtubules nucleate spontaneously in egg extract. The timing varies from 15 to 45 min post–calcium addition, depending on the extract and its age (older extracts have more spontaneous microtubules). Once initiated, nucleation tends to amplify rapidly, perhaps due to positive feedback (Ishihara et al., 2014). Activated *Xenopus* eggs, those released from CSF arrest by electrical stimulation, for example, also exhibit spontaneous microtubule assembly. Motors in the extract and egg can organize these microtubules into asterlike structures (Ishihara et al., 2014).

4. PREPARATION OF GLASS SURFACES FOR IMAGING

For imaging experiments, we typically squash small volumes of extract between two coverslips to provide a 5- to 25-μm thick preparation. Cleaning and preparation of the glass surface is important. Untreated glass tends to bind motor proteins such as dynein that influence microtubule organization, and it also prevents CPC recruitment to zones between asters. For cytokinesis signaling experiments we routinely passivate the glass using polyethylene glycol (PEG). Many PEG passivation schemes have been described in the literature. We discuss several here, starting with the most simple (Section 4.1), which is our current favorite. Passivation with poly-L-lysine-g-PEG was reported in (Lee & Vörös, 2005), now commercially available from SuSoS Chemicals. See Fig. 4 for examples of asters growing on different surfaces.

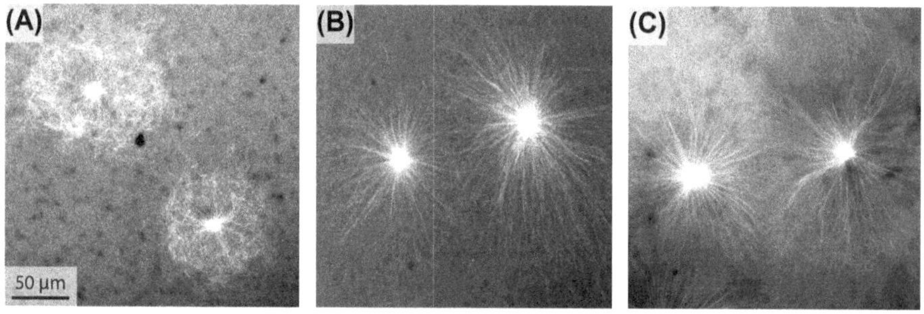

FIGURE 4 Surface passivation influences aster morphology.

These examples are nucleated from AurkA beads (Section 3.1) in I-phase extract. (A) Uncoated coverslip. Motor activity and nonspecific adsorption of microtubules generate disorganized asters. (B) Uncoated coverslip with 40 μg/mL CC1 added to inhibit dynein (Section 4.3). Dynein inhibition increases radial ordering of microtubules. (C) PLL-g-PEG–coated coverslips (Section 4.1). PEG treatment reduced nonspecific adsorption of motors and increased radial order.

4.1 PASSIVATION WITH POLY-L-LYSINE-g-POLYETHYLENE GLYCOL

4.1.1 Requirements

- Standard glass coverslips, No. 1.5 thickness, various sizes
- Parafilm
- Teflon rack for coverslips (Life Technologies C14784)
- Poly-L-lysine-g-polyethylene glycol (PLL-g-PEG) [SuSoS PLL(20)-g[3.5]-PEG(2)]
- VALAP, 1:1:1 mixture of wax, Vaseline, and lanolin
- Nitrogen or house air

4.1.2 Flamed coverslips

To flame-clean the coverslip surface, coverslips are individually dipped in 70% ethanol for a few seconds, pulled out, flamed with a Bunsen burner, cooled in air for a few seconds, then placed onto a droplet of 0.1 mg/mL PLL-g-PEG in 10 μM HEPES pH 7.4 on a piece of Parafilm. Coverslips 22 × 22 mm in size require 110 μL, and 18 × 18 mm coverslips require 90 μL. Allow coverslips to incubate at room temperature for 15–30 min. They are then washed three times in distilled water. The first two washes involve placing each coverslip PEG side down in a droplet of distilled water and allowing them to sit for 5 min. The last wash involves dipping each coverslip into a beaker of distilled water. Coverslips are then placed in slots in a Teflon rack so that they do not contact one another. They are dried with a jet of nitrogen, to prevent residue from water droplets. They are generally used the same day, but can be stored for up to a week at room temperature in the dark (Field et al., 2015).

4.1.3 Chemically cleaned coverslips

Deposit many (>50) coverslips in a beaker and clean by sonicating in 3 M KOH, rinsing in distilled water, then sonicating in 100% ethanol. Coverslips can be stored in 100% ethanol before use. Coverslips are removed from ethanol and incubated in drops of 0.1 mg/mL PLL-g-PEG solution in 10 mM K-HEPES pH 7.4 buffer for 15–30 min. After incubation, coverslips are placed in a Teflon drying rack and dried using house air (Ishihara et al., 2014).

4.2 CHEMICAL COUPLING OF SILANE-PEG TO GLASS SURFACES

This method was described in Bieling, Telley, Hentrich, Piehler, and Surrey (2010). We obtained similar or better results with the following simplifications: (1) Instead of cleaning coverslips with Piranha solution, we use oxygen plasma (Technics plasma etcher 500-II, oxygen gas at 100 mTorr, glow discharge at 200 W for 4 min). (2) Instead of a two-step silanization followed by PEG coupling procedure, we directly couple to the glass a silane-PEG (methoxy PEG silane, MW 5000, JenKem Technology USA # A3037). Silane-PEG is dissolved in 100% ethanol to 10% (w/v). Fully dissolving the silane-PEG requires heating. Coverslip sandwiches are made with this solution in between and incubated for 2 h at 75°C. These coverslips can be stored with a desiccant at −20°C for up to 1 year. We used these in Nguyen et al. (2014).

4.3 PASSIVATION WITH CASEIN

An advantage of casein passivation over PEG passivation is that growing microtubules track within a TIRF illumination field along the coverslip surface. This presumably reflects some adsorption of microtubule-associated proteins (MAPs) and motors. We made extensive use of kappa-casein coating in recent work, where we tracked single microtubules by TIRF (Ishihara et al., 2014; Nguyen et al., 2014). To clean the coverslips, use the protocol in Section 4.1.3. Then, instead of PLL-g-PEG solution, coverslips are incubated in droplets of 1 mg/mL casein in 10 mM HEPES pH 7.7 for 15–30 min. Coverslips are passed through two droplets of distilled water (no incubation), placed in a Teflon rack, and dried with a stream of house air.

As assayed by aster morphology and microtubule sliding, coverslips passivated with whole casein (Sigma C3400) bind less dynein but enrich for plus end–directed motors (Maloney, Herskowitz, & Koch, 2011). Coverslips passivated with kappa-casein (Sigma C0406) bind much less kinesin but more dynein, resulting in significant dynein activity on the glass surface. We obtained good TIRF images of dynamic astral microtubules while minimizing motor-generated forces from the substrate by coating the imaging coverslip with kappa-casein, coating the other coverslip with PEG, and adding a dynein inhibitor to the extract. For dynein inhibition we routinely use 0.04 mg/mL CC1, an easily expressed fragment of $p150^{Glued}$ that blocks dynein function (Quintyne et al., 1999). Vanadate usually has similar effects in extract systems, but may interfere with phosphatase activity.

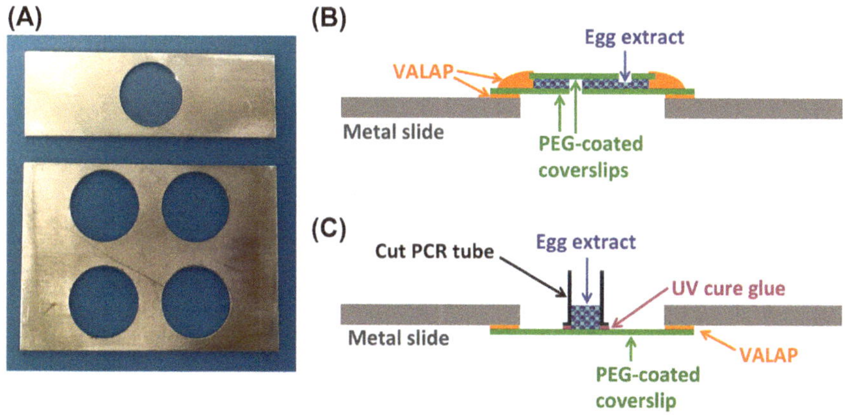

FIGURE 5 Metal support slides.

(A) Support slides made from aluminum or stainless steel. Single-hole slide is 240 × 750 × 2 mm, four-hole slide is 500 × 750 × 2 mm. Holes are 22 mm in diameter. (B) Coverslip mount for widefield and TIRF imaging of extract squash preparations. The same mount works on upright and inverted microscopes. (C) Coverslip and open well mount for supported lipid bilayer experiments. This mount is imaged from below by TIRF. See Figs. 6 and 7 for images.

4.4 METAL SUPPORT SLIDES

To hold the passivated coverslips we machined stainless steel or aluminum slides ~2 mm thick with 20 mm circular holes in the center. 22 × 22 mm square coverslips are attached to top of the metal slide using VALAP, with the passivated surface facing away from the slide. 4–12 μL of extract is pipetted to the center of the coverslip, then an 18 × 18 mm coverslip is placed on top with passivated side toward the extract. After the droplet spreads out fully (~5 s), we seal with VALAP so that the sample does not dry out. 7.5 μL extract gives a squash nominally 23 μm in thickness, but this may vary across the coverslip. These chambers can be viewed using both upright and inverted microscopes. We make metal slides with one, two, and four holes (Fig. 5A and B). With the four-hole slides, we can image four different reactions in parallel using an automated stage, visiting each reaction every 1–3 min.

5. PROTEIN IMMUNODEPLETION

Proteins can be specifically removed from egg extract by immunodepletion. This method has advantages when compared to gene knockdown in cultured cells, where expression of other proteins may change in response to target protein depletion over time. The degree of depletion can be monitored via Western blots and also by adding fluorescent antibody probes and imaging. Rescue experiments can be performed by adding back active fusion proteins. The most important requirement is high-quality IgG.

5.1 REQUIREMENTS

- 2.8 μm magnetic Dynabeads coated with Protein A (Thermo 10002D)
- IgG antibody, raised against protein of interest
- Random rabbit IgG for control (Jackson Immunoresearch 011-000-003)
- Rack to hold tubes with magnet to pull beads
  - For 1.5 mL Eppendorf tubes (Bangs Laboratories LS001)
- Wash Buffer, see Section 3.1
- Storage Buffer, see Section 3.1
- Rotating wheel for incubation

5.2 BEAD PREPARATION

1. Dynabeads are washed three times in Wash Buffer.
2. Saturate the beads with an affinity-purified polyclonal antibody raised against the protein of interest. In a typical preparation, we use 20 μg IgG per 50 μL of a 50% bead slurry in a 1.5 mL microfuge tube. Random rabbit IgG at an equivalent antibody concentration is used as a control.
3. Incubate the bead/antibody mixture while gently rotating overnight at 4°C or at room temperature for 30–60 min.
4. Wash three times with 1 mL Storage Buffer, and resuspend in 50 μL storage buffer. Beads with antibody attached can be stored at 4°C for 1–2 months. BSA reduces clumping and nonspecific adsorption of beads, and azide prevents bacterial growth.

5.3 PROTEIN DEPLETION

We optimize the immunodepletion protocol for particular target proteins.

Typically, two to three rounds of immunodepletion are required to remove proteins to <5%, but more IgG may be required to deplete abundant proteins or in cases where the IgG binds less efficiently. In some cases we have obtained more effective immunodepletion by adding IgG directly to extract, incubating 15 min at 0°C, then adding washed protein A (or G) beads, incubating 30 min, and removing beads.

1. Add *Xenopus* extract to prepared beads in a 1.5 mL tube at approximately a 1:20 dilution (typical would be 5 μL of beads and 100 μL of extract).
2. Incubate at 4°C for 45–60 min. Occasionally (every 5–10 min) flick the bead/extract mixture to disperse the beads and return to 4°C.
3. Pull beads to the side of the tube with a magnet, then move the extract to a second tube of prepared beads. Repeat the depletion step 2.

4. After the second depletion, separate the extract from the beads and assay the extract for activity.
5. Run an SDS PAGE gel and Western blot to determine extent of depletion.

6. SUPPORTED LIPID BILAYERS

The plasma membrane plays a central role in cell division. It defines the cell boundary, is shaped by the cortical actin cytoskeleton, interacts with microtubules, and ingresses during cytokinesis. There has been huge interest in reconstituting plasma membrane biology in the biophysics community, which led to several approaches that might be extended to egg extracts, notably giant unilamellar vesicles (GUVs) and SLBs. So far, we have focused on use of SLBs to model the plasma membrane in egg extract experiments, because these are technically easier than GUVs and better suited for microscopy experiments. In particular, SLBs are an excellent match to TIRF imaging, because they confine plasma membrane biochemistry to a thin layer near the coverslip which is ideally placed for TIRF imaging. By combining actin-intact extract, artificial MTOCs, and SLBs we were able to reconstitute spatially organized cytokinesis signaling in a cell-free system for the first time (Nguyen et al., 2014).

6.1 REQUIREMENTS

Lipids (all from Avanti Polar Lipids)

- L-α-Phosphatidylcholine (PC) (Brain, Porcine) (840053)
- L-α-Phosphatidylserine (PS) (Brain, Porcine) (840032)
- L-α-Phosphatidylinositol (PI) (Liver, Bovine) (840042)
- L-α-Phosphatidylinositol-4,5-bisphosphate ($PI(4,5)P_2$) (Brain, Porcine) (ammonium salt) (840046)

Equipment to create small unilamellar vesicles (SUVs)

- 10 μL Hamilton syringe, 1700 series, point style 2 (Sigma 20972)
- 100 μL Hamilton syringe, 1700 series, point style 2 (Sigma 20688)
- Glass vials (Borosilicate glass, phenolic screw cap) (VWR 66011-020)
- Nitrogen gas
- Argon gas
- Branson CPX1800 ultrasonic cleaner, or equivalent
- Two Teflon racks (Life Technologies C14784), to clean coverslips
- VALAP, 1:1:1 mixture of wax, Vaseline, and lanolin

Equipment for cleaning glass

- Glass coverslips 1.5 mm thickness
  - Option 1: 22 × 22 mm (VWR 48404-133)
  - Option 2: 22 × 60 mm (VWR 48367-059)
- Racks to wash 22 × 22 mm coverslips and keep them separate

    - Option 1: Polypropylene rack (Sigma–Aldrich Z688568)
    - Option 2: Teflon rack (Life Technologies C14784)
- Rack to wash 60 × 22 mm coverslips
    - Coplin staining jar (Thomas Scientific 8541K92)
- Glass beakers
    - 100 mL
    - 500 mL
- Plasma cleaning device
    - Option 1: Glow discharge unit (Electron Microscopy Sciences)
    - Option 2: Chemical vapor deposition system (Denton Vacuum)
    - Option 3: Other plasma cleaning devices (eg, Harrick Plasma Inc.)
    - Option 4: Handheld corona treater (Electro-Technic Products BD20AC). This option is less reproducible but less expensive and more portable than the other options
- UV light source
    - Compact UV lamp (UVP 95-0019-16)
    - Benchtop transilluminator (UVP 95-0449-01)
- Thin-walled PCR tubes, 0.2 mL (VWR 20170-012)
- Custom metal support slides (Section 4.4)
- Hot plate or temperature block

Buffers and reagents

- Extract buffer (XB) (Section 1.1)
- High-salt extract buffer (HS-XB): XB with 200 mM KCl
- VALAP (Section 4.1)
- Potassium hydroxide
- 100% ethanol (VWR 89125-186)
- Acetone
- Chloroform
- Norland Optical Adhesive 63, ultraviolet curing glue (Norland Products 6301)

6.2 PREPARATION OF SMALL UNILAMELLAR VESICLES

General information

Prepare SUVs at least 1 day prior to the experiment. Lipids are usually dissolved in organic solvents such as chloroform. To handle and store chloroform solutions, avoid plastic pipette tips and containers. Instead use glass or polytetrafluoroethylene (PTFE, eg, Teflon). We use Hamilton syringes and glass vials with PTFE screw cap lids.

To store organic solutions of phospholipids: prepare aliquots, cover the solution with a layer of argon to prevent oxidation of the lipids, close the cap tightly, and seal with Parafilm. Store at −20°C. More polar lipids, such as $PI(4,5)P_2$, are dissolved in more polar solvent mixtures, such as chloroform/methanol/water 60/33/7% (v/v). Furthermore, $PI(4,5)P_2$ is fairly unstable and volatile in organic solution at −20°C (lasting no longer than 1 week), thus it should be stored as SUVs in aqueous solution under argon at −80°C, as described in the following.

To mimic the inner leaflet of the plasma membrane of animal cells, we mixed two different mammalian lipids stocks.

- Stock A: three lipids in the molar ratio PC 0.6/PS 0.3/PI 0.1
- Stock B aka 5% PI(4,5)P_2: three lipids in Stock A plus PI(4,5)P_2 in the molar ratio of PC 0.6/PS 0.3/PI 0.05/PI(4,5)P_2 0.05

Depending on the quality of the extract, we occasionally added small amounts of PI(4,5)P_2 by combining the two stocks (Section 6.5.3).

1. In 4 mL clean glass vials, add lipid solutions in desired amounts and ratios, so the total concentration of lipids is 1−2 mM. An optimal volume of lipid per vial is 100−300 μL. Dry down lipids in a nitrogen gas stream, to obtain a translucent lipid film on the bottom of the glass vial.
2. Vacuum desiccate for 1−2 h to remove residual organic solvent.
3. Rehydrate lipid film with appropriate amount of HS-XB buffer to obtain a final 5 mM lipid suspension. Incubate buffer with lipid film for 30 min at 37°C, then vortex rigorously to obtain large multilamellar vesicles (MLVs). The result should be a milky solution.
4. To create SUVs freeze-thaw the MLVs five times using a dry ice/isopropanol mix. Vortex after each thaw step.
5. Sonicate suspension in bath sonicator for 2 min at a time until the solution is clear. Keep at 37°C when not sonicating. Take care not to overheat the lipid mixture during sonication. Some lipid suspensions, especially those made from mammalian lipids or with a high PE content, can take two to three rounds of 2-min sonication or longer sonication times. We found that freeze-thaw followed by sonication for each round is the fastest and most convenient method to break up larger MLVs into SUVs. This method produces SUVs with diameters in the range of 15−50 nm.
6. Aliquot SUVs from Stock A into desired volumes, typically 24 μL or larger. Store at −20°C. Aliquot SUVs from Stock B in smaller aliquots 5−10 μL and store at −80°C under argon, as PI(4,5)P_2 is more volatile.
7. Bilayers are assembled by mixing the appropriate amounts of Stocks A and B and adding the SUVs to small chambers affixed to coverslips (Section 6.4). For a larger reaction chamber prepared using a 0.5 mL microfuge tube, 10 μL is required to make one bilayer. For a smaller reaction chamber made from a PCR tube, 6 μL is sufficient for one bilayer.

6.3 CLEANING GLASS AND PREPARING CHAMBERS

We typically clean coverslips in alkaline solutions and organic solvents using sonication in a bath sonicator. These should be prepared the day before the experiment and can be stored for several days in ethanol at 4°C. Then we use air plasma for final cleaning and assemble the imaging chambers ~2 h prior to the experiment.

6.3.1 Chemical cleaning

1. Place coverslips into slots in a Coplin jar (for 22 × 60 mm coverslips) or a polypropylene rack (for 22 × 22 mm coverslips). It is important to keep each coverslip separate. Immerse coverslips into a beaker of 3 M KOH, and sonicate for 20 min.
2. Discard cleaning solution, and rinse five times with ddH_2O.
3. Immerse coverslips in ddH_2O, and sonicate for 20 min.
4. Discard ddH_2O, and rinse with 100% ethanol.
5. Immerse coverslips into 100% ethanol, and sonicate for 20 min.
6. Immerse into fresh ethanol. Proceed to plasma cleaning (Section 6.3.2), or store the coverslips in a sealed container for several days at 4°C.

Note: With certain batches of coverslips,[4] we found that a more rigorous pre-cleaning procedure was helpful:

1. Sonicate in 100% ethanol for 20 min, then rinse three times with ddH_2O.
2. Sonicate in 100% acetone for 20 min, then rinse three times with ddH_2O.
3. Incubate in 100% nitric acid (HNO_3) for 20 min, then rinse three times with ddH_2O.
4. Proceed to Step 1 in the cleaning procedure.

6.3.2 Plasma cleaning

1. Collect items required for chamber assembly before plasma cleaning. See Section 6.3.3. Everything must be ready before plasma cleaning begins, so chambers may be assembled as soon as cleaning is complete.
2. Use forceps to remove the rack with chemically cleaned coverslips from ethanol. Keep coverslips in the rack and blow dry using compressed nitrogen gas. Make sure each coverslip is dry.
3. We routinely used an EM Sciences Glow Discharge Unit. Place individual coverslips flat inside a Petri dish, then place the Petri dish into the chamber. Glow discharge at 25 mA for 8 min.
4. Immediately prepare the reaction chamber (Section 6.3.3) after plasma cleaning. Freshly cleaned and dried glass surfaces can adsorb dirt particles from unfiltered air, so we add the SUV solution within 30 min after plasma.

6.3.3 Preparation of reaction chambers

1. Cut either PCR tubes or 0.5 mL microfuge tubes to make chambers. Remove the lid and the lower conical part of the tube. See Figs. 5C and 6B.
2. Use metal slides (stainless steel or aluminum) with circular holes 20 mm in diameter. Attach a 22 × 22 mm plasma-cleaned coverslip to bottom of the slide using VALAP. The plasma-cleaned side should face up through the hole. Or use

[4]We have successfully used coverslips from various vendors. Most commonly we used #1.5 thickness, 24 × 60 mm or 22 × 22 mm square coverslips made of standard glass.

22 × 60 mm coverslip as a platform, again plasma-cleaned side up (Fig. 6C). The longer coverslips can be used in place of a slide in a microscope slide holder and can accommodate more chambers; however, they are less robust.

3. Cover the rim of a cut PCR tube with a thin layer of UV glue. Insert a pipette tip into the rim opening to ensure it is not occluded by glue.
4. Place tube with glue-covered rim down flat onto the surface of the freshly cleaned glass coverslip (Fig. 6B).
5. Expose to 365 nm UV light for 5–10 min to allow glue to polymerize.[5] It is crucial to load SUVs into the reaction chambers as soon as possible, so while the glue is polymerizing, thaw and mix lipid stocks, to form supported bilayers (Section 6.4).

6.4 SUPPORTED LIPID BILAYERS FROM MAMMALIAN LIPIDS

1. Thaw stocks of SUVs A and B prepared in Section 6.2. Seal tubes with Parafilm and sonicate 1–2 min until solution is clear. We found immersing the tubes in the center of the sonicator bath such that the liquid inside the tube is just below the surface of bath water works well. Make sure that lipids do not heat up.
2. Using the SUV stocks A and B make mixtures of the desired volumes and ratios of lipids. 6 μL of SUV is required for one PCR tube chamber. For our cytokinesis signaling experiments we typically made mixtures of 0%, 0.1%, and 0.5% $PI(4,5)P_2$ lipid mixtures in 60% PC, 30% PS, 10% PI and tested the extract for actin assembly properties (see Section 6.5).
3. Sonicate the mixtures for 1–2 min as described in Step 1.
4. Dilute SUV mixtures 5× in HS-XB buffer to 1 mM final lipid concentration.
5. Put 30 μL of diluted SUVs into each PCR tube well.
6. Add $CaCl_2$ to 4 μM final concentration (1.2 μL of 100 mM $CaCl_2$ per well).
7. Incubate on 42°C heated temperature block for 30–60 min. This incubation temperature approximates porcine body temperature (lipids are porcine derived).
8. Wash bilayers at room temperature:
   - 8.1. Add 170 μL of regular XB buffer into each well.
   - 8.2. Pipette up and down five times, then remove 150 μL buffer from each well.
   - 8.3. Add 150 μL XB buffer into each well, pipette up and down five times, then remove 150 mL of the buffer from each well. Repeat this until they have been rinsed five times. Do not touch bilayers. Make sure a layer of ~20 μL of buffer remains above the bilayer.
9. Keep slides with bilayers in a covered Petri dish at 4°C until use. They should last for several hours.

[5] To prevent loss of valuable lipids, one should practice gluing chambers to uncleaned coverslips, then adding water to check for leakage.

6.5 EVALUATION OF BILAYER INTEGRITY AND EXTRACT QUALITY

Before the experiment begins, optimize every aspect of the protocol, such as bilayer continuity, fluidity, and composition; extract quality; and concentrations of fluorescent probes and MTOCs.

6.5.1 Requirements

For bilayer integrity

- Fluorescent lipids
  - Option 1: 0.5 mol % Lissamine Rhodamine PE. 1,2-dipalmitoyl-sn-glycero-3-phosphoethanolamine-N-(lissamine rhodamine B sulfonyl) (Avanti Polar Lipids 810158C)
  - Option 2: 0.5 mol % Texas Red DHPE. Texas Red-1,2-dihexanoyl-sn-glycero-3-phosphoethanolamine (Biotium 60027)
  - Option 3: DiIC18 (Invitrogen D282) 10 mg/mL stock in DMSO (1000×)

For extract quality

- Small Petri dish
- Mineral oil (Sigma M5904)
- 1 M $CaCl_2$
- 20°C water bath
- Fluorescent probes (Section 2). In this protocol, labeled tubulin and Alexa 647-labeled anti-AurkB IgG
- MTOC (Section 3). In this protocol, anti-AurkA beads
- PEG passivated coverslips, 22 and 18 mm (Section 4.1)
- Metal imaging slides (Section 4.4)
- VALAP (Section 4.1.1)

6.5.2 Bilayer integrity

When first assembling SLBs, one should verify bilayer continuity and lateral fluidity by adding a low concentration of fluorescent lipids (Section 6.5.1). One can add Rhodamine PE, Texas Red DHPE, or the lipophilic dye DiIC18 (DiI) to the lipid mixture during the vesicle preparation. To confirm continuity, one can image the fluorescent bilayer, which should appear homogeneous. To confirm fluidity, one can perform fluorescence recovery after photobleaching (FRAP). Using the field aperture, photobleach a small region, then monitor diffusion of unbleached lipids into the region, which should take 1–5 min depending on the size of the bleached spot.

If slow fluorescence recovery reveals poor continuity or fluidity, one possible reason is aging of the lipids. For example, oxidized lipids do not form a fluid membrane. In this case, remake SUVs with fresh lipids. Another possible reason is that adsorbed vesicles did not coalesce to form a continuous bilayer, with various causes such as insufficient cleaning of the glass coverslips. Alternatively, one may increase the incubation time, the buffer ionic strength, or the divalent cations concentration to improve results (Nair, Salaita, Petit, & Groves, 2011).

After we tested a batch of lipids, we often omitted fluorescent lipids, to enable use of all fluorescence channels for experimental probes. Still, one can estimate bilayer integrity based on the actin probe Lifeact-GFP. Actin should form a uniform dynamic layer atop the bilayer, and gaps indicate holes in the bilayer.

6.5.3 Extract quality

Actin contractility in droplets

As soon as the extract has been harvested (Section 1.2.4), assay each batch for actin contractility in metaphase. Each batch of extract should not combine eggs from multiple frogs. As described in Section 1.3.2, pipette a droplet of extract under mineral oil, then observe the speed and extent of actin contraction. If no extract contracts well, abort the experiment.

Formation of aster–aster interaction zones

Convert each extract into I-phase (Section 1.3.1), and test for assembly of large microtubule asters that form AAIZs between glass coverslips (Nguyen et al., 2014). In addition to testing extract quality, this assay also verifies concentrations of

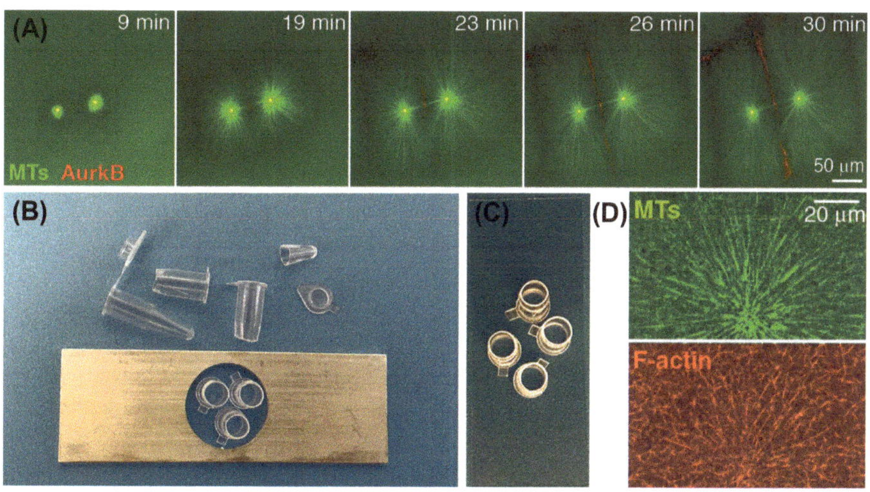

FIGURE 6 Evaluating extract quality and mounting wells for bilayers.

(A) Squash preparation containing AurkA beads (yellow), tubulin probe (green), and anti-AurkB IgG [marker for the chromosomal passenger complex (CPC)] (red). Widefield imaging. In a good extract, microtubule asters form zones of interacting microtubules termed aster–aster interaction zones (AAIZs). AAIZs recruit CPC. (B) Metal slide with chambers for preparation of supported lipid bilayers. (C) Bilayer chambers attached to 22 × 60 mm coverslip. (D) TIRF imaging of tubulin and actin on a bilayer to test actin morphology. Actin is aligned along the microtubule bundles in the aster. When the actin had this morphology, cytokinesis signaling was successful (Nguyen et al., 2014). If there is too much $PI(4,5)P_2$ for a given extract, we observed very dense F-actin not aligned with microtubules (not shown). If not enough, there was little F-actin on the coverslip (not shown).

fluorescent probes and MTOCs. One should add an optimal number of MTOCs to allow formation of aster–aster interaction zones. If asters are too far from one another, they cannot interact. If asters are too close to one another, they cannot recruit AurkB or form interaction zones (Field et al., 2015).

1. To 30 μL of extract, add tubulin and anti-AurkB IgG fluorescent probes, and anti-AurkA bead MTOCs.
2. Convert extract to I-phase with addition of calcium to 0.4 mM (Section 1.3.1).
3. Squash 7.5 μL of extract between two PEG passivated coverslips (Section 4.1) on a metal support slide, and seal with VALAP (Section 4.4).
4. Image using a widefield fluorescent microscope at 10× or 20×.
5. Robust microtubule asters should grow from each AurkA bead. When two asters are close enough to interact, a zone of low microtubule density should form between them. AurkB should localize to this zone within 15–20 min, depending upon the temperature (Fig. 6A).

$PI(4,5)P_2$ requirement

Egg extracts prepared from different frogs vary in their ability to promote actin polymerization at the cortical layer near the lipid membrane. To obtain the desired F-actin morphology at the cortical layer, we can vary the amount of negatively charged lipids in the supported bilayer. In general, more negatively charged lipid bilayers support more actin polymerization (Lee, Gallop, Rambani, & Kirschner, 2010). Routinely, we mix lipid stocks with (Stock B) and without (Stock A) $PI(4,5)P_2$ (Section 6.2) to prepare SLBs containing 0%, 0.1%, and 0.5% $PI(4,5)P_2$. Then, before each experiment, we assay for cortical actin morphology, such as alignment along microtubule asters (Fig. 6D).

6.6 IMAGING CYTOKINESIS SIGNALING

During cytokinesis in animal cells, the cleavage furrow is positioned by signals to the cell cortex that emanate from microtubule arrays interacting at the cell equator. An important protein complex that sends signals from microtubules is the CPC. Signals from the CPC help recruit additional proteins to the equatorial cortex resulting in cortical contraction and membrane ingression, separating the cell into two daughters. The key cortical signaling molecule is the active form of the small GTPase RhoA-GTP.

To reconstitute cytokinesis signaling, one requires high-quality extract and bilayers, MTOCs, and an appropriate collection of fluorescent probes. To visualize tubulin we use 250 nM tubulin-Alexa Fluor 488 or 568. To visualize RhoA-GTP we use 140 nM mCherry-rGBD. To visualize filamentous actin we use 140 nM Lifeact-GFP. To visualize the CPC we use 0.1–1 μg/mL anti-AurkB IgG labeled with Alexa Fluor 647 or a GFP-tagged version of the DasraA protein, another member of the CPC complex (Field et al., 2015; Nguyen et al., 2014). We have also imaged two cortical cleavage furrow proteins anillin and septin, using fluorescently labeled antibodies (Nguyen et al., 2014).

Timing and teamwork

Due to the simultaneous, time-sensitive tasks, we generally work in pairs. One or both people make the extract (approximately 1–2 h). Then one person plasma-cleans the glass and assembles chambers, then prepares the SLBs. Meanwhile, the second person tests extract quality and probe concentrations and selects the best extract for experiments (Fig. 6). Each process takes about 2 h with experienced hands. It is crucial to coordinate the two processes and to perform the experiments in a timely manner, as both the SLBs and the extracts last about 6 h at 4°C.

Final sample preparation

1. Prepare SLBs from mammalian lipids as described previously.
2. Convert extract to I-phase with addition of calcium to 0.4 mM (Section 1.3.1). Incubate at 20°C for 10 min, then return to ice for at least 2 min to depolymerize microtubules and filamentous actin.
3. To exchange out buffer, wash bilayers three times with 20 μL I-phase extract. Extract used for washing bilayers can be prepared earlier and frozen (Section 1.4.4). If frozen extract is used, wash two times with frozen extract, then a third time with fresh extract.

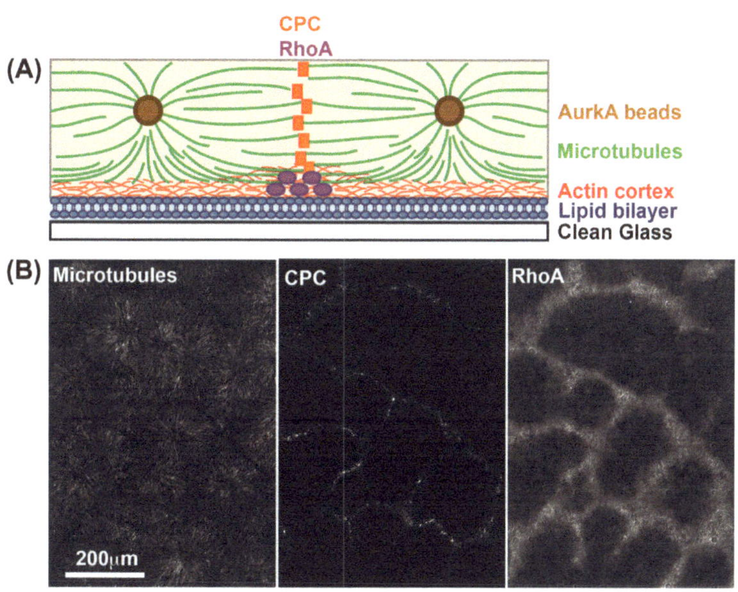

FIGURE 7 Reconstitution of cytokinesis signaling.

(A) Schematic of a bilayer experiment viewed from the side. (B) Representative experiment viewed via TIRF microscopy. Aster–aster interaction zones form where asters interact. The chromosomal passenger complex (CPC) is recruited and active RhoA is recruited to the bilayer (Nguyen et al., 2014).

4. Add desired probes and AurkA beads to 20 μL of I-phase extract and gently mix. As described in Section 6.5.3, use the beads concentration optimal for formation of aster–aster interaction zones.
5. Remove most of the last 20 μL I-phase extract used to wash the bilayer, and add the 20 μL extract containing probes and beads. At this point, we start a timer.
6. Monitor aster assembly reactions with an inverted TIRF or spinning disc confocal microscope. The temperature of the microscope room should be 18–22°C, though up to 25°C is tolerable. As the cell cycle progresses, microtubule asters will form and F-actin cables align with the tubulin polymer. Formation of AAIZs that recruit cytokinesis factors such as the CPC is observed starting at around 20 min. At the same time, activation of RhoA, as indicated by binding of the mCherry-rGBD probe to the bilayer, is also initiated. By 50–60 min, a stable active RhoA zone can be seen above CPC-positive zones (Fig. 7). By 90 min, most microtubules usually have disappeared, and we do not image beyond this point.

7. DROPLET MICROFLUIDICS

Microfluidic droplets are a versatile platform to study the effects of confinement and component limitation on the organization, size scaling, and dynamics of cytoskeletal structures. Extract without intact actin has been encapsulated in microfluidic droplets to investigate the organization of microtubule asters (Pinot et al., 2009) and size scaling of spindles (Good, Vahey, Skandarajah, Fletcher, & Heald, 2013; Hazel et al., 2013). Actin-intact extract in droplets has reconstituted dynamic actin networks in bulk (Pinot et al., 2012) and at the cortex (Abu Shah & Keren, 2014). This section assumes fully assembled microfluidic devices as a starting material, described in the above mentioned references. We present special considerations for encapsulating actin-intact extract in droplets, to study actin dynamics in confinement and interactions between actin and microtubules. Examples are shown in Fig. 8.

7.1 REQUIREMENTS

We most often use these materials, though substitutions can be made.

- Polydimethylsiloxane (PDMS) microfluidic devices, with inlet and outlet holes cored, bonded to glass coverslips (Fig. 8A)
- Oil/surfactant mixture: 2% Pico-Surf 2 surfactant in HFE7500 fluorous oil (Dolomite 3200281)
- Mineral oil (Sigma M5904)
- Syringes. Norm-Ject 1 mL (VWR 53548-001)
- Blunt needles. 30 g × 1/2″ (Brico BN3005)
- Long needles. 16 g × 1 1/2″
- Tubing. Tygon 0.010″ × 0.030″ ND-100-80 formulation (Cole–Parmer EW-06419-00)
- Two syringe pumps which can be mounted vertically

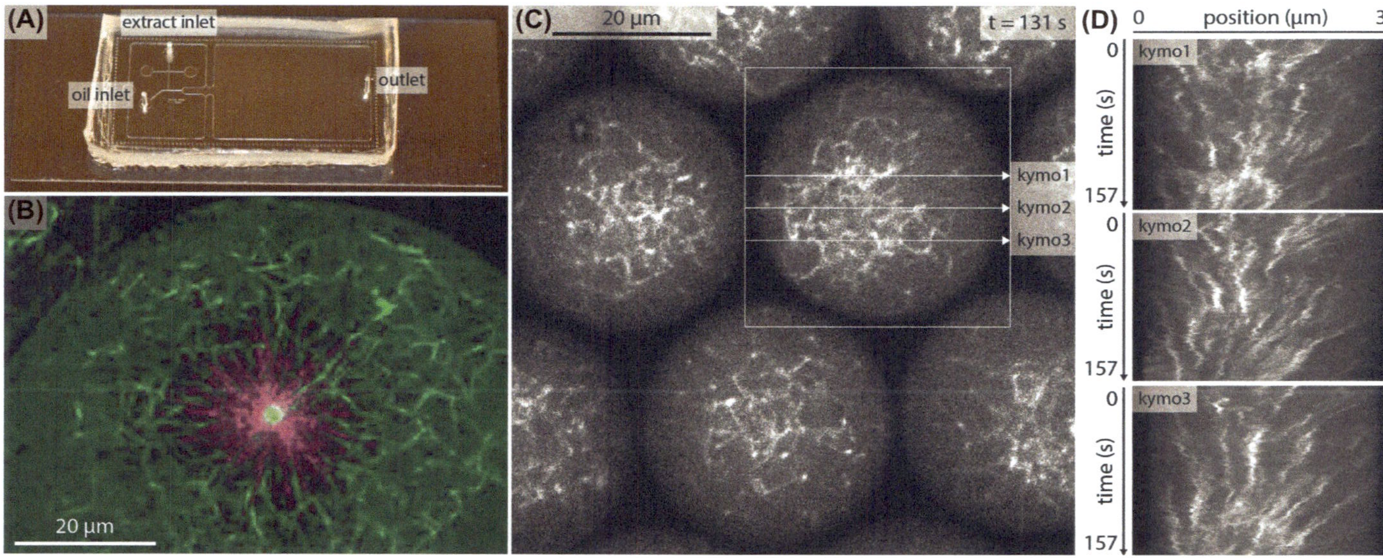

FIGURE 8 Cytoplasmic extract confined in microfluidic droplets.

(A) Each microfluidic device has an oil inlet, an extract inlet, and an outlet. Droplets are formed at the T-junction where the inlets meet. (B) I-phase extract encapsulated in microfluidic droplets. Green is actin (Lifeact-GFP), and magenta is tubulin (Alexa Fluor 647). An aster nucleated from an AurkA bead spans the droplet, and actin filaments align with microtubules near the aster center. (C) M-phase extract encapsulated in microfluidic droplets, packed into a hexagonal array. Fluorescence corresponds to Lifeact-GFP. (D) Kymographs of actin bulk contraction. The *arrows* in panel (C) correspond to the three kymographs in panel (D).

7.2 DROPLET GENERATION

Droplets must be formed at 4°C. At room temperature, gelation and contraction of the extract can clog the tubing. Furthermore, polymerization of microtubules before encapsulation in droplets can skew the concentrations of MAPs: at room temperature, droplets containing microtubule-nucleating centers contained significantly more MAPs than droplets without nucleating centers. If you have access to a cold room, omit the ice bath from the protocol.

1. Prepare microfluidic devices and materials.
   - 1.1. Soak devices in CSF-XB buffer (Section 1.2.3) to limit evaporation through the PDMS. Prevent buffer from entering the device, for example, by blocking the holes with tubing.
   - 1.2. Load 400 μL mineral oil into a 1 mL Norm-Ject syringe and chill on ice.
   - 1.3. Cut two lengths of tubing, long enough to reach from the syringe pump to the device. Press a blunt needle into one end, and cut the other end with a taper, which makes it easier to press tubing into PDMS holes.
   - 1.4. Prepare an ice bath in a small water basin.
   - 1.5. Load 300 μL oil/surfactant into a 1 mL Norm-Ject syringe.
2. Fill the microfluidic device with oil/surfactant at ~30 μL/min. Avoid air bubbles in the device, tubing, or syringe. Oil settles in the device, so tilting the device with the outlet uphill helps avoid air bubbles.
3. Submerge the device and tubing in the ice bath.
4. Finish preparation of the extract. For example, add fluorescent probes, MTOCs, or cycle to I-phase. Prepare at least 50 μL extract per device.
5. Using a long needle, aspirate extract into the syringe containing mineral oil. Avoid air bubbles. Keep the extract as cold as possible, so quickly connect the second tubing length and submerge in the ice bath.
6. Mount the extract syringe on the syringe pump, so the extract is in the ice bath.
7. Purge air from the extract tubing, then reduce the flow rate to 2 μL/min. At the same time, start the flow of oil at 2 μL/min, so the PDMS extract inlet is full of oil, and a droplet of oil forms atop the inlet. To prevent water from entering into the device, attach the extract with both extract and oil flowing.
8. Stop the oil flow, then flow extract at several microliters per minute. Monitor the device. When extract first enters the device, restart the flow of oil at several microliters per minute. Avoid filling the device with extract.
9. Form droplets. Start with 5 μL/min oil and 1 μL/min extract then adjust; the optimal extract to oil flow ratio depends on the geometry of the T-junction generator and the desired droplet size.
   - 9.1. If the droplets are too small, or the oil altogether outcompetes the extract at the T-junction, then decrease the oil flow rate or increase the extract flow rate.
   - 9.2. Conversely, if the droplets are too large, or the extract altogether outcompetes the oil at the T-junction, then increase the oil flow rate or decrease the extract flow rate.

10. After you make enough droplets, stop the extract flow but continue to flow oil to clear extract from the T-junction.
11. Stop the oil flow and remove the device from the ice bath.
12. Cut the tubing at the level of the PDMS and seal the holes with VALAP. Do not pull the tubes out of the PDMS, to avoid pressure fluctuations.
13. Mount on the microscope and image. If imaging for tens of minutes, place a Kimwipe wet with CSF-XB above the device to limit evaporation.

ACKNOWLEDGMENTS

Supported by GM39565 awarded to TJM. We thank Nikon Inc. for support of microscopy at MBL and the Nikon Imaging Center at Harvard Medical School (HMS). We thank the National Xenopus Resource at MBL for *Xenopus* animals and care. We thank Matt Sonnett and Keisuke Ishihara (HMS) for helpful discussions. We thank John Oakey (U Wyoming, Laramie, WY) for microfluidic devices and extensive training. JFP was supported by the Fannie and John Hertz Foundation, and the MIT Center for Bits and Atoms and Department of Physics. MBL work was supported by fellowships from the Evans Foundation, MBL Associates and the Laura and Arthur Colwin Summer Research Fund awarded to TJM and CMF.

REFERENCES

Abu Shah, E., & Keren, K. (2014). Symmetry breaking in reconstituted actin cortices. *eLife, 3*, e01433.

Applegate, K. T., Besson, S., Matov, A., Bagonis, M. H., Jaqaman, K., & Danuser, G. (2011). plusTipTracker: quantitative image analysis software for the measurement of microtubule dynamics. *Journal of Structural Biology, 176*, 168–184.

Bement, W. M., Benink, H. A., & von Dassow, G. (2005). A microtubule-dependent zone of active RhoA during cleavage plane specification. *The Journal of Cell Biology, 170*, 91–101.

Bieling, P., Telley, I. A., Hentrich, C., Piehler, J. J., & Surrey, T. (2010). Fluorescence microscopy assays on chemically functionalized surfaces for quantitative imaging of microtubule, motor, and +TIP dynamics. *Methods in Cell Biology, 95*, 555–580.

Burkel, B. M., von Dassow, G., & Bement, W. M. (2007). Versatile fluorescent probes for actin filaments based on the actin-binding domain of utrophin. *Cell Motility and the Cytoskeleton, 64*, 822–832.

Coue, M., Lombillo, V. A., & McIntosh, J. R. (1991). Microtubule depolymerization promotes particle and chromosome movement in vitro. *The Journal of Cell Biology, 112*, 1165–1175.

Field, C. M., Groen, A. C., Nguyen, P. A., & Mitchison, T. J. (2015). Spindle-to-cortex communication in cleaving, polyspermic *Xenopus* eggs. *Molecular Biology of the Cell, 26*, 3628–3640.

Field, C. M., Nguyen, P. A., Ishihara, K., Groen, A. C., & Mitchison, T. J. (2014). *Xenopus* egg cytoplasm with intact actin. *Methods in Enzymology, 540*, 399–415.

Field, C. M., Oegema, K., Zheng, Y., Mitchison, T. J., & Walczak, C. E. (1998). Purification of cytoskeletal proteins using peptide antibodies. *Methods in Enzymology, 298*, 525–541.

Field, C. M., Wühr, M., Anderson, G. A., Kueh, H. Y., Strickland, D., & Mitchison, T. J. (2011). Actin behavior in bulk cytoplasm is cell cycle regulated in early vertebrate embryos. *Journal of Cell Science, 124*, 2086–2095.

Gautier, J., Minshull, J., Lohka, M., Glotzer, M., Hunt, T., & Maller, J. L. (1990). Cyclin is a component of maturation-promoting factor from *Xenopus*. *Cell, 60*, 487–494.

Good, M. C., Vahey, M. D., Skandarajah, A., Fletcher, D. A., & Heald, R. (2013). Cytoplasmic volume modulates spindle size during embryogenesis. *Science, 342*, 856–860.

Groen, A. C., & Mitchison, T. J. (2016). Purification and fluorescent labeling of tubulin from *Xenopus laevis* egg extracts. In P. Chang, & R. Ohi (Eds.), *The mitotic spindle: Methods and protocols: Vol. 1413. Methods in molecular biology* (pp. 34–45). New York: Springer, Humana Press.

Hazel, J., Krutkramelis, K., Mooney, P., Tomschik, M., Gerow, K., Oakey, J., & Gatlin, J. C. (2013). Changes in cytoplasmic volume are sufficient to drive spindle scaling. *Science, 342*, 853–856.

Heald, R., Tournebize, R., Blank, T., Sandaltzopoulos, R., Becker, P., Hyman, A., & Karsenti, E. (1996). Self-organization of microtubules into bipolar spindles around artificial chromosomes in *Xenopus* egg extracts. *Nature, 382*, 420–425.

Hyman, A., Drechsel, D., Kellogg, D., Salser, S., Sawin, K., Steffen, P., … Mitchison, T. J. (1991). Preparation of modified tubulins. *Methods in Enzymology, 196*, 478–485.

Ishihara, K., Nguyen, P. A., Groen, A. C., Field, C. M., & Mitchison, T. J. (2014). Microtubule nucleation remote from centrosomes may explain how asters span large cells. *Proceedings of the National Academy of Sciences of the United States of America, 111*, 17715–17722.

Lee, K., Gallop, J. L., Rambani, K., & Kirschner, M. W. (2010). Self-assembly of filopodia-like structures on supported lipid bilayers. *Science, 329*, 1341–1345.

Lee, S., & Vörös, J. (2005). An aqueous-based surface modification of poly(dimethylsiloxane) with poly(ethylene glycol) to prevent biofouling. *Langmuir: The ACS Journal of Surfaces and Colloids, 21*, 11957–11962.

Lombillo, V. A., Stewart, R. J., & McIntosh, J. R. (1995). Minus-end-directed motion of kinesin-coated microspheres driven by microtubule depolymerization. *Nature, 373*, 161–164.

Maloney, A., Herskowitz, L. J., & Koch, S. J. (2011). Effects of surface passivation on gliding motility assays. *PLoS One, 6*, e19522.

Mitchison, T. J., & Kirschner, M. W. (1986). Isolation of mammalian centrosomes. *Methods in Enzymology, 134*, 261–268.

Murray, A. W. (1991). Cell cycle extracts. *Methods in Cell Biology, 36*, 581–605.

Nair, P. M., Salaita, K., Petit, R. S., & Groves, J. T. (2011). Using patterned supported lipid membranes to investigate the role of receptor organization in intercellular signaling. *Nature Protocols, 6*, 523–539.

Needleman, D. J., Groen, A., Ohi, R., Maresca, T., Mirny, L., & Mitchison, T. J. (2010). Fast microtubule dynamics in meiotic spindles measured by single molecule imaging: evidence that the spindle environment does not stabilize microtubules. *Molecular Biology of the Cell, 21*, 323–333.

Nguyen, P. A., Groen, A. C., Loose, M., Ishihara, K., Wühr, M., Field, C. M., & Mitchison, T. J. (2014). Spatial organization of cytokinesis signaling reconstituted in a cell-free system. *Science, 346*, 244–247.

Niethammer, P., Kueh, H. Y., & Mitchison, T. J. (2008). Spatial patterning of metabolism by mitochondria, oxygen, and energy sinks in a model cytoplasm. *Current Biology, 18*, 586–591.

Piekny, A., Werner, M., & Glotzer, M. (2005). Cytokinesis: welcome to the Rho zone. *Trends in Cell Biology, 15*, 651–658.

Pinot, M., Chesnel, F., Kubiak, J. Z., Arnal, I., Nedelec, F. J., & Gueroui, Z. (2009). Effects of confinement on the self-organization of microtubules and motors. *Current Biology, 19*, 954–960.

Pinot, M., Steiner, V., Dehapiot, B., Yoo, B.-K., Chesnel, F., Blanchoin, L., ... Gueroui, Z. (2012). Confinement induces actin flow in a meiotic cytoplasm. *Proceedings of the National Academy of Sciences of the United States of America, 109*, 11705–11710.

Quintyne, N. J., Gill, S. R., Eckley, D. M., Crego, C. L., Compton, D. A., & Schroer, T. A. (1999). Dynactin is required for microtubule anchoring at centrosomes. *The Journal of Cell Biology, 147*, 321–334.

Rankin, S., & Kirschner, M. W. (1997). The surface contraction waves of *Xenopus* eggs reflect the metachronous cell-cycle state of the cytoplasm. *Current Biology, 7*, 451–454.

Rauh, N. R., Schmidt, A., Bormann, J., Nigg, E. A., & Mayer, T. U. (2005). Calcium triggers exit from meiosis II by targeting the APC/C inhibitor XErp1 for degradation. *Nature, 437*, 1049–1052.

Reber, S. (2011). Isolation of centrosomes from cultured cells. *Methods in Molecular Biology, 777*, 107–116.

Riedl, J., Crevenna, A. H., Kessenbrock, K., Yu, J. H., Neukirchen, D., Bista, M., ... Wedlich-Soldner, R. (2008). Lifeact: a versatile marker to visualize F-actin. *Nature Methods, 5*, 605–607.

Rosenblatt, J., Agnew, B. J., Abe, H., Bamburg, J. H., & Mitchison, T. J. (1997). *Xenopus* actin depolymerizing factor/cofilin (XAC) is responsible for the turnover of actin filaments in *Listeria monocytogenes* tails. *The Journal of Cell Biology, 136*, 1323–1332.

Tsai, M. Y., & Zheng, Y. (2005). Aurora A kinase-coated beads function as microtubule-organizing centers and enhance RanGTP-induced spindle assembly. *Current Biology, 15*, 2156–2163.

Verde, F., Berrez, J. M., Antony, C., & Karsenti, E. (1991). Taxol-induced microtubule asters in mitotic extracts of *Xenopus* eggs: requirement for phosphorylated factors and cytoplasmic dynein. *The Journal of Cell Biology, 112*, 1177–1187.

Wühr, M., Freeman, R. M., Jr., Presler, M., Horb, M. E., Peshkin, L., Gygi, S. P., & Kirschner, M. W. (2014). Deep proteomics of the *Xenopus laevis* egg using an mRNA-derived reference database. *Current Biology, 24*, 1467–1475.

Index

'*Note:* Page numbers followed by "f" indicate figures and "t" indicate tables.'

C

D

E

L

M

Volumes in Series

Founding Series Editor
DAVID M. PRESCOTT

Volume 11 (1975)
Yeast Cells
Edited by David M. Prescott

Volume 12 (1975)
Yeast Cells
Edited by David M. Prescott

Volume 13 (1976)
Methods in Cell Biology
Edited by David M. Prescott

Volume 14 (1976)
Methods in Cell Biology
Edited by David M. Prescott

Volume 15 (1977)
Methods in Cell Biology
Edited by David M. Prescott

Volume 16 (1977)
Chromatin and Chromosomal Protein Research I
Edited by Gary Stein, Janet Stein, and Lewis J. Kleinsmith

Volume 17 (1978)
Chromatin and Chromosomal Protein Research II
Edited by Gary Stein, Janet Stein, and Lewis J. Kleinsmith

Volume 18 (1978)
Chromatin and Chromosomal Protein Research III
Edited by Gary Stein, Janet Stein, and Lewis J. Kleinsmith

Volume 19 (1978)
Chromatin and Chromosomal Protein Research IV
Edited by Gary Stein, Janet Stein, and Lewis J. Kleinsmith

Volume 20 (1978)
Methods in Cell Biology
Edited by David M. Prescott

Advisory Board Chairman
KEITH R. PORTER

Volume 21A (1980)
Normal Human Tissue and Cell Culture, Part A: Respiratory, Cardiovascular, and Integumentary Systems
Edited by Curtis C. Harris, Benjamin F. Trump, and Gary D. Stoner

Volume 21B (1980)
Normal Human Tissue and Cell Culture, Part B: Endocrine, Urogenital, and Gastrointestinal Systems
Edited by Curtis C. Harris, Benjamin F. Trump, and Gray D. Stoner

Volume 22 (1981)
Three-Dimensional Ultrastructure in Biology
Edited by James N. Turner

Volume 23 (1981)
Basic Mechanisms of Cellular Secretion
Edited by Arthur R. Hand and Constance Oliver

Volume 24 (1982)
The Cytoskeleton, Part A: Cytoskeletal Proteins, Isolation and Characterization
Edited by Leslie Wilson

Volume 25 (1982)
The Cytoskeleton, Part B: Biological Systems and In Vitro Models
Edited by Leslie Wilson

Volume 26 (1982)
Prenatal Diagnosis: Cell Biological Approaches
Edited by Samuel A. Latt and Gretchen J. Darlington

Series Editor
LESLIE WILSON

Volume 27 (1986)
Echinoderm Gametes and Embryos
Edited by Thomas E. Schroeder

Volume 28 (1987)
***Dictyostelium discoideum*: Molecular Approaches to Cell Biology**
Edited by James A. Spudich

Volume 29 (1989)
Fluorescence Microscopy of Living Cells in Culture, Part A: Fluorescent Analogs, Labeling Cells, and Basic Microscopy
Edited by Yu-Li Wang and D. Lansing Taylor

Volume 30 (1989)
Fluorescence Microscopy of Living Cells in Culture, Part B: Quantitative Fluorescence Microscopy—Imaging and Spectroscopy
Edited by D. Lansing Taylor and Yu-Li Wang

Volume 31 (1989)
Vesicular Transport, Part A
Edited by Alan M. Tartakoff

Volume 32 (1989)
Vesicular Transport, Part B
Edited by Alan M. Tartakoff

Volume 33 (1990)
Flow Cytometry
Edited by Zbigniew Darzynkiewicz and Harry A. Crissman

Volume 34 (1991)
Vectorial Transport of Proteins into and across Membranes
Edited by Alan M. Tartakoff

Selected from Volumes 31, 32, and 34 (1991)
Laboratory Methods for Vesicular and Vectorial Transport
Edited by Alan M. Tartakoff

Volume 35 (1991)
Functional Organization of the Nucleus: A Laboratory Guide
Edited by Barbara A. Hamkalo and Sarah C. R. Elgin

Volume 36 (1991)
***Xenopus laevis*: Practical Uses in Cell and Molecular Biology**
Edited by Brian K. Kay and H. Benjamin Peng

Series Editors
LESLIE WILSON AND PAUL MATSUDAIRA

Volume 37 (1993)
Antibodies in Cell Biology
Edited by David J. Asai

Volume 38 (1993)
Cell Biological Applications of Confocal Microscopy
Edited by Brian Matsumoto

Volume 39 (1993)
Motility Assays for Motor Proteins
Edited by Jonathan M. Scholey

Volume 40 (1994)
A Practical Guide to the Study of Calcium in Living Cells
Edited by Richard Nuccitelli

Volume 41 (1994)
Flow Cytometry, Second Edition, Part A
Edited by Zbigniew Darzynkiewicz, J. Paul Robinson, and Harry A. Crissman

Volume 42 (1994)
Flow Cytometry, Second Edition, Part B
Edited by Zbigniew Darzynkiewicz, J. Paul Robinson, and Harry A. Crissman

Volume 43 (1994)
Protein Expression in Animal Cells
Edited by Michael G. Roth

Volume 44 (1994)
***Drosophila melanogaster*: Practical Uses in Cell and Molecular Biology**
Edited by Lawrence S. B. Goldstein and Eric A. Fyrberg

Volume 45 (1994)
Microbes as Tools for Cell Biology
Edited by David G. Russell

Volume 46 (1995)
Cell Death
Edited by Lawrence M. Schwartz and Barbara A. Osborne

Volume 47 (1995)
Cilia and Flagella
Edited by William Dentler and George Witman

Volume 48 (1995)
Caenorhabditis elegans: Modern Biological Analysis of an Organism
Edited by Henry F. Epstein and Diane C. Shakes

Volume 49 (1995)
Methods in Plant Cell Biology, Part A
Edited by David W. Galbraith, Hans J. Bohnert, and Don P. Bourque

Volume 50 (1995)
Methods in Plant Cell Biology, Part B
Edited by David W. Galbraith, Don P. Bourque, and Hans J. Bohnert

Volume 51 (1996)
Methods in Avian Embryology
Edited by Marianne Bronner-Fraser

Volume 52 (1997)
Methods in Muscle Biology
Edited by Charles P. Emerson, Jr. and H. Lee Sweeney

Volume 53 (1997)
Nuclear Structure and Function
Edited by Miguel Berrios

Volume 54 (1997)
Cumulative Index

Volume 55 (1997)
Laser Tweezers in Cell Biology
Edited by Michael P. Sheetz

Volume 56 (1998)
Video Microscopy
Edited by Greenfield Sluder and David E. Wolf

Volume 57 (1998)
Animal Cell Culture Methods
Edited by Jennie P. Mather and David Barnes

Volume 58 (1998)
Green Fluorescent Protein
Edited by Kevin F. Sullivan and Steve A. Kay

Volume 59 (1998)
The Zebrafish: Biology
Edited by H. William Detrich III, Monte Westerfield, and Leonard I. Zon

Volume 60 (1998)
The Zebrafish: Genetics and Genomics
Edited by H. William Detrich III, Monte Westerfield, and Leonard I. Zon

Volume 61 (1998)
Mitosis and Meiosis
Edited by Conly L. Rieder

Volume 84 (2007)
Biophysical Tools for Biologists, Volume One: In Vitro Techniques
Edited by John J. Correia and H. William Detrich, III

Volume 85 (2008)
Fluorescent Proteins
Edited by Kevin F. Sullivan

Volume 86 (2008)
Stem Cell Culture
Edited by Dr. Jennie P. Mather

Volume 87 (2008)
Avian Embryology, 2nd Edition
Edited by Dr. Marianne Bronner-Fraser

Volume 88 (2008)
Introduction to Electron Microscopy for Biologists
Edited by Prof. Terence D. Allen

Volume 89 (2008)
Biophysical Tools for Biologists, Volume Two: In Vivo Techniques
Edited by Dr. John J. Correia and Dr. H. William Detrich, III

Volume 90 (2008)
Methods in Nano Cell Biology
Edited by Bhanu P. Jena

Volume 91 (2009)
Cilia: Structure and Motility
Edited by Stephen M. King and Gregory J. Pazour

Volume 92 (2009)
Cilia: Motors and Regulation
Edited by Stephen M. King and Gregory J. Pazour

Volume 93 (2009)
Cilia: Model Organisms and Intraflagellar Transport
Edited by Stephen M. King and Gregory J. Pazour

Volume 94 (2009)
Primary Cilia
Edited by Roger D. Sloboda

Volume 95 (2010)
Microtubules, in vitro
Edited by Leslie Wilson and John J. Correia

Kolibri 360 GmbH
Ferdinand-Jühlke-Straße 7
99095 Erfurt, Deutschland
produktsicherheit@kolibri360.de